ORGANIC FARMING
Scope and Use of Biofertilizers

AF739492

JDS Panwar
M.Sc., Ph.D, FISPP, FISPRD

Vice Chancellor, Venkateshwara Open University, Arunanchal Pradesh
Former Vice Chancellor, Mahatma Gandhi University, Meghalaya
Former Director, N K School of Nursing, Rampur (2013)
Former Head, Division of CPBM, IIPR, Kalyanpur, Kanpur
Former Head, Division of Plant Physiology, IARI, New Delhi

Amit Kumar Jain
M.Sc., Ph.D, D.Sc.
PGDNBT, PGDDE, FLS (London), FBS (India)

Assistant Regional Director
IGNOU Regional Centre Karnal, Haryana

Former Assistant Professor Department of Microbiology
Janta Vedic College, Baraut (Baghpat)

nipa®

NIPA® GENX ELECTRONIC RESOURCES & SOLUTIONS P. LTD.
New Delhi-110 034

NIPA® GENX ELECTRONIC
RESOURCES & SOLUTIONS P. LTD.

101,103, Vikas Surya Plaza, CU Block
L.S.C. Market, Pitam Pura, New Delhi-110 034
Ph : +91 11 27341616, 27341717, 27341718
E-mail: newindiapublishingagency@gmail.com
web: www.nipabooks.com

For customer assistance, please contact
Phone: + 91-11-27 34 17 17
Fax: + 91-11- 27 34 16 16
E-Mail: feedbacks@nipabooks.com

ISBN: 978-81-19235-88-9

Composed and Designed by NIPA®.

We
Whole Heartedly
dedicate the book to our
Respected Parents
who have been a source of inspiration
all throughout during this endeavour.

Preface

The present system of agriculture which we call conventional and practiced, which promoted an over riding quest for accumulation of wealth. One of the major challenges to agriculture in this decade will be to develop organic farming systems which can provide an additional avenue for climate change mitigation through such measures as enhanced soil carbon sequestration. Organic farming one of the widely used methods, is thought as the best alternative to avoid the ill effects of chemical farming. In a system's perspective organic farming primarily aims at cultivating the land and raising crops in such a way, as to keep the soil alive and in good health through use of organic wastes (crop, animal and farm wastes, aquatic wastes) and other biological materials along with biofertilizers to release nutrients to crops for increased sustainable production in an eco-friendly pollution free environment. It is becoming very popular and the farmers have now started adopting organic farming in a big way and as such the demand of organic fertilizers and biofertilizers is increasing in the market. In the light of the organic agriculture, this text book covers concept and relevance of organic agriculture in sixteen chapters as organic farming, basic information on biofertilizers, classification of biofertilizers, nitrogen fixation, *Rhizobium*, *Azolla*, *Azotobacter*, *Azospirillum*, Blue Green Algae, Phosphorus Solubilizing Micro-organisms (PSMs), *Mycorrhizae*, *Frankia*, Vermiculture and vermicomposting, liquid biofertilizers, production, quality and marketing of biofertilizers & some important media.

The book is written in a very simple form with up to date data and statistics. We have put the references of all chapters at the end of book that we hope will be useful to you. It is a mostly all-inclusive basic text book on organic farming system and will specifically meet out the requirement of the scientists, teachers, research scholars and students. We hope this book will be welcomed by the many, more people who live, work or simply enjoy non-urban areas. We do hope you enjoy reading Organic Farming: Scope and Use of Biofertilizers, as an essential reference book on the shelf. We would welcome feedback from readers on aspects of the book that we might amend in subsequent editions.

The authors would welcome suggestions from the readers to improve the text book.

(Authors)

Acknowledgement

The authors are thankful to well-wishers, colleagues and friends spanning all over the globe, stimulating us for writing this book. Their encouragement, invaluable suggestions and inclusion of photographs have given a real value addition to the book while keeping it updated with latest development on the subject. Our memory fell short of words to thank Dr. Srikant Mohapatra, Director (RSD), IGNOU, Delhi for his consistent cooperation and kind gestures. We cannot forget to thank our colleagues at the several other Organizations/Institutions and IGNOU Regional Centres namely Dr. A.K. Sharma, Dr. V P. Rupam, Dr. A.K. Srivastava, Dr. U.N. Tripathi, Dr. Vinita Katiyar and Dr. Dharam Pal for their encouragement and suggestions. The task could have not been complete if we don't name the authors of various books who helped us in updating of the subject material to the present status. Authors are thankful to the management and staff members of MGU, Meghalaya and VOU, Arunachal Pradesh for their kind co-operation.

Our special thanks to Dr. Y.S. Tomer, Dr. J. Mohan, Dr. Sudhir Kumar, Dr. A.K. Sharma, Dr. (Mrs.) Rajeshwari Sharma, Dr. V.K. Tomer at J.V. College Baraut and Prof. Y. Vimala, HOD Botany, CCS University, Meerut for their regular feedback, involvement, encouragement and suggestions .

We wish to thank to our friends especially Mr. Amit, Mr. Anuj, Sachin, Sh. Yogendra Kr. Jain and Dr. Surendera Sharma who directly and indirectly helped us and gave important suggestions from time to time for improving the presentation of this book. This entire tenure made us to compromise with many of our family commitments and whole heartedly thank our family members and nearer & dearer especially Smt. Santosh Panwar, Ms. Poonam Tomar, Ms. Neelam P. Tomar, Mr. Sanjay Panwar, Mr. Ajay Panwar, Prakhar Panwar, Smt. Indu, Sh. T.C. Jain, Er. R.C. Jain, Er. Navin Jain, Mr. Naveen Jain, Mr. Surender Jain, Mr. Sandeep Jain, Smt. Trishla Jain, Smt. Madhu Jain, Smt. Rita Jain, Ms. Babita Jain, Ms. Ritu Jain, Ms. Poornima Jain, Kanika, Riya, Naman, Rahul and Paras for bearing with us during difficult times and imparting unstinted support & encouragement.

We express our heartfelt thanks to our publisher for liberal use of the recent available expertise in the field of printing and publication of the book which enabled us for timely publication. Our readers such as students and teachers are welcomed for the suggestions that will be a source of encouragements to achieve the higher goals with improvement in the next edition.

Ultimately, we thank Almighty God "Master of this Universe".

Contents

Chapter - 1

Organic Farming

Agriculture is undoubtedly the largest livelihood provider in rural India. As we enter a new post WTO Scenario and IT revolution, new inventions are happening day by day in every sphere. Agriculture is not culture now, it is agri business! Agriculture constitutes over 25% of India's GDP. In recent years, much emphasis has been given by the Ministry of Agriculture on commercializing agricultural production in the country. Adequate production and distribution of food has become a high priority and global concern. In the fast changing world and increasing competition in a globalized economy, there is a need for exploiting the available resources to the maximum level and use of best technologies available world over, to cope up with domestic demand of food and also to target export market. This will be particularly relevant in South India, where farmers grow a number of crops, but have technical constraints in enhancing production and productivity because of inadequate exposure to high technology & inputs coupled with advanced production practices, logistics and marketing. In addition, shortage of labour is a major concern for progressive farmers of South India who are ready to offer more wages to labourers. Therefore, they are forced to become machinery-dependent in their farm operations. As a result, Southern India is emerging as one of the hottest destinations as well as a strong market with buying potential for farm machinery and equipment.

1.1 Govt. Policies and Programmes

New technological advances in agricultural practices have become all the more important against the backdrop of a series of programmes that have been launched by the Government of India to attain 4% annual growth in agriculture and to double the production of horticultural crops during the ongoing XII Five-Year Plan. The main thrust of all these programmes is on taking Indian agriculture to greater heights of excellence through spread of latest technology and modern practices on-farm and off-farm that would cover pre and post-harvest activities, marketing and exports.

1.2 Global Scenario of Agriculture/Farm Mechanization

The BRICS Nations i.e. Brazil, Russia, India, China and South Africa along with Japan and Turkey are joining the ranks of heavy weight agricultural machinery markets. According to Unacoma*, India and China were confirmed in 2011 as the countries with the greatest take-up of tractors with vast numbers of self-propelled machinery. India has emerged as New Agricultural Machinery mega market. After the grand success of our 5th AgriTech India 2013, where besides India, companies participated from 26 countries. Media Today Group have organized the 6th edition of AgriTech India 2014 during 22nd to 24th August 2014 at Bangalore International Exhibition Centre (BIEC), Bangalore, India, which emphasized to pump in more value additions to the existing technology from all over the world.

Organic agriculture is one among the broad spectrum of methodologies which are supportive of the environment. Organic production systems are based on specific and precise standard of production which aims at achieving optimal agro-ecosystem which are socially, ecologically and economically sustainable. Proper soil management without impairing soil health is the prerequisite for achieving higher productivity from agriculture land. The chemical inputs in crop production are becoming increasingly costly and in short supply. Moreover, the use of mineral fertilizers and pesticides doses of necessarily lead to better farming than the use of natural and organic methods of farming. There is a need to encourage more productive and environment friendly farming practices. Sustainable agriculture does not represent a return to pre-industrial revolution methods; rather it combines traditional conservation farming techniques with modern technologies and relies increasingly on renewable sources. Sustainable systems use modern equipment certified seed, soil and water conservation practices and the latest innovation in feeding and handling livestock crops rotating building up soil fertility, diversifying crops and livestock and controlling pest naturally. Another feature of sustainable farming is a regular addition of crop residues, manures/compost and sources of other organic materials to soil. The conventional method of farming helped the developing countries to raise the agricultural production but it has shown undesirable effects such as dominance of mineral fertilizers and chemical pesticides, lesser use of organic manures, compost, crop residues, biological inputs and neglect of mixed cropping.

---

* Unacoma : The Italian Agricultural Machinery Manufactures Federation, formed in 2012 to replace Unacoma (The Italian Farm Machinery Manufactures Association set up in 1945), brings together, and represents in Italy and abroad, the associations of Italian manufactures of implements (Assomao), self-propelled machines (Assomase), tractors (Assotrattori), components (Compacomp), gardening machinery (Comagarden), earth moving machinery (Comamoter) and enginess (Unamot).

The terms such as organic, natural, biological, biodynamic, alternative and sustainable farming have entered the agriculture vocabulary. In spite of these differences in terminology, basically identical ideologies, ideas and technologies are reflected. The common objectives of all the systems are to develop farming methods to restore, maintain or build up soil productivity, to protect the environment instead of causing pollution, produce safe and healthy food, and preserve or help the small farmer (Hooda, 2005; Sidhu, 2005). Soil micro-organisms and earthworms take care of the processes involved in decompositions and mineralization of the organic substrates, fix atmospheric nitrogen, solubilize insoluble forms of phosphates and produce natural growth promoting substance and antibiotics. Thus, agricultural practices should be converted from high-input agricultural systems into more sustainable organic agricultural systems (Jasem and Alireza, 2014).

1.3 Global Demand for Organic Produce

Organic farming is a holistic production management system that promotes and enhances health of the agro-ecosystem related to biodiversity, nutrient biocycles and soil biological and microbial activities. Organic farming emphasizes management practices involving the use of organic manures, green manures, biofertilizers, biopesticides and mechanical methods for sustainability in production. Organic farming works in harmony with nature rather than against it. It involves using techniques to achieve good crop yields without harming the natural environment or the people who live and work in it. Organic farming has been drawing attention from all over the world and the demand for the organic food is growing by leaps and bounds especially in developed countries.

During the last decades organic farming in India has attracted many farmers across the country and many of them have grown successfully crops like sapota, spices, banana, grape, coconut etc. with organic manures and biofertilizers. Organically grown spices such as black pepper, small cardamom, turmeric, ginger, cloves and maize are exported. There is increasing demand for organic tea, coffee, cashew nuts, fruits and vegetables from developed countries. International Federation of Organic Agriculture Movement (IFOAM) is an association which provides assistance to organic farmers to get their produce certified that is an organically grown produce. The certification is done by some of the international agencies that have also their officers in India. A standing and accredition committee on organic farming is constituted by the commerce ministry. The committee has finalized national standards for production of organic food. A steering committee under the chairmanship of secretary, commerce has also been formed to review the progress in this area. No doubt, the interest is the organic farming appears to be growing worldwide. How important, it can become in future is hard to assess, this will depends on the number of factors such as public awareness about environmental issues, advances in agricultural research viz. biotechnology, genetic engineering, microbiology etc, and attitude of governments/agencies. According to "The World of Organic Agriculture: Statistics and Emerging Trends 2014", published by the Research Institute of Organic Agriculture (FiBL)* and the International Federation of Organic Agriculture Movements (IFOAM), the area under organic cultivation increased from 11Mha in 1999 to 37.5Mha in 2012. The global market for organic produce grew from $15 billion in 1999 to $64 billion in 2012 (Anonymous, 2014a).

1.4 Definition and Options in Organic Farming

The concept of organic farming is not new to India as it has existed here before the introduction of modern methods of farming. It is actually based on the minimal use of off farm resources (inputs) and on management practices that restore, maintain and enhance ecological harmony. The International Federation of Organic Agriculture Movements (IFOAM) has defined it as **production system that sustains the health of soils, ecosystems and people**. It relies on ecological processes, biodiversity and cycles adapted to local conditions, rather than the sum of inputs with adverse effects. According to the Agricultural and Processed Food Products Export Development Authority (APEDA) of India **"Organic products are grown under a system of agriculture without use of chemical fertilizers and pesticides with an environmentally and socially responsible approach"**. This is a method of farming that works at grass root level preserving the reproductive and regenerative capacity of the soil, good plant nutrition, and sound soil management produces nutritious food rich in vitality which has resistance to diseases. As per the definition of the United States Department of Agriculture (USDA) "**Organic farming is a system which avoids or largely excludes the use of synthetic inputs (such as fertilizers, pesticides, hormones, feed additives etc.) and to the maximum extent feasible rely upon crop rotations, crop residues, animal manures, off-farm organic waste, mineral grade rock additives and biological system of nutrient mobilization and plant protection**". FAO suggested that "**Organic agriculture is a unique production management system which promotes and enhances agro-ecosystem health, including biodiversity, biological cycles and soil biological activity, and this is accomplished by using on-farm agronomic, biological and mechanical methods in exclusion of all synthetic off-farm inputs**".

* The Research Institute of Organic Agriculture (German) Forschungsinstitut fur biologischen Landbau (FiBL) is an independent, non-profit research institute with the aim of advancing cutting-edge science in the field of Organic agriculture.

There are three options in organic farming as-

a) **Integrated farming system:** Recycling of local organic resources generated from dairying, fisheries, poultry, goat farming, mushroom production, agro-forestry and agriculture system. This is low input organic farming.

b) **Integrated conventional and organic farming technology**: This involves integrated plant nutrients supply (IPNS) and integrated pest management (IPM). Bio-fertilizers could serve as alternate sources of nitrogen or phosphorus or both.

c) **Pure organic farming:** Use of organic manures/compost, recycling of plant residues, bio-fertilizers and bio-pesticides.

1.5 Key Characteristics of Organic Farming

- Providing crop nutrients indirectly using relatively insoluble nutrient sources which are made available to the plant by the action of soil micro-organisms.
- Nitrogen self-sufficiency through the use of legumes and biological nitrogen fixation, as well as effective recycling of organic materials including crop residues and livestock manures.
- Careful attention to the impact of the farming system on the wider environment and the conservation of wildlife and natural habitats.
- Weed, disease and pest control relying primarily on crop rotations, natural predators, diversity, organic manuring, resistant varieties and limited (preferably minimal) thermal, biological and chemical intervention.
- Protecting the long term fertility of soils by maintaining organic matter levels, encouraging soil biological activity, and careful mechanical intervention.
- The extensive management of livestock, paying full regard to their evolutionary adaptations, behavioral needs and animal welfare issues with respect to nutrition, housing, health, breeding and rearing.

1.6 Current Scenario of Organic Farming in India

At present organic farming is practiced in 162 countries and 37Mha of agricultural land are managed originally by 1.8 million farmers. The current market for organic foods in India is pegged at Rs. 2500 crore, which according to ASSOCHAM* is expected to reach Rs. 6000 crore by 2015. Thus a huge potential is seen in the nascent Indian organic sector. Organic products, which until now were mainly being exported, are now finding consumers in the domestic market also. The data of the organic farming in India is given in Table 1.1

Table 1.1: Data of Organic Products in India (2012-2013)

Number of Products Exported	135
Total Quantity Exported	164262 MT
Value of Total Export	US$ 374 Million
Total Certified Area including under Cultivation, Forest and Wild Harvest	5.21 Million hectare (Mha)
Organic Crops/Commodities/Products Produced in India	Basmati rice, Cotton, Coffee, Sugarcane, Pulses, Tea, Spices, Oil seeds, Fruits and their value added products, Organic Cotton Fiber, Functional Food Products etc.
Countries Importing Indian organic Products	EU, US, Canada, Switzerland, South East Asian countries and South Africa
Share of Indian Organic Products in Export	Oil Seeds- Soybean (41%), lead among the products exported followed by cane sugar (26%), Processed food products (14%), Basmati rice (5%), Other cereals and millets (4%), Tea (2%), Spices (1%), Dry Fruits (1%) and others.

Source: Anonymous (2014b)

A great diversity of climatic conditions exists in India, which may help in producing all varieties of organic products. In some parts of the country, the inherited tradition of organic farming seems to be an added advantage. Currently, India ranks 10th amongst top ten countries in terms of cultivable land under organic certification. GOI has implemented the National Programme for Organic Production (NPOP). This programme involves the accreditation of certification bodies, standards of India. Madhya Pradesh has covered largest area under organic certification followed by Rajasthan and Uttar Pradesh. The other important states are Uttarakhand, Kerala and many north-eastern states.

1.7 Need of Organic Farming

With the increase in population our compulsion would be not only to stabilize agricultural production but to increase it further in sustainable manner. The scientists have realized that the 'Green Revolution' with high input use has reached a plateau and is now sustained with diminishing return of falling dividends.

* Associated Chambers of Commerce and Industry (ASSOCHAM) is one of the apex trade association of India. It represents the interests of trade and commerce in India and acts as an interface between industry, government and other relevant stackholders an policy issues and initiatives.

Thus, a natural balance needs to be maintained at all cost for existence of life and property. The obvious choice for that would be more relevant in the present era, when these agrochemicals which are produced from fossil fuel and are not renewable and are diminishing in availability. It may also cost heavily on our foreign exchange in future. Modern, intensive agriculture causes many problems, including the following:

a) Artificial chemicals destroy soil micro-organisms resulting in poor soil structure and aeration and decreasing nutrient availability.

b) Artificial fertilizers and herbicides are easily washed from the soil and pollute rivers, lakes and water courses.

c) Artificial pesticides can stay in the soil for a long time and enter the food chain where they build up in the bodies of animals and humans, causing health problems.

d) The prolonged use of artificial fertilizers results in soils with a low organic matter content which is easily eroded by wind and rain.

e) Pests and diseases become more difficult to control as they become resistant to artificial pesticides. The numbers of natural enemies decrease because of pesticide use and habitat loss.

f) Dependency on fertilizers, greater amounts are needed every year to produce the same yields of crops.

1.8 Adverse Effects of Conventional Intensive Agriculture

Conventional intensive agriculture causes many adverse effects, which have been given as :

1.8.1 Decline of Soil Productivity

India has changed during the many decades from a country of scarcity to food sufficiency with the development of improved varieties of crops, irrigation facilities and use of conventional cultivation practices. During the period the cropping pattern in the country made a distinct shift in favour of rice wheat system. This has resulted in increased fertilizers, water and pesticides use. In the meantime, the use of organic matter/FYM/compost has declined.

1.8.2 Quality of the Soil

Soil quality is directly related to degradation, which can also be defined as the time rate of change in soil quality (Parr *et al.,* 1992). The maintenance or restoration of soil quality is highly dependent on organic matter and an array of beneficial micro-organisms. The proper and regular addition of different type of organic matter can effectively check many of these degradative processes. It is also the best and quicker means of developing a biological active soil that requires less energy for producing crops, increases the resistance to pests and diseases and enhances the decomposition of toxic substances. As the soil degradation processes intensify and increase, the soils productivity decreases. An integrated system of plant and animal production practices having site specific application that cover a long term goal is desirable. Organic wastes and residue offer the best possible means of restoring the productivity of eroded soil or reclamation of marginal soils.

1.8.3 Soil Pollution

The regular use of acid or acidogenic fertilizers causes a drop in pH of soil over a period of time, and it reduces crop yields drastically (Lal and Mathur, 1988). The long-term use of inorganic fertilizers without organic manures or crop residue damages the soil physical properties (Biswas *et al.,* 1917) and increase in soil erosion (USDA, 1978) and a decrease in water-soluble aggregates less than 0.25mm (Ketcheson and Beauchamp, 1978). The intensive cultivation with no crop rotation and organic materials application causes a depleting trend of soil organic matter leading to a decline in nitrogen reserves as well as low nutrient retention capacity owning to decreased Cation Exchange Capacity (CEC) (Lal and Kang, 1982). Over a period of time, it has become essential to fertilize the low organic matter soils not only with N, P, and K but also micronutrients. In order to minimize the above ill effects of intensive conventional farming, organic farming practices should be adopted by the farmers.

1.8.4 Environmental Pollution

The continued use of phosphatic fertilizers can result in-built of trace elements such as arsenic (Ar), cadmium (Cd), uranium (Ur) and radium (Ra) in soil. The transport of phosphorus fertilizers from terrestrial to aquatic reservoirs in the surface and sub-surface run off results in the deterioration of the water quality from the accelerated eutrophication (Sharpley and Menzel, 1987). Nitrogenous fertilizers create health and environment problems due to nitrate pollution of drinking water and the eutrophication of lakes and streams and depletion of stratospheric ozone due to nitrous oxide emission from nitrification and denitrification process (Keeney, 1982).

1.8.5 Cropping Patterns

The area under oil seeds and pulses has been decreasing. The inclusion of legumes in crop rotation as well benefits the following or companion crops. The rotations and intercropping involving many crops have been abandoned in the conventional agriculture. The levels of organic matter, total nitrogen and the microbial mass decease more with the monocropping than with crop rotations (McGill *et al.,* 1986). The structure and tilth in terms of total porosity and stable aggregation are also impaired (Ketcheson, 1980) in monocropping. Aggregate stability is high under crop rotations than under continuous monocropping (Toogood and Lynch, 1959).

1.8.6 Energy Consumption

In developing countries, chemical fertilizers account for about 70% of the commercial energy used in agriculture whereas in developed countries it is about 35%. But the total commercial energy used for production of chemical fertilizers in developed countries is more as compared to the developing countries, even though cultivated areas of the developed countries is only 37% of the total cultivated area of the world. Nitrogenous fertilizers are by far the most consumers of energy.

1.9 Organic System and Sustainable Agriculture

The ultimate of the sustainable agriculture is to develop farming system that are productive and profitable, conserve the natural resource base, protect the environment and enhance health and safety. Organic farming is system of food and fiber production that applies management skills and information to reduce the input costs, improve efficiency and maintain production levels through following principles and practices: **"Crop rotations and intercropping involving leguminous crops, recycling of organic resources of all kind (renewable sources of energy), nitrogen fixing and phosphate solubilizing micro-organisms, integrated plant nutrient supply (IPNS) system, integrated pest management (IPM), conservation tillage (maintenance of physical condition of soils)"**.

1.10 Composting

Composting is the natural process of 'rotting' or decomposition of organic matter by micro-organisms under controlled conditions. Raw organic materials such as crop residues, animal wastes, food garbage, some municipal wastes and suitable industrial wastes, enhance their suitability for application to the soil as a fertilizing resource, after having undergone composting. Composting may be defined as a biological process in which diverse and mixed group of micro-organism breakdown organic material to humus like substance. This product has lower bulk volume than the original raw material, is stable in nature, free of pathogens and plant seeds, has a slow decomposition rate and is an organic fertilizer. This can beneficially be applied to land. The quality of the product depends upon the nature and characteristics of organic raw materials. "**A mass of rotted organic matter made from waste is called compost**". The compost made from farm waste like sugarcane trash, paddy straw, weeds and other plants and other waste is called farm compost. The average nutrient contents of farm compost are 0.5% N, 0.15% $P_2O_5$and 0.5% K_2O. The nutrient value of farm compost can be increased by application of superphosphate or rock phosphate at 10 to 15 kg/t of raw material at the initial stage of filling the compost pit. Farm waste is placed in the trenches layer by layer. Each layer is well moistened by sprinkling cow dung slurry or water. Trenches are filled up to a height of 0.5m above the ground. The compost is ready for application within 5-6 months. Composting is essentially a microbiological decomposition of organic residues collected from rural area (rural compost) or urban area (urban compost).

Compost is a rich source of organic matter. Soil organic matter plays an important role in sustaining soil fertility, and hence in sustainable agricultural production. In addition to being a source of plant nutrient, it improves the physico-chemical and biological properties of the soil. As a result of these improvements, the soil (a) becomes more resistant to stresses such as drought, diseases and toxicity (b) helps the crop in improved uptake of plant nutrients; and (c) possesses an active nutrient cycling capacity because of vigorous microbial activity. These advantages manifest themselves in reduced cropping risks, higher yields and lower outlays on inorganic fertilizers for farmers.

1.10.1 Principle of Composting

Compost can be prepared from organic materials from different sources such as crop residues, straw, paddy, husk groundnut shell, wastes of safflower, sunflower and mustard, sugar cane trash, press mud and habitation wastes like city wastes/garbage, kitchen wastes, vegetables and fruit market wastes, sewage and sullage etc. In general some of the organic raw material contains 16.8 to 53.6% cellulose, 9.8 to 26.6% hemicellulose, 11.6 to 28.2% lignin, 1.8 to 23.5% protein, and 1.2 to 6.3 of ether soluble substances besides 2-30% of hot and cold water soluble fractions (sugars, starches, amino acids, urea, ammonium salts) (Table 1.2a, b & c).The ash contains 5-30% minerals (Gaur, 1997).The matter undergoes intensive decomposition under thermophilic and mesophilic conditions in pits, heaps or windrows. The decomposition process is carried out by bacteria, fungi, actinomycetes, protozoa, earthworms, termites, ants etc. The role of active cellulolytic and lignolytic organisms is of paramount importance in composting process. When the organic matter is broken down in presence of air (oxygen), the process is aerobic. In absence of oxygen, the process is known as anaerobic composting. This type of fermentation takes place in gobar gas plants, waste stabilization ponds, in marshy soils and composting process in Acharya's method of composting.

**Table 1.2a : ** Chemical Compositions of Some Organic Farm Wastes (%) on Dry Weight Basis

Sr.	Organic material (%)	Protein	Lignin	Cellulose	Hemicellulose
1.	Wheat Straw	2.4	18.8	37.8	26.9
2.	Barley straw	3.9	17.8	40.4	25.3
3.	Oat straw	2.7	18.0	36.4	26.6
4.	Rice Straw	3.1	11.6	35.8	22.9
5.	Groundnut shells	14.8	15.7	16.8	42.2
6.	Cattle dung	8.3	17.7	28.6	26.0
7.	Sheep dung	23.5	20.7	18.7	18.5
8.	Saw dust	1.8	28.2	53.6	9.8

Table 1.2b : Dung and Urine produced by Animals per day

Animal	Urine (ml/kg live wt.)	Quantity of Dung (Kg/day)
Pigs	5-30	3-5
Poultry	-	2.5-3.5
Horse	3-18	9-18
Cattle	17-45	18-30
Buffaloes	20-45	25-40
Sheep and goats	10-40	1-2.5

Table 1.2c : Nutritive Value of Animal Solid and Liquid Excreta

Animal	Dung (mg/g)			Urine (%)		
	N	P	K	N	P	K
Pig	20-45	6-12	15-48	0.38	0.1	0.99
Poultry	28-62	9-26	8-29	-	-	-
Cattle	20-45	4-10	7-25	1.21	0.01	1.35
Sheep and Goat	20-45	4-11	20-29	1.47	0.05	1.96

1.10.2 Necessity of Composting

a) These complex materials cannot be used as such as resource materials, and should be converted into simple inorganic element as available nutrient.

b) The rejected biological materials contain complex chemical compounds such as lignin, cellulose, hemicellulose, polysaccharides, proteins and lipids etc.

c) The material put into soil without conversion will undergo conversion inside the soil. These conversion processes take away all energy and available nutrients from the soil affecting the crop.

d) Hence conversion period is mandatory.

1.10.3 Methods of Composting

1.10.3.1 Coimbatore Method (Pits Method)

In this method, composting is done in pits of different sizes depending on available waste material. A layer of waste materials is first laid in the pit. It is moistened with a suspension of 5-10kg cow dung in 2.5 to 5.0 litres of water and 0.5 to 1.0kg fine bone meal sprinkled over it uniformly. Similar layers are laid one over the other till the material rises 0.75m above the ground level. It is finally plastered with wet mud and left undisturbed for 8 to 10 weeks. Plaster is then removed, material moistened with water, given a turning and made into a rectangular heap under a shade. It is left undisturbed till its use.The size of pit should be about 1m deep x 1.5-2.0m wide and of any suitable length (Plate 48). The site selected for the compost pit should be bear the cattle shed and water source at a high level so that no rain water gets in during monsoon season. The materials such as crop residues etc. for composting is spread in layers and each layer wetted with slurry made out of cattle dung, urine earth and inoculum either taken from a 15 day-old compost or inoculum of cellulolytic/lignolytic culture obtained from a commercial enterprise or a well-known institution. The moisture should be kept at 60-70% level initially which can be reduced at 50% level subsequently. The pit is filled in this ways layer by layer until reach 30 cm above ground level which should not take normally longer than one week. Care should be taken to avoid compacting of the material in the pit. Three turnings of compost upside down are given at 15, 30 and 60 days from filling the pit. At each turning, material is mixed thoroughly and moistened with water to maintain around 50% moisture.

1.10.3.2 Indore Method (Heap Method)

In this method of composting, organic wastes are spread in the cattle shed to serve as bedding. Urine soaked material along with dung is removed every day and formed into a layer of about 15cm thick at suitable sites. Urine soaked earth, scraped from cattle sheds is mixed with water and sprinkled over the layer of wastes twice or thrice a day. Layering process continued for about a fortnight. A thin layer of well decomposed compost is sprinkled over top and the heap given a turning and reformed. Old compost acts as inoculum for decomposing the material. The heap is left undisturbed for about a month. Then it is thoroughly moistened and given a turning. The compost is ready for application in another month.In the regions of heavy rainfall or during rainy season's heap method may be followed to prepare the compost. Indore method, the basic piles measure 2m wide at the base, 1.5m height and 2m long. The heap is laid with 20cm layers of carbonaceous material followed by another layer of 10cm of nitrogenous material such as fresh grass, weeds, garden plant residues/prunings, digested sewage sludge etc. The pattern of filling is continued until the pile is 1.5m high. The material is moistened keeping 60-70% moisture. The pile is covered with soil or dry straw so that heat developed is not dissipated. Alternately the heap may be covered with a thin plastic sheet to retain heat. Shredding of the material speeds up decomposition. Three turnings may be given at fortnightly intervals. The good quality compost is ready within 3 months.

1.10.3.3 Bangalore Method

This method of composting is partially aerobic, was developed at Bangalore, India (Acharya, 1993). It is useful where night soil and city garbage are used for composting and is suitable for areas with scanty rainfall. The method overcomes the disadvantages of Indore method but finished compost is obtained within 6 months. The trench or pits 1m deep are made; the breadth and length of the pit can be had depending on the availability of land and quantity of material to be composted. The trenches should have preferably sloping walls and a floor of 90cm slope to prevent water logging. Organic residues and night soil are put in alternate layers and later after filling the pit is covered with 15-20cm thick layers of refuse and a thin plastic on top of it. The materials are allowed to decompose without turning and watering for about 3 months. During this period, the material gets bio-composted, and reduced in volume, this may again be charged with refuse and liquid waste or nigh soil or cattle dung in alternate layers. Refuse is again placed on the top and plastered with or covered with mud or earth to prevent breeding of flies and loss of moisture. The material undergoes anaerobic decomposition at a very slow rate and it takes about 6-8 months to get the finished compost. The recovery of the finished compost is more as compared to aerobic composting and loss of nitrogen is negligible. Labour requirement is less as compared to Indore method. In Bangalore method of composting, dry waste material of 25cm thick is spread in a pit and a thick suspension of cow dung in water is sprinkled over for moistening. A thin layer of dry waste is laid over the moistened layer. The pit is filled alternately with dry layers of material and cow dung suspension till it rises 0.5m above ground level. It is left exposed without covering for 15 days. It is given a turning, plastered with wet mud and left undisturbed for about 5 months or till required.

1.10.4 Comparison among all Methods of Composting

In Coimbatore method, there is anaerobic decomposition to start with, following by aerobic fermentation. It is the reverse in Bangalore method. The Bangalore compost is not so thoroughly decomposed as the Indore compost or even as much as the Coimbatore compost, but it is bulkiest.

1.10.5 Advantages of Composting

- Final weight of compost is very less, excellent soil conditioner and volume reduction of waste.
- Composting temperature kills pathogen, weed seeds and seeds, and suppress plant diseases and pests.
- Matured compost comes into equilibrium with the soil.
- During composting number of wastes from several sources are blended together.
- Improves manure handling and reduces the risk of pollution.
- Reduce or eliminate the need for chemical fertilizers, and promote higher yields of agricultural crops.
- Facilitate reforestation, wetlands restoration, and habitat revitalization efforts by amending contaminated, compacted, and marginal soils.
- Cost-effective remediate soils contaminated by hazardous waste.
- Remove solids, oil, grease, and heavy metals from water runoff.
- Capture and destroy 99.6% of industrial volatile organic chemicals in contaminated air.
- Provide cost savings of at least 50% over conventional soil, water, and air pollution remediation technologies, where applicable.

1.10.6 Drawbacks of Using Composts

Agricultural use of composts remains low for several reasons:

- The product is weighty and bulky, making it expensive to transport.
- The nutrient value of compost is low compared with that of chemical fertilizers, and the rate of nutrient release is slow so that it cannot usually meet the nutrient requirement of crops in a short time, thus resulting in some nutrient deficiency.
- The nutrient composition of compost is highly variable compared to

chemical fertilizers.

- Agricultural users might have concerns regarding potential levels of heavy metals and other possible contaminants in compost, particularly mixed municipal solid wastes. The potential for contamination becomes an important issue when compost is used on food crops.
- Long-term and/or heavy application of composts to agricultural soils has been found to result in salt, nutrient, or heavy metal accumulation and may adversely affect plant growth, soil organisms, water quality, and animal and human health.

1.10.7 Lime Treatment -Composting Organic Materials with High Lignin Content

- By adding organic wastes such as sawdust, wood shavings, coir pith, pine needles, and dry fallen leaves, while preparing organic waste mixtures for composting, one can ensure that the compost produced contains sufficient and long-lasting humus. However, gardeners often find that where they use lignin-rich plant materials, the compost does not ripen rapidly. A technique for making good compost from hard plant materials involves mixing lime in a ratio of 5kg/1000kg of waste material. Lime can be applied as dry powder or after mixing with a sufficient quantity of water. Treatment with lime enhances the process of decomposition of hard materials.
- Liming can enhance the humification process in plant residues by enhancing microbial population and activity and by weakening lignin structure. It also improves the humus quality by changing the ratio of humic to fulvic acids and decreases the amount of bitumen, which interferes with the decomposition process. Instead of lime, powdered phosphate rock can be used in a ratio of 20kg/1000kg of organic waste. Phosphate rock contains a lot of lime. The phosphates and micronutrients contained in phosphate rock make composts rich in plant nutrients.

Composting Weeds

Composting weeds such as parthenium, water hyacinth (*Eichornia crassipes*), cyperus (*Cyperus rotundus*) and cynodon (*Cynodon dactylon*) are used.

Required Materials and Methods

- 250g of *Trichoderma viride* and *Pleurotus sajor-caju* consortia, and 5kg of urea.

  An elevated shaded place is selected, or a thatched shed is erected. An area of 500cm×150cm is marked out. The material to be composted is cut to 10-15cm in size. About 100kg of cut material is spread over the marked area. About 50g of microbial consortia is sprinkled over this layer. About 100kg of weeds are spread on this layer. One kilogram of urea is sprinkled uniformly over the layer. This process is repeated until the level rises to 1 m. Water are sprinkled as necessary to maintain a moisture level of 50-60%. Thereafter, the surface of the heap is covered with a thin layer of soil. The pile requires a thorough turning on the 21st day. The compost is ready in about 40 days.

1.10.8 Compost Enrichment

Farm compost is poor in P content (0.4-0.8%). Addition of P makes the compost more balanced, and supplies nutrient to micro-organisms for their multiplication and faster decomposition. The addition of P also reduces N losses. Compost can be enriched by:

- Application of superphosphate, bone meal or phosphate rock: 1kg of superphosphate or bone meal is applied over each layer of animal dung. Low-grade phosphate rock can also be used for this purpose.
- Use of animal bones: These can be broken into small pieces, boiled with wood ash leachate or lime water and drained, and the residue applied to the pits. This procedure of boiling bones facilitates their disintegration. Even the addition of raw bones, broken into small pieces and added to the pit, improves the nutrient value of compost significantly.
- Wood ash waste can also be added to increase the K content of compost.
- Addition of N-fixing and P-solubilizing cultures: The quality of compost can be further improved by the secondary inoculation of *Azotobacter*, *Azospirillum lipoferum, Azospirillum brasilense* (N-fixers); and *Bacillus megaterium* or *Pseudomonas* sp. (P solubilizers). These organisms, in the form of culture broth or water suspension of biofertilizer products, can be sprinkled when the decomposing material is turned after one month. By this time, the temperature of the compost has also stabilized at about 35°C. As a result of this inoculation, the N

content of straw compost can be increased by up to 2%. In addition to improving N content and the availability of other plant nutrients, these additions help to reduce the composting time considerably.

1.10.9 Benefits of Using Composts to Agriculture

Compost has been considered as a valuable soil amendment for centuries. Most people are aware that using composts is an effective way to increase healthy plant production, help save money, reduce the use of chemical fertilizers, and conserve natural resources. Compost provides a stable organic matter that improves the physical, chemical, and biological properties of soils, thereby enhancing soil quality and crop production. When correctly applied, compost has the following beneficial effects on soil properties, thus creating suitable conditions for root development and consequently promoting higher yield and higher quality of crops.

(1) Improves the Physical Properties of Soils

- Reduces the soil bulk density and improves the soil structure directly by loosening heavy soils with organic matter, and indirectly by means of aggregate-stabilizing humus contained in composts. Incorporating composts into compacted soils improves root penetration and turn establishment.
- Increases the water-holding capacity of the soil directly by binding water to organic matter, and indirectly by improving the soil structure, thus improving the absorption and movement of water into the soil. Therefore, water requirement and irrigation will be reduced.
- Enhances the soil temperature directly by its dark colour, which increases heat absorption by the soil, and indirectly by the improved soil structure.
- Helps bind the soil particles into crumbs by the fungi or actinomycetes mycelia contained in the compost and stimulated in the soil by its application, generally increasing the stability of the soil against wind and water erosion.
- Improves soil aeration and thus supplies enough oxygen to the roots and escapes excess carbon dioxide from the root space.
- Helps moderate soil temperature and prevents rapid fluctuations of soil temperature, hence, providing a better environment for root growth. This is especially true of compost used as surface mulch.

- Protects the surface soil from water and wind erosion by reducing the soil-dispersion action of beating raindrops, increasing infiltration, reducing water runoff, and increasing surface wetness. Preventing erosion is essential for protecting waterways and maintaining the quality and productivity of the soil.

(2) Enhances the Chemical Properties of Soils

- Enables soils to hold more plant nutrients and increases the anion exchange capacity (AEC), cation exchange capacity (CEC), and buffering capacity of soils for longer periods of time after composts are applied to soils. This is important mainly for soils containing little clay and organic matter.
- Builds up nutrients in the soil. Composts contain the major nutrients required by all plants N,P,K, calcium (Ca), magnesium(Mg), and sulphur (S) plus essential micronutrients or trace elements, such as manganese (Mn), copper (Cu), zinc (Zn), iron (Fe), boron (B), and molybdenum (Mo).
- Provides active agents, such as growth substances, which may be beneficial mainly to germinating plants.
- Stabilizes the volatile nitrogen of raw materials into large protein particles during composting, thereby reducing N losses.
- Buffers the soil against rapid changes due to acidity, alkalinity, salinity, pesticides, and toxic heavy metals.
- The nutrients from mature composts are released to the plants slowly and steadily. The benefits will last for more than one season.
- Adds organic matter and humus to regenerate poor soils.

(3) Improves the Biological Properties of Soils

- Supplies food and encourages the growth of beneficial micro-organisms and earthworms.
- Helps suppress certain plant diseases, soil borne diseases, and parasites.
- Successful competition for nutrients by beneficial micro-organisms.
- Antibiotic production by beneficial micro-organisms.
- Successful predation against pathogens by beneficial micro-organisms.

- Activation of disease-resistant genes in plants by composts; and high temperatures that result from composting kills pathogens.
- Reduces and kills weed seeds by a combination of factors including rotting, the heat of the compost pile and premature germination.

Research has shown that composts can help control plant diseases (e.g. Pythium root rot, Rhizoctonia root rot, chilli wilt, and parasitic nematode) and reduce crop losses. A major California fruit and vegetable grower was able to cut pesticide use by 80% after 3 years of compost applications as part of an organic matter management system. Research has also indicated that some compost, particularly those prepared from tree barks, release chemicals that inhibit some plant pathogens.

1.10.10 Economic and Social Benefits of Composting

The economic and social benefits of composting include the followings:

- Provides a source of plant nutrients and improves soil fertility; results in significant cost savings by reducing the need for water, pesticides, fungicides, herbicides, and nematodes.
- Produces marketable products and a less-cost alternative to standard landfill cover, artificial soil amendments, and conventional bioremediation techniques.
- Used as mulch for trees, orchards, landscapes, lawns, gardens, and makes an excellent potting mix. Placed over the roots of plants, compost mulch conserves water and stabilizes soil temperatures. In addition, it keeps plants healthy by controlling weeds, providing a slow release of nutrients, and preventing soil loss through erosion.
- Used as an alternative to natural topsoil in new construction, landscape renovations, and container gardens. Using composts in these types of applications is not only less expensive than purchasing topsoil, but it can also often produce better results when establishing a healthy vegetative cover.
- Reduces or avoids landfill or combustor tipping fees, and reduces waste disposal fees and long-distance transportation costs.
- Extends current landfill longevity and delays the construction of a more expensive replacement landfill or incinerator.
- Brings higher prices for organically grown crops.
- Composting can offer several potential economic benefits to communities.

- Offers environmental benefits from reduced landfill and combustion use.
- Creates new jobs for citizens.

1.10.11 Factors affecting Composting and their Regulation

Carbon/Nitrogen Ratio

The carbon nitrogen ratio of organic materials is the most important factor in composting and so it is influenced by the proportion of carbonaceous and nitrogenous components present in the residues to be composted. Micro-organisms require carbon for growth and nitrogen for protein synthesis. A C/N ratio 30 to 40 is optimal for efficient composting and ratios between 25-40 have been found satisfactory. If the organic material is low in nitrogen that is if C/N ratio is wide, biological activity decreases and several successions or cycles of microbial activities may be required to complete the process of composting. As the availability of stock, nitrogen is exhausted during composting, the atmosphere. With a C/N ratio of raw materials less than 30: 1, the proportion of nitrogen is in excess of the requirements of the micro-organisms. In such conditions although the bioprocess goes on, the inassimilable nitrogen is lost by volatilization as ammonia or by denitrification. Low C/N ratio under unfavourable conditions cause increasing loss of ammonia. About 20-40% nitrogen may be lost as ammonia during the preparation of compost from animal dung, night soil, water hyacinth and green immature organic materials. It is clear that simple nitrogenous compounds are more susceptible to loss and control of this for a greater retention of nitrogen in the final compost is of practical value in obtaining better qualify compost. The time of composting of high C/N ratio materials can be reduced by adding a mineral nitrogen sources or a nitrogen source such as activated sewage sludge, legumes residues, grass cuttings, aquatic weeds, slaughter house wastes, green leaves and sugar industry waster.

Shredding of Materials

The bioprocess can be accelerated if raw materials are shredded into smaller parts since then it becomes more susceptible to microbial action due to greater exposure of the surface area (Gaur and Sadasivam, 1980). The most desirable particle size for composting may be in the range of 3-5cm although larger sizes can be composted satisfactorily. The effect unchopped and chopped straw on the quality of compost has been reviewed (Gaur, 1999). The results showed that chopping helped in early composting of paddy straw and sugarcane trash. However, good enriched compost was also produced from unchopped farm wastes like crop residues by increasing the period of composting by over 30 days which is required by chopped materials.

Moisture Contents

Moisture content and degree of aeration are important factors in composting. If the amount of moisture in the compost pile is below 40% (w/w) decomposition will be aerobic but slow. Aerobic decomposition will occur at any moisture content between 30% and 100% if adequate turning is provided but higher moisture content should be avoided. The optimum moisture level for aerobic composting is 50-60%.However, a range of 40-80% may be quite satisfactory depending on the nature of material to be composted. Gaur *et al.* (1982) maintained moisture content of 80-85% for aerobic composting of fibrous materials including straw.

Temperature

High temperatures are essential for destruction of pathogenic organism and weed seeds. The thermophilic condition generally occurs within two to seven days of the start of operations. The temperature in the middle of the pile goes upto 55-70°C after which gradually cools to ambient temperatures. Decomposition is fastest in the thermophilic range. The optimum temperature for organic matter decomposition into carbon dioxide and water has been found to be 60°C (Wiley and Pierce, 1955) whereas Wiley (1957) reported the maximum temperature of 71°C to be the best but should not be allowed to exceed for long as decomposition will be slowed down due to thermal kill of micro-organisms. Only a few thermophilic organisms actively carry on decomposition above 70°C. While composting high C/N ratio materials such as wheat and rice straw, sorghum and jamun *(Syzqium cumine)* leaves maintaining C/N ratio of 48 to 50, it was observed by Gaur *et al.* (1982) that the temperature of compost did not go beyond 52°C. This indicated that extent of temperature rise will depend upon the type of organic material being composted as well as the size of the pile.

pH

Decomposition will occur satisfactorily in a neutral pH because most of micro-organisms grow best under neutral pH conditions. Under aerobic conditions initially a neutral raw material will experience a decrease in pH. The drop is due to activity of acid forming bacteria. It is followed by an increase in the pH due to proteolytic metabolism and liberation of ammonia from amino acids and accounts for subsequent rise in pH. Most materials decomposing aerobically will stay within a pH range that is conducive for microbial growth thus eliminating any need for pH control (Eberhardt and Pipes, 1972).

Micro-organism

Composting is dynamic process resulting in succession of group of microbial populations with specific metabolic functions, all of which are interrelated in holistic process. Most of the organic wastes and crop residues carry indigenous populations of bacteria, actinomycetes and fungi which contribute significantly in the composting process. The fungi and acid producing bacteria appear during mesophilic stage (Chang and Hudson, 1967 and Goleuk, 1954). As the temperature increases above 40°C, these organisms are replaced by thermophilic bacteria, actinomycetes and fungi. Spore forming bacteria develop at still higher temperatures and finally, mesophilic bacteria and fungi reappear as the temperature again fall. Many aerobic bacteria are initially present and multiply but after their increased activity as the temperature raise their numbers decreased. A minimum is reached at 55-65°C but their number increase again as temperature drops below 50°C. Mesophilic bacteria *Bacillus* spp. are especially involved in degradation of proteins, amino acid, peptones and blood meal. Actinomycetes degrade starch very actively and also water soluble fractions of raw materials. Thermophilic bacteria attack proteins, lipids and hemicelluloses. Lipids are degraded extensively whereas degradation of celluloses and hemicelluloses is comparatively slow. Lignin is most resistant compound to bio-decomposition. Actinomycetes such as *Thermonospora curvata* and many fungi are involved in cellulose decomposition. Mesophilic fungi are present in compost as it warms up below 40°C and they are quickly replaced by thermophillic fungi but reappear as the heaps cool to 40°C. They probably persist in the outer layers of the heap during thermophillic stage. Thermophillic fungi occur in the temperature ranges between 40°C and 60°C.

1.11 Coir Compost

The largest byproducts of coconut are coconut husk from which coir fibre is extracted. This extraction process generates a large quantity of dusty material called coir dust or coir pith. Large quantity of coir waste of about 7.5MT is available annually form coir industries in India. In Tamil Nadu state alone 5 lakh tons of coir dust is available. Coir pith has gained importance owing to its properties for use as a growth medium in horticulture. Because of wider carbon and nitrogen ratio and lower biodegradability due to high lignin content, coir pith is still not considered as a good carbon source for use in agriculture. Coir pith is composted to reduce the wider C:N ratio, reduce the lignin and cellulose content and also to increase the manorial value of pith. Composting of coir pith reduces its bulkiness and converts plant nutrients to the available form.

1.11.1 Coir Pith Composting Technology

Collection of Raw Material

Coir Pith Heap

Coir pith is collected from the coir industry without any fiber. If fibrous materials are present, it is removed by sieving at the source itself. Otherwise, it has to be removed at the end of composting at the compost yard. These fibrous materials will not get composted and it will hinder with composting process. It is advisable to bring fibre free coir pith for composting.

Site Selection for Composting

A separate area should be earmarked for composting. It is better to have an elevated place for composting. In between coconut trees, shade under any tree is good for composting. The shady area conserves the moisture in the composting material. The floor of the compost making area should be leveled. If earthen floor is available the floor can be made to hard by hard pressing and also by applying cow dung slurry. Presence of roof over the composting material is advantageous, since it protects the material from rain and severe sunshine.Coir pith is an aerobic composting; therefore it should be heaped above the soil. There is no need for pit or cement tub to make the compost. Coir pith should be spread to the length of 1.2m and breadth of 0.90m. Initially coir pith should be put up for 7.5cm height and thoroughly moistened. After moistening, nitrogenous source material should be added. The nitrogenous source may be in the form of urea or fresh poultry litter. If urea is applied, it is recommended that 5kg urea is required for one ton of coir pith. This 5kg equally divided into five portions and in alternative layer of coir pith 1kg of urea should be applied. If fresh poultry litter is applied, it is recommended @ 200kg for one ton of coir pith.

One has to proportionally divide and put the required amount of poultry litter over the coir pith. For example if one ton coir pith is divided into 10 portions, in the first layer, 100 kg poultry litter is added. After adding, the nitrogen source, the microbial inoculums *Pleurotus* and TNAU biomineralizer (2%) are added over the material. Over this one portion of coir pith is added and the same input mentioned above should be added. It is advisable to make a heap up to minimum of 1.2m height. But beyond 1.5m, it requires machinery to handle the materials. The increase in height retains the temperature generated in the coir pith compost process. If the height is low, whatever the heat generated will be dissipated easily.

Turning of Material

The compost heap should be turned once in 10days to allow the stale air trapped inside the compost material to go out and fresh air will get in. The composting process is an aerobic one, the organism performing the composting require oxygen for its metabolic activity. This turning of material indirectly aerates the substrate. The other way of giving aeration is inserting perforated unused PVC or iron pipe in the composting material both vertically and horizontally.

Maintenance of Moisture

Maintaining optimum moisture is the pre-requisite for uniform composting or waste material. Sixty percent moisture is to be maintained and the compost material should be always wet. But excess water should not be drained from the waste material is to take a handful of composting material and put in between the palms and squeeze it. If no water is coming out of the material, that moisture status is ideal for composting.

Maturity of Compost

The period of composting vary from substrate to substrate. If all the above said conditions are maintained in the composting, it will take sixty days (60 days) for some of the physical parameters to be observed in the compost. First observation is volume reduction of waste material. When the waste material is composted, the compost heap height will be reduced by 30%. The second observation is waste materials are turned to black in colour and the waste particle size is reduced. The third observation is that composted material emits earthy odour. The chemical observation for compost maturity is to be analyzed in the laboratory. The chemical observations are narrower C:N ratio (20:1), less oxygen uptake, less number of micro-organism, more amount of available nutrients and highly cation exchange capacity.

Harvesting of Compost

The composted material which is obtained from sieving is ready for use. If the composition is not used immediately, it should be stored in an open, cool place, to retain the moisture, so that the beneficial micro-organism present in the compost will not die. Once in a month, water is sprinkled over the compost material to maintain the moisture.

Nutritive Value

Nutritive value of the raw and composted coir pith compost is given in Table 1.3.

Table 1.3: Nutritive Value of Raw and Composted Coir Pith Compost

Parameters	Raw Coir Pith (%)	Composted Coir Pith (%)
Carbon	26.00	24.00
Nitrogen	0.26	1.24
Phosphorous	0.01	0.06
Potassium	0.78	1.20
Calcium	0.40	0.50
Magnesium	0.36	0.48
Iron(ppm)	0.07	0.09
Manganese(ppm)	12.50	25.00
Zinc(ppm)	7.50	15.80
Copper(ppm)	3.10	6.20
C:N ratio	112:1	24:1
Lignin	30.00	4.80
Cellulose	26.52	10.10

1.11.2 Benefits

- Addition of composted coir dust improves soil texture, structure and tilth, sandy soil become more compact and clayey soil becomes more arable.
- It improves the soil aggregation and water holding capacity (more than 5 times its dry weight) contributing towards increased soil moisture.
- The bulk density of both the sub surface (15-30cm) soil is reduced to considerable extent with the application composted coir pith.
- There is improvement in cation exchange capacity of soils, where composted coir pith is applied.
- Composted coir dust contains all plant nutrient elements and it can provide a supplemental effect along with inorganic fertilizers.
- Coir pith compost application increased ammonification, nitrification and nitrogen fixation are increased due to improved microbiological activity because of addition of humic materials.

1.11.3 Applications

- It is recommended that 5 tons of composted coir pith per hectare of land irrespective of the raised methods.

- It is advised that composted coir pith should be applied basally before take up the sowing.
- For nursery development in poly bags and in mud pots, while preparing the potting mixture 20% of composted coir pith can be mixed with the soil and sand before filling it in the poly bag or mud pot.
- For applying to the established trees like coconut, mango, banana and other fruit bearing trees, minimum 5kg composted coir pith is required.

1.11.4 Limitation

- It is not economical to buy composted coir pith and put in the farm for large areas. It is better to prepare compost in own farm.
- If immature compost is applied to the soil, even after entering into the soil, it will undergo decomposition inside the soil, by taking nutrients from the soil. Because of this, standing crop will get affected.
- Before buying composted coir dust, it should be ensured that the material is composted completely and quality analysis certificate is available with the material.

1.12 Sugarcane Trash Composting

Sugarcane produces about 10 to 12 tonnes of dry leaves per hectare per crop. The detrashing is done on 5th and 7th month during its growth period. This trash contains 28.6%organic carbon, 0.35-0.42% nitrogen, 0.04-0.15% phosphorus, 0.50-0.42% potassium. The sugarcane trash incorporation in the soil influences physical, chemical and biological properties of the soil. There is a reduction in soil EC, improvement in the water holding capacity, better soil aggregation and thereby improves porosity in the soil. Sugarcane trash incorporation reduces the bulk density of the soil and there is an increase in infiltration rate and decrease in penetration resistance. The direct incorporation of chopped trash increases the availability of nutrients leading to soil fertility. Sugarcane trash can be easily composted by using the fungi like *Aspergillus, Penicillium, Trichurus,* and *Trichoderma.* Addition of rock phosphate and gypsum facilitates for quicker decomposition.

1.12.1 Collection of Trash

The detrashed material has to be pooled together and transported to the compost yard. If no compost yard is available to farmer, anyone of the corner area in the sugarcane field itself can be used for making composting. There is no necessity to make a pit for composting. Composting can be done above the soil.Sugarcane trash is lengthy one. Handling and heaping the trash will be more cumbersome. It is recommended to shred the waste into small particles. This process reduces the volume of material, increases the surface area of the waste. If the waste material contains more surface area, more micro-organisms work effectively on the surface and degradation will be faster. Shredder is the ideal instrument to shred all the sugarcane trash. Chop cutter machine can also be used for this purpose. If no machinery is available manual shredding is recommended. Without shredding the composting process will take long time.

1.12.2 Inputs for Composting

(a) Microbial Consortium

Generally, Tamil Nadu Agricultural University (TNAU) biomineralizer is the consortium of micro-organism recommended for composting all the agro-wastes. For one ton of trash, 2kg inoculums are recommended. Without the inoculation of microbial consortium, the composting process will take its own time. The alternative source required micro-organism for composting is cow dung slurry. But in the cow dung slurry the required population of micro-organism for composting is low and they have to compete with other micro-organisms present in the cow dung for survival. In TNAU biomineralizer, only the required micro-organism meant for composting alone is present with high population. Therefore, it is recommended to go for TNAU biomineralizer.

(b) Animal Dung

Animal dung or fresh poultry litter can be used as a source of nitrogen to reduce the C:N ratio. For one ton of sugarcane trash 50kg fresh dung is recommended. The dung can be mixed with 100 litres of water and thoroughly mixed with sugarcane trash. Rock phosphate at 5kg/ton waste can be added to increase the phosphorus content of the compost.

Making Heap Formation

After mixing all the inputs with sugarcane trash, heap should be formed with a minimum height of 1.2m. This height is required to generate more heat in the composting process, and the generated heat will be retained long time inside the material.

Turning the Compost Material

The compost material should be turned periodically once in 15days to allow more aeration inside the material. In the turning process, bottom layer comes to top and top layer comes to bottom. So that uniform composting will occur.

Control of Moisture

Throughout the composting period 60% moisture should be maintained. If composting material is allowed to dry, all the established micro-organisms get killed and composting process will be terminated. Moisture maintenance is the critical factor for good composting.

Maturity of Compost

Volume reduction, earthy odor, brownish black colour and reduction in particle size are important parameters to be observed for assessing compost maturity. Once the compost attained the maturity, the compost heap should be disturbed and spread the material for curing. After 24hrs the composted material can be sieved through 4 mm sieve to get uniform compost material. The residues available after sieving will be recycled to the next composting batch for further composting.

1.12.3 Nutritive Value

Sugarcane trash compost contains 0.5% nitrogen, 0.2% phosphorus and 1.1% potassium, in addition to micronutrients traces. Sugarcane trash compost is a good source of nutrients for sugarcane crop.

1.12.4 Applications

The enriched compost can be applied at the rate of 5 tons/ha, as basal application to the field. Whatever the compost derived from sugarcane trash, it can be ploughed back into sugarcane field to enrich the soil.

1.12.5 Limitations

- Many farmers don't have separate land for composting the sugarcane trash. In that case shredding and *in-situ* composting inside the sugarcane field can be done.
- The detrashed material should be shredded into small pieces for quicker composting.
- If the detrashed materials are put as such for composting, it requires longer time and uniform composting cannot be obtained.

1.13 Composting of Poultry Wastes

1.13.1 Value Addition of Poultry Waste through Composting Technology:

Poultry industry is one of the largest and fastest growing livestock production systems in the world. In India, there are about 3430 million populations of poultry with a waste generation of 3.30MT per year. The localized nature of poultry production also means that it can represent a large percentage of the agricultural economy in many states or regions. Although economical and successful, the poultry industry is currently facing with a number of highly complex and challenging environmental problems, many of which are related to its size and geographically concentrated nature. From an agricultural perspective, poultry wastes playa major role in the contamination of ground water through nitrate nitrogen. Also, the eutrophication of surface water due to phosphorus, pesticides, heavy metals and pathogens present in the poultry wastes applied to soils are the central environmental issues at the present time.

Among the animal manures, poultry droppings have higher nutrient contents. It has nitrogen (4.55 to 5.46 %), phosphorus (2.46 to 2.82%), potassium (2.02 to 2.32%), calcium (4.52 to 8.15%), magnesium (0.52 to 0.73%) and appreciable quantities of micronutrients like Cu, Zn, Fe, Mn etc. In addition to this cellulose (2.26 to 3.62%), hemicellulose (1.89 to 2.77 %) and lignin (1.07 to 2.16%) are also present in poultry waste. These components upon microbial action can be converted to value added compost with high nutrient status. In poultry droppings, nearly 60%of nitrogen which is present as uric acid and urea is lost through ammonia volatilization by hydrolysis. This loss of nitrogen reduces the agronomic value of the product, besides causing atmospheric pollution. Composting with amendment seems promising in conservation of nitrogen in poultry droppings. Nitrogen in poultry waste can be effectively conserved by composting with suitable organic amendment. The technologies developed will be highly useful to the poultry farmers.

1.13.2 Technology for Composting of Poultry Wastes

1.13.2.1 Preparation of poultry waste compost using paddy straw

Inputs Required: Poultry droppings, Paddy straw and *Pleurotus sajor-caju*

A known quantity of fresh poultry droppings is to be collected and mixed thoroughly with chopped paddy straw (< 2cm size) @ 1:1.25 ratios so as to attain a C/Nratio of 25 to 30 which is considered to be optimum for composting. *Pleurotus sajor-caju* is inoculated @ 5 packets (250g each) per ton of substrate. The poultry waste and paddy straw mix should be heaped under shade. The moisture content of the heap should be maintained at 50 to

60%. Periodical watering should be done once in 15 days and turning should be given on 21st, 35thand 42ndday of composting (avoid turning during first 3 weeks of composting). Within a period of 50 days, materials are converted to matured compost with the following nutrient contents;

N :1.89%, **P** : 1.83%, **K** : 1.34% and **C/N** : 12.20%

1.13.2.2 Preparation of poultry waste compost using coir pith:

Inputs Required: Poultry droppings, Coirpith and *Pleurotus sajor-caju*

a. Collection of Poultry Waste from Caged System

A layer of 5cm sea sand and 10cm coir pith should be spread in the manure collection pit of caged system where the poultry droppings are allowed to settle. Dry coir pith should be applied periodically as per the Table 1.4. After a period of three months, the partially degraded coir pith and poultry droppings mix can be transferred to compost yard and heaped under shade (Fig 1.1).

Table 1.4: Amount of Application of Dry Coir Pith

Days	Quantity of Poultry Droppings (PD) excreted (kg)	Quantity of Coir Pith (CP) to be applied (for 1000 birds) (kg)	Application rate PD : CP ratio
1	70	105.0	1 : 1.50
1-7	490	735.0	1 : 1.50
7-14	490	735.0	1 : 1.50
14-21	490	612.5	1 : 1.25
21-28	490	612.5	1 : 1.25
28-35	490	490.0	1 : 1.00
35-42	490	490.0	1 : 1.00
42-49	490	367.5	1 : 0.75
49-56	490	367.5	1 : 0.75
56-63	490	245.0	1 : 0.50
63-70	490	245.0	1 : 0.50
70-77	490	122.5	1 : 0.25
77-84	490	122.5	1 : 0.25
84-91	490	-	-

b. Collection of Poultry Waste from Deep Litter System

Dry fiber free coir pith is spread as a layer to a height of 5-10cm on the floor of the poultry production unit. The birds are grown on this coir pith bed and the droppings are collected in the coir pith. After a period of three months, partially degraded coir pith containing poultry droppings and feathers are shifted to the compost yard and heaped under shade (Fig 1.2).

c. Method of Composting Poultry Waste with Coir Pith

A known quantity of the poultry waste as collected above along with coir pith is inoculated with *Pleurotus sajor-caju* @ 2 packets per ton of waste in order to speed up the composting process. This mixer should be placed under shade as heap. The moisture content of the heap should be maintained at 50 to 60%. Periodical turning must be given on 21st,28th and 35thdays of composting. Another two packets of *Pleurotus sajor-caju* is to be added during turning given on the 28thday of composting. Good quality compost will be attained after 45thday of composting. The nutrient contents of the composts of poultry litter collected from caged system and deep litter systems are given in Table 1.5.

Table 1.5: Nutrient Contents of the Composts of Poultry Litter

Nutrient	Caged System Manure	Deep Litter System Manure
Nitrogen (%)	2.08	2.13
Phosphorous (%)	2.61	2.40
Potassium (%)	2.94	2.03
C:N ratio	13:1	14:1

Fig. 1.1 : Caged Pit System

Fig.1.2 : Composting of Waste under Field Condition

Points to be Remembered

- Elevated shady place is highly suitable.
- Within a period of 10 to 15 days, the temperature of the heap will raise to maximum. If the temperature drops below 50°C, the heaps should be spread and moistened with water to bring the moisture content to 60%.
- Colour of the compost will turn from brown to black, and matured compost will be odourless.

- The volume of the compost heap will be reduced to 1/3.
- Temperature of the heap will be same as the ambient air temperature and stable.
- Matured compost will be light and fine textured.
- Moisture content of the heap can be measured using moisture meter or by taking handful of compost from the heap and squeezing it with the fingers. If excess water drips out from the compost, then it is considered to have >60% moisture. If small quantity of water oozes out as drops, then moisture content is considered to be optimum i.e., at 60%.
- Each compost heap should have a minimum of one ton to retain the heat for post decomposition.

1.13.3 Nutritive Value

Animal manures especially poultry manure are rich in N and the nutrient value of the manure is reduced by loss of N through ammonia volatilization and denitrification. Good quality poultry manure can be obtained by mixing the poultry waste with selective carbonaceous material such as coirpith and inoculation with suitable micro-organism. It can be used as an eco-friendly technique for the conversion of poultry waste into valuable compost.

1.13.4 Benefits

Poultry wastes contain higher concentrations of nitrogen, calcium and phosphorus than wastes of other animal species and the presence of nutrients provides more incentive for the utilization of this resource. The loss of nitrogen from poultry droppings can be effectively conserved by composting with coir pith and serves as a good source of organic nutrients to agricultural fields. To make the organic nutrients present in poultry waste available to plants, the waste has to be composted suitably to minimize the volatilization of ammonia.

1.13.5 Applications

This technology is widely suitable and applied to the poultry farmers to utilize the solid waste in an effective manner. The poultry waste compost will be a very good organic manure@6 ton/ha for all the crops.

1.13.6 Limitations

The uninterrupted availability of the raw materials has to be ensured for continuous production on a commercial scale.

1.14 Crop Residue Composting

Crop residues are the non-economic plant parts that are left in the field after harvest. The harvest refuses include straws, stubble, stover and haulms of different crops. Crop remains are also from thrashing sheds or that are discarded during crop processing. This includes process wastes like groundnut shell, oil cakes, rice husks and cobs of maize, sorghum and cucumber. The greatest potential as a biomass resource appears to be from the field residues of sorghum, maize, soybean, cotton, sugarcane etc. In Tamil Nadu 190 lakh tonnes of crop residues are available for use. These residues will contribute 1.0 lakh ton of nitrogen, 0.5 lakh ton of phosphorus and 2.0 lakh tons of potassium. However, crop residues need composting before being used as manure.

1.14.1 Collection of Waste

Crop residues accumulated in different locations are to be brought to compost yard. The compost yard is located in anyone corner of the farm with accessibility via good road. Water resource should also be available in sufficient quantity. The crop residues that are brought to compost yard should be heaped in one corner for further processing.

1.14.2 Shredding of Waste Materials

Particle size is one of the factors that influence the composting. It is advisable to shred all the crop residues that are used for composting. Shredding the waste manually is labour intensive. Shredder machine can be employed to shred all the crop residue biomass. Particle size of 2 to 2.5cm is recommended for quick composting.

1.14.3 Mixing of Green Waste and Brown Waste

Carbon and nitrogen ratio decides the initiation of composting process. If C:N ratio is wide (100:1) composting will not take place. Narrow C:N ratio of 30:1 is ideal for composting. To get a narrow C:N ratio, carbon and nitrogen rich material should be mixed together. Green coloured waste materials like glyricidia leaves, parthenium, freshly harvested weeds; sesbania leaves are rich in nitrogen, whereas brown coloured waste material like straw, coir dust, dried leaves and dried grasses are rich in carbon. In any composting process these carbon and nitrogen rich material is to be mixed together to make the composting quicker rather than putting green waste alone or brown color waste alone for composting. Animal dung is also a good source of nitrogen. While making heap formation, alternative layers of carbon rich material, animal dung and nitrogen rich material are to be heaped to get a quicker result in composting.

1.14.4 Formation of Compost Heap

Minimum 4 feet height should be maintained for composting. The composting area should be elevated one and have sufficient shade. While heap formation, all the crop residues should be mixed together to form a heterogeneous material rather than a single homogenous material. Alternate layers of carbon and nitrogen rich material with intermittent layers of animal dung are essential. After heap formation the material should be thoroughly moistened (Fig 1.3 &1.4).

1.14.5 Bio-inputs for Composting

One ton of crop wastes 2kg of biomineralizer are recommended for accelerating the compost process. This 2kg biomineralizer should be mixed with 20 liters of water and made slurry. When the compost heap is formed in between layers the slurry should be inoculated, so that it mixes with the waste material thoroughly for uniform coating of micro-organism on the waste material. Cow dung slurry is also a good source for microbial inoculum. But it carries unwanted micro-organisms also which may compete with composting organism. But when biomineralizer is not available, cow dung slurry is a good source material. For one ton crop residues 40kg fresh cow dung is required. This 40kg fresh cow dung is mixed with 100litres of water and it should be throughly poured over the waste material. Cow dung slurry acts as nitrogen source as well as source of microbial inoculum.

Fig. 1.3 : Bio-comopost Preparation with Farm Waste

Fig. 1.4 : Separation of Composted Materials through Sieving

1.14.6 Aerating the Compost Material

Sufficient quantity of oxygen should be available inside the compost heap. For this external air should be freely got in and comes out of the material.

Normally to allow the fresh air to get inside, the compost heap should be turned upside down, once in fifteen days. In this process top layer comes to bottom and bottom layer goes to top. This process also activates the microbial process and compost process is hastened. In some cases air ventilating pipe may be inserted vertically and horizontally, to allow the air to pass through. The wood chip that is available as waste in wood processing industry may also be used as bulking agent in the composting process. This bulking agent gives more air space to the compost material.

1.14.7 Maintenance of Moisture

Throughout the composting period 60% moisture should be maintained. On any situation, compost material should not be allowed to dry. If the material becomes dry, all the micro-organisms present in the crop residues will die and the compost process gets affected.

1.14.8 Maturity of Compost

Volume reduction, black colour, earthy odor, reduction in particle size is all the physical factors to be observed for compost maturity. After satisfying with the compost maturity index, the compost heap can be disturbed and spread on the floor for curing. After curing for one day, the composted material is sieved through 4 mm sieve to get uniform composted material. The residues collected after composting has to be again composted to finish the composting process.

1.14.9 Compost Enrichment

The harvested compost should be heaped in a shade, preferably on a hard floor. The beneficial micro-organisms like *Azotobacter* or *Azospirillum, Pseudomonas, Phosphobacteria* (0.2%) and rock phosphate (2%) have to be inoculated for one ton of compost. 40% moisture should be maintained for the maximum growth of inoculated micro-organism. This incubation should be allowed for 20 days for the organism to reach the maximum population. Now the compost is called as enriched compost. The advantage of enriched compost over normal compost is the quality manure with higher nutrient status with high number of beneficial micro-organisms and plant growth promoting substances.

1.14.10 Nutritive Values

The nutritive value of bio-compost varies from lot to lot because of varying input materials. But in general bio-compost contains all the macro and micro nutrients required for crops, which is given in Table 1.6. Even though the quantity available is low it covers all the requirements of the crop.

Table 1.6: Nutrient Content of Bio-compost Prepared from Different Crop Residues

Bio-compost	Nutrient Content (%)		
	Nitrogen (N)	Phosphorous (P)	Potash (K)
	Animal Refuse		
Cattle dung	0.3 - 0.4	0.1 – 0.2	0.1 – 0.3
Horse dung	0.4 – 0.5	0.3 – 0.4	0.3 – 0.4
Sheep dung	0.5 – 0.7	0.4 – 0.6	0.3 - 0.1
Night soil	1.0 – 1.6	0.8 – 1.2	0.2 – 0.6
Poultry manure	1.8 – 2.2	1.4 – 1.8	0.8 – 0.9
Sewage sludge	2.0 – 3.5	—	—
Cattle urine	0.9 – 1.2	Trace	0.5 – 1.0
Horse urine	1.2 – 1.5	Trace	1.3 – 1.5
Sheep urine	1.5 – 1.7	0.1 – 0.2	0.1 – 0.3
Wood Ash	—	—	—
Ash coal	0.73	0.45	0.53
Ash wood	0.1 – 0.2	0.8 – 5.9	1.5 – 36.00
	Habitation Waste and Factory Waste		
Rural compost	0.5 – 1.0	0.4 – 0.8	0.8 – 1.2
Urban compost	0.7 – 2.0	0.9 – 3.0	0.3 – 1.9
Farmyard manure	0.4 – 1.5	0.3 – 0.9	2.0 – 7.0
	Straw and Stalk		
Pearl millet	0.65	0.75	2.50
Cotton	0.44	0.10	0.66
Banana pseudo stem	0.61	0.12	1.00
Sorghum	0.40	0.23	2.17
Maize	0.42	1.57	1.65
Paddy straw	0.36	0.08	0.71
Tobacco	1.12	0.84	0.80
Pigeon pea	1.10	0.58	1.28
Sugarcane trash	0.53	0.10	1.10
Wheat	0.53	0.10	1.10
Tobacco dust	1.10	0.31	0.93

1.14.11 Benefits

- There is improvement in the physical, chemical and biological properties of the soil due to regular addition of bio-compost.
- Quality products will be obtained from the crop due to improvement in the soil fertility status.
- Quality and enriched manure from the crop and animal residues available in the farm. The manure contains both nutrients and beneficial micro-organisms.
- Soil organic matter content increased and soil biodiversity also improved due to enhanced soil organic matter content.

1.14.12 Application

Organic manures are highly regarded as good source of material to maintain soil health and increasing soil organic carbon content. Organic manures cannot be equated with inorganic fertilizers. But organic manures deliver all the nutrients to the soil but with little quantity. For one hectare of land 5 tons of enriched bio compost is recommended. It can be used as basal application in the field before taking up planting work.

1.14.13 Limitation

- It is better avoid woody material like heavy branches from pruned trees and other wooden materials. It will take long time and it interferes with other material for composting.
- While preparing the bio-compost, it should be ensured that the material is composted thoroughly.
- If the materials are not fully composted, the material should be sieved through 4mm sieve and sieved material will be taken as well as composted one. The residues will be put back for another round of composting.

1.15 Organic Certification Agencies

1) Aditi Organic Certifications Pvt. Ltd, No. 531/A, Priya Chambers, Dr. Rajkumar Road, Rajajinagar, 1st Block, Bangalore-560010, Tel.: +91-80-32537879, Fax: +91-80-23373083, Mobile: +91-9845064286 Email: aditiorganic@gmail.com, Website: www.aditicert.net
2) APOF Organic Certification Agency (AOCA) #3, 1st floor, 9th cross,

5th main, Jayamahal Extn, Bangalore–560046, Phone No: 080-55369888, Fax: 080-23430155, Email: aocabangalore@yahoo.co.in

3) Bureau Veritas Certification India Pvt. Ltd. (Formerly known as BVQI (India) Pvt. Ltd.), 6th Floor, Opp. Ansa Industrial Estate, Krishanlal Marwah Marg Off Saki-Vihar, Road, Andheri (East), Mumbai-400 072, (Maharashtra), Phone No.: 022-56956300, 56956311, Fax No. 022-56956302 / 10 , Email: scsinfo@in.bureauveritas.com

4) Chhattisgarh Certification Society, India (CGCERT), Raipur, A-25, VIP Estate, Khamhardih, Shankar Nagar, Raipur-492007, Chhattisgarh (India) Telefax : +91-771-283249, Email:cgcert@gmail.com

5) Control Union Certifications (Formerly known as Skal International (India)) "Summer Ville" , 8th Floor, 33rd – 14th Road, Junction Off Linking Road, Khar (West) Mumbai –400052 (Maharasthra) Phone 022-67255396/97/98/99 Fax 022-67255394/95 Email: cuc@controlunion.in; cucindia@controlunion. com controlunion@ vsnl.com

6) ECOCERT India Pvt. Ltd, Sector-3, S-6/3 & 4, Gut No. 102, Hindustan Awas Ltd. Walmi-Waluj Road, Nakshatrawadi Aurangabad – 431 002 (Maharashtra) Phone No.: 0240-2377120, 2376949 Fax No.: 0240-2376866 Email: ecocert@sancharnet.in

7) Food Cert India Pvt. Ltd, Quality House, H.No. 8- 2- 601/P/6, Road No. 10, Banjara Hills, Panchavati Colony, Hyderabad – 500 034, Tel. No.: +91- 40-23301618, 23301554, 23301582, Fax: +91-40-23301583, Email: foodcert@foodcert.in

8) IMO Control Pvt. Ltd. P.O. Aluva-683 105 Cochin, (Kerala) Telefax:0484-2630908-09/2620943 Email:Mathew.Sebastian@indocert.org

9) ISCOP (Indian Society for Certification of Organic Products), Rasi building, 162/163, Ponnaiyarajapuram Coimbatore–641 001, Tamil Nadu Mob. No.: 094432 43119

10) Lacon Quality Certification Pvt. Ltd Theepany, Thiruvalla - 689 101.(Kerala) Telefax: 0469 2606447, Email: laconindia@sancharnet.in

11) Natural Organic Certification Agency, House Ground Floor, Near P. N. Gadgil Showroom Pune-411 038 (Maharashtra), Phone No.: 020-25457869, 56218063 Fax: 020-2539-0096 Email: contact@nocaindia.com

12) One Cert Asia Agri Certification Pvt. Ltd. Agrasen Farm, Vatika Road, Vatika P.O., Off Tonk, Jaipur-303 905, (Rajasthan) Phone No. : - 0141-2770342 Telefax No: - 0141-2771101 Email: info@ onecertasia.in

13) Rajasthan Organic Certification Agency (ROCA), 3rd Floor, Pant Krishi Bhawan, Janpath, Jaipur 302 005, (Rajasthan),Phone No: 0141-2227104, Tele Fax: 0141-2227456 Email: dir_rssopca@rediffmail. com

14) SGS India Pvt. Ltd. Food, Retail & CSRS250 Udyog Vihar Phase – IV Gurgaon – 122 015 (Haryana) Phone No.: 0124-2399990-98, Fax No.: 0124-2399764 Email: namit_mutreja@sgs.com

15) Tamil Nadu Organic Certification Department (TNOCD) Coimbatore Thadagam Road, Coimbatore-641013, Tamil Nadu (India) Tel.: +91-422-2405080, Fax: +91-422-2457554. Email: tnocd@yahoo.co.in

16) Uttranchal State Organic Certification Agency (USOCA) Director 12/II Vasant Vihar Dehradun-248 006 (Uttaranchal) Phone No.: 0135-2760861 Fax: 0135-2760734 Email: uss_opca@rediffmail.com Tel. No.: 0422-2544199; 0422-6586060 E-mail: profdrkkk@yahoo. com; iscopcbe@yahoo.co.in

17) Vedic Organic Certification Agency Plot No. 55, Ushodaya Enclave, Mythrinagar, Miyanagar, Hyderabad – 500 050, Mobile No.: 09290450666, Tel. No.: 040-65276784, Fax:040-23045338, Email: voca_org@yahoo.com; Website: www.iscoporganiccertification.org *Source*: http://ncof.dacnet.nic.in/Training_manuals/Training_manuals_in_English/Organic_Agriculture_in_India.pdf

1.16 Microbial Inoculants

It has been observed in preparation of sawdust compost that inoculation with spores of cellulolytic fungus *Coprinus ephemerus* and amendment with plant nutrients (NPK) hastened the period of composting from the usual one or two years to three months and produced a compost that did not immobilize nitrogen as compared with the fresh sawdust (Davey, 1953, Wilde, 1958). The work done in India on microbial inoculants has shown their benefits in composting of farm wastes (Gaur, 1999).

1.16.1 Calcium Phosphate

The application of rock phosphate is important in composting when raw materials have wide C/P ratio (rice straw and dry leaves etc.). However, the quality of compost prepared from a mixture of rice straw, grass and water hyacinth was improved when rock phosphate was added both with and without pyrites. The enrichment of compost with rock phosphate besides increasing phosphorus content increased the content of calcium and micronutrients, particularly zinc, iron, manganese and copper. The benefits of rock phosphates in composting have been reported by several workers in India (Gaur *et al.,* 1995).The amendment of paddy straw with 2-3% levels of Mussorie rock phosphate (MRP) along with combined inoculation of *Aspergillus niger* and *Azotobacter chroococcum* accelerated the process of composting by reducing the organic carbon per cent and simultaneous enrichment in nitrogen and phosphorus content upto 3% of rock phosphate (Gaind and Gaur, 2000).

1.16.2 Destruction of Pathogens and Parasites

Composting temperatures in windrow's can easily rise to 70°-75°C and can be held for comparatively long periods. Up to 10 days is normal, and if, in windrowing plant, turning is done when the temperature has reached a plateau, and the temperature will drop sharply during the turning and restore itself to a slightly lower peak. This peak will get progressively lower with each turning till the compost is finally matured and the peak temperature is only slightly above the ambient. The destruction of pathogens and parasites of animal and human should be taken into consideration while composting materials containing sewage sludge and municipal wastes. The fact that fly larval breeding can effectively be controlled by composting solid wastes has been long known.

Acharya (1948) reported that pathogenic organisms in rural solid wastes which may lead to serious outbreaks of diseases can be destroyed by composting. Pathogenic micro-organisms from composted solid wastes could not be isolated (Van Vuren, 1950, Blair 1952). Gotaas (1956) recommended a temperature of 60°C for thermal kill of common pathogens. *Salmonella typhosa* does not survive beyond 46°C and death of these organisms occurs within an hour at 55-60°C and within twenty minutes at 60°C. *Shigella* sp. which causes dysentery and *Escherichia coli*common in sewage are similarly killed. *Mycobacterium tuberculosis* var. *hominis* does not survive at 66°C or above and their death occurs within less than twenty minutes at such temperatures. *Entamoeba histolytica* cysts which cause amoebic dysentery are destroyed within a few minutes at 45°C and only within a few seconds at 55°C. Tapeworm and Hookworm are quickly killed at 55°C. *Ascaris lumbricoides* eggs which develop as round worms and suck blood die within an hour at temperature above 50°C. Other pathogens are also killed within a few minutes at 55°C and above. Cooper and Goluek (1977) reported that the population of faecal coliforms, streptococci and coliphage were reduced drastically during the composting process. Bhaskaran (1957) reported that pathogens present in human excreta can be effectively destroyed by thermophilic stage of composting. Morgan and McDonald (1969) found that *Mycobacterium tuberculosis* did not survive composting process.

Nandkishore (1980) reported that pile height and the degree of aeration influenced significantly that death point of *Salmonella typhimurium* was found to be 65°C but an exposure at 50°C for twenty four hours was sufficient to kill the inoculated *S. typhimurium* in raw wastes or soil. De Bartoldi *et al.* (1980) observed that coliform bacteria, *Salmonella* sp., *Streptococcus* sp. and *Aspergillus fumigatus* were inactivated at varying temperature in the range of 58 to 75°C. *Streptococcus* and *A. fumigatus* were relatively more persistent. The maximum temperature of about 70°C was, however, obtained during the first ten days of composting.The extent and duration of high temperature in compost heaps as well as antibiosis as a result of mixed population of micro-organisms provide reasons to believe that pathogens and parasites are not able to survive the composting process. However, to achieve this physically, at least turnings of the heap are necessary to develop higher temperature.Anaerobic composting at mesophilic temperature range does not satisfactorily destroy the pathogens/parasites in a short time. Anaerobic composting of sewage or night soil should be preceded by aerobic composting for at least a week. The natural death of parasites and pathogens will occur in the anaerobic environment and microbial antagonism will eventually eliminate them but this will generally take a longer period of time, about six months.The composting process can serve useful purpose to obtain pathogen-free manure for use on agricultural or horticultural land. The full benefit of the process can be realized if the variable factors such as moisture, aeration, temperature, carbon-nitrogen ratio and nutrient balance are properly controlled.

1.16.3 Heavy Metals in Compost

It has long been recognized that plants require not only nitrogen, phosphorus and potash for their growth but also small amounts of other metals such as iron, calcium, magnesium, copper, zinc, sodium, manganese, chlorine, cobalt, molybdenum and selenium.The use of compost can increase the content of metals in the soil. Although, the total metal in soil is increased, but it may not be in available form for the uptake of these metals by plants. Their availability would depend on the soil organic carbon and pH. The higher is the humus content and pH, the lower is the availability of some metals to the plants.

The quantities of lead and cadmium present in municipal compost when it is applied to soil and mixed into the top 12-20cm of soil gets diluted. Moreover the plants would remove some of the metals for their growth so that the actual build-up would be slower.The only difference between an "essential trace element" and a; "heavy toxic metal" is one of the quantity and position. If zinc, lead and boron are naturally present in soil, then they are referred to as essential trace elements. If they are present is sewage sludges, their anticompost or anti-sludge lobbies refer to them as toxic metals. In practice, a vast amount of research has been carried out to investigate the complex relationship between municipal compost and plants by research workers of several universities and institutions and so for none have found any cause for concern in the use of municipal compost.

1.16.4 Cellulolytic and Lignolytic Micro-organisms

Cellulose and lignin are the most abundant single organic compounds in the biosphere. These make up the highest proportion of dead plant remains which are mainly degraded by micro-organisms. This degradation is ecologically of great importance since if these materials were not largely broken down, there might be a decrease in atmospheric carbon dioxide concentration. Partially degraded cellulose and lignin accumulate in soils as humus and under some conditions given rise to peat which may eventually be converted to lignite or coal.

1.16.5 Biodegradation of Cellulose

Cellulose is a β-1,4 linked linear homopolymer of anhydro-D-glucose which is of very high molecular weight. Cellulose molecules are linked together to form microfibrils. Cellulose is typically found in the walls of plant cells which have secondary thickening. These cell walls also contain pectin, lignin and hemicellulose. Cellulose is susceptible to biodegradation by means of hydrolytic enzymes called cellulases which convert it to glucose and oligomers thereof. The saprophytic micro-organisms (fungi, bacteria and protozoa) are predominant as cellulose degraders. The cellulolytic ability of protozoa was in doubt for many years due to the difficulty in isolating axenic, bacteria free cultures. Such cultures have now been isolated and demonstrated to produce carboxymethyl cellulase (Yamin and Trager, 1979). It is however still possible that *in vivo* some protozoa live in symbiosis with cellulolytic bacteria relying on bacterial cellulase. Cellulose is generally degraded extra-cellularly (except by protozoa) to soluble products which can diffuse to cellulase producing cells.

A variety of microbes are involved in its degradation. The main genera of bacteria are *Cellulomonas,* the non-ruling gliding bacteria (*Cytophaga* and *Sprorocytophaga*) the myxobacteria (*Angiococcus* and *Polyangium),* some strains of *Pseudomonas, Bacillus* and many of the actinomycetes (*Streptomyces, Micromonospora* and *Thermoactinomyces*). There are large numbers of cellulolytic fungi active in soil and leaf litter e.g. *Trichoderma, Chaetomium, Fusarium, Aspergillus, Penicillium, Rhizoctonia* and *Verticillium.* The fungi are more important in the ordinary soils, leaf litter and fairly dry crop residues. They are also important in cereal straw degradation as are actinomycetes. Both these groups are involved in aerobic composting of vegetable wastes. In this process there is usually a considerable increase in temperature upto 60 or 70°C which gives rise to successional changes in the microflora selecting for thermophiles such as *Thermoactinomyces* and the fungi *Chaetomium thermophile* (Lacey, 1979). Cellulose degradation is important in anaerobic habitats. Strictly anaerobic bacteria are usually the dominant cellulolytic organisms. *Cellulolytic clostrida* such as *C. cellobioparum* and *C. thermocellum* are found in anaerobic soils and sediments.

1.16.6 Factors affecting Cellulose Decomposition

a) **Available nitrogen level:** The application of inorganic nitrogen enhances cellulose breakdown in soil. The decomposition rate is proportional to the concentration of nitrogen added. The point at which additional quantities are no longer beneficial is at a ratio of 1 part of inorganic nitrogen for each 35 parts of cellulose.

b) **Temperature:** Cellulose decomposition can proceed from 5-65°C. Mesophiles dominate at moderate temperature. Thermophiles are active above 45°C.

c) **Aeration:** Aerobes dominate oxygenated environments and anaerobic bacteria being favoured by decreasing partial pressure of O_2. High soil moisture creates poor drainage and it is associated with proliferation of anaerobic cellulolytic bacteria.

d) **pH:** In environment of neutral to alkaline pH, many micro-organisms, are capable of growing and liberating the appropriate enzymes for cellulose hydrolysis. Under acid condition the disappearance of cellulose is mediated largely by filamentous fungi. Soils with lower pH degrade cellulose more rapidly.

e) **Presence of other carbohydrates:** The micro-organisms grow poorly in media containing purified cellulose as the sole source of carbon. Addition of readily metabolizable organic to soil accelerates cellulose decomposition.

1.16.7 Enzymology of Cellulolytic

Cellulose, being an insoluble polymer of very large size cannot permeate the microbial cell membrane. This means that in fungi and bacteria cellulose must be degraded extracellularly. In ciliate protozoa, cellulose particles can be taken up by phagocytosis, so degradation may be intracellular, which obviously reduces competition from other microbes for the released sugars. It is desirable to know whether enzymes are cell free or cell bound, but this can be difficult to assess as cellulases often have a high affinity for their substrate. If cellulase cannot be found in a culture filtrate, it may be either cell or substrate bound. This is probably the reason why some of the most vigorous cellulolytic microbes (*Myrothecium, B. succinogenes* and *Cytophaga*) actually seem to produce little detectable cellulase.

Reese *et al.,* (1950) postulated that for a cellulose to attack crystalline cellulose two factors were required C_1 and C_x. C_x was carboxymethyl cellulose able to attack soluble derivatives and amorphous cellulose but without activity against crystalline cellulose (e.g. cotton). C_1 was a factor which enabled C_X to act on cotton. C_1 is in fact an exo-acting β 1, 4 glucanase (often referred to as exo-cellulase or cellobiohydrolase, which removes cellobiose from the non-reducing end of the polymer). It is now widely (Gong and Tsao, 1979) but not universally (Reese, 1977) accepted that C_1 and cellobiohydrolase are identical. C_x is now more usually called endo β-1, 4 glucanase and is considered to be randomly acting enzyme hydrolyzing bonds within the polymer chain. The third member of the complex is β-1, 4 glucosidase (cellobiase) which hydrolyses cellobiose to glucose. Thus, cellulose degradation can be explained in the following chain reaction:

$$\boxed{\text{Native Cellulose}} \xrightarrow{C_1} \boxed{\text{Long Glucose Chain}} \xrightarrow{C_X} \boxed{\text{Cellobiose}} \xrightarrow{B.\ glucosidase} \boxed{\text{Glucose}}$$

1.16.8 Microbiology of Lignin Biodegradation

Lignin is highly resistant to biodegradation. As lignin is heterogeneous in bond type and positioning and also most of the bonds are not amenable to hydrolytic cleavage. Lignin is insoluble and difficult to wet. Lignin undergoes no significant decay in anaerobic environments and it is likely that molecular O_2 is required for its degradation. At one time it was considered that the only microbes capable of extensive lignin degradation were the white rot fungi and related litter degrading basidiomycetes. The ^{14}C-assay has allowed a reappraisal of microbial lignolysis. Brown rot fungi cause some modification of lignin, notably extensive demethylation and possibly limited oxidation. Soft rot fungi are preferentially cellulolytic. Bacteria were considered non-lignolytic but do cause slight modification. It is now known that a large number of species can cause degradation, notably species of *Nocardia, Streptomyces* and *Bacillus* but also *Pseudomonas, Flavobacterium, Aeromonas* and *Xanthomonas*. Other than wood rotters, fungi such as strains of *Penicillium, Fusarium* and *Aspergillus* have been shown to degrade lignin.

1.16.9 Microbial Physiology and Enzymology of Lignolysis

More information is available on the white rot fungi than other organisms. It has been shown that for *Phanerochaete chrysosporium* and *Coriolus versiocolor* growth, they cannot be supported by lignin alone and these functions require a co-substrate like a carbohydrate. *P. chrysosporium* requires O_2 for growth and high O_2, for lignolysis which is also very pH sensitive. The nature of N-source is unimportant but the concentration is crucial. Lignolysis being strongly inhibited on high nitrogen concentration, lignin degradation begins only after the cessation of linear growth, possibly in response to N starvation. It has also been suggested that it may be related to secondary metabolism. The presence of lignin is not essential for lignolytic activity to be produced (Chang *et al.*, 1980).

Analysis of degraded lignin makes it clear that the attack is oxidative and suggests some of the reactions involved. It is thought that demethylation occurs initially giving rise to dihydroxyphenol moieties which are oxidatively cleaved to give carboxyl rich residues. These modifications occur while the polymer is still intact. Further oxidative and hydrolytic reactions then degrade the new aliphatic chains. Lignin side chains are also attacked with hydroxy groups being oxidized to carbonyl and carboxyl groups and β aryl ether bonds may be cleaved. These reactions suggest that lignolysis is not a depolymerisation of the normal type, lignin undergoes an extensive modification which may ultimately lead to destruction in which basic units are degraded while still attached to polymer. A similar type of attack on soft wood lignin has been proposed for *Streptomyces viridosporus* by Crawford *et al.* (1982).

Due to its polymeric nature the enzymes involved in lignin degradation must be extracellular. It is however difficult to obtain culture filtrates with appreciable lignolytic activity, the enzymes being cell or substrate bound. To summarize, it is generally envisaged that lignin will require mono and dioxygenase enzymes to perform demethylation, hydroxylation, and ring cleavage and side chain oxidation. Tien and Kirk (1985) demonstrated that white rot fungus *Phanerochaete chrysosporium* can produce an extracellular enzyme capable of degrading lignin. The enzyme requires H_2O_2 for its activity. A mechanism for the mode of the action of this enzyme has been suggested (Schoemaker *et al.,* 1985; Harvey *et al.,* 1985). This proposes that the reaction proceeds via a single electron transfer oxidation and thus that the enzyme is a peroxidase rather that an oxygenase.

1.16.10 Azotobacter and Phosphate Solubilizer as Inoculant in Composting

After thermophilic phase of composting is over, an efficient strain of *Azotobacter chroococcum* and a phosphate solubilizers (*Aspergillus awamori*) can be inoculated in compost mass where organic matter, reaction and phosphorus content are optimum, During the process, the compost will be enriched with efficient strains of *Azotobacter* and phosphate solubilizers and the nitrogen and available phosphorus content of compost will be improved. It also plays an important role in ripening of compost and production of growth promoting substance and antibiotics can be additional gains of *Azotobacter* and phosphate solubilizers. Enriched compost will profitably increase the land fertility and crop yields. The role of PSF and nitrogen fixing bacteria in enrichment of phosphorous and nitrogen content of compost is now well recognized (Gaur, 1987).

1.16.11 Composting of Urban Wastes

The city solid waste is a major and voluminous resource which has increased tremendously with increasing urbanization. With increasing water treatment plants, sludge has become another potential resource to be processed by composting. The composted sludge has been reported to safe to handle and useful for application to soil. On an average, it has been found to contain 1.6%N, 1.0%P and 0.2%K on dry weight basis (Parr *et al.,* 1978). The composting of sludge is not being practiced so far in India. The available sludge is directly used as organic manure, without looking into the aspects of health and pollution hazards. Upto 1965, town-refuse composting was an organized practice in India and town compost was considered valuable as organic manure. However, subsequently the practice was discontinued probably due to increasing volumes of refuse on one hand and disinterest in the use of organic manure on the other.

1.16.12 Mechanical Compost Plants in India

The rural methods of composting are not proving suitable and convenient for handling large amounts of wastes from towns and big cities. The mechanism of composting has several advantages such as environmental sanitation, to minimize pollution recovery of discarded materials, handling of enormous quantities of solid wastes and production of compost in less time. The large metropolitan cities like Calcutta, Chennai, Delhi and Mumbai generate about 3,000-6,000 tonnes of urban wastes per day causing disposal problems. (Report-Ministry of Urban Development, 2003). The ideal size of compost plants may be those with a daily intake capacity of 150-200 tonnes/day of urban waste. The mechanical composting plants were commissioned in India in 1970's. Past experience in India shows that inspite of being heavily subsidized by the Government, mechanical plants have not been very successful for various reasons. A status report by the Ministry of Agriculture in 1985 showed that the capacity utilization of plants erected in several cities to be 8-43% with a majority of them not operating above one-third capacity.

From manurial point of view, compost produced at Mechanical plants has been reported to be quite useful for land application (Gupta *et al.,* 1986, Velayutham and Bhardwaj, 1994). However, generally compost produced from city wastes is relatively poor in plant nutrient and humus content.

1.16.13 Techno-economic Aspect

Setting up of compost plants involves large capital investments. The resources of the agencies setting up these plants are not enough to finance these from their own resources. Thus, they have to resort to financial assistance from banks. Due to high cost of interest, the cost of production of compost goes up beyond the reach of an average farmer. The consultancy service for proper maintenance and services are also not available to the desired extent. Potential marketing areas are not clearly identified and serious promotional measures are lacking.

1.16.14 Quality of Compost from City Wastes

The quality and plant nutrient content of compost from mechanical compost plants can be improved by using earthworms. Infact, very substantial role of earthworms has been described in the composting of sludge and city solid wastes (Haimi and Huhta 1987, Glogowski, 1992) Nutrient upgrading could be possible by addition of sludge or sewage water (Gaur, 1987).

1.17 Green Manuring

Green manuring is an excellent practice for adding substantial nitrogen and appreciable quantities of phosphorus, sulphur, potassium, etc. The process of green manuring also adds some humus fractions but due to rapid decomposition of organic matter, the build of humus substance is comparatively less than the residues of cereals. Both leguminous and non-leguminous plants can be used for green manuring. Green undecomposed material used as manure is called green manure. It is obtained in two ways: by growing green manure crops or by collecting green leaf (along with twigs) from plants grown in wastelands, field bunds and forest. Green manuring is growing in the field plants usually belonging to leguminous family and incorporating into the soil after sufficient growth. Plants those are grown for green manure known as green manure crops. The most important green manure crops are sunnhemp, dhaincha, pillipesara, cluster beans and *Sesbania rostrata* (Table 1.7 and 1.8 and Fig. 1.5).

a) ***Sesbania rostrata*** bears nodule on the stem as well besides root nodulation. It has been introduced in India from IRRI, Philippines. It is a tropical legume and thrives well under flooded and waterlogged conditions producing stem nodules. Nitrogen fixed by this green manuring crop is quite substantial. *Sesbania rostrata* is a stem nodulating green manure crop which is a native of West Africa. As it is a short-day plant and sensitive to photoperiod, the length of vegetative period is short when sown in August or September. A mutant (TSR-l) developed by Bhabha Atomic Research Centre (BARC), Bombay is insensitive to photoperiod, tolerant to salinity and waterlogged condition. Growth and nitrogen fixation is higher with TSR-l compared to the existing strains.

b) ***Sesbania aculeate*** is a quick growing succulent green manure crop. It adapts itself to varying conditions of soils and climate. It can be grown even in adverse conditions of drought, waterlogging, salinity etc. It grows satisfactorily even in alkaline soils and corrects alkalinity if grown repeatedly in such problem soils. Abundant root nodules are formed by indigenous or introduced rhizobia, within 2 to 4 months, green biomass in the range of 10 to 20 tonnes/ha are produced. Depending on the age of crop and harvest the stem gets woody and fibrous after three months of growth.

c) ***Crotalaria juncea*** is a very quick growing green manure-cum-fibre crop. It comes up well in loamy and heavy soils. It does not withstand continuous water logging or heavy irrigation. The crop even can be cut when it is about 45 days old. There are a number of varieties, growing duration varying from 75 to 150 days. The tall and late varieties are used for fibre extraction also.

1.17.1 Different Sources of Organic Wastes

Livestock and human wastes, crop residues, tree wastes and aquatic weeds, green manures, urban and rural wastes, agro-industrial wastes and marine wastes.

Fig. 1.5 : Some Most Important Green Manure Crops

Table 1.7: Biomass Productions and Nitrogen Accumulation of Green Manure Crops

Crop	Age(Days)	Dry matter(t/ha)	Nitrogen accumulated
Sesbania aculeata	60	23.2	133
Sunnhemp	60	30.6	134
Cow pea	60	23.2	74
Pillipesara	60	25.0	102
Cluster bean	50	3.2	91
Sesbania rostrata	50	5.0	96

Table 1.8: Nutrient Content of Green Manure Crops

Plant	Scientific name	Nutrient content (%) on air dry basis		
		N	P_2O_5	K
Sunhemp	*Crotalaria juncea*	2.30	0.50	1.80
Dhaincha	*Sesbania aculeata*	3.50	0.60	1.20
Sesbania	*Sesbania speciosa*	2.71	0.53	2.21

1.17.2 Advantages of Green Manures

Improves soil structure,increases water holding capacity and decreases soil loss by erosion.

1.18 Green Leaf Manure

Application of green leaves and twigs of trees, shrubs and herbs collected from elsewhere is known as green leaf manuring. Forest tree leaves are the main sources for green leaf manure. Plants growing in wastelands, field bunds etc., are another source of green leaf manure. The important plant species useful for green leaf manure are neem, mahua, wild indigo, Glyricidia, Karanji *(Pongamia glabra)* calotropis, avise *(Sesbania grandiflora),* subabul and other shrubs (Table 1.9 and Fig. 1.6).

Fig. 1.6 : Some Most Important Green Leaf Manure Crops

Table 1.9 : Nutrient Content of Green Leaf Manure

Plant	Scientific name	Nutrient content (%) on air dry basis		
		N	P_2O_5	K
Gliricidia	*Gliricidia sepium*	2.76	0.28	4.60
Pongania	*Pongamia glabra*	3.31	0.44	2.39
Neem	*Azadirachta indica*	2.83	0.28	0.35
Gulmohur	*Delonix regia*	2.76	0.46	0.50

Contd...

Peltophorum	*Peltophorum ferrugenum*	2.63	0.37	0.50
Weeds				
Parthenium	*Parthenium hysterophorus*	2.68	0.68	1.45
Water hyacinth	*Eichhornia crassipes*	3.01	0.90	0.15
Trianthema	*Trianthema portulacastrum*	2.64	0.43	1.30
Ipomoea	*Ipomoea*	2.01	0.33	0.40
Calotrophis	*Calotropis gigantea*	2.06	0.54	0.31
Cassia	*Cassia fistula*	1.60	0.24	1.20

1.18.1 Advantages

a) Growing of green manure crops in the off season reduces weed proliferation and weed growth.

b) Green manuring improves soil structure, increases water holding capacity and decreases soil loss by erosion.

c) Green manuring helps in reclamation of alkaline soils. Root knot nematodes can be controlled by green manuring.

1.18.2 Recycling of Farm Waste

Recycle of Garden Wastes

Scope

It has been estimated that organic resources available in the country alone can produce not less than 20MT of NPK plant nutrients. Vermicompost technology has promising potential to meet the organic manure requirement in both irrigated and rainfed areas. It has tremendous prospects in converting agro-wastes and city garbage into valuable agricultural input. Thus various economic uses can be obtained from organic wastes and garbage and prevent pollution. From vermiculture, we get well decomposed worm casts, which can be used as manure for crops, vegetables, flowers, gardens, etc. In this process, earthworms also get multiplied and the excess worms can be converted into vermiprotein which can be utilised as feed for poultry, fish, etc. Vermi-wash can also be used as spray on crops.

Feed Ingredients

Cow dung and agro wastes in the ratio of 1:1 to 1:3 may be mixed and allowed to predecompose for about two weeks in a separate tank adjacent to the vermicompost tank, before being fed to the earthworms.

Process (Fig. 1.7A-F)

1. Ideal tank size for small scale vermicompost production is 10'x6x2.5' (150ft.).
2. Create adequate no. of holes (about 8 holes of 5cm. diameter at the bottom) to facilitate drainage of excess water.
3. Bedding comprise of broken bricks, stone pieces, saw dust, sand and soil.
4. Release earthworms in this bedding.

A.

B.

C.

D.

E.

F.

Fig. 1.7 A-F : Recycling of Farm Waste

5. Pour the feed mix over this bedding of earthworms of about 15 to 20cm thick.
6. Pre decomposed material need not be added in layers but its depth should not be more than 1.5 to 2 feet.
7. Tank is covered with thatched roof to maintain the moisture of the tank feed (40% moisture should be maintained).
8. Under sub optimal moisture conditions the earthworms have a tendency to move downwards towards the bed of the tank. When the moisture, temperature and organic matter are optimum, the size, weight and cocoon producing capacity of earthworm's increases.

Conversion

1 Kg of worms numbering about 600 to 1000 can convert 25-45kg of wet waste/week. The compost recovery would be around 25kg/week under well managed conditions.

Harvest

The total decomposition may take about 75-100days depending on various factors. Therefore one tank may be used to 4 to 5 times in a year for vermicompost. A few days before the harvest watering of the tank are discontinued to allow migration of worms towards the bottom of the bed. The compost is then transferred outside without disturbing the bed and heaped on a plain open surface. The compost is sieved through a 3mm mesh and then packed in gunnies.

Hand Operative Sieve

About 1700kg of compost can be obtained from each cycle, while sieving the unhatched cocoons can also be retrieved. The excess worms can be retrieved and put in new tanks or sold or can be sun-dried to make vermin-protein. The cost of vermin-protein can be taken @ Rs 5/kg. The compost should be sun-dried and then bagged for sale/use.

Methods of waste treatment by Earthworms

1. Solid waste materials may be spread out over the soil surface, usually on pastures. But sometimes on crops or in forests. To be incorporated directly into the soil earthworms are important contributors to the burial and decomposition of the waste materials.

2. Wastes may be stacked into heaps or placed in bins, where they are treated like compost heaps-earthworm activity results in the production of large quantities of earthworm casts, which are widely sold as manure.

1.19 Organic Special Inputs

1.19.1 Panchakavya

Panchakavya or Panchagavya, an organic product has the potential to play the role of promoting growth and providing immunity in plant system. It consists of nine products viz. cow dung, cow urine, milk, curd, jaggery, ghee, banana, Tender coconut and water. When suitably mixed and used, these have miraculous effects as cow dung (7kg) and cow ghee (1kg). Mix these two ingredients thoroughly both in morning and evening hours and keep it for 3days as cow urine (10liters) and water (10liters). After 3days mix cow urine and water and keep it for 15days with regular mixing both in morning and evening hrs. After 15days mix the following and panchagavya will be ready after 30days.

- Cow milk-3liters
- Cow curd - 2liters
- Tender coconut water - 3liters
- Jaggery-3kg
- Well ripened poovan banana -12nos.

1.19.1.1 Preparation

All the above items can be added to a wide mouthed mud pot, concrete tank or plastic can as per the above order. The container should be kept open under shade. The content is to be stirred twice a day both in morning and evening. The Panchagavya stock solution will be ready after 30days (Care should be taken not to mix buffalo products). The products of local breeds of cow is said to have potency than exotic breeds). It should be kept in the shade and covered with a wire mesh or plastic mosquito net to prevent houseflies from laying eggs and the formation of maggots in the solution. If sugarcane juice is not available add 500g of jaggery dissolved in 3 liter of water.

1.19.1.2 Physico-Chemical and Biological Properties of Panchakavya

Physico-chemical properties of Panchakavya revealed that they possess almost all the major nutrients, micro nutrients and growth hormones (IAA & GA) required for crop growth. Predominance of fermentative micro-organisms like yeast and lactobacillus might be due to the combined effect of low pH, milk products and addition of jaggery/sugarcane juice as substrate for their growth (Table 1.10).The low pH of the medium was due to the production of organic acids by the fermentative microbes as evidenced by the population dynamics and organic detection in GC analysis. *Lactobacillus* produces various beneficial metabolites such as organic acids, hydrogen peroxide and antibiotics, which are effective against other pathogenic micro-organisms besides its growth. GC-MS analysis resulted in following compounds of fatty acids, alkanes, alconol and alcohols (Table 1.11).

Table 1.10 : Physico Chemical and Biological Properties of Panchakavya

Chemical Composition	
pH	5.45
EC (dSm^2)	10.22
Total N(ppm)	229
Total P(ppm)	209
Total K(ppm)	232
Sodium	90
Calcium	25
IAA(ppm)	8.5
GA(ppm)	3.5

Microbial Load	
Fungi	38800/ml
Bacteria	1880000/ml
Lactobacillus	2260000/ml
Total anaerobes	10000/ml
Acid formers	360/ml
Methanogen	250/ml

Table 1.11 : Compounds of Fatty Acids, Alkanes, Alconol and Alcohols in Panchagavya by GC-MS Analysis

Fatty acids	Alkanes	Alconol and Alcohols
Oleic acids	Decane	Heptanol
Palmitic acid	Octane	Tetracosanol
Myristic	Heptane	Hexadecanol
Deconore	Hexadecane	Octadeconol
Deconomic	Oridecane	Methanol, Propanol, Butanol and Ethanol
Octanoic	-	-
Hexanoic	-	-
Octadeconoic	-	-
Tetradeconoic	-	-
Acetic, Propionic, Butyric, Caproic and Valeric acids	-	-

1.19.1.3 Beneficial Effects of Panchakavya on Commercial Crops

Acid Lime: Continuous flowering is ensured round the year, Fruits are plummy with strong aroma and shelf life is extended by 10days.

Guava: Higher TSS and shelf life is extended by 5days.

Mango: Induces dense flowering with more female flowers, irregular or alternate bearing habit is not experienced and continues to set fruit regularly, flavour and aroma are extraordinary and enhances keeping quality by 12days in room temperature.

Banana: In addition to adding with irrigation water and spraying, 100 ml (3% solution) was tied up at the naval end of the bunch after the male bud is removed. The bunch size becomes uniform. One month earlier harvest was witnessed. The size of the top and bottom hands was uniformly big.

Turmeric: Enhances the yield by 22% and ensure low drainage loss, extra-long fingers and narrows the ratio of mother and finger rhizomes, helps survival of dragon fly, spider etc., which in turn reduce pest and disease load, sold for premium price as mother/seed rhizome, enriches the curcumin content and exceptional aroma and fragrance and no incidence of bud worm and continuous flowering throughout the year.

Vegetables: Yield enhancement by 18% and in few cases like cucumber, the yield is doubled, wholesome vegetables with shiny and appealing skin, Extended shelf life and very tasty with strong flavour.

Generally panchakavya is recommended for all the crops as foliar spray at 3% level (3litre panchakavya in 100 litres of water) (Table 1.12).

1.19.1.4 Recommended Dosage

1) **Spray system:** 3% solution was found to be most effective compared to the higher and lower concentrations investigated. Three litres of Panchagavya to every 100litres of water is ideal for all crops. The power sprayers of 10litres capacity may need 300ml/tank. When sprayed with power sprayer, sediments are to be filtered and when sprayed with hand operated sprayers, the nozzle with higher pore size has to be used.

2) **Flow system:** The solution of Panchakavya can be mixed with irrigation water at 50 litres per hectare either through drip irrigation or flow irrigation

3) **Seed/seedling treatment:** 3% solution of Panchakavya can be used to soak the seeds or dip the seedlings before planting. Soaking for 20 minutes is sufficient. Rhizomes of turmeric, ginger and sets of sugarcane can be soaked for 30 minutes before planting.

1.19.1.5 Seed Storage

3% of panchakavya solution can be used to dip the seeds before drying and storing them.

1.19.1.6 Periodicity

Table 1.12 : Time of Application of Panchakavya for Different Crops

Crops	Time schedule
Rice	10,15,30 and 50th days after transplanting
Black gram	Rainfed: 1st flowering and 15 days after flowering fermentation Irrigated: 15, 25 and 40 days after sowing
Green gram	15, 25, 30, 40 and 50 days after sowing
Jasmine	Bud initiation and setting
Vanilla	Dipping setts before planting
Groundnut	25 and 30th days after sowing
Tomato	Nursery and 40 days after transplanting: seed treatment with 1% for 12hrs
Onion	0, 45 and 60 days after transplanting
Bhendi	30, 45, 60 and 75 days after sowing
Moringa	Before flowering and during pod formation
Sunflower	30,45 and 60 days after sowing
Castor	30 and 45 days after sowing
Rose	At the time of prmuning and budding

1.19.1.7 Effect of Panchakavya

a) **Leaf:** Plants sprayed with Panchakavya invariably produce bigger leaves and develop denser canopy. The photosynthetic system is activated for enhanced biological efficiency, enabling synthesis of maximum metabolites and photosynthates.

b) **Stem:** The trunk produces side shoots, which are sturdy and capable of carrying maximum fruits to maturity. Branching is comparatively high.

c) **Roots:** The rooting is profuse and dense. Further they remain fresh for a long time. The roots spread and grow into deeper layers were also observed. All such roots help maximum intake of nutrients and water.

d) **Yield:** There will be yield depression under normal circumstances, when the land is converted to organic farming from inorganic systems of culture. The key feature of Panchakavya is its efficacy to restore the yield level of all crops when the land is converted from inorganic cultural system to organic culture from the very first year. The harvest is advanced by 15days in all the crops. It not only enhances the shelf life of vegetables, fruits and grains, but also improves the taste. By reducing or replacing costly chemical inputs, Panchakavya ensures higher profit and liberates the organic farmers from loan.

e) **Drought Hardiness:** A thin oily film is formed on the leaves and stems, thus reducing the evaporation of water. The deep and extensive roots developed by the plants allow withstanding long dry periods. Both the above factors contribute to reduce the irrigation water requirement by 30% and to ensure drought hardiness.

1.19.1.8 Panchakavya for Animal Health

Panchakavya is a living elixir of many micro-organisms, bacteria, fungi, proteins, carbohydrates, fats, amino acids, vitamins, enzymes, known and unknown growth promoting factors micronutrients trace elements antioxidant and immunity enhancing factors. When taken orally by animals and human beings, the living micro-organisms in the Panchagavya stimulate the immune system and produce lot of antibodies against the ingested micro-organisms. It acts like vaccine. This response of the body increases the immunity of animals and humans and thus helps to prevent illness and cures disease. It slows down the aging process and restores youthfulness. The other factors present in Panchakavya improve apetite, digestion, assimilation and elimination of toxins in the body. Constipation is totally cured. Thus the animals and humans become hale and hearthy with shining hair and skin. The weight gains are impressive.

- **Pigs:** Panchakavya was mixed with drinking water or feed at the rate of 10-50ml/pig depending upon the age and weight. The pigs became healthy and disease free. They gained weight at a faster rate. The feed to weight conversion ratio increased tremendously. This helped the piggery owners to reduce the feed cost and to get very good returns due to increased weight.
- **Goats and Sheep:** The goats and sheep became healthy and gained more weight in a short period after having administered 10-20ml Panchakavya/animal/day depending upon the age.
- **Cows:** By mixing Panchagavya with animal feed and water at the rate of 100ml/cow/day, cows become healthier with increased milk yield, fat content and SNF. The rate of conception increased. The retained placenta, mastitis and foot and mouth disease became things of the past. Now the skin of the cow is shiny with more hair and looks more beautiful. Instead of spraying urea on paddy straw (hay) before staking, a few farmers sprayed the 3% solution of Panchakavya, layer after layer during the staking and allowed it to ferment. The cows preferred such hay compared to unsprayed hay stock.
- **Poultry:** When mixed with the feed or drinking water at the rate of 1ml/bird/day, the birds became disease-free and healthy. They laid bigger eggs for longer periods. In broiler chickens the weight gain was impressive and the feed-to-weight conversion ratio improved.
- **Fish:** Panchakavya was applied daily with fresh cow dung in fish ponds. It increased the growth of algae, weeds and small worms in the pond, thus increasing the food availability to fish. The only precaution is that fresh water must be added to the ponds at frequent intervals. Otherwise, the growth of algae, weeds and other organisms will compete with the fish for available soluble oxygen in water. Alternatively, mechanical agitators can also be used to increase the oxygen content in the water. In ten months' time each fish grew to a weight of 2 to 3kgs. With reduced death rate of small fingerlings and increased weight of marketable fish, the fisheries became more profitable.

1.19.1.9 Panchakavya for Human Health

One lady with psoriasis all over the body was under allopathic treatment for over one and a half years. She happened to prepare Panchakavya for field use and stir the contents with her forearm. After 15days, the psoriasis in her fore arm got fully cured. Following her own intuition, she smeared panchakavya all over the body and to everyone's surprise; the psoriasis disappeared in 21days.

1.19.1.10 Dosage

50ml of filtered panchakavya mixed with 200ml of water, tender coconut water or fruit juice and taken orally in empty stomach in the morning and used for below given diseases.

- **AIDS/HIV:** AIDS/HIV patients regained lost appetite and digestion and put on weight. They slept better. Their fever, cough, diarrhoea and skin lesions disappeared within a month's treatment. Most of them are now working in the fields and others are pursuing their own professions. Even though the blood tests are still positive, they exhibit no symptoms of AIDS and lead a normal healthy life.
- **Psoriasis:** In Psoriasis, it is very effective and the lesions disappear within 6 months. In eczema and other allergic skin disorders, healing is even faster.
- **Neurological disorders:** When given to patients suffering from neurological disorders like convulsions and Parkinsonism, it helped to reduce the frequency of the attacks in convulsions and reduced shaking of the hands and head in Parkinsonism. They were able to reduce the regular medicines.
- **Diabetes mellitus:** When 50ml of filtered panchakavya/day was taken early in the morning on an empty stomach, it reduced the blood sugar and enabled the patients to reduce the dose of anti-diabetic drugs. Complaints like general weakness, indigestion, constipation and burning sensation in the feet disappeared within a month. They became active and healthy.
- **Pulmonary Tuberculosis:** It can be given in addition to the regular anti-TB drugs. Fever disappeared within a week and cough was controlled within two weeks. Appetite improved and the patients gained body weight. The duration of anti-TB treatment was reduced by one month.
- **Arthritis:** It completely relieves the joint pain, swelling and stiffness. Arthritis is cured within two months. Now even healthy people take it to become more healthy and energetic.

1.19.2 Dasakavya

Dasakavya is an organic preparation made from ten products in the form of panchagavya and certain plant extracts. "Gavya" is the term given to cow's products comprising of cow dung, cow urine, cow's milk, curd and ghee, which have miraculous effects on plant growth when suitably mixed. The Horticultural Research Station, Ooty has identified certain plant species for the temperate regions, *viz., Artemisia nilagirica, Leucas aspera*, *Lantana camera*, *Datura metal* and *Phytolacca dulcamera.* These are commonly available weed plants in the district, found abundantly along roadsides and in wastelands. The plants recommended for the tropical areas are neem (*Azadirachta indica*), erukam (*Calotrophis)*, Kolingi (*Tephrosia purpurea*), notchi (*Vitex negundo*), umathai (*Datura metel*), Katamanaku (*Jatropha curcas*), adathoda (*Adathoda vasica)* and pungam (*Pongamia pinnata*) (Fig 1.8). Since management of these can be made best use of in agriculture, as effective agents against certain pests and diseases.

Adathoda vasica

Azadirachta indica

Calotrophis

Datura metal

Jatropha curcas

Leucas aspera

Lantana camera

Pongamia pinnata

Vitex negundo

Fig. 1.8 : Some most important Dasakavya Crops

The plant extracts are prepared by separately soaking the foliage in cow urine in 1:1 ratio (1kg chopped leaves in 1 litre cow urine) for ten days. The filtered extracts of all the plants are then added @ 1 litre each to 5 litre of the panchakavya solution. The mixture is kept for 25 days and stirred well, meanwhile, to ensure thorough mixing of panchakavya and the plant extracts.

1.19.2.1 Mode of Use

The Dasakavya solution is filtered to avoid clogging of sprayer nozzles and is recommended as foliar spray at 3% concentration. Soaking of seeds or dipping the roots of seedlings in 3% solution of dasakavya for 20 minutes before planting enhances seed germination and root development.

1.19.2.2 Periodicity

Weekly sprays during crop growth for all vegetables and plantation crops.

1.19.2.3 Advantages

(i) Increases growth, yield and quality of the crops (ii) Controls pests like aphids, thrips, mites and other sucking pests, and (iii) Controls diseases like leaf spot, leaf blight, powdery mildew etc.

1.19.2.4 Effect of Dasakavya on Hill Crop Pests and Diseases

- **Rose:** Spraying 3% dasagavya in rose controls thrips and powdery mildew.
- **Gerbera:** Dasakavya as foliar spray is effective against gerbera powdery mildew.
- **Tea:** Regular spraying of 3% dasakavya at 15days interval is effective against blister blight disease.

CHAPTER - 2

Basic Information on Biofertilizers

2.1 Introduction

Microbes are the oldest form of the life on earth. These organisms date more than 3 billion years to a time when they were discovered with oceans. They are so tiny in size that million microbes can fit into the eye of a needle. Without microbes, we can't live. Thus, understanding microbes is vital to understanding past and future. Microbes play many roles in the earth's environment recycling dead plant and animal matter through the soil, removing CO_2 from the atmosphere by photosynthesis in the oceans and nitrogen from the atmosphere to form nitrogenous fertilizers plants. Microbes play an important role in nutrient recycling, management, organic matter, decomposition and increasing the crop productivity and quality and protection against diseases (Panwar *et al.,* 2009). Over the past century, many of the basic scientific principles of plant nutrition and soil fertility have been explained. In developed countries, inorganic chemical fertilizers have been widely accepted as a major source of improving and maintaining soil fertility.

Plants need sufficient nutrients in proper balance for normal growth and development. Seventeen plant nutrients are essential for proper crop development. Each is equally important to the plant, yet each is required in vastly different amounts. Our understanding of nutritional needs of crop plants begin from the famous five year willow tree experiment conducted by J.B. Van Helmont (1579–1644). He observed that willow tree had gained nearly 74.4kg but the loss of soil was only 56g, and concluded that the tree drew its nutrients from water not soil. Now we are aware that the soil is the repository for most of the plant nutrients, hence the concern for its continued health and sustainability (Brahmaprakash and Sahu, 2012). India had brought more than 2.5Mha land under certification of organics. In these systems production is based in synergism with nature, which makes systems of unending life i.e. sustainable. Deteriorative effects of synthetic chemical inputs are obvious, but, at the same time we need to revive soil health and living which support to sustainable production system.

Soil environment needs to be made congenial for living of useful microbial population, responsible for continuous availability of nutrients from natural sources.

2.2 Why to Explore Bio-fertilizers

Biofertilizers are inputs containing micro-organisms, which are capable of mobilizing nutritive elements from non-usable form to usable biological processes. They are less expensive, eco-friendly and sustainable. The beneficial microbes in the soil, which are the greater significance to horticultural crop, are biological nitrogen fixers, phosphate solubilizers and mycorrhizal fungi, which are the phosphate scavengers. Biofertilizer in conjunction with organic and inorganic fertilizers offers a great for sustainable crop production. Biofertilizers, as the name indicates are the formulation of living microorganisms are able to fix atmospheric nitrogen in the available form for plants (nitrate form) by living freely in the soil or associated symbiotically with plants fertilizers. However, in the scope the term biofertilizer is used for live preparations of microorganisms, alone or combination, which may help in increasing crop productivity by way of helping in nitrogen fixation, solubilization of insoluble plant nutrients and stimulate the plant.

Bio-fertilizers play a very significant role in improving soil fertility by fixing atmospheric nitrogen, both, in association with plant roots and without it, solubilise insoluble soil phosphates and produces plant growth substances in the soil. They are in fact being promoted to harvest the naturally available, biological system of nutrient mobilization (Venkatashwarlu, 2008a). The role and importance of biofertilizers in sustainable crop production has been reviewed by several authors (Biswas *et al.*, 1985, and Katyal *et al.*, 1994), but the progress in the field of BF production technology remained always below satisfaction in Asia because of various constraints. Large-scale use of chemical fertilizers is responsible for environment pollution and deterioration of soil structure. The cost has increased tremendously in recent past and farmers are constrained to apply the dose of plant nutrients in. their crops. At present, the terrestrial input of nitrogen from biological nitrogen fixation is held to be the range of 139-170 x 10^6t N/year as 65 x 10^6t N/year provided by fertilizer nitrogen (Panwar and Vijayluxmi, 2005).

The discovery of bacteria which had the ability to reduce non-usable (atmospheric nitrogen) into usable form had been a major breakthrough in the field of agricultural research. Of late, increasing attention is being paid to harmed potential benefits from these renewable sources of plant nutrients because of the following reasons:

a) Depleting the soil fertility due to widening gap between nutrient removal and supplies.

b) Depleting the feedstock fossil fuels and increasing cost of fertilizers.

c) Growing concern about environmental hazards.

d) Increasing the threat to sustainable agriculture.

e) Help in the establishment and growth of crop plants and tree, and enhance biomass production and grain yields by 10-20%.

f) Useful in the sustainable agriculture and suitable in organic farming.

g) They play an important role in agroforestry and silvipastoral systems.

h) Vermicompost helps in improving soil health and fertility.

2.3 Availability and Cost

Demand is much higher than the availability. It is estimated that by 2020, to achieve the targeted production of 321 million tones (MT) of food grain, the requirement of nutrient will be 28.8MT, while their availability will be only 21.6MT being a deficit of above 7.2MT (Mahdi *et al.,* 2010). Besides above facts, the long term use of bio-fertilizers is economical, eco-friendly, more efficient, productive and accessible to marginal and small farmers over chemical fertilizers (Venkataraman and Shanmugasundaram, 1992). The limited shelf-life, particularly of bacterial bio-fertilizers dictates that product streams must be with a quick delivery system at low temperatures.

2.4 Chronology of Major Developments related to Biofertilizers

Although the beneficial effect of legumes in improving soil fertility known since ancient times, the field of biological nitrogen fixation opened up with the discoveries by **Boussingault** and **Hellreigel** in 1886. The commercial history of biofertilizers began with the launch of nitrogen by **Nobbe and Hiltner**, laboratory culture of rhizobia in 1895 followed by discovery of *Acetobacter* and then Blue green algae and a host of other micro-organisms. In Indian, **N.V. Joshi** indicted first study on legume *Rhizobium* symbiosis and the first commercial production started as early as 1956. Biofertilizers offer a new technology for agriculture holding a promise to balance many of the shortcomings of the conventional chemical based technology. The main incentive for farmers to use biofertilizers seems to be that they hope to increase the yield or quality of their crops at a relatively low cost without a large investment of money.

Nobbe Hiltner produced for the first time a laboratory culture of Rhizobia under the name Nitragin in 1885. Starting with *Rhizobia,* a vigorous search for other N fixing micro-organisms began and soon it was found that there were non-symbiotic bacteria such as Azotobacter, which could fix atmospheric nitrogen. It was followed by a third group called blue green algae. Subsequently, micro-organisms capable of solubilizing phosphate were discovered. In 1970, a new group of bacteria called *Azospirillum* identified. Mycorrhizae (VAM) are the latest introduction in the list of biofertilizers which mobilize phosphorus.

In India, systematic research on biofertilizers started with the first study of N.Y. Joshi in 1920. This was followed by extensive research by Gangulee. Sarkaria and Madhok on the physiology of the nodule bacteria and inoculation for increasing crop yields. Important milestones in production development and promotion of biofertilizers in India are below given-

Some Milestones in Research, Production and Promotion of Biofertilizers in India

Year	Events
1834	Classical concept of Nitrogen Fixation by legumes "enunciated by French Agricultural Chemists J. B. Boussingault.
1886	Discovery of symbiotic nitrogen fixation by M. Hellriegel and H. Wilfarth, German Scientists production.
1888	Discovery that a bacterium (now named Rhizobium) is responsible for symbiotic nitrogen fixation in legumes by M. Beijerinck, a Dutch Scientist.
1895	Beginning of legume inoculants with the introduction of a biofertilizer named Nitragin in by F. Nobbe and L. Hitner in the USA.
1905	Discovery of *Azotobacter* by Beijerinck.
1905	Start of biofertilizer production in Canada.
1912	Initiation of quality control legislation of biofertilizer in USA.
1914	Start of biofertilizer production in Australia and Sweeden.
1920	First study on legume –*Rhizobium* symbiosis in India by NV Joshi.
1925	Discovery of *Azospirillum* by M. Beijerinck.
1927	Initiation of mass culture of *Rhizobium* in fermenter vessel with aeration by Matchettle in USA.
1930	Issue of license to a biofertilizer manufactured by the US Government on the basis of testing culture.
1932	Introduction of Yeast Extract Manittol Agar (YEMA) medium for *Rhizobium* by Fred.
1934	Earliest documented production of *Rhizobium* inoculants in India by M. R. Madhok.
1939	First report on the performance of *Azotobacter* in paddy/rice soil in India by B. N. Upal.
1939	Discovery of Blue Green Algae (BGA) as nitrogen fixer in paddy field by P. K. Dey.
1948	Recognition of peat as carrier in India.
1951	Use of charcoal as carrier in India.
1955	Use of *Azolla* as a biofertilizer in rice fields in North Vietnam.
1956	Use of Phosphate Solubilizing biofertilizers "Phosphobacterium" in the USSR.

1956	First commercial production of biofertilizers in India (New Delhi/Tamilnadu).
1957	Study on the solubilization of phosphate by micro-organisms in India by A. Sen and N.B. Pal.
1958	First attempt to standardize quality of legume inoculants in India by A. Sankaran.
1958	Recognition of *Azotobacter* culture as biofertilizers in the USSR with the trade name Azobacterin.
1960	First isolation of new non-symbiotic nitrogen fixing organism *Derxia gummosa by* P. K. Dey and Roma Bhattacharya of India.
1968	All India Co-ordinated Research Projects (AICRP) on Pulses improvement and soyabean set up by ICAR where *Rhizobium* study got priority.
1969	Use of Indian peat as carrier reported by V. Iswraan.
1970	Use of Charcoal, lignite and FYM as alternate carriers to peat reported by V. Iswraan.
1975	Coal as alternate carrier to peat reported by J.N. Dube.
1977	Indian Standard (ISI) for *Rhizobium* inoculants published.
1979	All Indian Co-ordinated Research Project on BNF initiated by Indian Council of Agricultural Research (ICAR), New Delhi.
1979	ISI standard for *Azotobacter* inoculants published.
1983	National Project on Development and Use of Biofertilizers by Ministry of Agriculture, Government of India.
1985	First national productivity award on biofertilizers given.
1988	Discovery of *Acetobacter* as Nitrogen Fixing bacteria in sugarcane by V.A. Cavalcante and J. Dobereiner in Brazil.
1988	National Facility for Blue Green Algae (BGA) Collection at IARI, New Delhi by Department of Biotechnology, Government of India.
1988	National Research Centre for BGA set up at IARI, New Delhi.
1991	National Facility of *Rhizobium* germplasm collection set up at IARI New Delhi.
1993	Mission Mode project on use and development of biofertilizers technology set up by the Government of India.
1993	Biofertilizer Newsletter published under the National Biofertilizers Project.
2001	National Bureau of Agriculturally Important Microorganisms (NBAIM) was established through a funded project sponsored by the Department of Agricultural Research and Education (DARE), Ministry of Agriculture, Government of India, Indian Council of Agricultural Research (ICAR), New Delhi.
2006	Inclusion of four Biofertilizers i.e. *Rhizobium, Azospirillum, Azotobacter* and Phosphate Solubilizing Bacteria (PSB) in the Fertilizers Control Order (FCO) in India.
2014	Technology for commercialization of Liquid Bio-Fertilizers.

2.5 Definition of Biofertilizers

Bio-fertilizers are natural fertilizers that are microbial inoculants of bacteria, algae, fungi alone or in combination. They are defined as a product containing carrier based (solid or liquid) living micro-organisms that are agriculturally useful in terms of nitrogen fixation, phosphorous solubilization or nutrient mobilization. They augment the availability of essential elements like nitrogen, potash, phosphorous, sulphur by directly supplying them or transforming them into soluble form. In addition, they also help plants to uptake several micronutrients. Biofertilizers more commonly known as microbial inoculants are artificially multiplied cultures of certain soil organisms that can improve soil fertility and crop productivity.

or

Bio-fertilizers being essential components of organic farming play vital role in maintaining long term soil fertility and sustainability by fixing atmospheric dinitrogen (N=N), mobilizing fixed macro and micro nutrients or convert insoluble P in the soil into soluble forms available to the plants, there by increases their efficiency and availability.

Biofertilizers have the ability to fix atmospheric nitrogen and mobilize phosphorus in soil from unavailable from the plant usable form. Biofertilizers include micro-organisms and their metabolites that are capable of enhancing soil fertility, crop growth and yield. These include both indigenous microbes and microbial inoculants that are microorganisms that replace fertilizers or increase a crop's fertilizer use efficiency. Thus, we can conclude that biofertilizers are ready to use live formulates of beneficial micro-organisms and their metabolites, which are capable of enhancing the soil fertility, crops growth and yield by mobilizing nutrients from no usable form to usable form through biological processes.

2.6 Necessity of Biofertilizers

An ideal fertile soil is characterized not only by optimum physical properties and chemical constituents conducive for plant growth, but also by a balance micro flora in the photosphere. With the introduction of green revolution technologies, there has been an increase in use of chemical fertilizers, pesticides, hybrid seeds, assured irrigation and agronomic practices that have disturbed the equilibrium of the ideal soil. Due to this, there is an ongoing attempt on the part of Government of India (GOI) to promote biofertilizers in Indian agriculture.

2.7 Effect of Chemical Fertilizers on Soil and Environment

a) Excessive and imbalanced use of chemical fertilizers has adversely affected the soil, causing decrease in organic carbon, reduction in microbial flora of soil, increasing acidity and alkalinity and hardening of soil.

b) Excessive use of nitrogen fertilizer are contaminating water bodies' thus affecting fish fauna and causing health hazards human beings and animals.

c) Production of chemical fertilizers adds to the pollution.

To overcome the deficit in nutrient supply and to overcome the adverse effects of chemical cultivation. It is suggested that effects should be made to exploit all the available resources of nutrients under the theme for integrated nutrient management. Under this approach, the best available option lies in the complimentary use of biofertilizers and organic manures in suitable combination with chemical fertilizers. This integrated approach of nutrient management not only ensures higher productivity but also ensures the good health of our soil and environment. Biofertilizers are essential components of this approach and are being promoted to harvest the naturally available, biological system of nutrient mobilization. Biofertilizers have important environmental and long term implications, negating the adverse effects of chemicals. At the farm level, the gains from increased use of the technology can spill over to other farms and sector through lesser water pollution then chemical fertilizers and even to extent organic manures can cerate (Rajasekaran *et al.*, 2009).

The gains from the new technology coming through the arrest of soil damage may not be perceived over a short span of time unlike for chemical fertilizers, which yield quick returns. At the sometime the farmer has to considerable initial cost in terms of skill acquisition, trail and failure and risk. In agrarian situations where agent's option operates with bounded nationality, adoption may be slow and influenced greatly by neighbors experience over time. Although, biofertilizers have been promoted as supplement of chemical fertilizers, in reality they are two alternative means of accessing plant nutrients. The strength of complementarily as against substitution between two inputs is open to empirical verification. The pricing of chemical fertilizers is from marginal cost based. The external cost of using chemical fertilizers though not measurable may also be taken into account when comparing with biofertilizers if the latter is to be promoted. It is good practice to promote biofertilizers as an input conjunctive to other forms of fertilizers, but keeping in view the protection given to chemicals, there is some ground for subsidizing the farmer to encourage their use.

2.8 Estimated Demand and Supply of Some Important Bio-fertilizers in India

The annual requirement and production of different bio-fertilizers have clearly shown tremendous gap in this area. Thus, a strategy for judicious combination of chemical fertilizers and biofertilizers will be economically viable and ecological useful. It should be recommended that biofertilizers are not a substitute, but a supplement to chemical fertilizers for maximizing not only the yield but also agro system stability.

2.9 Success of Biofertilizers Technology

Biofertilizers have various benefits, besides accessing nutrients, for current intake as well as residual, different biofertilizers also provide growth promoting factors to plants and some have been successfully facilitating composting and effective recycling of solid wastes. By controlling soil borne diseases and improving the soil health and soil properties, these organisms help not only in saving, but also in effectively utilizing chemical fertilizers and result in higher yield rates. GOI and the different state governments have been promoting use of bio-fertilizers through grants, extension and subsidies on sales with varying degrees of emphasis. With time farmers to learn about the technology forming their perception on the basis of agronomic realities of their regions, the knowledge gained from experiences of farmers around them and including themselves and the information provided by different disseminating agents and form their own decisions of adoption.

The commercial production of biofertilizers started more than a century ago. The first commercially available biofertilizer was *Rhizobium* inoculants produced in the USA in the year of 1885, which was sold under the trade name Nitragen. *Rhizobium* inoculants production was later started in several other countries as Canada in 1905 and Sweeden in 1914 and Australia in 1914. There are about 165 biofertilizers production units in India which are mainly producing the four type's inoculants apart a number of other bioinoculant as *Rhizobium, Azospirillum, Azotobacter* and Phosphate Solubilizing Bacteria (PSB), in which the organisms are used not specified in most reports. The production of biofertilizers was 1,000 tonnes in 1990 consisting largely of *Rhizobium* inoculants/ culture. By 2010, total biofertilizers production had reached 20,000 tonnes and an estimated 37,000 by 2011. The product pattern had become broader based. Not only has the total production increased over the years, the product pattern has also become more broad-based. In fact, quantitatively, products containing *Azotobacter* and P solubilizing micro-organisms have become much more important than *Rhizobium*. However, out of the total bio-inoculants production in 2009-10 the share of biofertilizers was only 30% and other products. Mainly bio-pesticides account for 70% of production as statistics compiled by NCOF (Bhattacharya and Tandon, 2012).

Production of biofertilizers started in India with significant government involvement with active participation of the public sector that is directed more by public policy and social objectives. Since biofertilizers are perishable and sensitive to quality of handling, the distribution of plants would be some extent reflects the regional distribution pattern. However, this is only partially valid as units with large distribution network do distribute over larger area (Table 2.1).

Table 2.1 : Biofertilizer production in India during the period from 2008-2009 to 2011-12

Sl. No.	State	Actual production of biofertilizers (MT)			
		2008-09	**2009-10**	**2010-11**	**2011-12**
1.	Andhra Pradesh	168.136	1345.28	999.60	1126.35
2.	Arunachal Pradesh	-	-	-	-
3.	Assam	129.3552	121.04	130.00	68.33
4.	Bihar	-	-	136.26	75.00
5.	Chhatisgarh	-	-	-	276.34
6.	Delhi	1165.1	1021.85	1205.00	1617.00
7.	Gujarat	1149.695	1309.19	6318.00	2037.35
8.	Goa	-	0	443.40	0
9.	Haryana	14.25	6.195	6.53	914.41
10.	Himachal Pradesh		8.5	9.00	1.29
11.	Jharkhand	15.0	15.0	0.00	8.38
12.	Karnataka	11921.057	3695.5	6930.00	5760.32
13.	Kerala	1187.001	1936.451	3257.00	904.17
14.	Madhya Pradesh	848.448	1587.6775	2455.57	2309.06
15.	Maharashtra	1249.87	1861.33	2924.00	8743.69
16.	Manipur	-	-	-	-
17.	Mizoram	1.996	2.5	2.00	-
18.	Meghalaya	-	-	0.00	-
19.	Nagaland	16.0092	18.25	21.50	13.00
20.	Orissa	405.03	289.867	357.66	590.12
21.	Punjab	1.14	301.232	2.50	692.22
22.	Pondicherry	561.7924	452.79	783.00	509.45
23.	Rajasthan	353.67	805.571	819.75	199.78
24.	Sikkim	-	-	-	-
25.	Tamil Nadu	4687.818	3732.5862	8691.00	3373.81
26.	Tripura	14.68	278.402	850.00	1542.85
27.	Uttar Pradesh	885.5174	962.6417	1217.45	8695.08
28.	Uttarakhand	48.23	32.00	45.00	263.01
29.	West Bengal	241.24	256.5	393.39	603.20
30	Total	25065.0352	20040.3534	37997.61	40324.21

Source: NCOF (Data as provided by Production Units/State Governments)

2.10 Plant Nutrients

These 17 nutrients are divided into three groups such as **(i)** major, **(ii)** minor and **(iii)** micro nutrients based on plant requirement. Major nutrients are nitrogen (N), phosphorus (P) and potassium (K); minor nutrients are calcium (Ca), magnesium (Mg) and sulphur (S), and the micro-nutrients are boron (B), chlorine (Cl), copper (Cu), iron (Fe), manganese (Mg), molybdenum (Mo), zinc (Zn) and nickel (Ni). In addition to the 13 nutrients listed above, plants require carbon (C), hydrogen (H) and oxygen (O), which are extracted from air and water. Most plants take up nutrients as inorganic ions irrespective of the form in which it is applied to soil. Soils analyses generally indicate that most of these elements are of these are bound in organic and inorganic complexes which are unavailable to plants for their uptake. Plants obtain most of their carbon and oxygen from air through photosynthesis and build up higher organic molecules. The required hydrogen is derived from water. About 95% of the plant tissues are made up of carbon, hydrogen and oxygen. The other elements are obtained by plants from soil. Of these, nitrogen, phosphorus, potassium, magnesium, calcium and sulphur are required in higher quantities than the rest of the elements. Nitrogen requirement is met from available forms of nitrogen viz. ammonium and nitrate which are fertilizers and organic manures. The micronutrients viz. boron, zinc, manganese, copper and chloride are taken up in traces by plants. Depending upon their availability in soil, they have to be applied through organic manures of chemical fertilizers. The balanced use of plant nutrients is needed for normal growth of plants. The requirements of plant nutrients can be met not only by chemical fertilizers but more advantageously by different techniques of organic recycling and large scale use of renewable sources of carrier based biofertilizers to change the system slowly from conventional to sustainable agriculture.

2.10.1 Nutrient Balance

While the fertilizer use over the last 45 years has risen from 0.066MT (NPK) in 1950 to 13.6MT in 1995, nutrient removal has not lagged behind on account of enormous increase in agricultural production. The net result has been a short fall of nearly 10MT of primary plant nutrients. More than half of this deficit exists in semi-arid regions alone. By 2020AD, when the estimated population will be around 1.5 billion, to provide minimum caloric requirement to her citizenry, India will need to produce at least 300MT of food grains at a cost of 30-35MT NPK from various sources. Overall nutrient needs will increase by another 15MT NPK, if the requirement of horticulture, vegetables and plantation crops, sugarcane, cotton, oilseeds and potato are also accounted for. Over and above NPK, the ion balance arising from other nutrient removals will be still wider due to a general lack of attention towards their utilization (Tilak and Singh, 1996).

2.11 Chemical Fertilizers vs Organic Farming

To meet the ever increasing demand of expanding population, agriculture production has been raised through optimum use of mineral fertilizers, adopting multicropping system and liberal application of agrochemicals. Though the use of fertilizers and chemicals has increased the yields dramatically, it has also resulted in rapid deterioration of land and water resources apart from wastage of scarce resources. This has adversely affected the biological balance besides resulting in the presence of toxic residues in food, soil and water; in addition it has imposed economic constraints on the developing countries. India is among the top four fertilizer consuming countries in the world and use of chemical fertilizer is increasing rapidly. Increase in nitrates in drinking water is believed to be due to excessive use of N fertilizers and animal manures. Nutrient enrichment, eutrophication and deterioration of surface water quality due to transportation of nutrients applied through fertilizers viz., leaching and/or run off and sediment erosion are other problems.

The appeal for biologically based systems of farming has fostered efforts in agriculture referred to by different names, viz. alternative agriculture, low-input agriculture or organic farming. The attention paid to organic agriculture and environment is associated with overall efforts towards making agriculture more sustainable. Organic agriculture is example of sustainable agriculture system with the strictest environmental requirements. The basic difference between a sustainable system and more conventional system is in nutrient resource supply. Conventional agricultural system grows crops using high inputs of inorganic fertilizers, whereas in more sustainable systems inputs are usually reduced and often become hetergenous and complex (organic manure). Chemical fertilizers supply one or two plant nutrients whereas organic fertilizers supply both macro- and micro-nutrients and humus substances particularly humic and fluvic acid (Gaur, 1984, 1986; Jarvis, 1996). Organic manures also showed residual effect on the succeeding crop. Moreover low-input or organically managed cropping systems were reported to produce more diverse and abundant soil microflora when compared with conventionally managed cropping systems. These systems minimize the use of inputs by substitution with farm-generated inputs, or exploit biological systems such as free-living/symbiotic nitrogen fixation to increase soil fertility and, potentially through use of PSM and VAM fungi to improve the phosphorus efficiency and use of other nutrients.

Mobilization of the possible organic sources and their incorporation in soil fertility management programme is a time-tested strategy to achieve balanced nutrition. Within soil, organic materials impart many important properties to the system and are involved in a variety of processes. Their continual breakdown provides mineral nutrients for new plant growth; they act as binding agent within the soil; they can increase pore space and alter cation exchange capacity of the soil. Also, there is evidence that organic system has a higher rate of biological activity and improved soil quality when compared with conventional systems. Organic matter decomposition and nutrient cycling are more known to improve the efficiency of chemical inputs. Therefore, soil micro-organisms are believed to play a crucial role in low-input sustainable agriculture.

All plant and animals except the organisms endowed within genes depend on fixed nitrogen to meet their nitrogen requirement. Chemical fertilizers are industrially manipulated and contain known quantities of nitrogen, phosphorus and potassium. On the other hand, biofertilizers do not come under the purview of the definition of term fertilizer because bio-fertilizers are carrier based living or latent cells of microorganisms when applied as seed or soil inoculant, multiply on seed and soil and participate in the nutrient cycling and benefit the crop productivity. They do not get depleted as the chemical fertilizers do after sometime. If the soil conditions remain optimum, the bio-fertilizer gets established and their frequent use may not be required.

Fritz Haber, a German chemist synthesized nitrogen and hydrogen into ammonia during early period of first Word Ward. This process, known as "**Haber-Bosech**", requires high temperature upto 426.66^{0}C, a catalyst and high pressure (above atmospheric pressure) for production of nitrogenous fertilizers. The commissioning of modern fertilizer plants is cost intensive and time taking. Interestingly certain microorganisms can fix inert atmosphere nitrogen ranging from a few kg to 250-300kg/ha, at ordinary temperature, and pressure. Selected strains of bacteria, fungi and actinomycetes can be used for preparation of carrier based inoculants in decentralized and small factories to supply available forms of plant nutrients like NPK and others.

Fertilizers are prepared from the raw materials viz. naptha, coal or fuel, oil which are non-renewable resources involving large sums of foreign exchange for activity of chemical fertilizer plants. Biofertilizers are produced from bacteria, fungi, actinomycetes, *Azolla* or BGA which are renewable sources and isolated from local soils without any foreign exchange cost. The contribution of other bio-fertilizers viz. phosphate solubilizing microorganisms and mycorrhizae in making available plants nutrients notably phosphorus and other plant nutrients such as sulphur, calcium, magnesium etc. is noteworthy. The fertilizer use efficiency is low and a cause for concern. In upland conditions, the fertilizer use efficiency of nitrogenous fertilizers is 40-45% in cereals and about 33% in rice fields. Losses of nitrogen through leaching, denitrification and ammonia volatilization are heavy. The efficiency of phosphatic fertilizer is much lower and is in the range of 20-30% because its major part is chemically fixed in soil and converted into unavailable form.

Chemical fertilizers is used indiscriminately cause pollution of soil and underground water leading to nitrate pollution and eutrophication of water bodies. Soil health is impaired on long term leases. Biofertilizers are environment friendly, build up soil health and improve soil quality.

- Chemical fertilizers are expensive and not easily accessible to small and marginal farmers. Biofertilizers are low cost inputs.
- Chemical fertilizers can be used in irrigated conditions whereas bio-fertilizers can be used both in rain fed (dry land agriculture) and irrigated conditions.
- Shelf life of chemical fertilizers is long whereas bio-fertilizers containing bacteria in particular have a short shelf life.

Chemical fertilizers are recommended for all crops based on soil tests. Nitrogenous biofertilizers are specific viz. legumes, cereals, cotton and sugarcane. However, phosphate solubilizing microorganisms are recommended for all crops. Promotion of agro-forestry and tree planting in waste lands and community holdings is most appropriate strategy to accomplish the twin objectives of having sufficient firewood and rejuvenate the degraded land. Therefore, both the research and extension have to take note of the increasing importance of IPNM in maintain of soil fertility and of plant nutrient supply to an optimum level for sustaining desired crop productivity through optimization of benefits from all possible sources of plant nutrients in an integrated manner. The cropping system rather than an individual crop, and the whole farm rather than an individual field should be to the focus of attention.

2.12 Concept of Conventional, Organic/ Biodynamic Farming/Eco-Friendly Systems

The practice of ecologically sound, economically viable and culturally appropriate agriculture is generally termed as sustainable agriculture. This not only benefits the farming community in particular but the society in general. The modern technology better known as conventional technology in the west is heavily dependent on non-renewal energy sources, viz. pesticides, and chemical fertilizers. The continuous, unbalanced and excessive use of agricultural chemicals on plant and soil is not only damaging the soil biodiversity and environment but also reduces the soil productivity potential. Moreover, the consumption of chemically contaminated food grains, fodder and animal products are jeopardizing the health of human beings and animals.

2.12.1 Conventional Farming

The technological advances made during 1950's had caused a dramatic change/shift in mainstream agriculture, creating a system that relied more on agro-chemicals viz. herbicides, insecticides, nematicides, chemical fertilizers, new varieties of crops and labor saving energy-intensive farm machinery. This system has come to be known as conventional farming. A growing cross section of people in the world is questioning the environmental, economic and social impacts of conventional agriculture. Consequently, many individuals are seeking alternative practices that would make agriculture more sustainable (Rupela, 2005).

2.12.2 Non-conventional Farming

Nature Farming: This farming was advocated in 1935 by Japanese Philosopher, Mokichi Okada (1882-1955). He paved the way for its practical use through his literature and results of his experiments in field cropping. In 1953, he set up a "Nature Farming Extension Society" as an organization to promote his newly advocated farming practices. Ohito Research Farm in Japan was established in 1982 which conducts Research and Development and has become headquarter for those practitioners. In 1992, MOA India and WSAA India branch was established in Bangalore. MOA nature Farming Model Farm was also established in Whitefield, near Bangalore in the same year.

The guidelines are based on the philosophy of Mokichi Okada Nature Farming practice in which soil itself consists of fire, water and soil elements as universal life giving powers. The focus of the Nature farming is to establish, linkage between humanity and nature. MOA Nature farming is not merely considered by their followers as alternative farming system, changing from conventional method to compost and natural pesticides but embodying both the philosophy that soil has the vital energy and the harmonious relationship between man and nature. The objectives of Nature Farming include development of agriculture, forestry and rural communities; promotion of healing dietary habits of consumers making available safe food both in quality and quantity; utilization of land effectively and other resources to increase soil productivity and conservation of energy and to reduce production cost. The basic methods to achieve the objectives are as follows:

- To ensure production of safe and quality produce by growing crops on soils' inherent power (Native fertility).
- Maintenance and improvement of soil fertility by use of crop rotation relay cropping, use of green manures, compost, natural materials rich in nutrients.

- To control disease, insects damage and weeds though ecological methods inclusive of utilization of companion crops and natural enemies.
- Shall not use chemical fertilizers, plant growth regulators and feed additives.
- Shall not apply undecomposed animal excrements and human faces to soil.

Conversion from Conventional to Nature Farming Methods: The conversion programme should begin with a fertility building phase and prohibited inputs may not be used at any stage during changeover. The period of conversion will be decided by the guidelines Administrative Committee after going through the technical information furnished in this context.

This includes plant materials existing in Nature and Compost prepared from such materials, bring earth from another place whose safety has been confirmed; can be used for mixing with soil, crop residues and compost made of such residues, charcoals and smoked-charcoals, oil-cakes, rice bran, fish meal, bone meal, guano phosphate, grass and wood ashes, limestone and dolomite, vermi-compost, peat moss, biofertilizers for restricted use, kitchen waste, fermented poultry excreta and rock phosphate.

2.12.3 Organic Farming

Organic farming as a philosophy has caught the imagination of farmers the world over and has been perceived as panacea for ills of conventional farming ever since Rachel Carson wrote "Silent springs" in the last 1960s. In the developed countries, the residues of chemical inputs such as pesticides and fertilizers entering the food chain are of major concern while under Indian conditions. It is the restoration of soil productivity and conservation of the basic resource-soil that assumes significance.

The basic principle of organic farming aims at the management of agro- and ecosystems. Under agro-system, the farmer has to manage the farm with coherent diversity by utilizing all the on-farm and adjacent resources. Such a practice helps to conserve the ecosystem rather than destroy it. Organic farming excludes use of agro-chemicals like chemical fertilizers and pesticides. However, organic farming would be successful in places where sufficient organic materials are available in surroundings and on farm and proper techniques of recycling of organic matter and use of renewable resources are employed for the purpose and also under conditions of scarce availability of costly chemical fertilizers. There are variations in organic farming as-

a) **Pure Organic Farming**: Use of organic manures/compost, recycling of plant residues, biofertilizers and biopesticides without any application of chemical fertilizers and pesticides.

b) **Integrated Farming System:** Recycling of local organic resources generated from dairying, fisheries, poultry, goat farming, mushroom production, agroforestry and agricultural system. This is low input organic farming.

c) **Integrated Conventional and Organic Farming Technology:** This involves integrated plant nutrient supply and integrated pest management. Biofertilizers could serve as alternate sources of nitrogen or phosphorus or both.

Conventional intensive agriculture causes many problems like soil becomes poor in organic carbon and plant nutrient, increasing rates of chemical fertilizers are required every year/ season to grow and get the same crop yields, control of pest and disease becomes more problematic and rivers, lakes, wells and water bodies get polluted with chemicals damaging the ecosystem. An organic farmer should have a healthy balance between nature and farming where plants and animals can grow and thrive. Organic farmers do not leave their farms to nature; they use all knowledge, techniques and products to work with nature. Organic farming avoids the use of pesticides which dissolve easily and can quickly find their way in food chains and water-bodies. The soil is living system and must be manured to feed the soil micro-organism for their growth and biochemical activities. Nutrients are recycled by composting crop residues and using farmyard manure, green manure etc. Rotation with legume crops helps in biological nitrogen fixation.

International Standards for Organic Farming: These were laid down by the International Federation of Organic Agriculture Movement in 1992 and are enumerated here.

a) To produce food of high nutritional quality in sufficient quantity.

b) To interact in a constructive and life enhancing way with all natural systems and cycles.

c) To encourage and enhance biological cycles within the farming system, involving micro-organisms, soil flora and fauna, plants and animals.

d) To maintain and increase long-term fertility of soils.

e) To use as far as possible renewable resources in locally organized systems.

f) To work as far as possible within a closed system with regard to organic matter and nutrient element.

g) To work, as far as possible with materials and substances this can be reused or recycled either on the farm or elsewhere.

h) To give all livestock living conditions this will allow them to perform the basic aspects of their innate behaviors.

i) To minimize all forms of pollution that may result from agricultural practices.

j) To maintain genetic diversity of agricultural system and its surroundings including protection of plant and wild life habitats.

k) To allow agricultural producers a living according to the United Nations Human Rights, to cover their basic needs and obtain an adequate return and satisfaction from their work, including a safe working environment.

l) To consider the wider social and ecological impact of the farming system.

2.12.4 Biodynamic Farming

The main objectives of biodynamic farming are as follows:

a) To produce high quality food and fibers out of potential individual farms.

b) To develop site adapted and balanced combination of plant and animal husbandry which will make farms independent of synthetic fertilizers.

c) To maintain and enhance the environment as well as the cultural life of the farm in co-operation with the social life of the region.

d) To take part in experimental research for the further development of agriculture that is attuned to the natural world.

Methods of Biodynamic Farming : The following steps are used in methodology as adopted for biodynamic farming.

a) Nutrients cycling and humus replacement are the foundation of a farm's wellbeing. It involves livestock husbandry, utilization of plant and animal residues, pasture management and crop rotation.

b) Intensive use of biodynamic preparations in the context allows, soil, plant and animal processes on the farm to be integrated in a healthy and productive way.

c) While biodynamic method of farming stresses the importance of macro- and micro-nutrients and advises the use of mineral and organic forms where appropriate (rock phosphate, dolomite, ground basalt, basic slag etc.). It sponsors the skillful use of organic matter as basic factor of soil life –hence fertility.

Biodynamic Preparations (B.P.) : All the preparations can be made on the farm or obtained by biodynamic farming and gardening association. Preparations, 500 and 501 are mixed in water and then sprayed on soil or plant and 502-507 are used in composting process and in preparation of liquid manures. Biodynamic preparation 500 soil spray is considered to promote humus formation processes and improves soil structure and root growth. BP 501 leaf spray is reported to promote photosynthesis, yield and nutrient quality. BP 502-507 is used in compost preparations which promote cycling of plant nutrients and maintains diverse biological pathways. Biodynamic farming has been practiced in Australian farms and in New Zealand on a small scale. There is needed to be re-examined and confirm such preparations and methods. There is a lack of data and authentic information on methods and products. It is yet to know whether the claims are reproducible and based on scientific principles.

2.12.5 Eco-friendly Systems in Sustainable Agriculture

Natural organic farming is not a single method but rather a variety of techniques according to a report of United States National Academy of Sciences (1989). These techniques have the common goal of reducing costs, preserving the environment, and protecting human health by lowering or eliminating use of toxic chemicals at farms. Natural farming usually requires harder work and greater management skills than chemically based agricultural practices. At present times, people are conscious about the issues viz. food safety and water quality. Sustainable agriculture involves several variants of non-conventional agriculture that are often called organic, alternative, regenerative, ecology friendly or low-input. However, just because a farm is organic or alternative does not mean that it is sustainable. For a farm to be sustainable, it must produce adequate amounts of high quality food, protect its resources and be both environmentally safe and profitable. Instead of depending on synthetic and expensive sources viz. pesticides and fertilizers, a sustainable farm relies as much as possible on beneficial natural process and renewable resources drawn from the farm itself.

The concept of sustainable agriculture is gaining acceptance in India because of the rapid degradation of natural resources, increase in cost of production in conventional farming and deterioration of the quality of rural life caused by environmental pollution. In India, most of the farmers are using natural products for production and protection of crops for two main reasons. The chemical fertilizers are too expensive to afford and it is risky to adopt conventional methods in over 70% of cultivated area which is presently under rainfed farming system. The most salient aspect of organic agriculture is emphasis on utilization of natural soil process viz. soil nutrients, water cycles and naturally occurring energy flow for producing nutritious foods which are free from chemical residues. The reliance is on crop rotations, crop residues, animal manures, green manures, vermicompost, gobar-gas plant spent slurry, municipal wastes, sludge, rock phosphate, enriched compost, tank silt etc. for maintenance of soil productivity and biopesticides for control of insects and diseases. A permaculture or natural approach where one relies on the productivity of permanent structure viz. trees, perennial species or other undisturbed plant soil system for food has been perceived to be ultimate goal of sustainable farming system. Thampan (1995) observed sporadic success in adopting organic farming by farmers of tree-based cropping system involving the combination of trees, shrubs and seasonal field crop and livestock. The multiple products so obtained and enhanced on-farm employment have strengthened the socio-economic security of the farmers.

2.13 Soils

Soil is an important basic resource and acts as sink for disposal of different types of wastes. It is made up of solid phase 50% by volume whose components are of mineral and organic matter, liquid phase, 25% by volume is water having solution of plant nutrients. The concentration of salt in soil on dry basis may range from 100 to 1000ppm. The content of calcium sulphate or calcium carbonate is often found more. Sulphate, nitrate, chloride, calcium, magnesium, potassium and sodium bicarbonate are also found in liquid phase. Besides, these elements other nutrients are found in traces. The gaseous phase has oxygen (20%), nitrogen (78.6%) Argon (0.9%) and carbon dioxide (0.50%) against atmospheric air which has 20.0% oxygen, 78.03% nitrogen, 0.94% Argon and 0.03% carbon dioxide. The proportion of carbon dioxide in soil air is about 16-fold higher than atmospheric carbon dioxide. This is due to soil respiration and heterotrophic activity in soil conditions.

Soil is essential to agriculture and is a complex living system besides the inert component. The loose but coherent structure of good soil holds moisture and invites airflow. Ants and earthworms mix the soil in natural way. Bacteria viz. *Rhizobium* living in the root nodules of legumes viz. pulses, soybeans and groundnuts, fix atmospheric nitrogen in association with the host. Other soil micro-organisms including fungi, actinomycetes and bacteria decompose organic matter, thereby, mineralizing more nutrients and actively participating in different plant nutrient cycles (Gaur *et al.,* 1971). Micro-organisms also produce polysaccharides and mucilaginous substances that bind the primary and secondary soil particles into aggregates (Gaur and Subba Rao, 1976; Gaur *et al.,* 1972). To improve the soil quality, soil must be fed with crop residues, green manures, oilcakes, farmyard manure etc. The use of efficient and competitive strains of biofertilizers may be of special significance to improve the nutritional quality of soil.

Soil is a favourable medium for the growth and development of soil micro-organisms. Bacteria are found in Indian soils in the range of 10 to 30 million, followed by actinomycetes (x10^5), fungi (x10^4), algae (x10^3), *Azotobacter* (x10^2). Most of them are principal decomposers of organic matter in soil and help in formation of humus substances. Few bacteria are involved in fixation of atmospheric nitrogen and some of the microbes are instrumental in phosphate solubilization. They participate in different biological cycle's viz. carbon, nitrogen, phosphorus, sulphur and transformations of iron, manganese, potassium, zinc, arsenic, selenium etc. Sustainable agriculture does not represent a return to old methods but it combines traditional conservation-minded techniques with modern technologies. Emphasis is placed in crop rotation, building of soil fertility, and use of soil microorganisms as biofertilizers to improve soil health, diversifying crops and livestock and controlling pests naturally. Whenever possible external sources viz. commercially procured chemicals and fuels are replaced by resources found on or near the farm. These internal sources include solar, wind or biogas as energy, bacterial or fungal biofertilizers, fixing atmospheric nitrogen, phosphate solubilization or accelerating the nutrient mineralization process. In some cases, external resources may be essential for reaching sustainability. As a result, such farming systems can differ considerably from one and another because each tailors its practices to meet specific production, environmental and economic needs.

2.13.1 Microbial Biodiversity in Soil

The biosphere or part of the earth and its atmosphere where life is found is made of different size of ecosystem regulated by solar energy. Ecosystems have both biotic and abiotic components. The biotic component comprises closely interacting groups of flora and fauna having varying energy needs which comprise of heterotrophic, chemotrophic and photoautotrophic microorganisms. Biodiversity refers to the richness of life as manifested by their morphological, cultural and biochemical characteristics of microorganisms and their interdependence in a habitat. It is because of biodiversity that each ecosystem functions as the basic life support system of the planet earth. Biodiversity sustains the food and nutrition security of human beings and mediates the ecological process by resisting environmental perturbations in a given soil and climatic conditions. Presently greater emphasis is being given on the biodiversity of above-ground biotic components belonging to plant and animal kingdom. The living organisms in soil and their diversity in relation to their contribution to sustainable agriculture often do not receive adequate attention. Without taking into consideration the magnitude and biodiversity of soil biota which is more or less equal to above ground mass, the understanding of the global diversity will be fragmentary. Soil is a natural habitat of an array of diverse groups of organisms related to the plant and animal kingdoms. They vary in morphology and are found from microscope to large bodies belonging to flora and fauna. There are organisms which are beneficial to improve soil properties and plant growth but some pathogenic types damage the agricultural crop by causing plant diseases viz. root-and stem-rot. The decomposition of plant and animal residues in soil, mineralization (conversion of organic nutrients to inorganic form) of plant nutrients, solubilization of insoluble forms of nutrients, biological nitrogen-fixation, mediation in natural cycles etc. are processes in which soil biota activity participate. In the absence of soil organisms, the biological process will stop, making life of plants and animals impossible on the earth. The soil organisms comprise of fauna and flora.

2.13.1.1 Soil Fauna

Their biomass is smaller as compared to soil flora. Soil fauna or soil animals comprise large varieties of organisms with different shapes and sizes and adaptive strategies. Based on their sizes and adaptive strategies, they are divided into microfauna, mesofauna and macrofauna (Fig 2.1).

Microfauna: Organisms belonging to this group are very small with an average size of less than 0.2mm. They are found in soil pores and water films that cover soil particles and feed on microflora and other microfauna. Many are parasitic on higher plants and some persists on plant residues. Nematodes and protozoans are the main representatives of this group.

Protozoans are one-celled organisms confined mostly to surface horizon of soil. They are slightly bigger than bacteria and usually the most varied groups' viz. flagellates, amoebae and ciliates. Aeration and availability of food materials are the facts influencing their population. Their population in normal soils varies between 10^4 and 10^5/g of dry soil. Protozoans are important predators of bacteria resulting in partial sterilization of soil exerting a regulatory control on the bacterial population of soil. They may prove beneficial in control of pathogenic bacteria in the root zone. The flagellated protozoa belonging to the class Mastigophora are predominant in the soil. They are uni-nucleate and their flagella are terminal, even if there are 40 or more of them,. Important genera are *Allantion, Bodo, Cercobodo, Entosiphon, heteromita, Oikomonas, Sainouran, spiromonas, Spongomonas* and Termites, Amoebae included the class Sarcidina are characterized by their movement through pseudopodia. The protoplasm may be naked or, as in some genera, encased in shells. Important genera of this class are *Amoeba, Biomyxa, Diffugia, Euglypha, Hartamanella, Lecythium, Naglerai, Nucleria* and *Trinema*. Ciliates belonging to the class Ciliata are characterized by the presence of cilia aroundthe bodies. They also possess two dissimilar nuclei, the macromolecules and micronutrients. Important soil inhabitants of this class are *Balantiophorus, Colpidium, Enchelys, Colpodo, Halteria, Pleurotricha* and *Vorticella* etc.

Nematodes are found in all types of soils and are among the most abundant soil fauna. Their numbers are highest ranging from 10^4 to $10^6/m^2$ in cultivated soils. Nematodes are composed of those that are predatory on bacteria, fungi and protozoa, parasite on higher plants and those that feed on decaying organic matter. Nematodes often cause serious damage to plants and over 500 species of plant parasitic nematodes are known to exist in soils. These round and spindle-shaped organisms have adaptable mouthparts and pointed form. As such, it is easy for them to penetrate plant tissues. Though major crop pests, nematodes also possess the potential to suppress the population of fungi and bacteria in the root zone.

Mesofauna: Organisms of this group range in length from 0.2 to 2mm. Population densities in favourable soil conditions can raise upto $10^6/m^2$ of soil surface. Mites and Collembola are of common occurrence in soil. These organisms degrade plant litter and are predators of other mesofauna and fungi. Although present in all soil layers, maximum population is recorded in the rhizosphere of plants.

Macrofauna: This group comprises of small mammals, insects, mites, centipedes, millipedes, slugs, snails, spiders, earthworms etc. They exceed 10mm in size and include predators and feeders of plant materials. They contribute to soil fertility by mixing organics in soil and improve physical condition of soils. On death, their substantial residues are degraded in soil and release plant nutrients. The most important members of this group from point of view of soil fertility are termites, ants and earthworms.

Termites and ants are common in tropical, subtropical and mediterranean soils. Through their activities, they modify the organic substances of soil either by digestion or translocation. Termites utilize undecomposed organic matter as food resource and thus play a role in their degradation in soil. Their biomass in soil is much less than earthworms. Termites may damage the growing plants when present in large numbers in soil, depleted of all organic matter on the surface soil making it vulnerable to erosion.

Earthworms are well known representative of soil macrofauna having the maximum influence on the physical and biological properties of the soil, they are found in large numbers in soil which are moist and rich in organic matter, exchangeable calcium and phosphates. In rich cultivated soils, their population may range between 0.6 and 2.5million/ha with live weight ranging between 300 and 1250kg/ha. In unfavourable soils, their population may drop to even less than 30,000/ha with a biomass of about 15kg/ha.

Earthworms are classified into detrivores and geophages. The detrivores include epigeic and anecic forms, and geophages comprise the endogenic earthworms. The epigeic bring the fragmentation of organic materials restriction on the soil surface. This activity has no effect on soil structure. The anecics feed on litter mixed with soil particles and produce surface casts. The endogenic earthworms live within the soil deriving nutrition from organically rich soil they ingest. The common examples of detrivores which are involved in humus formation and vermicomposting are *Perionyx excavatus, Eisenia foetida, Eudrilus euginae, Lempito mauriti, Octachaetona serrate* and *Octachetana surensis*. The common example of humus feeders (Geoghages) is *Octachaentonathustoni*. Earthworms ingest organic debris alongwith sufficient soil particles. The ingested material is subjected to enzymatic and grinding action before excretion as 'casts'. In rich soils, the casts may work out to more than 15 tonnes/ha. They organisms may pass through bodies annually around 40 tonnes soil and organic matter/ha. It is claimed that earthworms cast are richer in organic matter, available plant nutrients, and microbial population. The accelerated microbial population and their activities are helpful in improving the physical, chemical and biological properties of the soil. However, data substantiating these claims are fragmentary, hence, systematic investigation are required to confirm the claims of the promoters of vermicomposting.

It is represented by diverse type of microorganism viz. bacteria, actinomycetes, fungi and algae. The living roots of higher plants are also important component of soil. The biomass of soil flora is higher than the biomass of soil fauna. This group performs important functions viz. soil granulation, organic matter mineralization, fixation of atmosphere nitrogen production of plant growth promoting substances etc. (Dahiya and Srivastava, 2005). The organic transformations initiated by soil fauna are carried forward by microflora and cumulative effect of interaction viz. synthesis of microbial polysaccharides, binding effect of fungal mycelia and plant roots bring about significant improvements in soil structure and physical condition of soils. Soil enzymes, vitamins, amino acids and hormones etc. are produced in soil due to microbial processes.

Bacteria: Soil bacteria are the most abundant group of all micro floras present in soil. Indian soils in normal conditions have 10 to 25million/g bacteria in dry soil. These are simplest and smallest forms of life, present in the soil. Their growth rate is faster than other microflora of soils. Most of the soil bacteria are heterotrophic and depend on their needs of carbon and energy on native and added fresh organic matter. Heterotrophic bacteria alongwith other microorganism extensively participate in organic matter decomposition, release of plant nutrients and humus formation. The chemoautotrophs are of special significance because they participate in nitrification process and nutrient cycling viz. sulphur, iron etc. soil bacteria in general are involved in cycling of plant nutrients.

Soil bacteria as a group are quite prominent in fixation of atmospheric nitrogen and is of great relevance in sustainable agriculture. Nitrogen-fixing bacteria form symbiotic, associative symbiotic or asymbiotic associations with higher plants. The symbiotic bacteria form associations with higher plants of the leguminous group. The organisms belong to different species of *Rhizobia*. In some legumes viz. *Sesbania rostrata* nodules are formed on stem as well. The inclusion of leguminous trees, pulses and oilseed crops in the farming system or crop rotation is of great importance in substantial gains of atmospheric nitrogen.

The associative and asymbiotic bacteria are categorized into aerobic, microaerophilic, facultative and anaerobic. The nitrogen-fixing bacteria in this sector are *Azotobacter, Clostridium, Beijerinckia, Azospirillum* etc. The soil health, namely the population of beneficial bacteria is influenced by soil conditions and farming system. Organic matter, moisture, temperature, pH and nutrient content affect the growth and activity of soil bacteria.

Algae: Algae are present in diverse locations viz. land, fresh water and sea. The organisms present in upper layers of soil are chlorophyll bearing and derive their energy from photosynthesis. Algae are also found in sub-soils as spores or cysts or as vegetative forms that do not depend on photosynthesis. These forms, however, represent only a small fraction of the total algae biomass. These microorganisms are not as numerous as bacteria, actinomycetes or fungi. Their development on the surface of cultivated or virgin moist lands is frequently noted by naked eye but isolates can be obtained from lower depths as well. The counts of algae usually range from about 100 to 50,000/g of soil samples, taken just below, the surface of arable lands. They are unicellular, Filamentous, or colonial. Blue green algae, green algae and diatoms are numerous in the soil. Owing to the presence of photosynthetic pigments in their cell, algae are phototrophic and use CO_2 from the atmosphere and give out O_2. Green algae are the dominant algal flora in acid soils; they are also numerous in neutral and alkaline soils. *Chlorella, Chlamydomonas Chlorococcum, Oedigonium* and *Protosiphom* are the most common green algae found in the soil. Diatoms have highly silicified outer layer and are common in neutral to slightly alkaline soils. The prominent genera in soil include *Achnanthes, frangilaria, Pinularia* and *Synedra*.

Blue Green algae are prokaryotic with their pigments localized in chromatophores. They have phycocyanin pigment in addition to chlorophyll. They do not have flagella and do not reproduce sexually. They prefer neutral to alkaline environments. *Oscillataria, Phormidium, Cylindrospermum, Anabaena, Scytomea, Tolypothrix* and *Fischerella*. Some of the blue green algae possess specialized cell known as heterocysts which are involved in nitrogen fixation.

Fungi: Different groups of fungi are present in soil. They are active in presence of organic materials in general and in acidic soils in particular. The population density of fungi in normal cultivated soils is less compared to bacteria and actinomycetes. However, their biomass is more when compared with bacteria. Fungi are active decomposers of cellulose lignin, starch, gums etc. in soil. They are efficient in carbon assimilation from decomposing organic residues into fungal hyphae, which serve as source of energy and nutrients for bacteria and other microflora to carry forward the biochemical transformations. In unfavourable conditions viz. drought weather, they can survive by forming heat resistant spores. Their role is of much importance than bacteria especially in acidic forest soils and in mulched soils. In cultivated Indian soils, their population is found in 10^4/g of dry soil. A wide range of soils under many different types of vegetation and from many different geographical areas have been examined for the presence of fungi. *Aspergillus, Mucor, Trichoderma, Penicillium, Alternaria, Fusarium, Rhizopus, Verticillium, Botrytis, Zygorhynchus, Pullularia, Gliocladium, Monilia, Chaetomium* and *Phytium* are among the most commonly encountered genera of fungi in the soil.

Certain species of fungi viz. *Glomus* form symbiotic association with the roots of higher plants. Their associations are known as mycorrhizae, resulting in increase of nutrient absorption area of the root system. Mycorrhizae are beneficial for higher plants as they increase the availability of phosphorus, zinc, copper and molybdenum. Mycorrhizal associations improve greater water uptake and nitrogen-fixation by *Rhizobium* and *Frankia* in leguminous and non-leguminous plants and better resistance to disease by inhibiting root pathogens. Ectomycorrhiza and vesicular arbuscular mycorrhiza (VAM) showed maximum potential in forestry and associated only with tree roots with their mycelia developing on the surface of the roots to form a sheath. The occurrence of VAM fungi have been found in wide range of cultivated plants and trees where the mycelia do not form a sheath on the root surface but penetrate into the root. Some fungi are pathogenic and cause root rot, damping off, wilt, blight etc. in plants resulting in reduced yields and damage to the plants.

Actinomycetes: They are unicellular and filamentous organisms. They are smaller in size and closer to bacteria. They occur in large numbers in well aerated moist soils, rich in organic matter. They are highly sensitive to acidic conditions and do not function at a soil pH below 5. Their population density in India is found to range between 1 and 3 million/g of dry soil. These organisms are active as antibiotic producers and participate in decomposition of resistant plants material components like lignin and cellulose. Frankia, an actinomycete, is known to possess ability to fix atmospheric nitrogen in association with non-leguminous shrubs and trees of *Causuaria, Alnus rubra, Myricagale, M. faya* etc. The quantity of nitrogen fixed has been found to vary from as low as 12kg to as high as 320kg/ha/year.

In the order of abundance in the soils, the commonest genera of actinomycetes are *Streptomycetes* (up to 70%) followed by *Nocardia* and *Micromonospora,* although *Actinomycetes, Actinoplanes, Microbiospora, Streptosporangium* are also generally encountered in the soil.

2.13.2 Formation of Different Soil

Soils are formed when geologic materials are exposed. This may occur after a dramatic event such as a volcanic eruption or from a simple disturbance. Soil formation is the result of the combined action of weathering and colonization of geologic material by microbes. The microbial community that initially colonizes a newly disturbed environment will bring changes to the local environment by degrading and recycling organic material. For example, if extreme soil erosion has removed top soil, phosphorus may be present, but nitrogen and carbon must be imported by physical or biological processes. Under these circumstances, autotrophic, nitrogen fixing cyanobacteria are active in pioneer stage nutrient accumulation. This primary microflora will eventually be replaced by a new community of microbes that further alter the ecosystem. This set in motion successive waves of microbial communities until a community that is sufficiently diverse and physiologically well-suited is established and a stable climax ecosystem is developed. The major types of plant soils systems are shown in Fig. 2.2.

2.13.2.1 Tropical and Temperate Region Soil

In warm, moist tropical soils, organic matter is decomposed very quickly and the mobile inorganic nutrients can be leached out of the surface soil environment, causing a rapid loss of fertility. To limit nutrient loss, many tropical plant root systems penetrate the rapidly decomposing litter layer. As soon as organic material and minerals are released during decomposition, the root takes them up to avoid losses in leaching. Thus it is possible to recycle nutrients before they are lost due to water movement through the soil (Fig. 2.2a), with deforestation, there is no leaf litter so nutrients are not recycled, leading to their loss from the soil and decreased soil fertility.

Tropical plant soil communities are often used in slash-and-burn agriculture. The vegetation on a site is chopped down and burned to release the trapped nutrients. For a few years, until the minerals are lost from these low organic matter soils, the farmer must move to a new area and start over by again cutting and burning the native plant community. this cycle od slash and burn agriculture is stable if there is sufficient time for the plant community to regenerate before it is again cut and burned, if the cycle is too short, rapid and almost irreversible degradation of the soil occurs.

In many temperate region soils, in contrast the decomposition rates are that of primary production, leading to litter accumulation, Deep root penetration in temperate grasslands results in the formation of fertile soils, which provide valuable resources for the growth of crops in intensive agriculture (Fig. 2.2b). The soils in many cooler conferous forest environments suffer from an excessive accumulation of organic matter as plant liter (Fig. 2.2c). In winter, when moisture is available, the soils are cool and this limits decomposition. In summer, when the soils are warm, water is not as available for decomposition. Organic acids are produced in the cool, moist litter layer, and they leach into the underlying soil.

These acids solubilize soil components such as aluminium and iron, and a leached zone may form. Litter continues to accumulate and fire becomes the major means by which nutrient cycling is maintained. Controlled burns have become part of the environmental management in this type of plant soil system. Bog soils provide a unique set of conditions for microbial communities (Fig. 2.2d). In these, decomposition is slowed by the waterlogged, predominantly anoxic conditions, which leads to peat accumulation. When such areas are drained and become more toxic and SOM is degraded, resulting in soil subsidence. Under oxic conditions, the lignin cellulose complexes of the accumulated organic matter are more susceptible to decomposition by filamentous fungi.

2.13.2.2 Cold Moist Soils

Soils in cold environments, whether in Arctic, Antarctic, or Alpine regions, are of extreme interest because of their wide distribution and impacts on global level processes. The colder mean soil temperatures at these sites decrease the rates of both decomposition and plant growth. In these cases SOM accumulates, and plant growth can become limited due to the immobilization of nutrients. Often, below the plant growth zone, these soils are permanently frozen. These permafrost soils hold about 11% of the earth's soil carbon and 95% of its originality bound nutrients. These soils are very sensitive to physical disturbance and pollution, and the widespread exploration of such areas for oil and minerals can have long term effects on their structure and function. In water saturated bog areas, oxygen limitation means that bacteria are more important than fungi in decomposition processes, and there is decreased degradation of lignified materials. As in other soils, the nutrient cycling processes of nitrification, denitrification, nitrogen fixation and methane synthesis and utilization although occurring at slower rates, can have major impacts on global gaseous cycles.

2.13.2.3 Desert Soils

Soils of hot and cold arid and semi-arid deserts are dependent on periodic and infrequent rainfall. When it rains, water can puddle in low areas and be retained on the soil surface by microbial communities called desert crusts. These consist of *Cyanobacteria* and associated microbes including *Anabaena, Microcoleus, Nostoc* and *Scytonema*. The depth of the photosynthetic layer is perhaps 1mm and the filaments and slime link the sand particles, which damage the surface soil albedo (the amount of sunlight reflected), water in filtration rate, and susceptibility to erosion. These crusts are quite fragile, and vehicle damage can be evident for decades, after a rainfall, nitrogen fixation begins within approximately 25 to 30hrs, and when the rain evaporates or drains, the crust dries up and nitrogen is released for use by other microorganisms and the plant community.

2.13.2.4 Geologically Heated Hypothermal Soils

Geologically heated soils are found in such areas as Iceland, the Kamchatka peninsula in eastern Rudsssia. Yellowstone National Park, and at many mining waste sites. These soils are populated by bacteria and archea, many of which are chemolithoautotrophs. A wide variety of chemoorganotrophic genera also are found in these environments: these include the aerobes. *Thermomicrobium, Thermoleophilim*, and also the anaerobes *Thermosipho* and *Thermotaga*. An important microorganism found in heated mining wastes is Thermoplasma. Such geothermal soils have been of great interest as a source of new microbes to use in biotechnology, and the search for new, unique microorganisms in these are is intensifying all over the world.

2.13.3 Rhizosphere Effect and R: S Ratio

Rhizosphere is central to microbial and nutrient dynamics and describes the zone of soil surrounding roots of legumes in relation to symbiotic nitrogen-fixing bacteria. It includes both, the area of influence and the physical location around the root. This definition is now extended to the zone of soil in which microbial activity is affected by the roots of plant species which release organic substances (Dommergues, 1978). The rhizosphere has been further subdivided into (i) Ectorhizosphere (outer rhizosphere), and (ii) Endorhizosphere (a zone where invasion and colonization of root cortical cells by soil microorganisms occurs). The distinct boundary of the root surface with the soil has been termed the 'rhizosphere'. This term was coined by Hiltner in 1904 based on the observations. Rhizoplane refers to the zone of attachment of microorganisms to the root surface. However, microbial colonization across the soil-root interface is often difficult to distinguish and, therefore, it is more reasonable to consider it as a continuum. Microbial colonization is minimal at the region around the root cap and is greatest in the zone of elongation behind the root tip (Lynch, 1990).

Microscopic examination reveals the presence of a vast microbial community surrounding on the surfaces of roots and roots hairs. Bacteria are found to be localized in colonies and chains of individual cells. Filamentous fungi and actinomycetes are not observed as frequently. Protozoa particularly flagellates are situated in the water films on the root hairs and on the epidermal tissues. The root influence on soil bacteria as measured by plating techniques is often expressed as rhizosphere effect. This can be expressed based on R: S ratio. Rhizosphere effect is governed by root variation in plant species, its age and is also influenced by cultural, soil and environmental conditions.

Katznelson (1965) suggested rhizosphere: non-rhizosphere ratio of the population to find out the degree of influence of plant roots on soil micro-organisms. It is defined as the ratio of microbial numbers per unit weight of rhizosphere soil (R) to the numbers in a unit weight of the adjacent non-rhizosphere soil (S). The rhizosphere effect is consistently greater for bacteria than for the other microorganisms. Those responding in a significant way are short, Gram negative which generally make use a larger percentage of rhizosphere soil than the normal soil microflora. The percentage occurrence of short, Gram positive rods, coccoid rods and spore forming bacteria (*Bacillus* spp.) declines, Arthrobacter group, cocci, or long non-spore forming rods are not generally influenced.

The genera of *Pseudomonas, Flavobacterium, Alcaligenes* and occasionally *Agrobacterium* were commonly reported. The other bacteria viz. *Arthrobacter, Brevibacterium, Corynebacterium, Micrococcus, Bacillus*, etc. showed their presence in this zone but the environment was not so favourable. Anaerobic bacteria have also effected by the root in view of reduced O_2 tension due to enhanced root and microbial respiration. Soil microflora is activated by growing plants and the nutrient cycling is more rapid in the rhizosphere than in the non-rhizosphere soil. The plant roots release number of low and high molecular weight organic substances which are easily available to organisms. The most important contribution of plant is the excretion of root exudates and sloughed off tissues as source of energy and nutrients. The products excreted by plant under aseptic conditions are amino acids, organic acids, carbohydrates, pentose and hexoses and oligosaccharides, growth factors, enzymes and auxins. The plant begins to excrete organic compounds in the environment from the beginning of seed swelling and germination. Generally the germinating seed exudes more nitrogen as protein and peptides than in amino acids whereas the roots release more of free amino acids than in proteins and peptides. Other quantitative difference includes that reducing compounds are released more by seedlings than germinating seeds.

Rovira (1965) reported 3 possible ways by which the rhizosphere micro-organisms can affect plant growth **(i)** the production of metal chelating compounds, solubilizing some important plant nutrients **(ii)** formation of the certain growth promoting substances, and **(iii)** protection against plant pathogens, Krasilnikov (1963) reported the production of biologically active substances by rhizosphere micro-organisms considered to be an important factor in soil fertility. There are evidences about uptake of organic compounds by plant roots by different workers (Winter, 1961; Miller and Schmidt, 1965). Both positive and negative effects of bacteria on plant growth were observed by Hussain and Vancura (1970). The negative effect might be due to high concentration of b-indolyl acetic acid which was produced by strains No. 5 and 10 of *Pseudomonas fluorescens*. Favourable influence on plant growth was observed when the above strains were used as inoculant in non-sterile soil. The strains producing b-indolyl acetic acid and gibberellin-like substances enhanced the growth of plants. *Pseudomonas fluorescens* produced both these compounds in addition to biotin and pantothenic acid. The production of b-indolyl acetic acid was reported by *Azotobacter* (Vancura and Macura, 1960, Jackson *et al.,* 1964) *Pseudomonas solanacearum* (Sequeera and Williams, 1964), *Arthrobacter* and *Nocardia* (Katznelson and Sirois, 1961). The production of gibberellic acid (A_3) by *Azotobacter* was also reported (Vancura, 1961).

The influence of microbial metabolites on plant growth is most effective under conditions when the effect of microorganisms on plants is continuous e.g. microbial inoculants applied to seeds survive and multiply in the rhizosphere during the vegetative and productive part of plants. *Pseudomonas* and probably other Gram negative *bacilli* dominated among microorganisms utilizing amino acids, organic acids and other early available substances. It is known that the compounds exuded by plant root were utilised preferentially by Gram negative bacilli from the root surface and rhizosphere soil. In General, Gram (-), non-spore forming rods with simple nutritional requirements which respond to amino acids, are stimulated to a greater degree by roots than the coccoid and Gram (+) spore forming rods (Glick, 1995). The greater proportion of non-sporulating rods among the rhizosphere isolates and their ability to respond to nutrients viz. amino acids reflects on the governing mechanism(s) by which root and the rhizosphere are colonized.

2.13.4 Rhizodeposition

The extent of 'C' loss from the roots has been described as "rhizodeposition" and plays a significant role in the rhizosphere effect. It must be emphasized, however, that the living root itself serves as a substrate for the endorhizosphere inhabitants (Curl and Truelove, 1986). The 'C' flow from roots is released into the soil environs as **(a)** water soluble exudates, **(b)** secretions, **(c)** lysates and **(d)** gases. Not all compounds are released at anyone stage of plant growth and development but during the complete cycle from seed emergence to senescence. Detailed experiments based on the use of radioactive tracer techniques showed that the values obtained for per cent of C lost as rhizodeposition can be quite substantial. It is estimated that nearly 30-60% of net fixed 'C' is transferred to roots in annual plants of this; 40-90% is lost as rhizodeposition. However, complexity of the soil system makes it difficult to draw accurate carbon budgets for various plants (Kraffczyk *et al.,* 1984). Various estimates suggest that the yield of microorganisms utilizing the products of rhizodeposition is approximately 0.35 g dry wt/g substrate consumed. The categories of microbial products released in the rhizosphere are rather wide of which some are absorbed by the roots whereas others remain bound. On microbial dynamics in the rhizosphere, Bowen (1980) had suggested alternative approach towards study of the behaviour of microorganisms in rhizosphere which included use of "microbial probes" as against the usual methods of collecting "root exudates". Through this one can relate 'C' loss directly to microbial population, be it mycorrhizae or growth promontory rhizobacteria. The only problem would however, be the diffusion of the substrate into soil but the latter can be overcome by comparing the analysis of the root exudates.

2.13.5 Substrates in the Rhizosphere

A wide variety of organic substances are found in the rhizosphere which can be broadly categorized as under (Rovira *et al.,* 1979).

- ***Exudates:*** Low molecular weight compounds, i.e., sugars and amino acids that leak from intact cells.
- ***Secretions:*** Compounds actively released from root cells.
- ***Lysates:*** Materials released through the lysis of older epidermal cells.
- ***Mucigel:*** Gelatinous material on the root surface comprising of plant mucilage, bacterial cells, metabolic products and colloidal organic and mineral materials.
- ***Plant mucilage:*** It is released from four different sources viz. by golgi bodies of the root cap cells, hydrolysate of the primary cell wall located between the root cap and epidermis, secretions from the epidermal cells and roots hairs with primary walls and compounds resulting from the microbial degradation and modification of the dead epidermal cells.

Experimental study of microbial growth on parts of the root indicated the amounts of substrate for growth at different sites (Fig. 2.3). Growth curves on roots in soil suggest a rapid logarithmic increase in microbial population over the first 3-4 days, followed by a leveling of the curve (Bowen, 1979). The rapid phase was equated with the richness of the substrate, and the latter period with the carrying capacity of the system. i.e., the amount of energy able to support the population that has developed on the root. In fact, phase III is not strictly stationary because it shows a rise due to growth of micro colonies along the root. Generation time in the slower phase may be 100 hrs or more but only 5-6hrs in the rapid phase.

Root exudates are crucial to rhizosphere micro-organisms and comprise of large variety of molecules viz. sugars, amino acids, organic acids, fatty acids and sterols, growth factors, nucleotides. flavoures, enzymes and a range of other miscellaneous compounds. Some of these compounds in root exudates serve as chemoattractants for the rhizosphere bacteria at very low concentrations, e.g., attraction of *Azospirillum* by benzoate at 10^{-9} to 10^{-12}M when compared with *Agrobacterium tumefaciens* and *Rhizobium meliloti* by luteolin at 10^{-9}M. At relatively higher concentrations (10^{-2} to 10^{-3}) amino acids viz. glutamine, threonine, serine, cysteine and arginine attract *Pseudomonas* sp. Furthermore roots release agglutinating factors that cause agglutination of saprophytic free-living pseudomonads (Li and Alexander, 1988). About 30% of the non-rhizosphere bacteria showed positive agglutination to pea root whereas the level for rhizosphere bacteria was nearly 85%, suggesting adaptation of the latter group to the surrounding environment (Anderson *et al.,* 1988). Special recognition mechanism and relationship was found for *Rhizobium*-host roots wherein lectins and agglutination reactions are only one phase. Several workers have distinguished between a non-specific attachment phase with the root hairs and symbiotic specificity phase accompanied by the formation of extra cellular microfibrils. Similar microfibrillar attachment was observed with other rhizosphere micro-organisms (Thomashow *et al.,* 1990).

Recognition events in rhizobia-host system result in increased porosity of root hair walls and alteration of plasmalemma and proton flux. While non-homologous bacteria also cause somewhat similar changes, the intensity of reaction is low. To monitor *in situ* performance of PGPR, van Overbeck and van Elsas (1995) developed a strain RIWE8 of *Pseudomonas fluorescens* which expresses itself from root induced promoter with *Lac Z* as reporter gene. A proline-controller promoter is present in this modified strain which offers opportunity for ecologically safe and efficient use in areas of bio-control and growth promotion.

The flavonoids released by roots and seeds are important plant signals to microbes that form root nodules and later reduce free N_2. Several such inducers were identified in legumes and their specific role for rhizobia defined. While these were suggested to be important for free-living forms as well, specific functions are not clarified as yet. Another important group of compounds are the plant and microbial produced phytohormones (Penrose and Glick, 1997).

Hormones are defined as a group of naturally occurring organic substances that influence the physiological processes at low concentrations. There is strong evidence that supports the hypothesis that plant growth and development in response to inoculation may be due to microbial production of plant growth regulatory substances. *Pseudomonads* in particular were studied for phytohormones on account of their growth promotory activity. Both, IAA and gibberellins were reported from *Pseudomonas fluorescens,* especially GA, transformation of tryptophan to IAA was observed for these bacteria and seed inoculation found to improve plant growth and productivity. Some pseudomonad strains were also found to secrete *biotin* and pantothenic acids, besides to IAA (100 to 200 mg/g cell dry mass). Although cytokinin like substances too has been detected, their production is largely restricted to species of *Azotobacter, Agrobacterium* and *Rhizobium.*

2.13.6 Microbial diversity in Rhizosphere

Detailed investigations of microbial diversity showed that short, Gram(-) rods viz. *Alcaligens, Flavobacterium* and *Pseudomonas* Predominate in the rhizosphere (Glick, 1995). However, the initial root colonizers are associated with soil organic matter. The ability to grow on both, roots and residue was demonstrated with root-colonizing pseudomonads which make up for a dominant population in soils alongwith others viz. *Agrobacterium, Alcaligens, Arthrobacter, Azotobacter, Bacillus, Clostridium, Rhizobium and Serratia;* collectively these were termed as *Rhizobacteria.* The general influence of these bacteria on plant varies from beneficial to deleterious to near neutral. Some specific strains that were pathogenic to sugarbeet seedlings were termed deleterious rhizobacteria (DRB). PGPR's especially the fluorescent pseudomonad act directly or indirectly on major and minor plant pathogens either by antibiosis, competition with the pathogen for essential nutrients or through the release of growth hormones. Growth promotion by fluorescent pseudomonad results directly either through their metabolites or as a consequence of the release of siderophores, HCN, antibiotics and others. In the latter circumstances, inhibition of the activity of phytopathogens is responsible for improved plant growth. Indirect growth promotion can alter the physiology of the plant thus improving the host defense to pathogen attack i.e. induced resistance (Cook *et al.,* 1995; Johri *et al.,* 1997).

The successful establishment of PGPR in an ecological niche viz.

rhizosphere is dependent upon colonization, competition, antibiosis, HCN production and chemotaxis. Among these, colonization is an active process and involves survival of bacteria onto seeds, in soil, multiplication in response to seed exudates rich in carbohydrates and amino acids, attachment to the root surface and proliferation in soils containing indigenous microorganisms. However, an effective colonizer should be able to sustain under the stress viz. water potential, temperature, pH, soil type, composition of root exudates and soil texture to be successful in the milleu of the rhizosphere. A good colonizer adheres strongly to the roots and possesses short doubling time thus achieving a selective advantage. For example, fluorescent pseudomonads show a generation time of 27hrs on barley during first four days of plant growth but the duration increases to 10hrs when the plants are 7 to 16 days old. Glycoproteins of root origin (agglutinins) showed to be involved in the colonization process in *Pseudomonas putida.* More recently, it was possible to identify the genes responsible for root colonization but main colonization gene (s) are yet to be elucidated. Competitive advantage in case of fluorescent pseudomonads arises as a consequence of higher levels of iron-chelating siderophores, secretion of antibiotics and release of hydrogen cyanide (HCN). Considerable data on each of these components exists which tends to suggest that each microbial strain can behave in a different manner and that growth promotion and bio control action represents two sides of a coin.

2.13.7 Biomass Changes in Rhizosphere

The bacterial community in the soil primarily lives in smaller soil pores, while the fungal component tends to bridge between aggregate because of its ability to translocate nutrients from and between different regions of the soil. This difference in strategies of fungi and bacteria suggests that rhizosphere bacteria depend primarily on soluble exudates from the root for nutrients while many of the filamentous fungi obtain their substrates from outside the root zone. The increased availability of readily utilizable organic matter from root exudates modifies the location of microbial development in soil. These areas of newly, developing populations are, therefore, more accessible for predation by soil protozoans.

Another critical factor that influences microbial response to the rhizosphere relates to the level and flux rate of mineral nitrogen and possibly phosphorus. In more intensively managed and fertilized soils, nitrogen availability will decrease the population of filamentous fungi, which have the functional advantage of being able to translocate nitrogen to the site of available organic carbon. This will in turn increase the relative amount of carbon processed through the bacterial biomass. The improved nitrogen availability results in decreased demand on native soil organic matter as a source of nitrogen for microbial growth. Other factors that are known to shift the relative bacterial to fungal dominance in the utilization of plant root exudate are the levels of various inhibitors, e.g. heavy metals, pesticides, herbicides and fungicides in this environment.

2.13.8 Beneficial Vs Deleterious Rhizobacteria

Micro-organisms, present in the rhizosphere or "rhizobacteria", are named by the specific activity shown: PGPR (Plant Growth Promoting Rhizobacteria), NPR (Nodulation Promoting Rhizobacteria), EPR (Emergence Promoting Rhizobacteria), PHPR (Plant Health Promoting Rhizobacteria), DRB (Deleterious Rhizobacteria), DRMO (Deleterious Rhizosphere Microorganisms), and YIB (Yield increasing Bacteria). Rhizosphere research has been dramatic changes in approach and methodologies during the last 90 years and has currently entered a phase where in situ detection of microbial activities is feasible through modern analytical techniques and tools. It is also being realized that a single root often carries a multitude of symbioses viz. mycoorhizal, actinorhizal, and bacterial and, therefore, the net gain by a plant is something more than the simple one to one relationship. This fact is receiving further support from the observations that common signal molecules might be operating as attractants and that the host response to colonization by an alien may be manifested through expression of genes that produce proteins that exhibit serological cross-reactivity amongst widely differing mutualistic partners. All of these statements are based on the data available widely in literature but is often not viewed as a common facet of rhizosphere biology. The reason is not lack of apprehension but often, non-availability of suitable tools to tackle a complex situation (Chanway *et al.,* 1991). In the context of introduced biofertilizers, there is an increased emphasis on developing microbial consortia of nitrogen-fixes, 'P' solubilizers, growth promoters and siderophore/ antibiotic activity so that growth promotion is more or less guaranteed and inter-conversion and utilization of nutrients is not limited to one group with the resultant change in the rhizosphere equilibrium. After this, entire zone is most critical in nutrients release, mobilization and competition.

2.13.9 Phyllosphere Effect

Ruinen (1961) showed the importance of the external leaf surface as an environment for micro-organisms. Since, then Sinha and his co-workers have reported on the interrelationship between the plant foliage and the micro-organisms found on the surface. The quantity and quality of microbes in the region named as phyllosphere vary with the plant species, and its morphological, physiological and environmental factors. The age of the plant, its leaf spread, morphology and maturity level and the atmospheric factors under which it grows, greatly influence the phyllosphere microflora. Both native or resident organisms and casual depositors are found on the leaf surface. More organisms are retained on a surface which is hairy, waxy or highly circulate than on a smooth dry surface. Leaf diffusates and exudates have been reported to encourage growth and multiplication of the phyllosphere micro-organisms. Since large variety of microbes, both pathogens and saprophytes are found on the leaf surface, the stimulation of growth in the micro-organisms varies widely, and is often selective. Thus the phyllosphere effect seems to be specific to a given set of plant foliage and its microbial load. An change in the phyllosphere effects affects plant growth, which in turns affects the physiological activity of the root system. Such changes in root result in an altered pH and spectrum of chemical exudations, causing a change in the rhizosphere microflora. Thus, there exists a link between phyllosphere microflora and rhizosphere microflora.

2.14 Concept of IPNM

Five basic management principles for sustaining cropping system such as replenishment of removed plant nutrients, maintenance of humus level in soil, avoidance of weeds, pests and diseases, control of soil acidity, alkalinity and toxicity and soil erosion control were suggested (Greenland, 1985). These principles for proper management point out on the possible meager contribution of soil biota wherein rhizobacteria contribute significantly towards soil organic matter dynamics, nutrient cycling and control of pest and diseases. A comprehensive understanding of the biological processes in Cropping Systems can provide models for judicious and compatible use of organic materials of different sources, microbial inoculants and chemical fertilizers.

The basic concept underlying the IPNM is the maintenance of adjustment of soil fertility and of plant nutrient supply to an optimum level for sustaining the desired crop productivity through optimization of the benefits from all possible sources of plant nutrients in an integrated manner. The appropriate combination of mineral fertilizers, organic manures, crop residues, compost of nitrogen fixing crops varies according to the system of land use and ecological, social and economic conditions and decrease in subsidy on fertilizers by Government have become the cause of concerns to the government, fertilizer industry and farmers.

Import of fertilizers to meet the growing demand has imposed a heavy foreign exchange burden on the country. In this context, it was considered necessary for an alternative renewable source of nutrient supply to the crops and as such emphasis has been given to the fertilizers of the biological origin. The biofertilizers and green manures are used in agriculture to combat the ill effects of chemical fertilizers. It is necessary to adopt an integrated nutrient supply system by means of judicious combination of fertilizers, organic manures and biofertilizers (Table 2.2).

Table 2.2: Nutrient Potential of Organic Sources in India (MTs)

Sources	Quantity (MT)	N %	P %	K%	NPK
Cow/Buffalo Dung	200	3.4	1.3	2.2	6.9
Crop Residue	300	6.0	0.3	1.6	7.9
Total	500	9.4	1.6	3.8	14.8

Source : Anonymous (2012)

2.14.1 Integrated Plant Nutrient Management (IPNM)

The conventional system of farming involving the use of high-yielding varieties, mineral fertilizers and pesticides results not only in mining of plant nutrients but also in deterioration of environment and food quality. The exploitive agriculture to get higher and higher output exerts excessive demand on natural resources. The estimated food grain target for 2000AD is placed at 240MT from limited non-expandable soil resource on account of habitation, urban and industrial pressures on land. The future Indian scenario points out for making serious efforts and a greater attention to increase food production following integrated approach for balanced and combined use of organic manures and mineral fertilizers along with the use of specific biofertilizer depending on the Cropping or Farming system.

Sustainable agriculture requires the management of resources in a way to fulfill changing human needs without damaging order deteriorating the quality of environment and conserving vital natural resources. For the country, the compulsion of increasing crop yield per hectare is simply a matter of necessity to achieve production target of atleast 225MTs by the turn of century to feed over one billion people in the ever deteriorating land: man ratio: Indian is fourth largest user of chemical fertilizers (12.5MTs of NPK nutrients) in the world, its soils are still being depleted of their inherent nutrients reserve as a result of wide gap between additions (12.5MTs) and removal (18.5MTs). One ton produce remove on an average 32kg N: 12 kg P and 58kg K.

The strategy for sustainable satisfactory yield levels envisages nutrients balances and efficient nutrient cycling. This can be successfully achieved through integrated plant nutrients supply system. Integrated nutrients management involves tapping of all the major sources of plant nutrients are it soil nutrient reserve, mineral fertilizers, bulky organic manures, compost, green manures, biological inoculants etc. in a judicious way and to ensure their efficient use. One important feature of this approach is that the fertilizer recommendation should take into account the cropping system as a whole rather than individual crop in a system. The nutrient supply from soil can be enhanced by adopting appropriate soil management and conservation practices to reduce losses of nutrient through leaching, erosion and run off. Amelioration of problem soils and improving soil physical conditions insure maximum possible efficiency of applied nutrients.

There is great need to use available fertilizers most efficiently. This will ensure not only clean environment but also higher return by using a somewhat lower fertilizers dose. Fertilizer use efficiency is of high order when nutrients are applied in needed quantities in a balanced proportion at appropriate time in a proper from with suitable method in right way to a crop. Besides, soil moisture, weed control also has a considerable impact on response of fertilizers and efficient use of added plant nutrients. At present, the nutrient use in India is much less compared to that of other countries. Against world average consumption of 95Kg/ha we are using 74 kg/ha with national productivity level of merely 1.1 metric ton per hectare. Under the emerging situation of the global energy crisis and escalating high cost of inorganic fertilizers dependence on chemical fertilizers alone as a source of plant nutrients has to be supplemented by some other sources. India has a great potential for the production and use of organic manures in cropping systems.

Organic manures are valuable not only as supplier of nutrient elements such as NPK but even more so for their ability to mobilizer native soil phosphorus and render it available to crops. They even enhance fertilizer use efficiency when applied in conjunction with mineral fertilizers. Organic residues however are bulky and have low nutrient contents. The estimated contrary to the Low External Inputs (LEI) and organic Farming, the IPNM involves a low to medium external inputs approach taking into holistic view of soil fertility and plant nutrient management for a targeted yield based not only on cropping and farming systems but also distinct geographical areas or villages as dynamic system.

With almost twice the quantity of plant nutrients being removed from soil than what is added through fertilizers, the growing nutrient imbalance pose a major threat to sustain soil health and crop productivity. The recent fall in fertilizers consumption due to unprecedented hike in the prices of K and P fertilizers has further aggravated the problem and has underlined the need for adoption of Integrated Nutrient Supply System (INSS) which involves the combined use of different nutrient sources such as fertilizers, organic manures and biofertilizers etc.

Biofertilizer are the products containing living cells of different types of microbes which have an ability to mobilize nutritionally important elements from non-usable to usable from through biological process. Although the advent of the phenomena is as old ad a century, the need of its conventional exploitation was not felt in traditional agriculture. In recent years, biofertilizers have emerged as an important component of INSS and hold a promise to improve the crop yields and nutrient supplies.

Food production of a country is closely associated with its population. Increased production from a unit area brings abundant pressure on land. This leads to denudation of forests and grass land cover and increase of soil erosion. Energy resources are necessary for intensive agriculture and investments are not only involved in highly mechanized farm operations but also in the production of fertilizers, herbicides, pesticides and other materials commonly used in modern agricultural practices. Energy and fertilizer constitute a serious problem and farming does not remain economically feasible. Moreover, increased use of fertilizers may lead to health hazards apart from soil erosion. Increasing population has therefore compelled many nations to take necessary steps to increase food production by alternative means.

The most important of the biological processes existing on earth are photosynthesis and nitrogen fixation. BNF is second to photosynthesis. Nitrogen is one of the major limiting factors responsible for primary production in agro-ecosystems. The process of photosynthesis involves the fixation of C and N through BNF. The atmosphere contains about 350 ppm of CO_2 and 79% N_2. However, nitrogen gas is not available in the form of that most plants or animals can use. They require nitrogenous components such as nitrate or ammonium and are usually supplied to them as fertilizers. The required forms of nitrogen are rather expensive and beyond the reach of the most of the poor farmers.

BNF may instance help to alleviate heavy fertilizer demand which could be better used in various parts of the world. A good inventory of the process of nitrogen for agriculture crop production indicates that BNF is predominant. Approximately 175 million metric tonnes are fixed annually. Approximately 90million metric tonnes coming from legumes crop and 10 from non-legumes crops with 45million coming from pastures and grasslands.

The global values of BNF run in several dollars. Nitrogen economy is different. In the recent years, the cat is well illustrated by quoting two estimates varying by approximately 100%. The estimates of Dahwiche (Germany) suggested that about 100 million metric tonnes of N_2 is turned over on the earth each year whereas; Burns and Hardy (USA) in 1975 suggest that the amount is closed to 200 MTs annually. These doses appear to be agreement that perhaps two third of the nitrogen cycling annually on the earth comes from biological rather than chemical nitrogen fixation. There is various food crops existing on the earth comes from biological rather chemical nitrogen fixation. There are various food crops existing on the earth of which 95% of the world's food supply comes directly or indirectly from plants. 60% of the calories and 50% of the proteins consumed by people comes from the cereal grains, while 20% of the protein is provided by the seed legumes. This protein comes directly or indirectly from either biologically or chemically fixed nitrogen.

Integrated Plant Nutrient Supply System : Soil is the most important resources for sustainable agriculture since it contains essential nutrients, stores water and provides the medium in which plants grow. Thus, an integrated plant nutrient supplies (IPNS) system may play a vital role in sustaining both soil health and crop production on a long term basis. An innovative idea, for overcoming the problems of inadequate availability of chemical fertilizers, their escalating prices and the issues related to environmental hazards have resulted in greater demand on plant nutrients resources viz. to be used in an integrated way (Fig 2.4).

The IPNS system can thus be described as the combined use of chemical fertilizers, organic manures and biofertilizers with the objective of obtaining high yields, and ensuring environmental safety. Swaminathan (1974) stated "What we hence need is an integrated nutrient supply system involving fertilizer, organic manures of different kinds and biological nitrogen fixation through the introduction of legumes in crop rotations". Commenting further on the IPNS he stated, "Legumes will help in restoring soil fertility but here again the amount of nitrogen in legumes grains as a proportion of that in all arable produce has declined from 24.3% in 1961-65 to 14.8% in 1971-72". It is based on this approach that the Bio-village programme at the M.S. Swaminathan Research Foundation is being operated.

The studies conducted in India and abroad showed that neither the organic manures nor the mineral fertilizers alone can sustain high productivity under modern intensive farming, where the nutrient turnover in the soil-plant systems is quite high. Organic manures alone may suffice for lower nutrient demand under low to medium intensity cropping but combination of organic manures and chemical fertilizer become imperative for sustainable producing. Nutrient balance is a key factor in increased yield and quality of the produce, maintenance of soil productivity and health and protection of environment. Moreover, it takes into account the availability of the nutrients already present in the soil, crop requirement, crop removal of the nutrients, the economics of fertilizers and profitability, farmer's investment ability, agro-techniques, soil moisture regimes, weed control, plant protection, seed rate, sowing time, soil salinity/ alkalinity, physical environment, soil health involving, microbiological conditions of soil, and status of available nutrients, etc. It is not a static but a dynamic system. Thus, every time a crop is grown, all the nutrients should be applied in a particular proportion rather the fertilizer application should be tailored to the crop needs keeping in view the capacity of soil to fulfill these requirements. To achieve this, it is necessary to keep on overall balance in the total cropping system. This may indicate the necessity for the application of different nutrient at specific times, in a particular order to derive the maximum benefit from the application of a given quantity of nutrients; pinpointed focus on balanced fertilization can thus be placed through soil testing, IPNS system and land development programme, etc.

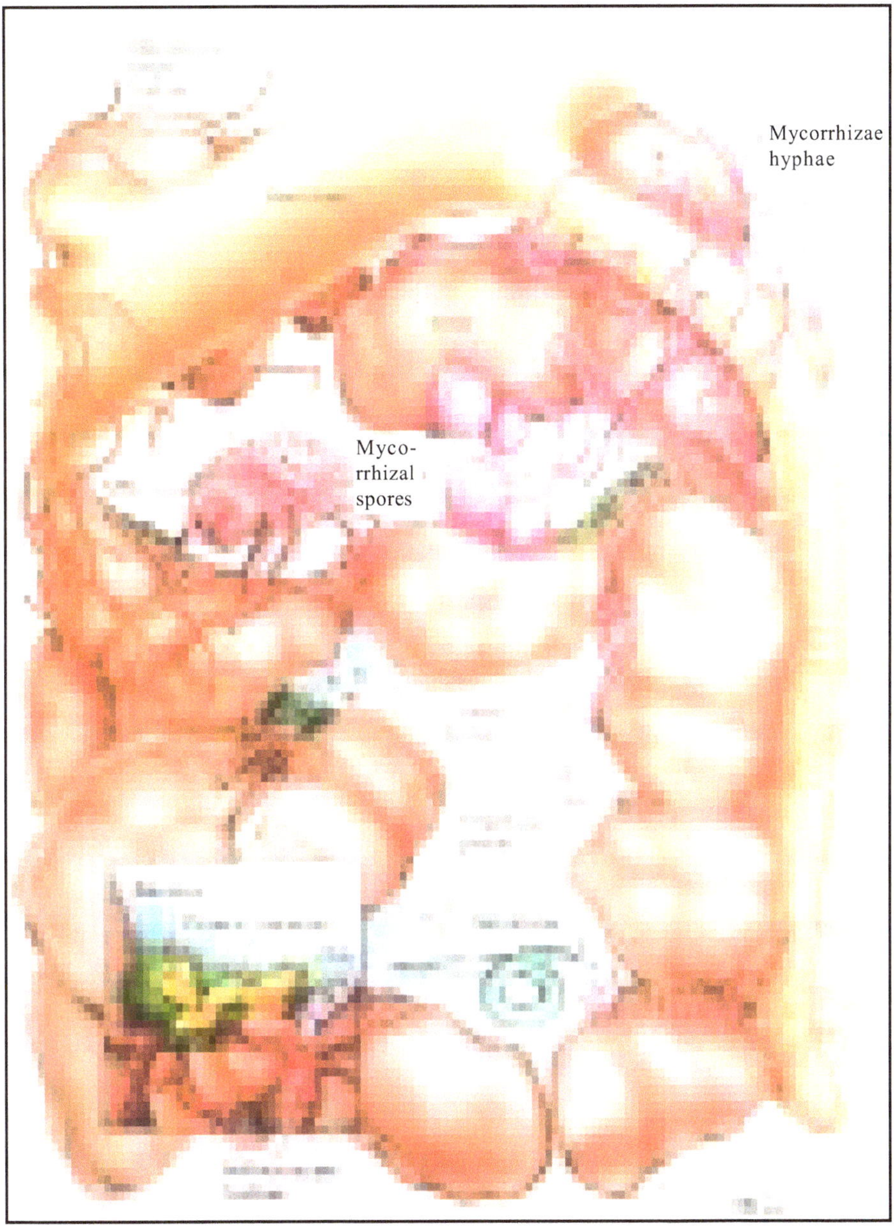

Fig. 2.1 : The Soil Habitat.

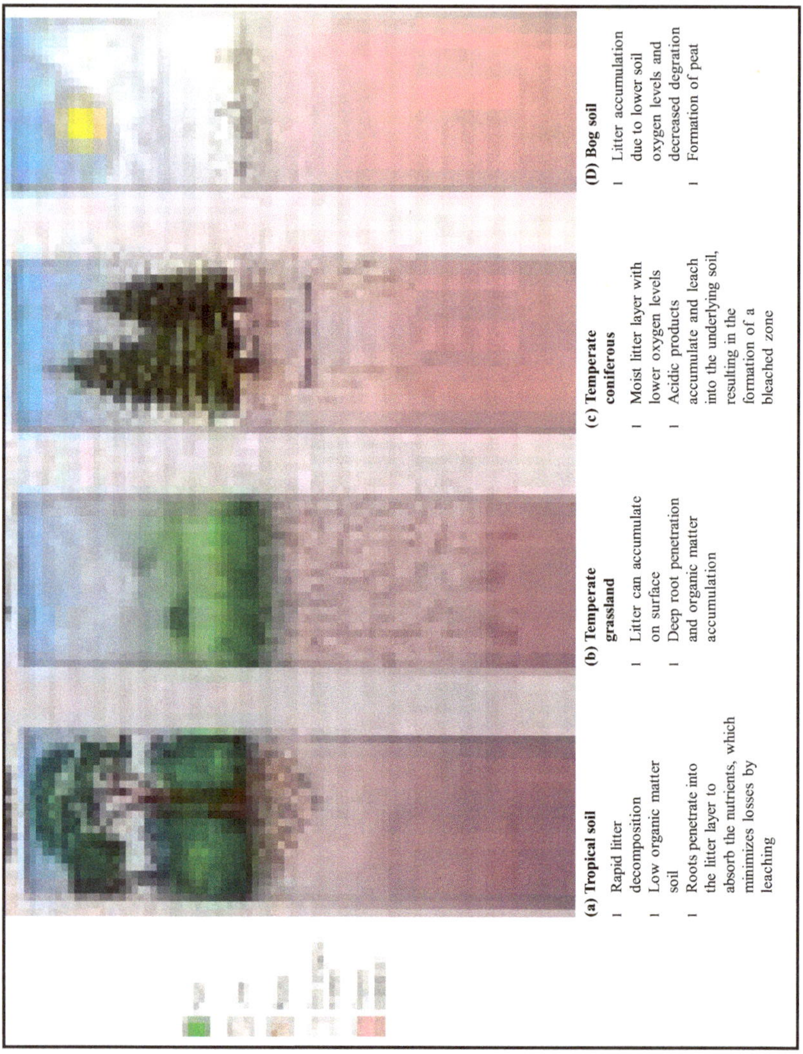

Fig. 2.2 : Examples of Tropical and Temperate Region Plant-Soil Biomes.

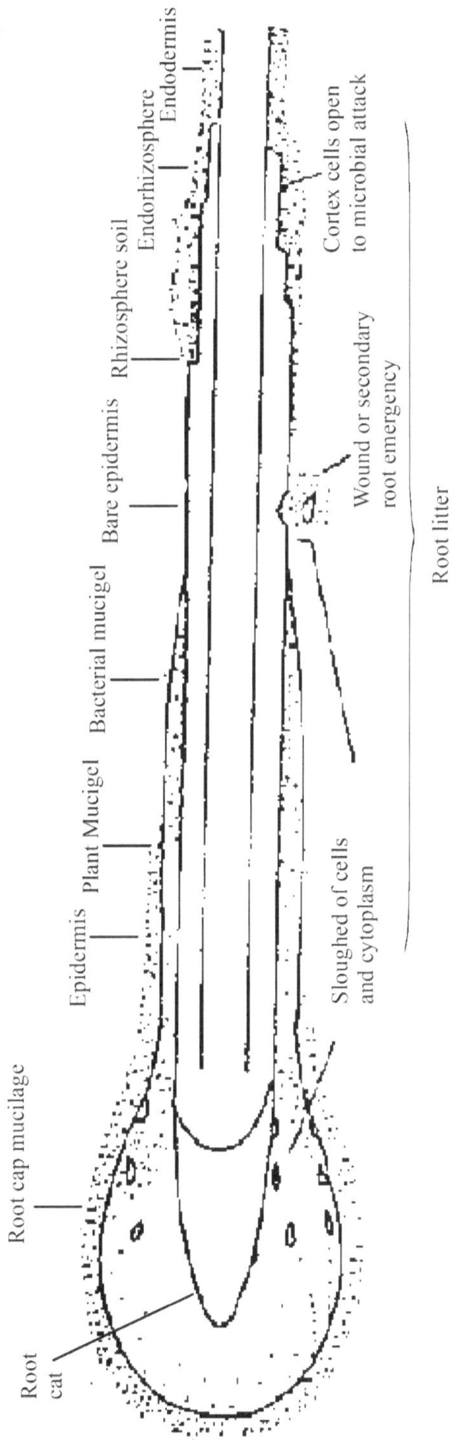

Fig. 2.3 : Root tip region showing the relationship between various components

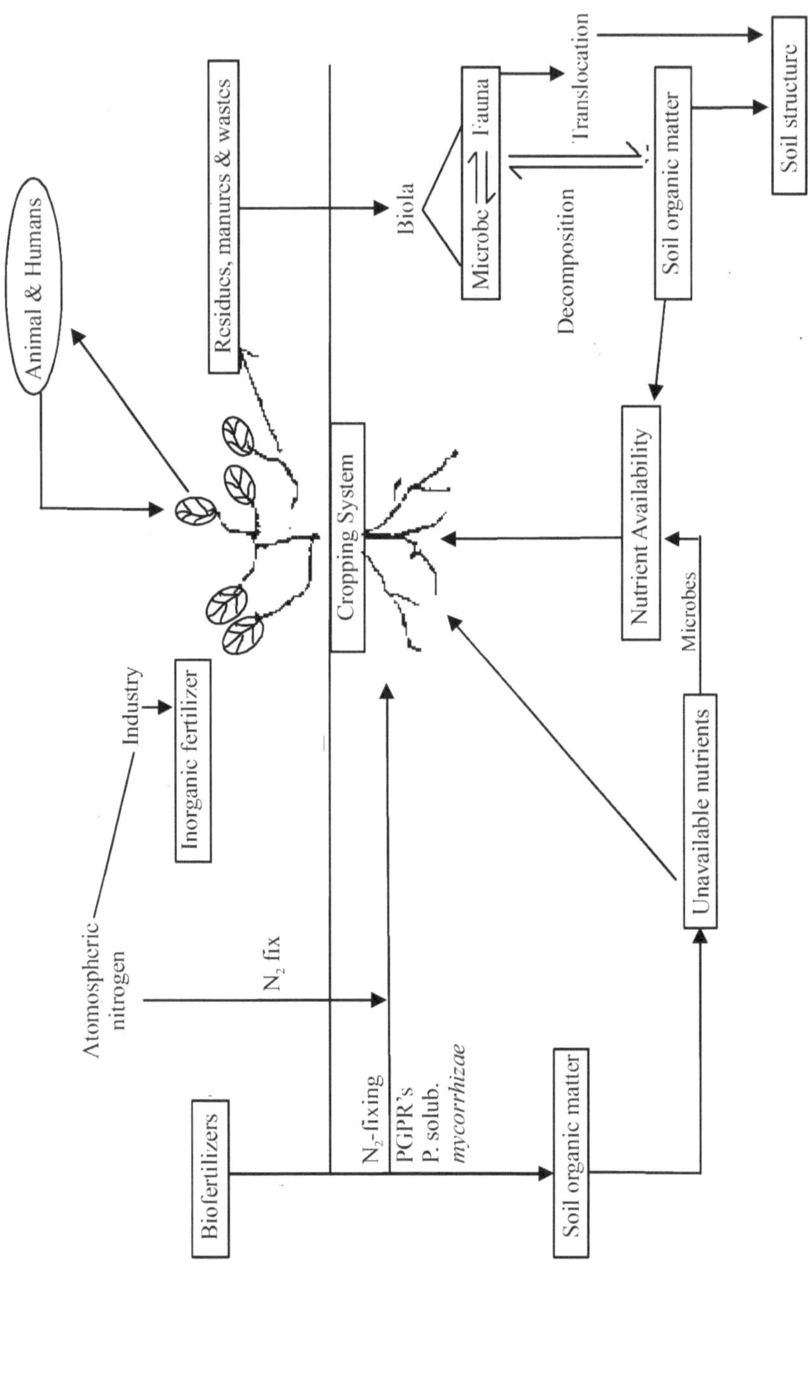

Fig. 2.4 : Schematic diagram shows the involvement of biofertilizers in integrated nutrient management.

CHAPTER - 3

Classification of Biofertilizers

Biofertilizers can be defined as the preparations containing strains of micro-organisms which can augment the microbiological process viz. nitrogen fixation, phosphate solubilization or mineralization, excretion of plant growth promoting substances or cellulose or lignin biodegradation in soil, compost or other environments. Biofertilizers can be divided into nitrogen fixing bacteria, phosphate solubilizing and mobilizing microorganisms and organic matter decomposers as (Table 3.1 and Fig. 3.1):

On the basis of Nitrogen Fixation: On the basis of N_2 fixation, they are divided in to two sub categories as

a) Symbiotic nitrogen fixers (*Rhizobium, Azolla* and *Frankia*)

b) Non-symbiotic nitrogen fixers (*Azotobacter, Azospirillum, Clostridium,* Blue Green Algae (BGA) as *Nostoc* , *Anaebaena* and *Calothrix*)

On the basis of Phosphate: On the basis of Phosphate, they are divided in to two sub categories as :

a) Phosphate Solubilizing Biofertilizers (*Bacillus, Pseudomonas, Aspergillus, Penicillium*)

b) Phosphate Mobilizing Biofertilizers (VAM, *Glomus, Acaluspora, Gigaspora, Scleroigha*)

On the basis of Organic Matter: On the basis of decomposition, they are divided in to two sub categories such as:

a) Cellulolytic biofertilizers

b) Lignolytic biofertilizers

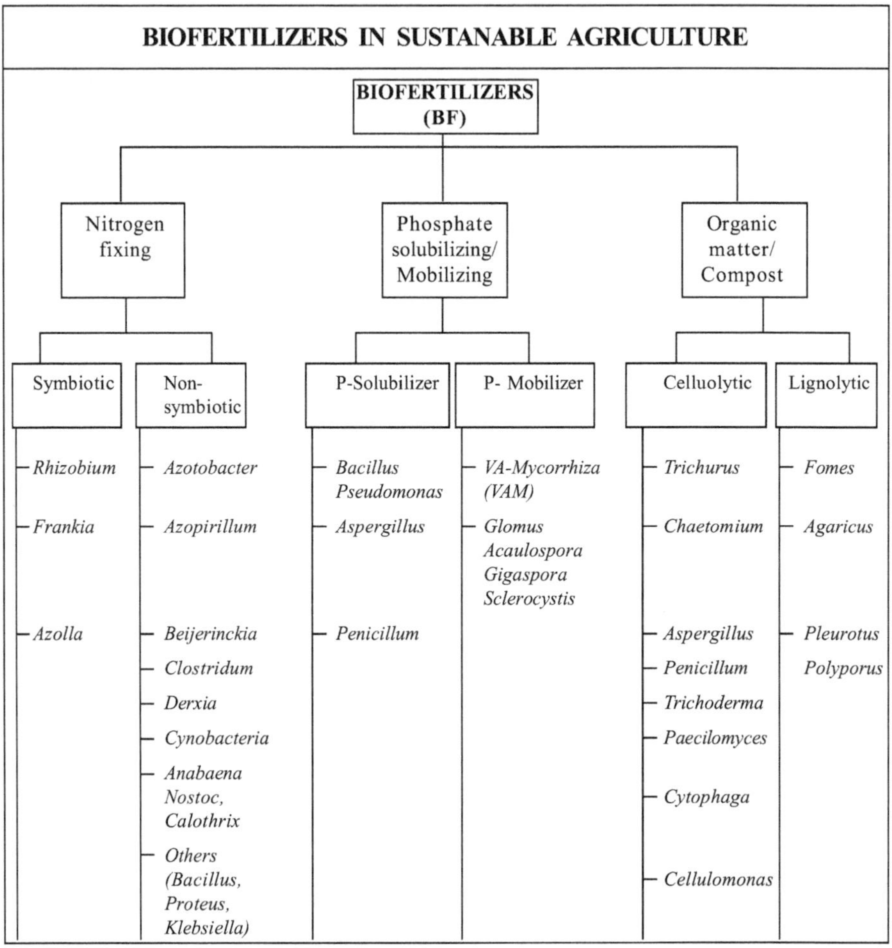

Fig. 3.1 : Classification of Biofertilizer

Table 3.1: Different Groups of Biofertilizers

S. No.	Groups	Examples
	N_2 Fixing Biofertilizers	
1.	Free-living	*Azotobacter, Clostridium, Anabaena, Nostoc,*
2.	Symbiotic	*Rhizobium, Frankia, Anabaena azollae*
3.	Associative Symbiotic	*Azospirillium*
	P Solubilizing Biofertilizers	
1.	Bacteria	*Bacillus megaterium var. phosphaticum*
		Bacillus circulans, Pseudomonas striata
2.	Fungi	*Penicillium* sps, *Aspergillus awamori*
	P Mobilizing Biofertilizers	
1.	Arbuscular mycorrhizae	*Glomus*sp., *Gigaspora*sp., *Acaulospora*sp.,
		*Scutellospora*sps. and*Sclerocystis*sp.
2.	Ectomycorrhiza	*Laccaria*sp., *Pisolithus*sp., *Boletus* sp., *Amanita* sp.
3.	Orchid mycorrhiza	*Rhizoctoniasolani*
	Biofertilizers for Micro-nutrients	
1.	Silicate and Zinc Solubilizers	*Bacillus sp.*
	Plant Growth Promoting Rhizobacteria	
1.	*Pseudomonas*	*Pseudomonas fluorescens*

3.1 Nitrogen Fixing Biofertilizers

They are the preparations generally used as seed inoculants for the supply of nitrogen to the crops. Several heterotrophic bacteria have capacity to fix nitrogen from atmosphere and make it available for plant use.

3.1.1 Symbiotic Biofertilizers

3.1.1.1 Rhizobia

*Rhizobia,*Gram (-) soil bacteriabelongstofamilyRhizobiaceae, symbiotic in nature, fixnitrogen50-100kg/ha withle gumesonly. It is useful for pulse legumes like chickpea, red-gram, pea, lentil,blackgram, etc., oil-seed legumes like soybean and ground nut and forage legumes(berseem and lucerne).They are able to enter into symbiotic relationship with pulses and forms root nodules through the symbiosis by providing nitrogen to the host plants. The *Rhizobia*-legume association could fix up to 40- 200kg N/ha/season able to meet up to 80-90% of the crop nitrogen needs. They fix atmospheric nitrogen, and thus not only increase the production of the inoculated crops, but also leave a fair amount of nitrogen in the soil; which benefits the subsequent crop different legumes fix different amount of nitrogen depending on the crop, soil and environmental conditions. The efficiency of *legume Rhizobium* symbiosis depends upon the host crop and its compatibility with the *Rhizobium* bacteria, crop management and several soil factors like available soil nitrogen, soil moisture, soil pH and soil temperature etc. Successful nodulation of leguminous crops by *Rhizobium* largely depends on the availability of compatible strain for a particular legume as in mungbean (Jain *et al.,* 2007, 2008) and black gram (Rathi *et al,* 2009). It colonizes the roots of specific legumes to form tumor like growths called root nodules, which acts as factories of ammonia production. *Rhizobium* has ability to fix atmospheric nitrogen in symbiotic association with legumes and certain non-legumes like Parasponia. A strain of *Rhizobia* that no dulates and fixes alarge amount of nitrogen in association with one legume species may also do the same in association with certain other legume species and enhance the cereals production (Baset Mila and Shamsuddin, 2010). Following seven cross inoculation groups of *Rhizobia* have been recognized according to their host: (i) *Bradyrhizobium spp. Parasponia* (a non-legume, formally called Trema), (ii) *Bradyrhizobium japonicum* (Slow growing) Soybean (*Glycine max*), (iii) *Sinorhizobium fredii* (fast growing type) Alfalfa, (*Medicago sativa*), (iv) *Azorhizobium* (forms both root and stem nodules; the stem have adventitious roots) Sesbania (Aquatic), (v) *Rhizohium leguminosarum* by. *Phaseoli, Beans* (*Phaseolus*) (vi) *Rhizohium leguminosarum* by. *Viciae* Clovers (vii) *Photorhizobium* (Photosynthetically active).

3.1.1.2.Azolla—Anabaena symbiosis

Azolla is a small-leaf floating fern, which contains an endosymbiontic community living in the dorsal lobe cavity of the leaves. The presence in this cavity of anitrogen fixing filamentous *cyanobacteria* into the only fern-*cyanobacteria* association that presents agricultural interest by the nitrogen input that this plant could introduce in the fields. Normally one kg *Azolla* can fix 1.9 to 2.5kg nitrogen in about one month under optimum growth conditions. The nitrogen content of *Azolla* on dry weight basis is generally 5.0%. The fern can be used in rice crop by two ways as (i) green manure or (ii) dual crop. In case of green manuring, the fern is grown in the field or in separate shallow ponds and in corporate into soil prior to transplanting of rice. As a dual crop it is intercropped with rice and then mixed in the soil while crop is growing. It can fix about 30-40 kg N/ha/crop of *Azolla* that takes about 253 days. The limitation in use of *Azolla* as biofertilizer is scarcity of water and high temperature in north India. Higher temperature than 40°C is harmful for the fern maintenance of

Azolla cultures in pond during winter is also a big problem. *Anabaena azollae* is a cyanobacterium, forms symbiotic association with a water fern *Azolla*. In India, for paddy fields *Azolla*is apromising bio fertilizer and extensively used by farmers insever alrice growing countries such as Philippines, China, Vietnam, Thailand and Sri Lanka. It is grown in the slow flowing creeks or water beds and applied to crop between planting of rice. After a certain period of growt hit is incorporated to soil before transplanting or left to be shaded out as rice can opybuilds up. They are rapidly mineralized due to the narrow C:N ratio (as we use it in a succulent stage when there is not much lignification of cell walls and is easily degradable) and provide nitrogen to the plants. Apart from nitrogen fixation, *Azolla* is also known to suppress weed population in wet land rice and hence provides an additional economic advantage to rice cultivation. Recently, *Azolla* was also used as feed to enhance milk production in milching animals. *Azolla microphylla* (15t/ha) increases grain yield by 29.2% with neem cake reported by Sundaravarathan and Kannaiyan (2002).

3.1.1.3 Frankia

An actinomycetes is known to develop association with non-leguminous plants viz. Alnus, Casuarina, Ceanothus, Eleagnus, Myrica, Hippophae and Purshia etc. and the causal organism is named as *Frankia*. It showed non-leguminous symbiosis. Nearly 120 species, mainly trees and shrubs are now known to harbour N_2 fixing bacteria in their root nodules which are not rhizobia. The non-leguminous nitrogen fixing systems are of modest agricultural importance, because their potential is not widely known. The extent of nitrogen gain by such Angiosperms varies enormously depend on soil type, climatic conditions and plant age. Nitrogen gains in the range of 12-200kg/ha/yr was reported for *Alnus* species, and 27-179kg/ha/yr for *Hippophae* species. Tests of individual location suggest the nitrogen fixation of 58kg/ha/yr for *Casuarina,* 60 kg/ha/yr for *Ceanothus* but only about 10kg/ha/yr for *myrica.* Plants in the group have little agricultural potential as grain crops or forage, but *Alder* could be available in agro-forestry. They may be used for upgrading poor soils. Its role as biofertilizer is insignificant.

3.1.2 Non-Symbiotic Nitrogen Fixers

3.1.2.1 Acetobacter inoculant

Acetobacter diazotrophious is one of the newly discovered endophytic diazotrophs isolated from leaves, stems and roots of sugarcane. In the eighties Dobereiner and coworkers in Brazil discovered this association and named *Acetobacter diazotrophicus*. These live in xylem vessels, intercellular space of root, shoot or leaf, ensuring proper supply of nutrients for nitrogen fixation. It was widely studied and used as a model system to assess the bacterial endophyte plant interaction. After its discovery, it was reported from a variety of crops like coffee (Jimnez-Salgado *et al.*, 1997), ragi (Loganathan *et al.*, 1999) and pineapple (Herandez*et al.*, 2000). and a latest report throw light on *Gluconacetobacter* sp as a natural colonizer of the wild rice (*Porteresia cocarctata Tateoka,* formerly *Oryza coarctata Roxbi*) and a salt tolerant pokali rice variety (Loganathan *et al.*, 2003). In India, its nitrogen fixing ability and plant growth promoting rhizobacteria (PGPR) productionability have been described, tested and found promising (Saravanan*et al.*, 2008). Besides, the bacterium secretes plant growth promontory substances such IAA and favours the germination and root development, which in turn help in the absorption of plant nutrients effectively from soil. The application of *Acetobacter* biofertilizers can benefit the sugarcane growing areas in the country by increasing sugar contents in cane from 0.3 to 0.5%. *A. diazotrophious* could enhance- rice growth and that growth promotion- might be related to the transfer of biologically fixed N although other factors such as auxin production could be involved reported by Sevilla and Kennedy *et al.*, (2000).

Gluconoacetobacter diazotrophicus colonizes the internal root tissue of sugarcane and fix nitrogen. Since it occupies vascular tissue, it has the obvious advantage of being first in line and thus solves the problem of competition by non diazotrophs. *Gdiazotrophicus* is also found in *Pennisetum purpureum, Ipomoea batatas* and *Arabica* (non-legume plants).The largest effect in this group was obtained with sugarcane, which can obtain up to150kg N/ha from BNF (Dobereiner, 1997). In Tamilnadu, 24 strains of *G.diazotrophicus* were isolated from sugarcane (root, stem and leaves) and screened for nitrogenase activity. The strain isolated from sugarcane stem showed maximum acetylene reduction assay, 410.92n moles of C_2H_4/hr/mg cell protein followed by SoL3 from sugarcane leaf and hence the strain from sugarcane stem was recommended as an effective biofertilizer (Meenakshisundaram *et al.*, 2010).

3.1.2.2 Azospirillum

Azospirillum, a member of Spirillaceae, is associative N_2 fixing micro-organism beneficial for non-leguminous plants. Associative diazotrophs are those in which some in independence between the partners but they can grow satisfactory apart. They are found in association with cereals, grasses, vegetables, and oil seeds either freely in soil or indeed the root system. They not only fix nitrogen but also benefit plants by supplying growth hormones and vitamins. *Azospirillum* with farmyard manure (FYM), led to saving of 15.25kg equivalent of nitrogen also the above ground portion of plant through associative symbiosis. *Azospirillum* inoculation helps the plants in better vegetative growth saving nitrogenous fertilizers by 25-30% and fix nitrogen from 10 to 40kg/ha. *Azospirillum are* micro-arophilic in nature (Panwar and Sirohi, 1989). Following three species of *Azospirillum* have been identified as **(i)** *A. lipoferum,* ***(ii)*** *A. brasilense, and* ***(iii)*** *A. amazonenes. Azospirillum brasilense* is a bundant in Indian soil. *Azospirillum* is often used with My corrhizal fungi for optimum growth and yield. One of the characteristics of *Azsopirillum* is ability to reduce and denitrify, therefore it is these characters. Inoculations with *Azospirillum* have registered increased in different vegetable crops.

It also secretes phytohormones in the plant root regions, which inturn enhances the root growth. In developing countries like India, the use of *Azospirillum* as a biofertilizer would not only to the nitrogen supplementation to crops but also help in improving the fertility of soil in the long run. Its application is common practice in vegetables in Tamil Nadu. Increased yield in pearl millet obtained by inoculation of *Azospirillum* was due to production of IAA, gibbelellins, any cytokine like substances by the bacterium and their subsequent effect on the plant (Kumar *et al.,* 2010). Researchers have demonstrated the feasibility of *Azospirillum* inoculation to mitigate negative effects of NaCl on plant growth parameters. This beneficial effect of *Azospirillum* inoculation was previously observed in wheat (*Triticum aestivum* cv.'Buck Ombú') seeds, where a mitigating effect of salt stress was also evident (Creus *et al.,* 1997). *Azospirillum* inoculated wheat(*T. aestivum*) seedlings subjected to osmotic stress developed significant higher coleoptiles,with higher fresh weight and better water status thannon-inoculated seedlings (Alvarez *et al.,* 1996; Creus *et al.,* 1998). *Azospirillum* spp.is not considered to be a classic bio-control agent of soil-borne plant pathogens. However, there have been reports on moderate capabilities of *A.brasilense* inbio-control of crown gall-producing *Agrobacterium* (Bakanchikova *et al.,* 1993),bacterial leaf blight of mulberry (Sudhakar *et al.,* 2000) and bacterial leaf and/or vascular tomato diseases (Bashan and Bashan 2002b). In addition, *A. brasilense can* restrict the proliferation of other nonpathogenic rhizosphere bacteria (Holguin and Bashan, 1996). These antibacterial activities of *Azospirillum* could be related to its already known ability to produce bacteriocins (Oliveira and Drozdowicz 1987) and siderophores (Tapia-Hernández *et al.,* 1990). *A. brasilense* can synthesize phenylacetic acid (PAA), an auxin like molecule with antimicrobial activity (Somers *et al.,* 2005). Biofertilizers made from *Azospirillum* issuitable for C_4 crops such as sugarcane, maize, bajra and sorghum.

3.1.2.3 Blue Green Algae (BGA)

BGA are also known as *Cyanobacterium* as species of *Anabena, Aulasira, Gioeoirieha, Nostoc, Plectonema* etc. *Cyanobacteria* enter into a much wider range of association than *Rhizobia* due to their ability to fix nitrogen independently of the environment provided by the plant; they can be seen as more likely candidates than Rhizobia to form productive associations with cereals. *Cyanobacteria* were also found in the intercellular spaces of paranodules tissue in agreement with findings for *Azorhizobium* (Yu and Kennedy, 1995) and *Azospirillum* (Zemen*et al.,* 1992). Many of them are nitrogen fixing and synthesize their own energy source for nitrogen fixation and release it into surrounding in the form of amino acids, proteins and other growth promoting substances. Lowland rice is an ideal ecosystem for their survival, growth and nitrogen fixation and therefore. They can fix about 20-30-kg N/ha/season. Application of 1kg BGA culture in rice can increase the yield up to 10%-under optimum conditions. These belongsto eight different families, photo trophicin nature and produce auxin, indoleacetic acid and gibberellicacid, fix20-30kgN/hainsubmergedricefieldsastheyareabundantinpaddy, so also referred as '**Paddy Organisms**'.Nisthekey input required in large quantities for lowl and rice production. Soil N and BNF by associated organisms are major sources of N for low land rice. The 50-60% N requirement is met through the combination of mineralization of soil organic N and BNF by free living and rice plant associated bacteria (RogerandLadha, 1992). Most N fixing BGA are filamentous, consisting of chain of vegetative cells including specialized cells called heterocyst which function asmicro no dule for synthesis and N fixing machinery. Several cyanobacteria such as *Anabaena, Nostoc, Cylin-drospermum, Aulosira, Tolypothrix* are excellent nitrogen fixers.

3.2. On the Basis of Phosphate Solubility/Mobility

Phosphorus differs fundamentally from nitrogen in that no natural channels exist for return of large amount of losses occurring annually. Hence, the supply is inevitable shrinking and deposits are limited. Farmers are constrained in using phosphorus owing to in high cost of phosphatic fertilizers. In such situations, the release of insoluble phosphorus in soil and fixed phosphorus in clay minerals by microorganisms assumes great significance. They may be broadly classified in two groups namely phosphate solubilizers and phosphate absorbers.

3.2.1 Phosphate Mobilizing Biofertilizers

P mobilizers facilitate the mobilization of soluble phosphorus from distant places in soil, where plant roots cannot reach and thus increase availability of P to plants. Mycorrhizae are prominent P mobilizers. Mycorrhizae are a symbiotic association between plant roots and a few fungus. The fungal partner is benefited by obtaining its carbon requirements from host's photosynthates and the plant in turn gains the much needed nutrients especially phosphorus, calcium, copper and zinc which would otherwise be in accessible to the host. This uptake of nutrients is facilitated with the help offine absorbing hyphae of the fungus. These fungi are associated with majority of agricultural crops. There are seven genera of these fungi that produce Arbuscular mycorrhizal symbiosis with plants. They are *Glomus, Gigaspora, Scutellospora, Acaulospora Entrophospora, Archaeospora* and *Paraglomus.* They account for 5-50% of the biomass of soil microbes (Olsson *et al.,*1999). Hyphal biomass of AM fungi may amount to 54-900kg/ha (Zhu and Miller, 2003), and some products formed by them may account for another 3000kg (Lovelock *et al.,* 2004). Pools of organic carbon such as glomalin produced by AM fungi may even exceed soil microbial biomass by a factor of 10–20 (Rillig *et al.,* 2001). Approximately 10-100m mycorrhizal mycelium can be found per cm root (Mc Gonigle and Millerv, 1999). The mechanism that is generally accepted is a wider physical exploration of the soil by mycorrhizal fungi (hyphae) rather than by roots. A speculative mechanism to explain Puptake by mycorrhizal fungi involves the production of glomalin (Lovelock *et al.,* 2004) The micro-organisms always participate in cycling of phosphorus in the environment hence there is recycling and not exhaustion.

AM fungi play an important role in water economy of plants. Their association improves the hydraulic conductivity of the root at lower soil water potentials, and this improvement is one of the factors contributing towards better uptake of water by plants (Mahdi *et al.,* 2010). A few proposed mechanisms by which AM fungi also help in activation of plant defense systems include changes in exudation patterns and concomitant changes in mycorrhizosphere populations, increased lignifications of cell walls, and competition for space for colonization and infection sites (Kasiamdar *et al.,*2001).

3.2.2 Phosphate Solubilizers

Phosphorus is second only to nitrogen in mineral nutrients most commonly limiting growth of crops, and an essential element for plant development and growth making up about 0.2% of plant dry weight. Plants acquire P from soil solution as phosphate anions. However, these are extremely reactive and may be immobilized through precipitation with cations such as Ca^{2+}, Mg^{2+}, Fe^{3+} and Al^{3+} depending on the particular properties of a soil. In these forms, P is highly insoluble and a large portion of soluble in organic phosphate applied to the soil as chemical fertilizer is immobilized rapidly and becomes unavailable to plants (Goldstein, 1986). Hence, the amount available to plants is usually a small proportion of this total application.

Several bacteria, particularly as *Bacillus polymyxa, Bacillus subtilis* and *Pseudomonas striata,* and fungi *as Penicillium digitatum* and *Aspergillus awamori* have the ability to convert the insoluble inorganic phosphorus into the soluble form, which can be utilized by crop plants. The inoculants of these microorganisms are called PSB (Phosphate solubilizing bacteria) or PSM (Phosphate solubilizing microorganisms) inoculants. They also produce siderophores (iron chelating substances, e.g. Pseudobactin), which chelate with iron and make it unavailable to harmful fungi as *Eriwina* in rhizosphere, and produce plant growth hormones like indole acetic acid and gibberellic acid etc. These cultures can be used in all crops of legumes, cereals and vegetables. Normally these bacteria can solubilize about 15kg P ha/season. Their inoculation was found to increase the yield of crops by 10 %.

Phosphate rock minerals are often too in soluble to provide sufficient P for crop uptake. Use of phosphate solubilizing microorganisms increase crop yields up to 70% (Verma, 1993). Combined inoculation of AM and PSB enhanced uptake of both native P froms oil and P coming from the phosphatic rock (Goenadi *et al.,* 2000; Cabello *et al.,* 2005). Microorganisms with phosphate solubilizing potential increase the availability of soluble phosphate and enhance the plant growth by improving biological nitrogen fixation (Kucey *et al.,* 1989; Ponmurugan and Gopi, 2006).

Phosphorous Biofertilizers

Phosphorous is very often present in the soil in unavailable form. Several soilbacteria particularly those belonging to the genera *Bacillus* and *Pseudomonas* posses the ability to bring insoluble phosphates in the soluble forms by secreting organicacids. These acids lower the pH and bring about the dissolution of bound forms of phosphorous. These bacteria are commonly known as phosphobacteria. They can be applied either through seed or soil application. Phosphorous is one of the important elements for plant growth. It is the second major chemical fertilizer, being applied for crop production. For optimum crop production such as wheat, soil level should be maintained above 30ppm available P in the top 15cm. It is anticipated that only 10-15% recovery is obtained for applied P fertilizers by plants. The remainder of the fertilizers-P is accumulating in the soil. This is as two major forms: inorganic soil 'P' which comprises adsorbed phosphate and precipitated Fe, Al, $MnFePO_4$, $AlPO_3$, $Ca_3(PO_4)_2$ acidic soil and Ca salts e.g. $Ca_3(PO_4)_2$ alkaline soils. Another general form of unavailable soil phosphorus is organic soil P, which comprises in phosphates, phospholipids, nucleic acids etc. It occursin in organic as well as organic form, inorganic forms are compounds of calcium, iron, and aluminium in rock phosphates and organic forms are phytin, phospholipids, nucleicacids, mositol phosphates which are added to the soil by decaying vegetation. Rock phosphate is the raw material for the production of phosphate fertilizers. Phosphate biofertilizers were first prepared by USSR using *Bacillus magaterium* var. *Phosphaticum* as phosphate solubilizing bacteria and product was termed as phosphobacter.

Many countries are studying the direct utilization of rock phosphate. Australia has developed "Biosuper" as pellets composed of rock phosphate, sulphur and sulphur oxidizing bacteria. Heterotrophic phosphate solubilizing microorganisms need a large amount of organic matter before they can excrete organic acids. Even if phosphate is solubilized, and phosphate ions are incorporated into the microbial biomass, so roots cannot absorb enough of them. Once the organic matter becomes exhausted, the microbial biomass decreases and releases phosphate into soil. The death of microbial biomass can be accelerated by various soil treatments, including tillage, drying and sterilization. Plants can absorb phosphate after microbial proliferation has ceased. The absorption of phosphate by plants can be accelerated by inoculation with AMF. Through biological nitrogen fixation, solubilization of insoluble phosphates and mobilization of plant nutrients more quantities of nutrients are made available for crop plants by the root associated organisms. Increased nitrogen, phosphorus and potassium content of inoculated plants at different stages of crop growth have been recorded (Rajasekaran and Sundaramoorthy, 2010).

3.2.2.1 Mycorrhizae

The term Mycorrhizae denotes "fungus roots". It is a symbiotic association between host plants and certain group of fungi at the root system, in which the fungal partner is benefited by obtaining its carbon requirements from the photosynthates of the host and the host in turn is benefited by obtaining the much needednutrients especially phosphorus, calcium, copper and zinc etc., which are otherwise inaccessible to it, with the help of the fine absorbing hyphae of the fungus. These fungi are associated with majority of agricultural crops, except with those crops/plants belonging to families of *Chenopodiaceae, Amaranthaceae, Caryophyllaceae, Polygonaceae, Brassicaceae, Commelinaceae, Juncaceae* and *Cyperaceae.* They are ubiquitous in geographic distribution occurring with plants growing in artic, temperate and tropical regions alike. VAM occurs over a broad ecological range from aquatic to desert environments (Mosse *et al.*, 1981). Of 150 species of fungi that have been described in order *Glomales* of class *Zygomycetes,* only small proportions are presumed to be *mycorrhizal.* There are 6 genera of fungi that contain species, which are known to produce *Arbuscular mycorrhizal fungi* (AMF) with plants (Panwar and Thakur, 1995). Two of these genera, *Glomus* and *Sclerocytis* produce chlamydospores only. Four genera form spores that are similar to *Gigaspora, Scutellospora, Acaulospora* and *Entrophospora.* The oldest and most prevalent of these associations are AM symbioses that first evolved 400 million years ago, coinciding with the appearance of the first land plants. In comparison, crop domestication is a relatively recent event, beginning 10,000 years ago (Sawers *et al.*, 2007). *Mycorrhizal* fungi can increase the yield of a plant by 30-40%. Plants that suffer from nutrient scarcity, especially, phosphorus develops m*ycorrhizae.* Association of mycorrhizae is wide spread in bryophytes a large number of pteridophytes, most or all species of *Gymnosperms* and some 90% or more of angiosperms. In recent years use of artificially produced inoculums of my corrhizal fungi has increased due to its significance and multifarious role as

- Benefits more than 99% of earth's plants.
- Increase absorptive surfaces or root systems up to 700%.
- Occupy 100 times more soil volume than a non-mycorrhizal plants entire root system.
- Extend through the soil up to 30ft away from a plant host. Depress many root diseases caused by pathogenic fungi and nematodes. Develop tolerance to climatic and edaphically stresses including high soil temperature and heavy metal toxicity.

Ectomycorrhizae

It is also called as *Ectotrophic mycorrhizae* and found in approximately 10% of the world flora. It colonizes outside of plants cells anda root such as the fungus completely encloses each feeder rootlet in a sheath of mantle of hyphae, which penetrate only between the cells of root hairs. But due to the presence of layer of hyphae it almost looks like a host tissue and this layer is known as pseudoparenchymatous sheath. From this sheath hyphae enter the cortex and remain only in the outer cortical cells to form a network called as Hartig net. All the nutrients are absorbed by the fungal mantle and transported to the root through the Hartig net. The fungi involved in *ectomycorrhizae* come under *Basidiomycetes* and they grow best at pH5-6. The excess of inorganic fertilizers suppress *ectomycorrhizal* development and these results in the stunted growth and chlorosis. When the defense reaction of the higher symbiont diminishes, as it is likely to happen in senescent or diseased trees, the lower symbiont may become endotropic. Such instances have been designated as *ectoendomychorrhizae*.

Endomycorrhizae

Endomycorrhizae, the fungus does not form and external sheath but colonizes inside the plant root cells and establishes direct connections between the cells of the roots and surrounding soils. The fungi involved in endotropic association belong to either to the *Basidiomycetes* (possessing aseptate hyphae)or to the *Basidiomycetes* or fungi Imperfecti (possessing septate hyphae). Root colonization involves the formation of intracellular and highly branched hyphae scattered throughout the root system. The establishment of an intracellular symbiosis between the fungus and the host roots requires the penetration of the cell by fungal cell well hydrolyzing enzyme. Fungal hyphae enters the cells of the host plant and thus penetrates the cell wall at the site of contact with the aid of various hydrolytic enzymes like xylanases, pectinases and celluloses, the fungi can break down lignin and cellulose and thus contribute to the decaying of organic matter. In this respect, they differ from *ectomycorrhizal* fungi on the host for their carbon nutrition.

For *ectomycorrhizal* fungi the basidiospores, pure mycelial culture, chopped sporocarp, sclerotia and fragmented *mycorrhizal* roots or soil from *mycorrhizosphere* region can be used as inoculum. The inoculum is mixed with nursery soils and seeds are sown. Now a day's commercially available mycelial inoculum are also available in the market.

3.2.2.2 VAM (Vesicular-Arbuscular Mycorrhizae)

VAM fungi form symbiotic association with a number of economically important crop plants. These fungi can improve the plant growth through increased uptake of phosphorous, sulphur, calcium and zinc. Further these fungi are known to enhance resistance to diseases and help the host plant to absorb more water under moisture stress conditions. VAM fungi are the ubiquitous soil microbes found associated with most of the angiosperms, pteridophytes, and bryophytes but absent in plants which from only *ectomycorrhizae* (Pinaceae and Betulaceae) or the two other specific types of *endomycorrhiza* of Ericales and Orchidales.

VAM develop special structures called as arbuscles and vesicles that help in the transfer of the nutrients especially phosphates from the soil into the root system. VAM has been reported from 1000 genera of plants representing from 200 families. Fungus associated with VAM is generally non-septate Zygomycetes belonging to the genera Clomus, Gigaspora, Acaulospora, Seerocystis and Endogone. The fungi being obligate biotrophs don't growon synthetic media and hence are classified according to the morphological characteristics of the spore formed in the soil. The efficient VAM fungal strains are the potential candidates as biofertilizers for their use in crop production. VAM increases the plant growth mainly by increasing the soil volume from which plant absorbs relatively less mobile nutrients such as Zn, P, and Cu etc., which are very important for optimum plant growth. However, response to VAM fungi may vary depending upon crop species, soil P levels and the presence of other micro-organism in the root regions.

The potential of Arbuscular mycorrhizal fungi as biofertilizers and bioprotect ors to enhance the micro propagated plantlets. One of the major implements to the success of micro propagation is the very high mortality rate of tissue culture plantlets either during acclimatization phase or during transfer to field conditions attributed mainly to desiccation and microbial infection. It appears that the advantages of VAM association, such as improved uptake and transport of water and other nutrients, tolerance to root pathogens resulted in reduction of the mortality rate of tissue culture plant lets. Besides VAM can be used as a disease control agent as the fungus is said to be releasing such chemical compounds which have an inhibitory effect on the pathogens. Studies have shown that VAM formed by *Gigaspora* exhibits the inhibitory effect on the development of pigeon pea blight caused by *Phytophthora drechsleri.*

Zinc Solubilizers

The nitrogen fixers like *Rhizobium, Azospirillum, Azotobacter,* BGA and Phosphate solubilizing bacteria like *Bacillus magaterium, Pseudomonas striata,* and phosphate mobilizing mycorrhizae have been widely accepted as bio-fertilizers (Subba Roa, 2001a). However, these supply only major nutrients but a host of microorganism that can transform micronutrients are there in soil that can be used as bio-fertilizers to supply micronutrients like Zn, Fe, Cu etc. Zinc being utmost important is found in the earth's crust to the tune of 0.008% but more than 50% of Indian soils exhibit deficiency of zinc with content must below the critical level of 1.5ppm of available zinc (Katyal and Rattan, 1993). The plant constraints in absorbing zinc from the soil are overcome by external application of soluble zinc sulphate ($ZnSO_4$). But the fate of applied zinc in the submerged soil conditions is pathetic and only 1-4% of total available zinc is utilized by the crop and 75% of applied zinc is transformed into different mineral fractions (Zn-fixation) which are not available for plant absorption (crystalline iron oxide bound and residual zinc). There appears to be two main mechanisms of zinc-fixation, one operates in acidic soils and is closely related with cat ion exchange and other operates in alkaline conditions where fixation takes by means of chemisorptions, (chemisorptions of zinc on calcium carbonate formed a solid-solution of $ZnCaCO_3$, and by complexation by organic ligands (Alloway, 2008).The zinc can be solubilizedby micro-organisms viz., *B. subtilis, Thiobacillus thioxidans* and *Saccharomyces* sp. These micro-organisms can be used as bio-fertilizers for solubilization of fixed micronutrients like zinc. The results have shown that a *Bacillus* sp. (Zn solubilizing bacteria) can be used as bio-fertilizer for zinc or in soils where native zinc is higher or in conjunction with insoluble cheaper zinc compounds like zinc oxide (ZnO), zinc carbonate($ZnCO_3$) and zinc sulphide (ZnS) instead of costly zinc sulphate (Mahdi *et al.*,2010).

3.3 Organic Biofertilizers

Soil organic matter plays important role in the maintenance and improvement of soil properties. It is a dynamic material and is one of the major sources of nutrient elements for plants. Soil organic matter is derived to a large extent from residues and remains of the plants together with the small quantities of animal remains, excreta, and microbial tissues. Soil organic matter is composed of three major components i.e. plants residues, animal remain and dead remains of microorganisms. Various organic compounds are made up of complex carbohydrates, (Cellulose, hemicellulose, starch) simple sugars, lignins, pectins, gums, mucilages, proteins, fats, oils, waxes, resins, alcohols, organic acids, phenols etc. and other products. When plant and animal residues are added to the soil, the various constituents of the soil organic matter are decomposed simultaneously by the activity of microorganisms and carbon is released as CO_2, and nitrogen as NH_4 —>NO_3 for the use by plants. Other nutrients are also converted into plant usable forms. This process of release of nutrients from organic matter is called mineralization. The insoluble plant residues constitute the part of humus and soil organic matter complex. The final product of aerobic decomposition is CO_2 and that of anaerobic decomposition are Hydrogen, ethyl alcohol (CH_4), various organic acids and carbon dioxide (CO_2). Soil organisms use organic matter as a source of energy and food.

The process of decomposition is initially fast, but slows down considerably as the supply of readily decomposable organic matter gets exhausted. Sugars, water-soluble nitrogenous compounds, amino acids, lipids, starches and some of the hemicellulases are decomposed first at rapid rate, while insoluble compounds such as cellulose, hemicellulose, lignin, proteins etc. which forms the major portion of organic matter are decomposed later slowly. Thus, the organic matter added to the soil is converted by oxidative decomposition to simpler substances which are made available in stages for plant growth and the residue is transformed into humus.

The microbiology of decomposition/degradation of some of the major constituents (viz. Cellulose, hemicellulose, lignin, proteins etc.) of soil organic matter/plant residues are discussed in brief in the following paragraphs.

3.3.1 Decomposition of Cellulose

Cellulose is the most abundant carbohydrate present in plant residues/ organic matter in nature. When cellulose is associated with pentosans (eg. xylans & mannans) it undergoes rapid decomposition, but when associated with lignin, the rate of decomposition is very slow. The decomposition of cellulose occurs in two stages as :

a) In the first stage, the long chain of cellulase is broken down into cellobiase and then into glucose by the process of hydrolysis in the presence of enzymes cellulase and cellobiase, and

b) In second stage, glucose is oxidized and converted CO_2 and water.

$$1.\ \text{Cellulose} \xrightarrow[\text{Hydrolysis}]{\text{Cellulase}} \text{Cellobiose} \xrightarrow[\text{hydrolysis}]{\text{Cellobiase}} \text{Glucose}$$

$$2.\ \text{Glucose} \xrightarrow{\text{Oxidation}} \text{Organic Acids} \xrightarrow{\text{Oxidation}} CO_2 + H_2O$$

The intermediate products formed/released during enzymatic hydrolysis of cellulose (Cellobiose and Glucose) are utilized by the cellulose-decomposing organisms or by other organisms as source of energy for biosynthetic processes. The cellulolytic microorganisms responsible for degradation of cellulose through the excretion of enzymes (Cellulase and Cellobiase) are fungi, bacteria and actinomycetes.

3.3.2 Decomposition of Hemicelluloses

Hemicelluloses are water-soluble polysaccharides and consist of hexoses, pentoses, and uronic acids and are the major plant constituents second only in quantity of cellulose, and sources of energy and nutrients for soil microflora.When subjected to microbial decomposition, hemicelluloses degrade initially at faster rate and are first hydrolyzed to their component sugars and uronic acids. The hydrolysis is brought about by number of hemicellulolytic enzymes known as "hemicellulases" excreted by the microorganisms. On hydrolysis hemicelluloses are converted into soluble monosaccharide/sugars (xylose, arabinose, galactose and mannose) which are further convened to organic acids, alcohols, CO_2 and H_2O and uronic acids are broken down to pentoses and CO_2. Various microorganisms including fungi, bacteria and actinomycetes both aerobic and anaerobic are involved in the decomposition of hemicelluloses.

3.3.3 Lignin Decomposition

Lignin is the third most abundant constituent of plant tissues, and accounts about 10-30% of the dry matter of mature plant materials. Lignin content of young plants is low and gradually increases as the plant grows old. It is one of the most resistant organic substances for the microorganisms to degrade however certain Basidiomycetous fungi are known to degrade lignin at slow rates. Complete oxidation of lignin result in the formation of aromatic compounds such as syringaldehydes, vanillin and ferulic acid. The final cleavages of these aromatic compounds yield organic acids, carbon dioxide, methane and water. Microorganisms involved in the decomposition of organic matter are listed in the Table 3.2.

Table 3.2 : List of Micro-organisms, those are involved in decomposition of organic matter

Constituents		**Microorganisms**	
	Bacteria	**Fungi**	**Actinomycetes**
Cellulose	*Achromobacter,*		
	Bacillus, Cellulomonas,	*Aspergillus,*	*Micromonospora*
	Cellvibrio, Clostridium, Cytophaga, Vibrio	*Chaetomium, Fusarium, Pencillium, Rhizoctonia,*	*Nocardia Streptomyces*
	Pseudomonas,	*Rhizopus, Trichoderma,*	*Thermonospora*
	Sporocytophaga etc.	*Verticillttm.*	
Hemicellulose	*Bacillus, Achromobacter,*	*Aspergillus, Fusarium,*	
	Cytophaga Pseudomonas,	*Chaetomium,*	*Streptomyces*
	Erwinia, Vibrio,	*Penicillium,*	*Actinomycetes*
	Lactobacillus	*Trichoderma, Humicola*	
Lignin	*Flavobacterium,*		
	Pseudomonas,	*Humicola, Fusarium*	
	Micrococcus,	*Fames, Pencillium,*	*Streptomyces*
	Arthorbacter,	*Aspergillus, Ganoderma*	*Nocardia*
	Xanthomonas		

3.4. Mixed Biofertilizers

Mixed biofertilizers containing a consortium of nitrogen fixers, phosphate solubilizers and plant growth promoting *Rhizobacteria (*PGPR) were found to promote the growth of cereals, legumes and oil seeds better and save nitrogen and phosphate fertilizers in crops. Response was better and there was 50% nutrient substitution in crops like rice and legumes depending up on soil type and yield level. The response of bio fertilizers was better when used along with 75% chemical fertilizers.

3.5 Other Biofertilizers

3.5.1 Potash Biofertilizers

Some microbes can be used as potash biofertilizer as *Bacillus mucilaginous*, *Frateuria aurantia*, *Trichoderma harzianum* etc. Release of K through solubilization of muscovite, orthoclase and microline by*Bacillus mucilagenosus* has been reported (Sugumaran and Janathanam, 2007).

***Bacillus mucilagenosus*:** Slime froming micro-organisms have been found to be useful for weathering of K bearing minerals. *B. mucilagensous*, belonging to Bacilliaceae family is gram +ve, rod shaped slime producing bacteria which produce nuclease, endonucleases, cellobiase, protease, ribonuclease and phosphomonoestoarase Phylogenic analysis indicates that *B. mucilagenosus* and *B. edaphicus* are member of the genus Paenibacillus. Some good strains are KNP4/3, KNP 414, AS1.153 (China) etc. They secrete organic acids by oxidation of glucose. It has been observed that *B. mucilagineous* increase K availability in soils which then leads to an increase in the K content of Plant. The effect is more when it is co-inoculated with a PSB.

Frateuria aurantia:*Frateuria aurantia* belongs to the Pseudomonaceae family.This bacteria is gram-ve, rod shaped, obligatory anaerobe. It has been found to be capable of mobilizing potash, and may enhance crop productivity in K-deficient soils (Ramarethinam and Chandra, 2006). *F. aurantia* is also known to solubilize calcium. Even though K-solubilizing/mobilizing biofertilizers are being produced and marked, their mode of action is still not fully clear and much more research is needed in this area.

3.5.2 EM-based Biofertilizers

The concept of effective micro-organisms (EM) was developed by Prof. TeruoHiga in Okinawa, Japan. EM is a multi-culture of major micro-organisms and a consortium of photosynthetic bacteria, lactic acid bacteria, yeast, actinomycetes and fermenting fungi with those found in nature and capable of co-existing with other microorganisms. EM microbial solution can convert all waste into very good fertilizers within a short time. EM is an alternative technology which is useful in recycling and management of organic wastes.

3.5.3 Nano-based Biofertilizers

Nano science is now contributing to our knowledge in different area including agriculture. Efforts are being made to produce 'Nano Biofertilizers' based on Nano biotechnology. It has been observed that some nanoparticles (Particles size of 10^{-9}m). If mixed with rock phosphate can enhance the solubilizing capacity of P solubilizing microbes. Efforts are also on going to utilize nanoparticles to achieve higher mobilization of Zinc mediated by rhizobacteria for crop production. Nano-biofertilizers are being produced and marked in Malaysia (Bhattacharya and Tandon, 2012).

Table 3.3: Summary of the Microorganisms and their functions used as Biofertilizers

Micro organisms	Function/Contribution	Limitation	Use for crops
Rhizobium	a) Estimated fixation of 30-200kg N/ha. b) 10-35% increase in yield of field crops c) Fixed N also benefits next crop	a) Fixes N only with legumes. b) Not very effective in traditional area. c) Needs optimum supply of P and Mo	All pulses, beans Legume oilseeds (soybean, groundnut) Forage legumes like clover, luceme, others
Azotobacter and *Azospirillum*	a) Fixation of 20-25kg N/ha. b) 10-15% increase in yield. c) Production of growth promoting substances such as IAA, GA, vitamins etc.	a) Demands high organic matter b) Azospirillum is effective C_4 plants	Cereals, millets Cotton, Sugarcane Vegetables/others
Acetobacter (*Gluconoacetobacter*)	a) Can fix 100-150kg N/ha. b) 10-20% increase yield. c) Can solubilize P also.		
Blue Green Algae/Cyanobacteria	a) Fixation of 20-30kg N/ha. b) 10-15% increase in yield. c) Production of growth promoting substances.	Demands and high organic matter and specific C substrate.	Only for sugarcane
Azolla-Anabaena	Can fix 30-100 kg N/ha 10-25% increase in yield	a) Poor survival at high temperature b) High demand for P,	Only for flooded rice or as a green manure
	a) Can solubilize 20-30% of insoluble phosphate.	Demand high organic matter.	For all soils and crops

Contd...

	b) 10-20% yield increase		
P mobilizing BF (Mycorrhizae)	a) Can mobilize P for plant b) 10-20% increase in yield	Demand organic matter free crops.	Potentially for all upland field and
K biofertilizer *Frateuria aurantia* *Bacillus muvilagenosus*	Can release 10-15 % k from K bearing minerals	Details under research soils and crops	Potentially for all

Source : Bhattacharya and Tandon (2012)

Biofertilizers production received a major impetus after the GOI, Ministry of Agriculture, launched the National Project on Development and uses of Biofertilizers in 1983. Under this project, the government provided non-recurring grants in aid up to 20 lakhs for setting up of biofertilizers production units of 150 tonnes/annum capacity each by the agro industries/co-operatives /public sectors undertaking /NGOs and private agencies. Since, inspection 77 units of production of biofertilizers have been established with central assistance creating a production capacity of 9.625 tonnes/annum under this project.In October 2004, the national project on biofertilizers became a part of the project National Project on Organic Farming (NPOF) under this scheme, there is no provision to set up more biofertilizers production units through NABARD/NCDC as credit linked back ended subsidy release pattern at 25% of the total project cost (Rs. 160 lakhs) or Rs. 40 lakhs whichever is less the project is continuing (Bhattacharya and Tandon, 2012). Based on the latest available data for 2011-12 from NCOF, the total biofertilizers production of 69178.09 tonnes was made up of 19385.12 tonnes of P solubilizing biofertilizers, 10621.96 tonnes of *Azotobacter*, 6006.51 tonnes of *Rhizobium* and 4193.37 tonnes of *Azospirillum* (Table 3.4). In case of biofertilizers the production and supply of microbial cultures, the quality of the cultures and the lack of publicity are affecting their popularity as nutrient sources. The government has no control over manufactures of biofertilizers for any of the state of India. Only a few entrepreneurs possess ISI mark for their products and most of the products are of substandard quality. Due to these *laxities* on the part of Govt. the farmers are confused about their rates, availability and expiry dates. The necessary action by government and its policies will certainly go to a long way in the further development of the biofertilizers.

Table 3.4: State wise Production of various biofertilizers during 2011-12

Sl No.	State	Name of the Biofertilizers						
		AZB	AZS	RZB	PSB	TOTAL BF	OTHER INOCULANTS*	GRAND TOTAL
1.	Andhra Pradesh	121.99	319.98	101.57	581.66	1126.35	133.20	1259.55
2.	Assam	20.48	12.56	6.86	28.43	68.33	0.00	68.33
3.	Bihar	28.00	0.00	16.60	30.40	75.00	0.00	75.00
4.	Chhattis-garh	11.91	0.00	99.65	164.78	276.34	0.00	276.34
5.	Delhi	1157.00	27.00	47.00	386.00	1617.00	1582.00	3199.00
6.	Gujarat	433.35	314.93	219.67	1069.40	2037.35	12.05	2049.40
7.	Goa	0.00	0.00	0.00	0.00	0.00	592.14	592.14
8.	Haryana	503.81	0.00	84.60	326.00	914.41	50.00	964.41
9.	Himachal Pradesh	0.34	0.15	0.20	0.60	1.29	0.00	1.29
10.	Jharkhand	0.42	0.00	4.61	3.35	8.38	0.11	8.48
11.	Karnataka	756.40	1015.33	1114.87	2873.72	5760.32	14343.48	20103.80
12.	Kerala	31.83	161.87	164.30	546.17	904.17	4122.63	5026.80
13.	Madhya Pradesh	246.43	3.00	565.03	1494.60	2309.06	67.53	2376.59
14.	Maharashtra	2262.39	731.03	2376.65	3373.62	8743.69	622.90	9366.59
15.	Mizoram	0.00	0.00	0.00	0.00	0.00	0.00	0.00
16.	Nagaland	2.50	1.50	2.50	6.50	13.00	0.00	13.00
17.	Orissa	72.45	36.42	144.76	336.49	590.12	0.00	590.12
18.	Punjab	94.20	13.96	18.65	565.41	692.22	0.00	692.22
19.	Pondi-cherry	11.12	85.39	73.22	339.72	509.45	1304.34	1813.79
20.	Rajasthan	27.76	0.00	25.02	147.00	199.78	0.00	199.78
21.	Tamil Nadu	197.66	926.45	385.04	1764.66	3373.81	5941.29	9315.10
22.	Tripura	537.58	463.00	0.04	542.24	1542.85	0.01	1542.86
23.	Uttar Pradesh	3871.79	14.92	439.41	4352.96	8695.08	46.54	8741.63
24.	Uttara-khand	128.25	4.37	6.77	123.62	263.01	0.00	263.01
25.	West Bengal	104.30	61.60	109.50	327.80	603.20	35.67	638.87
	Total	10621.96	4193.37	6006.51	19385.12	40324.21	28853.884	69178.09

* Other include compost enriches (*Trichoderma* and *Paceliomycetes* etc.) PGPRs, BGA, *Azolla,* ZBS/KMB.

3.6. Constraints in the use of Biofertilizers

3.6.1 Production Constraints

Despite significant improvement/refinement in BF technology over the years, the progress in the field of BF production technology is below satisfaction due to the followings:-

a) **Unavailability of appropriate and efficient strains**

Lack of region specific strains is one of the major constraints as bio-fertilizers are not only crop specific but soil specific too. Moreover, the selected strains should have competitive ability over other strains, N-fixing ability over a range of environmental conditions, ability to survive in broth and in inoculants carrier.

b) **Unavailability of suitable carrier**

Unavailability of suitable carrier (media in which bacteria are allowed to multiply) due to which shelf life of bio-fertilizers is short is a major constraint. Peat of a good quality (more than 75% carbon) is a rare commodity in India. Nilgiri peat is of poor quality (below 50% carbon). According to the availability and cost at production site, choice is only with lignite and charcoal in India. As per the suitability the order is peat>lignite>charcoal>FYM>soil>rice husk. Good quality carrier must have good moisture holding capacity, free from toxic substances, sterilisable and readily adjustable pH to 6.5-7.0. Under Indian conditions where extremes of soil and weather conditions prevail, there is yet no suitable carrier material identified capable of supporting the growth of bio-fertilizers. Better growth of bacteria is obtained in sterile carrier and the best method is gamma irradiation of sterilization (while using autoclave, lime mixed lignite is filled up to two third capacity of steel trays for 1-2hrs for three days and sterilized at 121^0C for carrier material.

c) **Mutation during fermentation**

Bio-fertilizers tend to mutate during fermentation and there by raising production and quality control cost. Extensive research work on this aspect is urgently needed to eliminate such undesirable changes.

3.6.2 Market Level Constraints

a) **Lack of awareness of farmers**

Inspite of considerable efforts in recent years, majority of farmers in

India are not aware of bio-fertilizers, their usefulness in increasing crop yields sustainably.

b) **Inadequate and Inexperienced staff**

Because of inadequate staff and that too not technically qualified who can attend to technical problems. Farmers are not given proper instructions about the application aspects.

c) **Seasonal and unassured demand**

The bio-fertilizer use is seasonal and both production and distribution is done only in few months of year, as such production units' particularly private sectors are not sure of their demand.

d) **Lack of quality assurance**

The sale of poor quality bio-fertilizers through corrupt marketing practices results in loss of faith among farmers, to regain the faith once is very difficult and challenging.

3.6.3 Resource Constraint

a) **Limited resource generation for BF production**

The investment in bio-fertilizer production unit is very low. But keeping in view of the risk involved largely because of short shelf life and no guarantee of off take of bio-fertilizers, the resource generation is very limited.

3.6.4 Field Level Constraints

a) **Native microbial population**

Antagonistic microorganism already present in soil competes with microbial inoculants and many times do not allow their effective establishment by out-competing the inoculated population.

b) **Soil and climatic factors**

Among soil and climatic conditions, high soil fertility status, unfavorable pH, high nitrate level, high temperature, drought, deficiency of P, Cu, Co, Mo or presence of toxic elements affect the microbial growth and crop response.

c) **Faulty inoculation techniques**

Majority of the marketing sales personals do not know proper inoculation techniques. Bio-fertilizers being living organisms required proper handling, transport and storage facilities.

3.7 Plant Growth Promoting Rhizobacteria (PGPR)

PGPR were first defined by **Kloepper** to describe soil bacteria that colonize roots of plants following inoculation into seed and that enhance plant growth. Beneficial effects of PGPR are as follows:

- Reduces deleterious effects of pathogens on crop growth by protection against pathogens by production of antibiotics.
- Solubilization of mineral nutrients by inducing specific ion flux in plant cell.
- Production of plant hormones like IAA, GA_3, cytokinin and induce formation of ethylene
- Production of siderophores that chelate iron and make it available to the plant root.
- Induce systemic resistance in plant.
- Fixation of atmospheric nitrogen that is transferred to the plant.

The effect of PGPR on seed germination, seedling growth and yield of field grown maize were to be significantly enhanced (Gholami*et al.,* 2009) and in chickpea (Mishra *et al.,* 2010). Stimulation of different crops by PGPR has been demonstrated in both green house and field trials. Strains of *Pseudomonas putida* and *P. fluorescens* have increased root and shoot elongation in canola, lettuce, and tomato; yields in radishes, potato, rice, sugar beet, tomato, lettuce, apple, citrus, beans, ornamental plants, and wheat (Suslov, 1982; Kloepper, 1994). Wheat yield increased by 30% with *Azotobacter* inoculation; 43% with *Bacillus* inoculants (Kloepper, 1992) and 10-20% in field trials using a combination of *Bacillus megaterium* and *A. chroococcum.* (Brown, 1974). There is been a large body of literature describing potential uses of plants associated bacteria as agents stimulating plant growth and managing soil and plant health (Glick,1995). PGPB are associated with many, if not all plant species, and are commonly present in many environments and protect plant against pathogens. Under iron-limiting conditions PGPR produce low-molecular-weight compounds called siderophores to competitively acquire ferric ion. The basis of antibiosis as a bio control mechanism of PGPR has become increasingly better understood over the past two decades (Whipps, 2001). A variety of antibiotics have been identified, including compounds such as amphisin, 2, 4-diacetylphloro-glucinol (DAPG), hydrogen cyanide, oomycin A, phenazine, pyoluteorin, pyrrolnitrin, tensin, tropolone, and cyclic lipopeptides produced by Pseudomonads (Defago, 1993)

A variety of microorganisms also exhibit hyperparasitic activity, attacking pathogens by excreting cell wall hydrolases (Chernin and Chet, 2002). Chitinase produced by *Serratia. plymuthica* C48 inhibited spore germination and germ-tube elongation in *Botrytis cinerea*(Frankowski*et al.,* 2001). The ineffectiveness of PGPR in field appears to be because of their inability to colonize plant roots (Benizri, *et al.,* 2001). PGPR has dual role in enhancing crop production. The main role is production of growth hormones. However, many PGPR have a potential role as bio-control agents. Further researches are required to examine the replacement of insecticides.

Root colonizing bacteria like, *Azospirillum* and *Pseudomonas* sps. are known to produce growth hormones which often leads to increase root and shoot growth. Plants differ in the leaves and ration of the hormones required to maintain normal growth and development. Industrial production of inorganic fertilizers a costly process dependent on energy derived from fossilfuel, which is getting depleted at a faster rate. On the contrary use of microbial inoculants is not only a low cost technology but also takes adequate care of soil health and environmental safety. Intensive search a number of micro-organisms have been recognized as nitrogen fixers. This is no doubt a low cost technology capable of bringing rich-dividends to the farmers. However, transferring a technology to the farmer's field is of paramount importance (Panwar and Swarnalaxmi, 2005). Generally the effect of biofertilizers on crop growth and yield is not as sticking as that of chemical fertilizers. Since it is a living system and the influence is subject to environmental, biological and nutritional stresses.

3.7.1 Microbes as PGPR in Biotic and Abiotic Stress

PGPR strains *Bacillus, Azospirillum Azotobacter* and *Pseudomonas* spp. etc. are found to improve seed germination, root/shoot length, nodulation, biomass, yield and nutrient mobilization in various crops (Table. 3.5). The beneficial bacteria also protect the crops from the attack of soil borne pathogen among which *Fusarium spp., Rhizoctonia solani, Rhizoctonia bataticola* and *Scierotium rolfsii* are highly destructive. Pyrrolnitrin, Phenazine-1-carboxylic acid and 2,4 produced by *Pseudomonas* spp. were the major antibiotics involved.

3.7.1.1 PGPR : Usefulness under biotic stress

Combined inoculation of other PGPR with rhizobia was observed to exert positive effects on the growth of legumes. Co-inoculation of PGPR along with *Rhizobium* increased the root nodulation. The beneficial effects of co-inoculation are strain dependent. The basic mechanisms involved in this synergistic activity were not completely known and remains a challenge. One possibility is that PGPR, by altering the host's secondary metabolism and/or creating antibiosis in the rhizosphere, out compete the pathogens and eliminate competition of *Rhizobium* with deleterious micro-organisms for colonization of the plant parts. Alteration of the plant flavonoid metabolism was proposed as another mechanism of synergistic activity of PGPR and rhizobia. Inoculation of *Azospirillium* sps. produces large quantities of auxins and stimulates the formation of epidermal cells that become root hair cells or additional infection sites for rhizobial colonization. Unlike chemical fertilizers, PGPR exert a beneficial effect on rhizobial nodulation of legumes and should be exploited for the economic benefit of subsistence farming systems. A complete understanding of the mechanisms of the synergistic activity between PGPR and *Rhizobium* facilitates the better utilization and improvement of bio-inoculants in legume based cropping systems.

Table 3.5: Plant growth promoting rhizobacteria (PGPR) in enhancing the growth of crop plants.

Crop	Bacterial isolate	Effects on growth and yield
Wheat	*Bacillus subtilis*	Seed germination and grain yield
Pigeonpea	*Bacillus substilis* AF1	Increase in shoot & root length and biomass
Mungbean, Pea	*Azotobacter*	Increase in yield
Chick pea	*Azospirillum brasilense* *Azotobacter, Pseudomonas* sp.	Enhanced seed germination, root and shoot dry weight, yield and nodulationNumber of shoots plant, yield, seed protein
Soybean	*Bacillus cereus* UW 85	Increase in root growth, nodulation and yield
Faba bean	*Azospirillum brasilense*	Increase in growth of root and shoot and nodulation
Groundnut	*Bacillus subtilis* *Bacillus amyloliquefaciens* *Pseudomonas aeruginosa,* *Pseudomonas fluorescens*	Increase in root and shoot length, biomass and yield.
Sunflower	*Azospirillum, Azotobacter* *Pseudomonas*	Increase in yield
Blackpepper, ginger, turmeric, spice crop	*Pseudomonas fluorescens* *Phosphobacteria*	Nutrient mobilization

PGPR's and their effectiveness has been established on foot rot and slow decline of black pepper and rhizome rot of ginger and dump rot of cardamom. Strains of *Pseudomonas fluorescens* were found to increase the growth and vigor of these crops apart from suppressing the soil borne diseases (Sarma *et al.,* 2000).

3.7.1.2 PGPR : Usefulness under Abiotic Stress

Microbial colonization of plant root is affected by biotic and abiotic factors as root exudates, competition, inorganic nutrients, soil pH and temperature etc.

Among the agriculturally important abiotic stress, drought, heat, soil salinity and alkanity as the most detrimental and cause loss of crop productivity all over the world.

Kulkarni and Nautiyal (2000) conducted a study to examine the growth response of a *Rhizobium* sp. NBR1330 isolated from root nodules of *Prosopis juliflora* growing in alkaline soil. The individual stress survival limit of the strain in a medium containing 32% NaCl was 8hrs and at 55°C up to 3 hrs also conducted an ecological survey (Kulkarni *et al.,* 2000) to characterize 5000 *Rhizobium* sp., *Sesbania* strains of diverse geographical origin, isolated from the root nodules of *Sesaania aculeata* growing in neutral (pH 7) and alkaline (pH 8.5 and above) soils. The rhizobia from the alkaline soil showed the significantly higher salt tolerance than those isolated from neutral soil. High temperature was tolerated efficiently by *Rhizobium* sp. NBRI 2505 Sesbania, in the presence of salt at higher pH. Several soil bacteria produce osmolytes to protect themselves against frequent fluctuations in osmotic conditions e.g. *Paenibacillus polymyxa* (previously *Bacillus polymyxa*) and *Bacillus subtilis* produces glycine betaine (Lucht and Bremer, 1994). This compound lowers the water potential outside the cell wall and inhibits intercellular spaces. The increased concentration of these compounds could be sensed by the plant as a local dehydration. Consequently, dehydration genes such as ERD 15 and RAB 18 would be activated in plants in order to tolerate drought.

Tripathi *et al.,* (1998) have studied in detail salinity stress responses in the plant growth promoting rhizobacteria *Azospirillum* sps. They have shown that in order to adapt to the fluctuations in soil salinity/osmolarity, the bacteria accumulate compatible solutes such as glutamate, proline, glycine betaine, trehalose etc. Proline seems to play a major role in osmoadoptation. With increase in osmotic stress the dominant osmolyte in *Azospirillum brasilense* shifts from glutamate to proline. Accumulation of proline in *Azospirillum brasilense* takes place both uptake and synthesis.

Rangarajan *et al.* (2001) analyzed populations of *Pseudomonas* for their biochemical characters and genetic diversity using molecular tools including RAPD and PCR-RFLP. It was observed that increasing salinity caused a predominant selection of salt tolerant species, in particular *Pseudomonas pseudoalcaligenes* and *Pseudomonas alcaligenes,* irrespective of the host rhizosphere. Rangarajan *et al.,* (2003) have recently isolated indigenous Pseudomonas strain from rhizosphere of rice cultivated in the coastal agriecosystem and screened for *in vitro* antibiosis against *Xanthomonas oryzae* p.v. *oryzae* and *Rhizoctonia solani.* PSB that could tolerate high salinity have also been isolated (Kumar, 2003). P-starvation inducible quinoprotein glucose dehydrogenase (Gcd) has been shown to be responsible for the high gluconic acid secretion resulting efficient P solubilization.

3.8 Constraints in Biofertilizer Technology

Though the biofertilizer technology is a low cost, ecofriendly technology, several constraints limit the application or implementation of the technology the constraints may be environmental, technological, infrastructural, financial, human resources, unawareness, quality, marketing, etc. The different constraints in one way or other affecting the technique at production, or marketing or usage.

3.8.1 Technological Constraints

- Use of improper, less efficient strains for production.
- Lack of qualified technical personnel in production units.
- Production of poor quality inoculants without understanding the basic microbiological techniques.
- Short shelf life of inoculants.

3.8.2 Infrastructural Constraints

- Non-availability of suitable facilities for production.
- Lack of essential equipment and power supply etc.
- Space availability for laboratory, production and storage etc.
- Lack of facility for cold storage of inoculant packets.

3.8.3 Financial Constraints

- Non-availability of sufficient funds and problems in getting bank loans.
- Less return by sale of products in smaller production units.

3.8.4 Environmental Constraints

- Seasonal demand for biofertilizers.
- Simultaneous cropping operations and short span of sowing/planting in a particular locality.
- Soil characteristics like salinity, acidity, drought and water logging etc.

3.8.5 Human Resources and Quality Constraints

- Lack of technically qualified staff in the production units.
- Lack of suitable training on the production techniques.

- Ignorance on the quality of the product by the manufacturer.
- Non-availability of quality specifications and quick quality control methods.
- No regulation or act on the quality of the products.
- Awareness on the technology.
- Unawareness on the benefits of the technology.
- Problem in the adoption of the technology by the farmers due to different methods of inoculation.
- No visual difference in the crop growth immediately as that of inorganic fertilizers.

CHAPTER - 4

Nitrogen Fixation

Under agricultural conditions, the nitrogen removed is usually greater than the nitrogen input. To maintain fertility, nitrogen must be returned to the soil. This takes place mainly by biological nitrogen fixation, which is responsible for the major reduction of dinitrogen to ammonia that is used by the plants. Biological nitrogen fixation accounts for most of the fixation of atmospheric nitrogen in to Ammonium, thus representing the key entry point of molecular nitrogen into the biogeochemical cycle of nitrogen (Fig.4.1). Some bacteria can convert atmospheric nitrogen into ammonium. Most of this nitrogen fixing prokaryote is free living in the soil. A few form symbiotic association with higher plants in which the prokaryotes directly provides the host plant with fixed nitrogen in exchange for other nutrients and carbohydrates (Table 4.1). Such symbiosis occur in nodules that form on the roots of the plant and contain the nitrogen fixing bacteria. The most common type of symbiosis occurs between members of the plant family Leguminosae and soil bacteria of the genera *Azorhizobium, Bradyrhizobium, Protorhizobium, Rhizobium* and *Sinorhizobium,* these are collectively called as *Rhizobia* (Table 4.1 and Fig.4.2). Another common type of symbiosis occurs between several woody plant species such as alder trees, and soil bacteria of the genus *Frankia*. Still other types involve the South American herb viz. *Gumera* and the tiny water Fern *Azolla*, which form association with the *cyanobacteria, Nostoc* and *Anabaena*, respectively (Table 4.2).

4.1. N_2 Fixation Requires Anaerobic Conditions

It requires anaerobic conditions because oxygen irreversibly inactivates the nitrogenase enzymes involved in nitrogen fixation, nitrogen must be fixed under anaerobic conditions. Thus each of the nitrogen fixing organisms either functions under natural conditions or can create an internal anaerobic environment in the presence of oxygen. In cyanobacteria, anaerobic conditions created in specialized cells heterocyst (Fig.4.2). Heterocysts are thick walled cells that differentiate when filamentous cyanobacteria are deprived of NH_4^+. These cells lack photosystem II, the oxygen producing photosystem of chloroplast, so they do not generate oxygen (Burris, 1976).

Heterocysts appear to represent an adaptation for nitrogen fixation, in that they are wide spread among anaerobic cyanobacteria that fix nitrogen.

Cyanobacteria that lack heterocysts can fix nitrogen only under anaerobic conditions such as those that occur in flooded fields. In Asian countries, N_2 fixing cyanobacteria of both the heterocysts and non-heterocysts types are a major means for maintaining an adequate nitrogen supply in the soil of rice fields. These micro-organisms fix nitrogen when the field is flooded and disease as the fields dry, releasing the fixed nitrogen to the soil. Another important source of available nitrogen in flooded rice fields is the water fern *Azolla*, which associates with the *Cyanobacterium anabaena*. The *Azolla-Anaebaena* association can fix as much as 0.5kg of atmospheric nitrogen per ha/day, a rate of fertilization that is sufficient to attain moderate rice fields. Free living bacteria that are capable of fixing nitrogen are aerobic, facultative or anaerobic (Table 4.1).

- Aerobic nitrogen fixing bacteria such as *Azotobacter* are thought to maintain reduced oxygen conditions (microaerobic conditions) through their high levels of respiration (Burris, 1976). Others such as *Gleothece*, evolve O_2 photosynthetically during the day and fix nitrogen during the night.
- Facultative organisms, which are able to grow under both aerobic and anaerobic conditions, generally fix nitrogen only under anaerobic conditions.
- For anaerobic nitrogen fixing bacteria, oxygen does not pose a problem, because it is absent in their habitat. These anaerobic organisms can be either photosynthetic (*Rhodospirillum*) or non-photosynthetic (*Clostridium*).

Table 4.1: Examples of the Micro-organisms that can carry Nitrogen Fixation

Symbiotic Nitrogen Fixation	
Host Plant	**Nitrogen Fixing Symbionts**
Leguminous, Legumes, *Parasponia*	*Azorhizobium, Bradyrhizobium, Protorhizobium, Rhizobium, Sinorhizobium*
Actinorhizal: alder (tree), Ceanothus (shrub),	
Casuarina (tree) and Datisca (shrub)	*Frankia*
Gunnera	*Nostoc*
Azolla (water fern)	*Anabaena*
Sugarcane	*Acetobacter*

Contd...

Free Living Nitrogen Fixation	
Types	**Nitrogen Fixing Genera**
Cyanobacteria (BGA)	*Anabaena, Colothrix, Nostoc*
Other bacteria	
• Aerobic	*Azospirillum, Azotobacter, Beijerinckia, Dexia*
• Facultative	*Bacillus. Klebsiella*
Anaerobic	
• Non-photosynthetic	*Clostridium, Methanococcus (Archaebacterium)*
• Photosynthetic	*Chromatium, Rhodospirillum*

Table 4.2: Associations between Host Plants and *Rhizobia*

Plant Host	*Rhizobial* Symbiont
Parasponia	*Bradyrhizobium*spp.
Soybean	*Bradyrhizobium japonicum* (slow growing type) *Sinorhizobium-fredii* (fast growing type)
Alfalfa (*Medicago sativa*)	*Sinorhizobium meliloti*
Sesbania (Aqautic)	*Azorhizobium* (forms both root and stem nodules; the stems have adventitious roots)
Bean (*Phaseolus*)	*Rhizobium leguminosarum* bv. *phaseoli Rhizobium tropicii, Rhizobium etli,*
Clover (*trifolium*)	*Rhizobium leguminosarum* bv. *trifolii*
Pea (*Pisum sativum*)	*Rhizobium leguminosarum* bv. *viciae*
Aeschnomene (Aquatic)	*Photorhizobium* (Photosynthetically active rhizobia that form stem nodules, probably associated with adventitious roots)

4.2 Symbiotic Nitrogen Fixation Occurs in Specialized Structure

Symbiotic nitrogen fixing prokaryotes dwell within nodules, the special organs of the plant host that enclose the nitrogen fixing bacteria (Fig. 4.1). In the case of *Gumera*, these organs are existing stem glands that develop independently of the symbiont. The case of legumes and actinorhizal plants, the nitrogen fixing bacteria induce the plant to form root nodules.

Grasses can also develop symbiotic relationship with nitrogen fixing organisms, but in these associations root nodules are not produced. Instead, the nitrogen fixing bacteria seem to colonize plant tissues or anchor to the root surfaces, mainly around the elongation zone and the root hairs (Reis *et al.*, 2000). For example, the nitrogen fixing bacterium *Acetobacter diazotrophicus* lives in the apoplast of stem tissues in sugarcane and may provide its host with sufficient nitrogen to grant independence from nitrogen fertilization (Dong *et al.,* 1994). The potential for applying *Azospirillum* seems to fix little nitrogen when associated with plants (VandeBroek and Vanderleyden, 1995).

Legumes and actinorhizal plants regulate gas permeability in their nodules, maintaining a level of oxygen within the nodule that can support respiration but is sufficiently low to avoid inactivation of the nitrogenase (Kuzma*et al.,* 1993). Gas permeability increases in the light and decreases under drought or upon exposure to nitrate. The mechanism for regulating gas permeability is not yet known.

Nodules contain an oxygen-binding heme protein called leghaemoglobin. Leghaemoglobin is present in the cytoplasm of infected nodules cells at high concentrations (700μM in soybean nodules) and gives the nodules a pink color. The host plant produces the globin portion of leghaemoglobin in response to infection by the bacteria (Marschner 1995); the bacterial symbiont produces the heme portion. Leghaemoglobin has a high affinity for oxygen (a K_m of about 0.01 μM), about ten times higher than the ß chain of human hemoglobin. Although leghaemoglobin was once thought to provide a buffer for nodule oxygen, recent studies indicate that it stores only enough oxygen to support nodules respiration for a few seconds (Denison and Harter, 1995). Its function is to help transport oxygen to the respiring symbiotic bacterial cells in a manner analogous to hemoglobin transporting oxygen to respiring tissues in animals (Ludwig and de Vries, 1986).

4.3 Establishing Symbiosis Requires an Exchange of Signals

The symbiosis between legumes and rhizobia is not obligatory. Legume seedlings germinate without any association with rhizobia, and they may remain unassociated throughout their life cycle. *Rhizobia* also occur as free-living organisms in the soil. Under nitrogen-limited conditions, however, the symbionts seek out one another through an elaborate exchange of signals. This signaling, the subsequent infection process, and the development of nitrogen-fixing nodules involve specific genes in both the host and the symbionts.

Plants genes specific to nodules are called nodulin (Nod) genes; rhizobial genes that participate in nodule formation are called nodulation (nod) genes (Heidstra and Bisseling, 1996). The nod genes are classified as common nod genes-nodA, nodB, and nodC are found in all rhizobial strains; the host-specific nod genes-such as nodP, nodQ, and nodH; or nodF, node, and nodL differ among rhizobial species and determine the host range. Only one of the nod genes, the regulatory nodD, is constitutively expressed, and as we will explain in details, its protein product (NodD) regulates the transcription of the other nod genes.

The first stage in the formation of the symbiotic relationship between the nitrogen-fixing bacteria and their host is migration of the bacteria toward the roots of the host plant. This migration is a chemotactic response mediated by chemical attractants, especially (iso) flavonoids and betaines, secreted by the roots. Theses attractants activate the rhizobial NodD protein, which then induces transcription of the other nod genes (Phillips and Kapulnik, 1995). The promoter region of all nod operons, except that of nod, contains a highly conserved sequence called the nod box. Binding of the activated NodD to the nod box induces transcription of the other nod genes.

4.4 Nod Factors Produced by Bacteria Act as Signals for Symbiosis

The nod genes activated by NodD code for nodulation proteins, most of which are involved in the biosynthesis of Nod factors. Nod factors are lipochitin oligosaccharide signal molecules, all of which have a chitin ß-1-4linked N- acetyl-D-glucosamine backbone (varying in length from three to six sugar units) and a fatty acyl chain on the C-2 position of the non-reducing sugar (Fig. 4.3).Three of the nod genes (nodA, nodB, and nodC) encode enzymes (nodA, nodB, and nodC, respectively) that are required for synthesizing this basic structure (Stockkermans *et al.,* 1995)

1. NodA is an N-acyltransferase that catalyzes the addition of a fatty acyl chain.
2. NodB is a chitin-oligosaccharide deacetylase that removes the group from the terminal non-reducing sugar.
3. NodC is a chitin-oligosaccharide synthase that link N-acetyl-D glucosamine monomers.

Host- specific nod genes that vary among rhizobial spices are involved in the modification of the fatty acyls chain or the addition of groups important in determining host specificity (Carlson *et al.,* 1995).

- NodE and nodF determine the length and degree of saturation of the fatty acyl chain; those of *Rhizobium leguminosarum* bv. *viciae* and *R. meliloti* result in the synthesis of an 18:4 and a 16:2 fatty acyl group, respectively.
- Other enzymes, such as NodL, influence the host specifically of Nod factors through the addition of specific substitutions at the reducing or nonreducing sugar moieties of the chitin backbone.

A particular legume host responds to a specific Nod factor. The legume receptors for Nod factors appear to be special lectins (sugar-binding proteins) produced in the root hairs (Van Rhijn *et al.,*1998; Etzler *et al.,* 1999). Nod factors activate these lectins, increasing their hydrolysis of phosphoanhydride bonds of nucleoside di-and triphosphates. This lectin activation directs particular rhizobia to appropriate hosts and facilitates attachment of the rhizobia to the cell walls of a root hair.

4.5 Nodule Formation Involves Several Phytohormones

Two processes- infection and nodule organogenesis- occur simultaneously during root nodule formation. During the infection process, rhizobia that are attached to the root hairs release Nod factors that include a pronounced curling of the root hair cells (Fig. 4.4A and B). The rhizobia become enclosed in the small compartment formed by the curling. The cell wall of the root hair degrades in these regions, also in response to Nod factors, allowing the bacterial cells direct access to the outer surface of the plant plasma membrane (Lazarowitz and Bisseling, 1997).

The next step is formation of the infection of the infection thread (Fig. 4.4C), an internal tabular extension of the plasma membrane that is produced by the fusion of golgiderived membrane vesicles at the site of infection. The thread grows at its tip by the fusion of secretory vesicles to the end of the tube. Deeper into the root cortex, near the xylem, cortical cells dedifferentiate and start dividing, forming a distinct area within the cortex, called a nodule primordium, from which the nodule will develop. The nodule primordia form opposite the protoxylem poles of the root vascular bundle (Timmers *et al.,* 1999). Different signaling compounds, acting either positively or negatively controls the position of nodule primordia. The nucleoside uridine diffuses from the stele into the cortex in the protoxylem zones of the root and stimulates cell division (Lazarowitz and Blesseling, 1997). Ethylene is the synthesized in the region of the pericycle, diffuses into cortex, and blocks cell division opposite the phloem poles of the root.The infection thread filled with proliferating rhizobia elongates through the root hair and cortical cell layers, in the direction of the nodule primordium. When the infection thread reaches specialized cells within the nodule, its tip fuses with the plasma membrane of the host cell, releasing bacterial cells that are packaged in a membrane derived from the host cell plasma membrane (Fig. 4.4D). Branching of the infection thread inside the nodule enables the bacteria to infect manyFig. 4.4E and F (Mylona *et al.,* 1995).

At first the bacteria continue to divide, and the surrounding membrane increase in surface area to accommodate this growth by fusing with smaller vesicles. Soon thereafter, upon an undetermined signal from the plant, the bacteria stop dividing and begin to enlarge and to differentiate into nitrogen-fixing endosymbiotic organelles called bacteroids. The membrane surrounding the bacteroids is called the peribacteroid membrane.The nodule as a whole develops such features as vascular systems (which facilitate the exchange of fixed nitrogen produced by the bacteroids for nutrients contributed by the plant) and a layer of cell to exclude O_2 from the root nodule interior. In some temperate legumes e.g., peas, the nodules are elongated and cylindrical because of the presence of a nodule meristem. The nodule of tropical legumes, such as soybeans and peanut, lack a persistent meristem and are spherical (Rolfe and Gresshoff, 1988).

4.6 Historical Review

Ancient peoples acknowledged the benefits of annually rotating leguminous and non-leguminous crop plants. The first explanation for this effect was made in the 1830s by the French scientist Boussingault, who suggested that leguminous plants could utilize atmospheric nitrogen for growth. Boussingault's work met the harsh criticism of the distinguished German organic chemist, Liebig, and consequently Boussingault's claim fell into disregard with the scientific elite of the period. However, during the next 50 years others sought to verify Boussingault's results.

The three decades following the experiments of Hellriegel found researchers seeking to identify not only the organisms responsible for symbiotic N_2fixation but also free-living N_2-fixing microbes. Winogradsky demonstrated fixation by the strictly anaerobic *Clostridium* spp., Beijerinck showed the obligatory aerobic *Azotobacter* spp. were capable of N_2 fixation, and Drewes reported fixation by the algae viz. *Nostoc* and *Anabaena,* Fred, Baldwin thoroughly reviewed the work on the leguminous symbioses in their treatise. That monograph describes the morphological, cultural and physiological characteristics of the rhizobia, cross-inoculation groups, the nodule formation process, inoculation methods, the quantification of fixation rates under various conditions, and so on. In the late 1920s, investigations on the biochemistry of the N_2 fixation process began in earnest. Progress was made in several major areas such as:

a) Effects of different partial pressures of N_2 on the fixation process were investigated.

b) H_2 was reported to be a specific competitive inhibitor of nitrogen fixation.

c) Ureides were found to be the major nitrogen assimilate in certain leguminous plants, and

d) The inter-relationships of carbon and nitrogen metabolism during the N_2 fixation process were actively investigated.

It was during this period, before active, cell-free extracts or pure enzyme became available, that the first stable product of nitrogen, fixation was identified. Virtanen supported hydroxylamine, whereas Burris considered ammonia to be the first stable product. In the early 1940s, a new technique was introduced which aided in establishing ammonia as the first stable product, of nitrogen fixation. Burris developed the use of the stable isotope $^{15}N_2$ for nitrogen fixation research. By the early 1950s, it was firmly established from many additional lines of evidence that ammonia was the first stable product of nitrogen fixation.In 1960, both the group at Wisconsin reported reproducible, active, cell-free extracts from BGA and *Clostridium pasteurianum* respectively. These reports were quickly followed by others describing active extracts from *Klebsiella pneumoniae, Azotobacter vinelandii, Azotobacter chrococcum, Rhodospirillum rubrum, Bacillus polymyxa, Chromatium* spp. and *Rhizobium* bacteroids. These extracts were used to define the requirements for the nitrogen-fixing reaction and also pointed out why most previous attempts had failed. Nitrogenase, the enzyme catalyzing the reduction of N_2 to NH_4^+, was found to require large amounts of ATP, to be strongly inhibited by ADP, to require a constant supply of low potential electrons and to be irreversibly inactivated by O_2.

In these early studies on the cell-free systems, ATP was supplied via the phosphoroclastic metabolism of pyruvate was later replaced by an exogenous ATP-regenerating system. The utilization of $Na_2S_2O_4$ as an electron donor, as suggested by Bulen, coupled with the ATP-regenerating system, provided a readily available and reliable *in vitro* method for measuring nitrogenase activity. These surrogate systems permitted investigation of the ATP and electron requirements, the H_2 evolution activity of nitrogenase, the discovery of alternative substrates and the purification of nitrogenase.

In the 1970s, highly purified nitrogenase component proteins suitable for protein chemistry and enzymological characterization were obtained from *A, vinelandii, A.* chroococcum, *B. polymyxa, C. pasteurianum, K. pneumoniae* and *R. rubrum.* The investigation on the nature of nitrogenase structure and function advanced rapidly, utilizing such classical tools as enzyme kinetics and protein chemistry, such state-of-the art physical methodologies as electron paramagnetic resonance (EPR), Mossbauer spectroscopy and extended X-rays absorption fine structure.

Nitrogenase requires the functioning of two distinct proteins for the reduction of atmospheric nitrogen. Neither of these component proteins separately displays any activity characteristic of nitrogenase itself. The smaller of the two proteins, dinitrogenase reductase (also called Fe protein having two components) is reduced by an appropriate reductant and binds two molecules of ATP. Dinitrogenase reductase then transfers one electron to dinitrogenase (also called Mo-Fe protein, component 1) with the concomitant hydrolysis of both ATP molecules. The electrons are transferred within dinitrogenase to the various iron-sulfur centers and the iron-molybdenum cofactor (Fe-Mo-Co). The latter is believed to be the substrate-reduction site.Hageman recently proposed a nomenclature for nitrogenase based on a functional role of the component, proteins (dinitrogenase and dinitrogenase reductase), rather than on a physical property or an arbitrary designation. This nomenclature, though not accepted by all, will be utilized here since it does describe the major functional role for these proteins (as far our present understanding permits) and it provides the reader unfamiliar with the field with a way of more easily comprehending the biochemistry of the process.

4.7 Mechanism of Nitrogen Fixation

Nitrogen fixation is a reductive process which requires a strong reducing agent (electron donor) as substrate. The electron donors vary in the different physiological groups of nitrogen fixing organisms.

4.7.1. Pyruvate

It is produced from the fermentation of sugars, is the electron donor for anaerobic bacteria such as *Clostridia*, Pyruvate supports active nitrogen fixation in extracts of *C. pasteurianum, Bacillus polymyxa, Klebsiella pneumoniae, Chromatium* and some cyanobacteria. The first nitrogen fixing extracts were prepared from *C. pasteurianum* with pyruvate as substrate. The fermentation of pyruvate was extremely rapid, and occurred along with nitrogen fixation. A large volume of CO_2 and H_2 accumulated in the reaction vessel. Earlier workers had described a complex enzyme system capable of converting pyruvate to acetyl phosphate with the evolution of CO_2 and H_2. This reaction was termed the phosphoroclastic reaction because inorganic phosphate was the final acetyl acceptor. Strong reductants are produced during pyruvate metabolism. Nitrogen fixation by oxidative cleavage of pyruvate involves a number of steps:

a) Pyruvate is decarboxylated by pyruvate dehydrogenase enzyme containing TPP.

b) Electrons are passed from the reduced (bleached) enzyme to ferredoxin. This reaction yields the dark oxidized form of the enzyme and leuco ferredoxin.

c) Clastic reaction (Fig.4.5): the acetyl moiety of pyruvate is transferred to Coenzyme A (CoA), yielding acetyl CoA.

d) The thioester bond energy of this enzyme is conserved as ATP via the intermediate acetyl phosphate.

e) The phosphoroclastic reaction supplies nitrogenase enzyme with both reduced ferredoxin and ATP. Nitrogenase catalysis the reduction of H_2 to NH_4^+.

f) Excess reduced ferredoxin is converted to molecular hydrogen by an active hydrogenase. The reduction of a variety of substrates is catalyzed by ferredoxin linked reductase.

4.7.2 Formate

It is a strong reducing agent which supports nitrogen fixation in extracts containing ferredoxin linked formate dehydrogenase enzyme. It functions as an electron donor in *C. pasteurianum, B. polymyxa* and *K. pneumoniae.*

4.7.3 Hydrogen

Gaseous hydrogen is also a strong reducing agent. It supports nitrogen fixation in extracts containing ferredoxin linked hydrogenases. The H_2-nitrogenase electron transport chain of *C. pasteurianum* can be reconstructed from pure components; Purified hydrogenase contains 4 atoms each of nonhaeme iron and acid liable sulphur, and is therefore an iron sulphur enzyme. Its MW is 60,000. It is composed of two identical subunits each of MW 30,000.

4.7.4 Krebs cycle metabolites

In strictly aerobic bacteria such as *Azotobacter*, the reducing power for nitrogen fixation is developed during Krebs cycle respiration. Organic acids, alcohols and sugars are utilized. Salts of malic, succinic and lactic acids supports vigoursly growth and acetylene reduction in pure cultures of *Spirillum* and *Azospirillum*sps, *Azotobacter paspali* and *A. lipoferum* can utilize glucose and sucrose as the sole source for growth and nitrogen fixation. By contrast, these two sugars are poor substrates for *A. brasilense.* In aerobes, Krebs cycle is the major route for generating adenosine-5'triphosphate. In *Azospirillum brasilense*, respiration generates adenosine-5'triphosphate, which is required for the operation of nitrogenase. The TCA cycle is also important for generating reducing power.

4.7.5 Pyrimidine nucleotides

In *A. brasilense*, electrons from TCA cycle intermediate are transferred to oxidized NAD. Reduced NAD is one of the sources for N_2 to NH_3 reduction by nitrogenase. In *A. vinelandii*, $NADPH_2$ is the electron donor in nitrogen fixation. The substrates most active in generating $NADPH_2$ in cell free extracts are isocitrate, malate and glucose-6-phosphate. The electrons carriers in this organism are *Azotobacter* ferredoxin and azotoflavin, a flavodoxin. $NADPH_2$ (NADH) support nitrogen fixation in extracts of aerobic, facultative fermentative and photosynthetic bacteria. They are the most important carrier for linking cellular reducing power to nitrogenase.

4.7.6 Photosynthetic Donors

There have been several reports of the coupling of a native photosynthetic ETC to nitrogen fixation in cyanobacteria and photosynthetic bacteria. Nitrogen fixation in legumes depends upon the supply of photosynthate from the host to the bacteroids inside the nodules. In the cyanobacteria, photo synthetically derived electrons drive nitrogenase activity *in vivo*. Electrons from carbon substrates also drive nitrogenase activity.

4.7.7 Electron Carriers

Electron carriers of the ferredoxin and flavodoxin type are the natural reductant of nitrogenase.

4.7.7.1 Ferredoxins

These are iron and sulphide containing proteins which functions as electron carriers, but have no enzymatic functions of their own. Ferredoxin from various nitrogen fixing organisms differ in molecular weight, Fe and Sulphide content, redox properties and biological activity. All ferredoxins isolated from nitrogen fixing organisms react with their homologous nitrogenase. *A. vinelandii* ferredoxin has a 14,500MW and contains 8 atoms of Fe and 8 sulphide residues. It is distinct from other iron sulphur proteins purified previously from the organism. *Rhizobium* ferredoxin has a 9,400MW. It is difficult to purify because of its oxygen sensitivity. Cyanobacteria ferredoxin has about 10,000MW and contains 2 atoms of Fe and 2 of sulphide. It thus resembles the ferredoxin of higher plants. The ferredoxins of *Chlorobium* (green photosynthetic bacteria) and Chromatin (Purple photosynthetic bacteria) resembles those of clostridia. Chromatium ferredoxin has a 9,000 MW and contains 8Fe atoms and 8 sulphide groups. The red bacterium *Rhodospirillum rubrum* is unique in having two types of ferredoxin. Type I Ferredoxin (8,700MW) has 6 Fe and 6 sulphide, while Type II (7,500MW) has 2 Fe and sulphide.

The prosthetic group of ferredoxin may be a dimeric iron sulphur cluster (Fe_2S_2) or a tetrameric iron-sulphur clusters (Fe_4S_4). The sulphur is inorganic or acid labile sulphur, so called because acid treatment liberates it as H_2S during denaturation of the protein (Fig.4.6). It has a valence of 2. An individual protein may contain one or more copies of the basic Fe-S structures. According to one model of the tetrameric structure, 4Fe and 4S atoms occupy alternate corners of a cube. Projecting in a tetrahedral fashion from the iron atoms are 4S atoms of cysteine residues which anchor the iron-sulphur clusters to the protein.

4.7.7.2 Flavonoids

They have been isolated and crystallized from a number of bacteria, including C. *pasteurianum, A. vinelandii* and *Rhodospirillum rubrum*. Their molecular weight ranges from 14,600 to 23,000. The flavodoxin of *A. vinelandii* is called azotoflavin (Fig.4.7). It functions as an electron carrier for nitrogen fixation. The ETC of *A. vinelandii* consists of Azotobacter ferredoxin, insoluble heat labile component, and azoflavin. These components constitute a chain between NADPH and nitrogenase. Flavodoxins do not have metals or liable sulphide. Their chromophore is FMN, and they contain one mole per mole of protein. Flavodoxins are yellow in the oxidized form and blue in partially reduced preparations. They can accept two electrons, each at a different level of oxidation. Flavodoxins have a high percentage of acidic amino acids. The probable binding site of FMN in flavodoxin is cysteine. The active center is probably located in -COOH region of flavodoxin. Most flavodoxins contain 120-150 amino acid residues, corresponding to a molecular weight of about 15,000. Flavodoxins from different species have similar, but never identical, amino acid compositions.

4.7.8 Nitrogenase

Nitrogenase is the central and unique catalyst in the nitrogen fixation process. It binds molecular nitrogen and converts it into free ammonium ion. Nitrogenase is very similar in all organisms from which it has been isolated. It has very broad substrate specificity. The enzyme catalyzes the reduction of several compounds containing triple bonds (Table 4.3).

Table 4.3: Some Compounds Reduced by Nitrogenase Enzyme

Substrate	Structure	Products
Dinitrogen	$N\equiv N$	NH_3
Acetylene:	$HC\equiv CH$	$H_2C= CH_2$
Methylene acetylene	$CH_3\text{-}C\equiv CH$	$CH_3CH= CH_2$
Hydrogen cyanide	$H\text{-}C\equiv N$	$CH_4 + NH_3$
Methyl cyanide	$CH_3\text{-}C\equiv N$	$C_2H_6 +NH_3$
Methyl Isocyanides	$CH_3N\equiv C$	$CH_3NH_2 +CH_4$
Hydrogen azide	$H\text{-}N\text{-}N\equiv N$	$NH_3 +N_2$
Nitrous oxide	$N\equiv N\text{-}O\text{-}$	$NH_3 +H_2O$
Hydrogen ions	H+	H_2
	C_2H_2	C_2H_4

4.7.8.1 Structure of Nitrogenase Components

The nitrogenase system is present only in prokaryotes. It consists of two dissociating protein components (Fig. 4.8).

Component I: it is also called as Enzyme I, molybdoferredoxin, azofermo and Mo-Fe protein, and contains Mo, Fe and S^{2-}

Component II: It is also called as enzyme II, azoferedoxin, azofer and Fe protein, and contains Fe and S^{2-}. Each component by itself has no activity of its own. Combinations of the two inactive components reconstitute nitrogenase activity. The two components interact as a complex *in vivo*, and one component is required for the stability of the other. The two inactive components have similar physical and chemical properties.

4.7.8.1.1 Fe Protein

The Fe protein contains 4 iron and 4 acid labile sulphur atoms. It is a dimer of MW ranging from 57,674 (*Clostridium pasteurianum*) to 73.000 (*Corynebacterium* sp). In *C. pasteurianum*, two subunits of the Fe protein are identical. The dimers contain one Fe_4S_4 cluster. All Fe proteins isolated are dimers with 4 Fe atoms. It is likely that two cysteine thiols from each subunit are liganded to the Fe_4S_4Centre. The Fe protein from all organisms examined has (a) MW from 57,000 to 73,000 (b) 4Fe atoms and 4 liable S atoms, (c) 2 identical subunits and (d) extreme sensitivity to oxygen.The amino acid sequence of the Fe protein from clostridium pasteurianum has been determined. The Fe protein consists of two identical monomers each with 273 amino acid residues. The cysteine residues are in position 37, 82, 94, 129, 181 and 231 from the NH_2 end in each peptide (monomer). The Fe_4S_4 cluster is apparently liganded to two cysteine thiols of each peptide. Other arrangements may, however be possible. The Fe protein does not show any sequence homology with any other iron-sulphur protein. It has the highest content of Gly-Gly (1/39 residues) amongst all proteins examined. Four of the Gly–Gly sequences are predicted to form beta bends. The Fe protein contains several dipeptide and tripeptide sequences. It is noteworthy for the lack of tryptophan; there is an overall negative charge at neutral pH. The isoelectric point is about 4.4. The protein contains 8 regions with 5 or more hydrophobic residues. The reduced protein undergoes different types of conformational charges on binding to MgATP and MgADP. The protein is designed to be flexible.The Fe_4S_4 cluster can exist in four possible oxidation states $(X)^{-4}$, $(X)^{-3}$, $(X)^{-2}$ and $(X)^{-1}$, where (X)= $(Fe_4S_4(Cys)_4)$. The Fe protein supplies electrons one a time to the Mo-Fe protein. This transfer takes place when the Fe protein is complexed with $MgATP^{—}$. The Fe protein like the 4Fe and 8Fe ferredoxins, operates between the $(X)^{-2}$ and $(X)^{-3}$ states. The Fe protein has a midpoint potential of -250 to-295 mV in the absence of $MgATP^{-}$, when completed with $2MgATP^{—}$, the Em is about -400mV. Binding of $MgATP^{2-}$to the Fe protein increases its oxygen sensitivity. The binding of MgADP is greater than that of MgATP.

4.7.8.1.2 Mo-Fe Protein

The Mo-Fe Protein is a tetramer of MW 220,000-245,000. It contains (i) two molybdenum (Mo) atoms, (ii) 28-34 Nonhaeme iron (Fe) atoms, and (iii) 26-28 labile sulphur atoms (S). Fe and S are present in about equal amounts. This suggests that Fe is present in the typical iron-sulphur cluster. The Mo-Fe protein of C. *pasteurianum*, *K. pneumonia* and A. *vinelandii* is composed of two copies each dissimilar subunit ($\alpha_2\beta_2$). The subunits have MW of about 50,000 and 60,000. The two subunits are distinct proteins rather than alternations of one protein. Similarities between the subunits are probably the result of gene duplication that may have occurred during the evolution of nitrogenase. One subunits is code by the *nif K* gene and the other by the *nif D* gene. Each subunit requires the other for stability in vivo. When a mutation in either gene makes its product unstable, the other gene product is also rapidly degraded.The Mo-Fe protein of the bacterial species mentioned above is complexes with:

(i) Up to 4 Fe_4S_4 centers (16 Fe atoms),

(ii) Two centers that appear to contains $MoFe_6S^*_6$ called the *Fe-Mocofactor* or FeMo-Co (12 Fe atoms), and

(iii) Possibly also a Fe_2S_2 center (2Fe atoms).

The Mo-Fe cofactor of Mo-Fe proteins contains 8 Fe atoms and 6 acid-labile sulphides for each atoms of Mo. The Fe-Mo cofactor appears to have a MW of less than 5,000. FeMo-co is an active site of nitrogenase. Carbon monoxide (CO) is a potent inhibitor FeMo-co catalyzed acetylene reduction.

4.7.9 Mechanism of Enzyme Nitrogenase Action

Mechanism of enzyme nitrogenase is completed in following steps as given in Fig.4.9.

(i) In most organisms, the physiologically functional reductant of nitrogenase is *ferredoxin*. Other natural reductant includes *flavodoxin* and *NADPH*. Artificial reductants include $Na_2S_2O_4$ and reduced methylviologen. *Reduced* ferredoxin, the electron donor, reduces the *Fe protein* of nitrogenase.

Fe Protein (oxidized)+> → Fe protein (reduced)

(ii) The reduction of N_2 to NH_4 is exothermic. Yet nitrogen fixation requires energy in the form of 5- triphosphate (ATP), because of high activation energy. The Fe protein of nitrogenase specifically binds to MgATP and lowers it redox potential. A complex containing both Fe and Mo-Fe proteins and MgATP is assembled. In the absence of MgATP the midpoint potential (Em) of Fe protein is about – 250 to 295mV. After biding with MgATP the Em is about – 400mV.

Fe protein + 2MgATP + Mo-Fe protein

Fe protein. 2MgATP. Mo-Fe protein

(iii) The Fe protein transfers an electron to the Mo-Fe protein. This results in the oxidation of the Fe protein and the reduction of the Mo-Fe protein. This reaction is coupled to ATP to ADP hydrolysis. ATP is not hydrolyzed to ATP until the Fe protein transfers an electron to the Mo-Fe protein. There is 12 or more ATP hydrolyzed from each N_2 reduced, or 4 strate. Thus, there appear to be 2 ATPs hydrolyzed for each electron transferred. *In vivo* and *in vitro* ATP requirements are not necessarily the same. Growth yield experiments indicate that in *Azotobacter* only 4 or 5 ATPs are required for each N_2 fixed. On the other hand 29 ATPs are required in *K. pneumoniae* and 20 ATPs in *C. pasteurianum.*

Fe protein 2 MgATP.Mo-Fe Protein →

e^-

Fe protein (oxidized) (2MgADP +Pi). Mo- Fe protein (reduced)

(iv) The reduced Mo-Fe protein can in turn reduce the substrate. A number of substrates other than N_2 can be reduced by nitrogenase. Both Fe and Mo-Fe proteins are required for all these reductions which are coupled to MgATP hydrolysis.

e^-

Fe protein (oxidized) (MgATP). Mo-Fe Protein (reduced) N_2 →
or H^+
or C_2H_2

Fe protein (oxidized) (MgATP). Mo-Fe Protien(oxidized) + NH_3
or H_2
or C_2H_4

4.7.10 Hydrogen Evolution

If no suitable electron acceptor is available, nitrogenase forms molecular hydrogen. Hydrogen is evolved by many nitrogen fixing bacteria during nitrogen fixation. No hydrogen is, however, evolved when they are grown on NH^{4+}, because nitrogenase synthesis is repressed. Nitrogenase itself can be responsible for H_2 evolution. Nitrogenase receives electrons during nitrogen fixation and transfers them to the dinitrogen molecule. Three electrons € associate with each atom of nitrogen. Three protons (H^+) are therefore withdrawn from the intracellular medium to neutralize the negative charge of the electrons. Thus, each dinitrogen molecule yields two molecules of ammonia.

Electron transfer from nitrogenase to molecular nitrogen has a side reaction which is energetically wasteful. In the absence of N_2 and in the presence of ATP and electrons, the electrons combine with protons to yield H_2. ATP is hydrolyzed during hydrogen evolution. Even when gaseous nitrogen is available, some electrons are evolved as H_2, the evolved hydrogen is of no apparent use to the plant or the bacteria. Energy derived from metabolism is used in separating free protons (H_+), the energy is wasted. Under in vivo conditions, 50-65% of the electrons are lost as H_2*nif* mutants are unable to evolve H_2. Carbon monoxide inhibits N_2 fixation but not H_2 evolution. H_2 is a competitive inhibiter of N_2 Fixation.

Some strains of *Rhizobium* carry the *hup* gene in the plasmids (Fig 4.10). Such strains are able to synthesize hydrogen oxidizing uptake hydrogenase. This enzyme converts molecular hydrogen back to electrons and protons, which are then reused by nitrogenase. The increased bacterial efficiency would enable the host plant to direct less energy for supporting the bacteria, and more for increasing seed yield. Wild type *Rhizobium* strains capable of oxidizing H_2 through hydrogenase appear to be more efficient in nitrogen fixation than strains without hydrogenase.

R. japonicum mutants defective in hydrogenase are less efficient that the wild type. Experiments with field grown soybeans inoculated with hydrogenase producing strains of *Rhizobium* give higher seed yields than plant having *Rhizobium* strains which do not synthesize hydrogenase. *K. pneumonia* mutants deficient in hydrogenase activity, but having nitrogenase, give out more H_2 than the wild type. Nitrogen fixation and hydrogen oxidation do not appear to be coregulated. *nif* mutants of *Azotobacter* cannot fix nitrogen, but retain the ability to oxidize hydrogen. Hydrogen uptake also place with NH^{4+} grown cells.Since ATP and electrons are wasted during hydrogen evolution, the process would appear to be harmful to cells. One would except that natural selection would eliminate hydrogen evolving strains having active nitrogenase. It has been suggested that the evolved hydrogen protects nitrogenase from oxygen inhibition. This would account for survival of hydrogen producing nitrogen fixers. However, the strictly anaerobic *C. pasteurianum* also evolves hydrogen from nitrogenase. Differences between Symbiotic and Non-symbiotic nitrogen fixation are given in Table 4.4.

Table 4.4: Differences between Symbiotic and Non-symbiotic Nitrogen Fixation

Sl. No.	Symbiotic Nitrogen fixation	Non-symbiotic Nitrogen Fixation
1.	It is completed with the help of association of bacteria with protein.	It is completed without the help of any association.
2.	Sugar (Glucose) is directly utilized and converted into the phosphogluconic acid.	Sugar (Glucose) is not directly utilized. Firstly, it is converted into the pyruvic acid and then into phosphogluconic acid.
3.	Electron donor is glucose	Electron donor is pyruvic acid.
4.	Glucose is converted into Phosphogluconic acid and gives the electron to NADP, then to Fd, then Fd to NHI and at last ammonia format ions takes place.	Hence, electron comes from pyruvic acid and further process same like symbiotic nitrogen fixation.

4.7.11 Energy Requirements

The nitrogen-nitrogen triple bond is exceedingly strong requiring 941 kJ/mol to completely dissociate the molecule. The large energy requirement for the reduction to the double-bonded intermediate (523 kJ/mol) may explain the difficulty in reducing dinitrogen even though the free energy for the overall reaction is negative.

$$N_2(gas) + 3H_2(gas)\ U \longrightarrow 2\ NH_3(aq),\ 25°C \qquad \Delta G = -53.34\ kJ\ mol^{-1}$$

The energy of drive biological N_2 fixation is supplied by ATP. Since two ATPs are hydrolyzed for each electron transferred from dinitrogenase reductase to dinitrogenase, a minimum of 12 ATPs are needed to reduce N_2. However, for each mole of N^ reduced, a minimum of one mole of H_2 is evolved from nitrogenase. The evolution of H_2 requires the same energy input (two ATPs per electron transferred) as other substrates, thus raising the minimum number of ATPs to 16. Also, the eight low potential electrons (six for N_2; two for H^+) can be assumed to possess energy as they could alternatively be used for oxidative phosphorylation or some other energy yielding reactions. Evans assumed each pair of electrons is equivalent to three ATPs and, therefore, the apparent minimum requirement for N_2 reduction is approximately 28 ATPs.The ATP for biological N_2 fixation must originate from the metabolism of carbon compounds (although photosynthetic microorganisms may derive energy directly from light). Theoretically, 0.11 mol of glucose is needed to produce 1 mol of ammonia. Commonly, in the leguminous symbioses, 12g of glucose must be metabolized to produce the energy needed to reduce 1g of N_2. This results in an overall efficiency of 12%. The efficiencies of the free-living heterotrophs are considerably lower.

4.7.12 Catalytic Mechanism

Currently the most detailed mechanisms of nitrogenase catalysis are those offered by Thorneley. The flow of electrons through nitrogenase is well established; reduced flavodoxin/ ferredoxin transfers an electron to dinitrogenase reductase, then the single electron is transferred to dinitrogenase-with the hydrolysis of two molecules of MgATP to MgADP. The mechanism by which the electrons are transferred or stored within dinitrogenase and the manner by which the electrons are allocated to substrates is poorly understood.The oxidation/ reduction of dinitrogenase reductase can be monitored by EPR, Mossbauer, potentiometery and visible absorption spectroscopy. The Fe_4S_4 center operates between the $[Fe_4S_4(Cys)_4]^{2-}$ state and the $[Fe_4S_4(Cys)_4]^{3-}$ state. The binding of MgATP can be followed by EPR, thiol group reactivity, O_2 sensitivity, chelation susceptibility and, of course, ligand binding methods.

It is known if there is an obligatory in *vivo* order in which dinitrogenase reductase is charged with MgATP and then reduced or if this processes occur independently. The rate of reduction by $Na_2S_2O_4$ of oxidized dinitrogenase reductase in the absence of MgATP is approximately 1000 times faster than the rate observed during catalytic turnover. Thus, dissociation of the MgADP-oxidized dinitrogenase reductase complex may greatly affect the rate of re-reduction. However, $Na_2S_2O_4$, which is normally utilized in these investigations, is not a very effective reductant for dinitrogenase reductase and the use of the natural election donors may yield different results. Theoretically, reduced dinitrogenase free in solution may transfer electrons to a molecule of oxidized dinitrogenase reductase that is complexed to dinitrogenase. The reduced, MgATP-complexed dinitrogenase reductase rapidly associates with dinitrogenase. The rate constant for this association has been estimated at greater than $10^7 m^3 mol^{-1} s^{-1}$ with a dissociation constant of the order of 0.5 μmol^{-3}. Ratios of dinitrogenase reductase to dinitrogenase of both 2 to 1 and 1 to 1 have been reported as optimal for nitrogenase activity. Hageman reported that a 1:1 ratio of dinitrogenase reductase: dinitrogenase produced full activity of the ATP hydrolysis and electron transfer reactions for *A. vinelandii* nitrogenase. A report on a heterologous nitrogenase complex (*A. vinelandii* dinitrogenase plus *C. pasteurianum* dinitrogenase reductase) indicates that *A. vinelandii* dinitrogenase does have two binding sites for dinitrogenase reductase. *C. pasteurianum* nitrogenase requires a 2 to1 ratio of dinitrogenase reductase to dinitrogenase for full catalytic activity. These differences in binding ratios may reflect suitable distinctions in the electron transfer reactions between various nitrogenases. Although nitrogenase is frequently referred to as a complex, there is little evidence to suggest a given dinitrogenase reductase molecule remains bound to a particular dinitrogenase molecule for a complete catalytic cycle (in terms of substrate reduction). Recently,

Hageman have presented persuasive evidence that during proton reduction dinitrogenase reductase and dinitrogenase associate and dissociate with each electron transfer event. Mortenson and Thorneley suggest that perhaps for acetylene or N_2 reduction a longer-lived complex may be necessary.

4.7.13 Inhibitors

4.7.13.1 Classical Inhibitors

Classical inhibitors are those compounds that also serve as substrates for nitrogenase. These inhibitors can be classified into five different groups depending upon their inhibition patterns *versus* the other substrates, (i) H_2, N_2O and cyclopropene: these compounds are competitive *versus* N_2. (ii) CN, N_3 and CH_3NC: these three are mutually competitive but non-competitive versus N_2. (iii) C_2H_2: acetylene is non- competitive versus N_2 and N_3. N_2 is competitive versus acetylene, (iv) CO; carbon monoxide is competitive with all substrates but is unable to block H^4 reduction, (v) H_2 evolution: H^+ reduction is not inhibited by the presence of H_2 nor blocked by CO. Originally these results were interpreted as five different sites or perhaps modification of a single site. There is no active site data to support this interpretation and these results may simply reflect differences in the allocation of electrons to substrates.Several new substrates have been investigated to pursue the conformation of the substrate reducing active sites(s) of dinitrogenase. McKenna have synthesized and utilized cyclopropene and 3,3-difluorocyclopropene as substrates and active site probes, 3,3-Difluorocyclopropene inhibits H_2 evolution, acetylene reduction and N_2 reduction. Diazurine (a strained-ring diazene analog) is a substrate and also an inhibitor of acetylene reduction. Dilworth and Thorneley reported that N_2, CO and N_2O inhibited the reduction of azide to hydrazine. H_2 was not an inhibitor of hydrazine formation from azide.

4.7.13.2 Regulatory Inhibitors

The regulation of nitrogenase is governed by energy, nitrogenous compounds and oxygen. Normally, nitrogen fixation activity is expressed when it is required for the growth of organisms. Once the nitrogen fixing system has been genetically turned on, it can either genetically or biochemically turned off. In this section we will deal only with the biochemical considerations.

4.7.13.3 Energy

Nitrogen fixation imposes a considerable energy drain on the energy metabolism of an organism. Nitrogenase requires ATP for catalysis and is inhibited by ADP. The ratio of ADP/ATP affects electron transfer and thus allocation of electrons to substrates as well as total activity. *In vivo,* N_2-fixing cells have

ADP/ATP ratios of 0.3-0.5, whereas organisms under non-N_2-fixing conditions have ratios of 0.8-0.9. The *invivo* data, if considered exclusively, would imply nitrogenase is 80% inhibited under normal conditions. Mortenson and Upchurch (1981) utilized p/2e ratios as a measure of the efficiency of nitrogenase m *vitro* and found that the best apparent efficiency is at an ADP/ATP ratio of 0.3-0.5. Thus many other factors in addition to ADP/ATP ratios contribute to optimize rates of N_2 fixation *in vivo.*

4.7.13.4 Ammonia

The effect of added nitrogenous compounds, particularly ammonia, to cultures of actively fixing cultures is perhaps the most often studied aspect of nitrogenase regulation. Addition of ammonia to the culture media causes a rapid switch-off of nitrogenase, simple repression of nitrogenase synthesis or a combination of these effects. The observed effects depend upon, the organism, the carbon and fixed nitrogen sources and degree of membrane energization, The mechanism by which ammonia represses nitrogenase synthesis is not known. However, during depression of *K. pneumoniae,* the m/mRNA possesses a half-life of about 20 min. After the addition of ammonia to a depressed culture, the half-life of these mRNAs was reduced to approximately 9min. apparently, there is no need for post-transcriptional control of nitrogenase by ammonia in *K. pneumoniae.*

4.7.13.5 Covalent Modification

The nitrogenase of the photosynthetic bacteria *Rhodospirillum rubrum, Rhodopsuedomonas capsulata* and *R. palustris* exist in one of the two different forms are actually interchangeable by the addition or removal of a group covalently bound to dinitrogenase reductase containing phosphate, a sugar (presumably ribose) and an adenine-like molecule. Cultures grown on glutamate or N_2 are poised so that modification occurs during harvesting and express little or no nitrogenase activity upon isolation. Cultures grown under NH_4^+ limited conditions lack this potential and possess active cell-free N_2-fixing preparations. The inactive dinitrogenase reductase can be converted into the active form by removal of the covalently linked adenine-like molecule. The deactivation is not well understood, but during this process ribose, phosphate and the adenine-like molecule are attached to dinitrogenase reductase. The inactive form of dinitrogenase reductase is incapable of transferring electrons to dinitrogenase. Glutamine synthetase may play a role in the interconversions of dinitrogenase reductase.

4.7.13.6 Oxygen

In vitro, oxygen irreversibly inactivates both nitrogenase components. The half-lives of dinitrogenase and dinitrogenase reductase in air are approximately

10 min and less than one min, respectively. *In vivo,* it is not known if O_2-denatured nitrogenase proteins can be reactivated or must be completely re synthesized. Based on investigations of putaredoxin and other Fe-S proteins, O_2 denaturation of the nitrogenase proteins is most probably due to the oxidation of the labile sulfide of the Fe_4S_4 clusters and/or FeMo cofactor to the zero oxidation state.The physiology of each nitrogen-fixing organism must have the capacity to maintain the activity of the O_2~sensitive nitrogenase proteins as well as provide adequate sources for energy and reducing equivalents. The physiology of each nitrogenfixing organism must be compatible with its respective ecological niches. For example, the *Azotobacter* possesses a unique mechanism to protect nitrogenase components during O_2 stress. A complex is formed between Fe_2S_2 protein and dinitrogenase reductase. This Fe_2S_2 protein has a molar mass of $14x10^3$ daltons in *A. chroococcum* and $23x10^3$ in *A. vinelandii.* Distress may initially oxidize the Fe_2S_2 protein, thereby increasing its affinity for dinitrogenase reductase. Reduction of the Fe_2S_2 protein (removal of O_2 stress) causes dissociation of the complex and restores nitrogenase activity.Conversely, sub-optimal partial pressures of O_2 can severely affect nitrogenase activity in those organisms which generate ATP via oxidative phosphorylation. Low extensions reduce ATP levels and thus lower the nitrogenase activity.

4.7.13.7 Ammonia Assimilation

Ammonia, the first stable product of nitrogen fixation, must be assimilated for transport to symbionts and/or conversion into proteins. Classical enzyme studies, ^{15}N and more recently ^{13}N have been employed to trace the assimilatory pathways of N_2-fixing organisms. Several reviews appeared recently describing the ammonia assimilation process in prokaryotes plants and legume nodules. There are two possible primary routes for the assimilation of ammonia into amino acids viz. (a) glutamate dehydrogenase and (b) glutamine synthetase-glutamate synthase. Glutamate dehydrogenase provides the major assimilatory pathway when ammonia is abundant. However, during nitrogen fixation when nitrogen-limiting conditions prevail, it operates predominately in. a catabolic mode rather than a synthetic one. Glutamate dehydrogenase has a high Michaelis constant for ammonia and furthermore the specific activities in crude extracts of nitrogen limited cells are usually quite low. Conversely, less have reported the kinetic parameters for soybean and lupine glutamate dehydrogenase, respectively, and suggest that this enzyme may have a more important role in ammonia assimilation than implied by the Michaelis constant for ammonia. Other amino acid dehydrogenase activities, such as alanine or aspartate, are too low to be of physiological importance.

The glutamine synthetase-glutamate synthase pathway is the major assimilatory pathway under nitrogenfixing conditions in prokaryotes. The

Michaelis constants for all the substrates are at physiological levels and specific activities can account for the observed rates of ammonia assimilation. This pathway requires ATP in addition to reductant and thus may represent a mechanism for regulatory control in free-living organisms.Once glutamine and glutamate are formed, all the other amino acids and other nitrogenous compounds can be formed by transamination or similar processes. However, in some symbiotic associations more complex pathways are required. In nodulated leguminous plants, the ammonia produced *via* nitrogen fixation within the *Rhizobium* bacteroids is excreted into the plant cytosol. Although two glutamine synthetases have been reported in free-living cultures of rhizobia only one has been reported within bacteroids. These evidences suggest glutamine synthetase of *Rhizobium* bacteroids performs an insignificant role in symbiotic ammonia assimilation. Glutamine synthetase of the plant cytosol is the major pathway of ammonia assimilation in nodulated leguminous plants. In certain leguminous plant species such as lupine, the principal form of nitrogen assimilates that are translocated to other plant tissues are amides. In other leguminous species, for example soybeans, allantoin and allantoic acid, commonly referred to as the ureides, are the major transport forms of nitrogen. The ureides are synthesized entirely in the plant host cell. Schubert has shown that the synthesis of the purines takes place entirely within plant cytosol, whereas the catabolism of the purines involves sequential steps within the cytosol, peroxisome and endoplasmic reticulum.

4.8 Genetics

Until the 1970s, the molecular genetics of N_2 fixation was essentially unexplored. Earlier progress was confined to the isolation and biochemical characterization of mutants unable to use N_2 as a nitrogen source. The primary which had received the most biochemical attention viz. *Clostridium* and *Azotobacter.* Because many of the genetic tools available for *E. coli* could be applied to *K. pneumoniae,* it became clear that this nitrogen-fixing cousin of *E. coli* offered the greatest advantages for genetic analysis. As a result of this recognition, two major advances were made. First, Streicher reported on the intra strain transfer of *K. pneumoniae* genes for nitrogen fixation *(nif)* by bacteriophage PI and showed their linkage to genes for histidine biosynthesis. Almost simultaneously, Dixon demonstrated that genes for nitrogen fixation could be transferred from *K. pneumoniae* to *E. coli* by conjugation and, amazingly, expressed there to form a functional nitrogenase S3^{r}stem. These advances established the approximate chromosomal location of *nif,* identified an easily-selected, linked marker and demonstrated that some *nif* genes were clustered on the .chromosome. These successes, a political climate favorable to the support of this research and major advances in molecular technology, provided the basis for the exponential acquisition of genetic and molecular information which has taken place in the ensuing decade.The genetic tools and current results of analysis will be outlined for several types of diazotrophs after a general discussion of the available approaches and methods.

4.8.1 Approaches and Techniques Available

An excellent review of the techniques used to explore the *nif* genes of *K. pneumoniae* has recently been published. Because it is written for use of the laboratory and is part of a more comprehensive methods volume, it is certainly a critical reference for those working in the area of nitrogen fixation. Here we will attempt to point out approaches which can be used with diazotrophs not as genetically accessible as *K. pneumoniae* and initially confine our remarks to free-living organisms.First, a bank of independent mutations blocking the ability to fix dinitrogen can be generated. Although NTG (N-methyl-N-nitro-N-nitrosoguanidine), a powerful mutagen, is often used, it readily results in multiple lesions which may complicate the interpretation of pleiotropic phenotypes. Other mutagens such as alkylating agents, hydroxylamine, nitrous acid or UV may offer fewer, problems. Because all strains are not equally sensitive to these agents, a dose-response curve should be established for each before use.After mutagenesis of an appropriately genetically marked strain, *nif* mutants are isolated as those unable to grow with N_2 but able still to grow well with ammonium, salts. Next some phenotypic characterization of the *nif* mutant strains can be made. If the wild-type phenotype(s) is restored spontaneously at a frequency of 10^{-9} or higher, the original lesion is likely to be a point mutation, although the possibility of suppression should be considered. Generally loss of a plasmid, deletions or multiple mutations does not revert at these frequencies. In addition, any pleiotropic effects of the mutations should be examined, *e.g.* altered growth rates on nitrogen sources other than N_3.

A biochemical and physical analysis of the m/components of the mutants can proceed without regard to the genetic capabilities of the organism. However, the establishment of the number of genes represented among the mutations greatly decreases the amount of biochemistry which must be done. Residual *in vivo* and *in vitro* nitrogenase activity should be measured to determine suitability for further biochemical or genetic studies. Furthermore, if activity is present *in vitro* but not *in vivo,* a block in endogenous electron flow to the nitrogenase complex is likely. Additional functional characterization can be made if antibodies to purified protein components of the nitrogenase complex are available. Purified components can sometimes to be used to restore activity to inactive extracts of mutants, thus identifying the altered protein.Because the three polypeptides of the dinitrogenase and the dinitrogenase reductase can be easily visualized by a comparison of the protein patterns of repressed and depressed cells on denaturing polyacrylamide gels (SDS-PAGE), a rapid screen of mutantsrepresentative of each linkage group or gene will assess the gross integrity of the nitrogenase complex. Additional *nif*- specific polypeptides can be identified by pulse-labeling proteins during depression of nitrogenase, separation by SDS-PAGE, followed by autoradiography. Finally, as a further refinement, pulselabelled extracts can be subjected to two-dimensional gel electrophoresis and subsequent autoradiography. These experiments can give gene-protein relationships as well as operon structure when used to analyze the appropriate mutations.

If a genetic exchange mechanism such as transduction, transformation or conjugation is available, mapping can proceed. For Gram (-) organisms lacking indigenous transferrable plasmids, conjugation has often been obtained with the promiscuous drug resistance plasmids of the PI incompatibility group. Two factor crosses of nif^{1}x nif^{2}can be made to determine the relative linkage of *nif* lesions. This procedure, called the ratio test, can be performed using an unlinked marker to standardize the results of each cross. Thus the closer two mutations are in the chromosome, the fewer prototrophic recombinants will be obtained relative the number of recombinants for the unlinked marker. If a linked marker is available, such as his D in *K. pneumoniae,* three point crosses can be carried out in a similar manner. It should be remembered that the size of the DNA transferred, *e.g.* the capacity of the phage head in transduction, determines the physical limits for establishing linkage between markers. In addition, fine structure mapping requires a relatively small genetic vector. If a linked marker is not available for mapping purpose, it is possible to generate such a mutation by the use of NTG. A *nif*strain is mutagenized with NTG and plated for *nif* revertants. Most of these revertants will have been the result of the action of NTG and, therefore, will have a high probability of having an additional mutation(s) nearby. Temperature sensitive mutations can be isolated and used, even without knowledge of the genes altered, or specific phenotypic alterations can be sought.

To investigate the number of genes represented by the mutational clusters, complementation analysis is required. The complementation test simply determines whether two mutations affect the same gene. For this analysis it must be possible to establish a relatively stable merodiploid(a strain with two copies of the chromosomal segment covering the mutations). This is generally brought about by the introduction (conjugation or transformation} of a plasmid containing the appropriate region of DNA into a recombination-deficient recipient strain (Rec^{+}). Although complementation has been successfully performed in a Rec^{+} background, an absolute interpretation of results is questionable.

The construction of a plasmid containing may be accomplished either in *vivo* or in vitro. To form an F-prime or an R-prime in *vivo* carrying *nif*, conjugation into a *nif*⁻ Rec⁻ recipient can be carried out and a *nif*⁺ recipient selected. Since recombination is prevented, the incoming chromosomal DNA can be stably maintained only if it is a part of the plasmid replicon. Other ways of restricting recombination are to use a recipient with a large deletion of the area coding for *nif* proteins, or to use another, closely related species as recipient so that DNA homology is low but one in which fit/expression can be monitored.

In vitro construction of plasmids carrying *nif* genes can be carried out using recombinant DNA techniques, which have been adequately reviewed elsewhere. The highly conserved sequence of *nif* H and *nif* D genes among diazotrophs allows the identification of nitrogenase structural genes from almost any species with an appropriately labelled probe (pSA30 from *K. pneumoniae*).Once a plasmid containing the *nif* genes has been obtained, mutations can be introduced either by mutagenesis of the plasmid, by homogenization with chromosomal mutations or by cotransduction of *nif* lesions with a selectable marker into the plasmid. Following the derivation of a bank of plasmid and chromosomal mutations, all pair-wise crosses can be constructed and the *nif* phenotypes observed. Information concerning operon structure can be obtained from insertion mutations induced by transposable genetic elements such as drug resistance transposons or bacteriophage Mu. These insertions have been shown to produce strongly polar effects on genes operatordistal to the point of the insertion. Therefore, non-complementarity between an insertion mutation and point mutations in several clustered genes suggests an operon or transcriptional unit. Another advantage of insertion mutation is their, usefulness in the construction of deletions. Since these elements excise imprecisely, deletions are often observed among strains selected for loss of the insert.

A third type of genetic insertion resulting in the fusion of the gene for β-galactosidse to the operator-promoter region of the gene of interest has proved to be especially valuable in studying *nif* regulation. Because of the extreme O_2 sensitivity of some *nif* products and the lack of discriminatory assay systems for most of these products, observations of the control of expression are quite difficult. By fusing the gene for â-galactosidase to the control region of the various m/operons, fluctuations in expression are readily followed. The limitation to this procedure, as for the other transposable elements, is to obtain a suitable vector for introducing the element into the organism of choice. Unless the bacterium of interest is sensitive to or μ bacteriophages, a conjugable plasmid or transformation must be used which may have limited efficiency.

4.8.2 Nif genes in ***K. pneumoniae***

The results of numerous investigations aimed at the elucidation of the number, arrangement and control of the genes essential for nitrogen fixation by

K. pneumoniae have been reviewed well and frequently in the literature. Here,we will attempt to summarize briefly the information with updates. Fig.4.11 the order, operon organization, direction of transcription and approximate molar mass of the produces of the genes presently identified in the *nif* regulon of *K. pneumoniae.*

hts Q B A L F (W) M V S U X NE Y K D H J shi A

? ? 55 45 23 ? 28 42 45 22 18 50 46 19 60 56 32 120 (kd)

Fig. 4.11 : Order, operon organization, direction of transcription and approximate molar mass of the products of the *nif* regulon genes of *K. pneumoniae*

The order was established relative to *hisD by* three-point crosses, deletion mapping and physical analysis with cloned fragment. The additional use of transposon mutagenesis of the cloned fragments has allowed a more detailed physical map to be established. It is now apparent that the *nif* regulon is contained in a 23kb segment and that all the essential genes in this segment have been identified.The number of operons and the direction of transcription have been identified by complementation between insertion mutations and point mutations. This data suggest seven or eight transcriptional units read in the leftward direction toward the genes for histidine biosynthesis.

Although the protein products of only five of the *nif genes* have been purified to homogeneity, the functions of most are known but not the actual enzymatic reactions catalyzed.The dinitrogenase reductase and the dinitrogenase, which have been purified, are coded by *nif* Hand *nif KD,* respectively.The protein products of both genes involved in electron transport to the nitrogenase complex, *nif F* and *nif J,* have been purified, nif codes for a flavoprotein which is essential for physiological electron transport. In contrast, the J protein is a dimer of *ca.* 245, 000 daltons containing 30mol Fe and 24mol labile S/mol protein and is probably an oxidoreductase. Three of the genes whose products have not been purified are involved in the FeMo cofactor synthesis, *nif B, nif N* and *nif E.* Because extracts of strains with mutations in m/M which have inactive nitrogenase can be restored to activity by the addition of dinitrogenase reductase, it appears that M protein is involved in a post-transcriptional modification of the product of *nif H.* A recent analysis of mutations in *nif V* indicated that the V product may also be involved in post-transcriptional modification but of the dinitrogenase protein. Although dinitrogenase in a *nif* V background reduces some substrates, it is not capable of reducing N_2 with physiological electron fluxes. Two other gene products, those of m/S and *nif U,* also appear to be involved -in the maturation of dinitrogenase since strains containing Mu insertions in S or Fe/were shown to lack normal levels of this protein.

The last two genes for which functions are known are those comprising the *nif LA* operon. The *nif A* product is required for the expression of all *nif* genes except for its own operon. In contrast, the *nif L* product is not essential for *nif* expression and acts as a repressor for all transcripts (except its own operon) in response to O_2 NH_4^+ and temperature.The remaining genes, *nif Q, nif X, nif* T and *nif* W have not been assigned a specific function. The gene designed *nif Q* was inferred from the observation that a deletion from *his D* which recombined with all known nif mutations resulted in a leaky *nif* phenotype. In addition, the strain containing this deletion was not dramatically altered in acetylene reduction, suggesting that protein may influence substrate selection. By analysis of the protein products of cloned, fragments of the m/regulon, genes X and Y were identified. These two genes have not been shown to be necessary for nitrogen fixation.Only limited analysis has been carried out on *nif W.* It was assigned on the basis of reversion of a Mu induced mutation which was polar on to *nif F* and may represent an operator proximal region of *nif* F that is non-essential. Finally a mutation between *nif J* and *nif H* which appeared to complement a *nif J* mutation was assigned the designation *nif C.* Additional mapping studies have demonstrated that this mutation lies between well characterized *nif J* lesions and that intracistronic complementation occurs which confused the interpretation of earlier results.In summary, 18 to 19 gene assignments have been made in the m/regulon. All the *nif* W been shown to produce a protein product, 15 by genetic experiments and two, to three by examination of cloned fragments. It is clear that the commitment to nitrogen fixation is a major investment for a bacterium.

4.8.3 *Regulation of nif Gene*

A large number of environmental factors are involved in the regulation of expression of nitrogen fixing activity, among which are ammonia concentration, dissolved oxygen tension, presence of amino acids, availability of molybdenum and temperature. It appears that control may operate at transcription, the stability of transcripts and the stability of the enzyme complex. In most cases, evidence for regulatory effectors still rests with the physiological descriptions rather than detailed molecular mechanisms. Ammonia, the end product of nitrogen fixation, is assimilated by the successive action of glutamine synthetase (GS) and glutamate synthase. As a result, the investigations of the mechanism of ammonia repression of nitrogenase have focused on these enzymes. Because the levels of GS and the state of its covalent: modification was seen to fluctuate in response to the supply of fixed nitrogen available to the cell, attention began to focus on the importance of this enzyme in a generalized nitrogen control system. Concomitant with the changes in GS activity and content, changes in the levels of other enzymes capable of supplying NH% or glutamate to the cell were also observed and a cause and effect relationship was proposed. The description of two types of mutants mapping in the region of the structural gene for GS, **(i)** those leading to glutamine auxotrophy, Gln^-, which also resulted in a *nif* phenotype and **(ii)** those designated Gln which were constitutive for *nif* expression in the presence of NH^+_4, confirmed the involvement of the assimilatory system. The model which grew out of these studies suggested that the activation of the synthesis of the enzymes of nitrogen fixation and nitrogen metabolism required an increase in the cellular level of GS and its conversion to the unadenylylated form, which could then function as a positive regulatory element for transcription.Since this model was proposed, there has been an accumulation of evidence showing that there is no apparent correlation between the adenylylation state of GS and the level of other enzymes of nitrogen metabolism. The discovery of additional genes involved in nitrogen regulation tightly linked to gln A has made a re-evaluation of nitrogen regulation necessary. Genetic and physical data presently support the model proposed by McFarland for *Salmonella* and elaborated on by Merrick as illustrated in Fig 4.11.

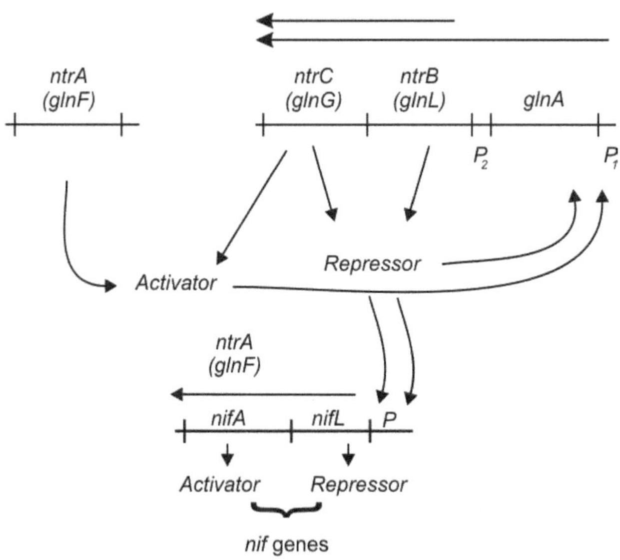

Fig. 4.11 : Model for Regulation of Nitrogen Metabolism

(Gene designations in parentheses are those used in *E. coli*. The *ntr* designation derives from nitrogen regulation indicate transcriptional units and directionality-indicate regulatory functions)

Loss of *ntr A* or *ntr C* results in loss of ability to express nitrogen controlled genes at high levels. Thus, when nitrogen is limiting, *ntrA* protein or an enzymatic product of that protein interacts with *ntr C* protein to form an activator. The *ntr C* product also can interact with the *ntr B* protein form a repressor when ammonia levels are high. As a consequence of the loss of *ntr C,* both repression and activation of *gln A* are eliminated resulting in a low *glnA* expression insensitive to nitrogen availability. In contrast, mutations in *ntrB* allow a high constitutive level of expression *of glnA.*This model now appears to hold several enteric bacteria. In *E. coli* and *K. pneumonias the* P_{11} protein, a component of the adenylylation system for glutamine synthetase, has been reported to be a corepressor capable of interacting with the *ntrBC* complex. Mutants incapable of converting P_{11} to its uridylylated form in response to low levels of ammonia were unable to activate nitrogen assimilation genes including nitrogenase. Mutations occurring within the *nif* regulon which render the expression of that regulon insensitive to fixed nitrogen are at the operator-promoter site of *nifLA* or within the *nifLA* operon. These results suggest that the transcription of *nifLA* is controlled by the general nitrogen-regulatory system and the products of that operon in turn regulate the remaining nif operons. Recent experiments have pointed out the remarkable similarities between the *nifA* and *ntr C*protein products. Both are approximately 55 000 daltons, have similar isoelectric points, and require a functional *ntrA* gene for regulatory activity, Both have been shown to be capable of activity with the *nifLA* operon, *glnA ntrB ntrC* operon, and the *Rhizobium melioti nifH* promoter in addition to those genes governing the catabolism of certain amino acids. These similarities have led to the suggestion that the two genes are evolutionarily related. Significantly the *K .pnemoniae nif*H prompter is responsive only to *nifa* protein. As yet the physiological meaning of the ability of *nifA* protein to substitute for *ntrC* protein is unclear. Perhaps it is simply a sparing effect for *ntrC* or the vestiges of the evolutionary antecedent.The *nifL*product which is the Nif regulon repressor affords a more immediate response to fixed nitrogen than is possible through nitrogen regulation. Since *nif L*and *nifA* coding for the activator protein are transcribed as a single operon, the *nifL* protein must be maintained in an inactive state during active nitrogen fixation and be converted into a repressor state by molecular signals sensing fixed nitrogen or O_2.

Synthesis of most polypeptides in *K. pneumoniae* has been shown to be temperature sensitive. It is as yet unresolved whether A protein and/or L protein are responsible for thermosensitivity.

The mechanisms involved in the control mediated by amino acids or by nitrate remain to be elucidated. However, both repressive effectors are still functional in mutants of *K. pneumoniae* derepressed for nitrogenase biosynthesis in the presence of ammonium.The involvement of molybdenum (Mo) in the regulation of biosynthesis of nitrogenase in *K. pneumoniae* has been brought into question with results from more recent studies. Under Mo deprivation some dinitrogenase apoprotein is made and can be activated by the addition of molybdate to the cells. Thus Mo most not be essential for *nif* polypeptide synthesis in this bacterium. Only in *A. vinelandii* does Mo appear to be required for conventional dinitrogenase synthesis.A from of control at the level of mRNA destabilization operates in *Klebsiella* for m/expression. Under nitrogen fixing conditions, nif mRNA appears to be remarkably stable with a half-life of 18 min or longer. When these diazotrophs were shifted to repressive conditions, the half-life became significantly shorter. However after addition of rifampicin plus NH_4^+, mRNA hybridization experiments suggested that *nif* mRNA was detectable well after *nif* protein synthesis was terminated. This result would suggest that a translational control was superimposed on the transcriptional regulation and mRNA stability in *Klebsiella* orthat a specific nuclease was required to shorten the half-life of *nif* mRNA upon repression.

4.8.4 Azotobacter Species

The members of the Gram-negative *Azotobacter* genus are nitrogen-fixing obligate aerobes. They are widely distributed in soil and some species have been found' associated with the rhizosphere of tropical grasses. There are reports of increased plant growth following inoculation with *Azotobacter,* possibly due to the production of plant growth hormones. Because of the potential for coupling ammonia excretion from N_2 reduction with plant-hormone production in an organism that has an associative growth mode, this genus deserves considerable attention. Although *A. vinelandii has* been reported to have 10 times the amount of DNA that *E. coli* contains, mutants lacking the ability to fix nitrogen were obtained rather readily. While purine and pyrimidine auxotrophs of *A. vinelandii* have been isolated, amino acid auxotrophs have not been reported even though considerable effort has been made to obtain them. Therefore, a very limited array of genetic markers is presently available for extensive mapping efforts in this genus.Several *nif* mutations have been characterized with respect to activities for dinitrogenase and dinitrogenase reductase, antigenic cross-reacting material (CRM) and electron paramagnetic resonance signals. Among the mutants described were all the predictable classes for a system of two proteins which could be assayed separately. Two additional classes of special interest were found: one which was *nif* but hyper produced the reductase and one which produced nitrogenase in the presence of ammonia.

Although a transformation system was described 10 years earlier, a reliable procedure was not established until 1976. Refinements of this procedure have been published. The plate-transformation system was used for the determination of a rough linkage map for several *nif* mutant strains. Because the size of DNA transferred was not determined and congression (simultaneous transfer of markers on separate fragments of DNA) was relatively high, no physical interpretation of distance could be made from the ratio test crosses. However, the results did demonstrate that the *nif* genes in *A. vinelandii* not fall into one cluster. Subsequent DNA-DNA hybridization studies with cloned *K. pneumoniae nif* genes have shown that the highly-conserved structural genes for nitrogenase *nif H* and *nif D* do occur on a single, small fragment.

Regulation of the nitrogen-fixing complex of *Azotobacter* is similar to other free-living diazotrophs in being ammonia with the involvement of the ammonia assimilatory system. Two classes of regulatory lesions have been described as (a) presumed point mutations resulting in the loss of activity and CRM for both nitrogenase proteins, as well as (b) a mutation overproducing the dinitrogenase reductase and simultaneously not producing dinitrogenase. The existence of these phenotypes has been interpreted to mean that a common regulatory gene is required for expression of the structural genes for the nitrogenase complex.Surprisingly, an extended analysis of *A. vinelandii nif* mutants which included two-dimensional PAGE of the proteins of several *nif*$^+$ revertants revealed that the 'conventional' dinitrogenase polypeptides were missing and four additional, ammonia-repressible proteins were present. Thus the *nif*$^+$ revertants were pseudorevertants. Further investigation has led to the conclusion that this diazotroph may possess an alternative nitrogen fixing system which is Mo repressed. While the description of this alternative system awaits further confirmation, the survival potential for a bacterium forced to fix nitrogen in an environment depleted of Mo is clear. The biochemistry of N_2 reduction without the involvement of Mo should prove extremely interesting.

4.8.5 *Cyanobacteria*

The cyanobacteria can be roughly divided into three main groups, the unicellular, filamentous non-heterocystous and filamentous heterocystous forms. Although nitrogen fixing species occur in all classes, all the filamentous heterocystous species have been shown to possess this capacity. The heterocyst is the differentiated cell that functions as an anaerobic site specifically designed for N_2 reduction and release of fixed nitrogen to vegetative cells.The only reliable and efficient genetic exchange system for the cyanobacteria to date has been developed for *Anacystis nidulans,* a non-nitrogen fixing species. The consequences of the absence of a genetic exchange system in the diazotrophic cyanobacteria was brought into clear focus by their omission from a recent review of the genetics of nitrogen fixation by Roberts. However, progress is being made towards genetic analysis through the generation of defined mutations and cloning technology.Although a few mutant strains had been isolated in several species of filamentous cyanobacteria, until the work of Currier there had not been a concerted effort to obtain well characterized mutants in one species. Now a bank of mutant strains has been isolated in *A. variabilis,* including auxotrophs and those altered in nitrogen fixation and heterocyst development. To accomplish mutant isolation from a filamentous organism, it was necessary to separate the rare mutant cell from the filament. To rupture ceils randomly and fragment the filaments, cavitation was used and the short filaments or single cells were then subjected to penicillin enrichment, and selection for auxotrophs.

The strongly conserved nature of the structural genes for nitrogenase, *nif H* and *D* was demonstrated by heterologous hybridization of *K. pneumoniae* m/genes carried on the plasmid pSA 30 with fragments of *Anabaena* DNA, In addition, homology between *nif* genes other than the structural genes for the nitrogenase was reported for DNA from *Anabaena* and *K. pneumoniae* which may be a unique feature among the diazotrophs. These studies have indicated that the *nif* structural genes in *Anabaena* are rearranged from the order found in *K. pneumoniae* and are probably separated on the chromosome.During the course of the hybridization experiments, what appears to be a second copy of the *nifH* gene was found. Because the restriction pattern was different from that of the *nif* H gene located near *nifD,* it is known whether this is a second copy of *nifH,* a nonfunctional pseudogene, or another gene which by chance has homology. A role for such a gene in a hypothetical alternative nitrogen-fixing system such as that found in *A. vinelandii was* also suggested.The regulation of the capacity for nitrogen fixation by cyanobacteria differs from other free-living diazotrophs in a few areas and is discussed in several reviews. Ammonia repression (and probably O_2 repression) operates in the filamentous cyanobacteria. Supporting the involvement of a generalized nitrogen regulatory system in nitrogenase control were results which showed that the GS inhibitor MSX (methionine~SR-sulfoximine) caused a relief of inhibition of heterocyst formation by exogenous ammonia and the excretion of newly fixed ammonia from cells of N_2-fixing *Anabaena cylindrica.* The majority of the nitrogenase in filamentous cyanobacteria occurs within the heterocysts of a photo synthetically active culture. The elegant experiments of Wolk and coworkers have demonstrated that ammonia generated by nitrogenase in *A. cylindrica* is assimilated *via* the glutamine synthetase-glutamate synthase enzyme couple. Therefore, as might have been expected, the GS activity, which is the result of a single enzyme in *Anabaena,* is slightly higher in heterocysts, although the glutamate synthase activity is lower as compared to vegetative cells. These findings support the contention that glutamine is produced from newly fixed N_2 within the heterocyst and exported to the vegetative cell where glutamate is made.The GS in the cyanobacteria does not enjoy large fluctuations in activity in response to nitrogen source nor is it covalently modified, In addition, the excretion of ammonia from *Anabaena azollae* when in association with the water fern *Azolla caroliniana* has been shown to be the result of the absence of GS activity and antigen from the cyanobacterium.

4.8.6 Photosynthetic Bacteria

The purple, non-sulfur phototrophs have been most intensively studied among the free-living photosynthetic bacteria, excluding the cyanobacteria. Within the Rhodospirillaceae, the nitrogenase system of *R. rubrum* has received most biochemical attention, while genetic transfer systems have been established in *R. capsulata* and *Rhodopseudomonas sphaerodies.* Although some metabolic and morphological differences exist, it is assumed here that the nitrogen fixing function is conserved in these organisms, an assumption which has received support from a number of physiological studies.Interest in the nitrogen fixing capacity of these bacteria has been rekindled by the demonstration of regulation of nitrogenase activity by covalent modification' of the dinitrogenase reductase. This mechanism of inactivation and subsequent activation suggest that the number of genes essential for nitrogen fixation may be increased in these organisms. Several genetic tools are available in the Rhodospirillaceae and a recent review has an excellent description of each. Conjugation mediated by PI incompatibility group plasmids appears to be ubiquitous among these Gram-negative phototrophs. Plasmids transfer was first used to promote chromosomal transfer and demonstrate linkage in *R. sphaerodies.* Subsequently, transfer of several P and W group plasmids into this species has been shown as well as low chromosomal mobilization. Similar experiments of conjugational transfer have been reported for *R. capsulata.* The introduction of RP4 :: Mu *cts* has been accomplished in both species. Although Mu was not thermo inducible in *R. sphaerodies,* phages were produced, In contrast, *R. capsulata* strains differ in Mu expression; 37bP containing RP4 :: Mu *cts* was reported to be thermo inducible while no Mu expression was observable i B100 . These experiments suggest that in *vivo* engineering with Mu may be practical in *R, sphaeroides* and *R. capsulata* 37b4. In addition, the introduction of R751 :: Mudlac (Ap^R, lac*)* into *R. sphaeroides* has been accomplished and should allow operon fusions in this organism. By the introduction of the mercury resistance transposon Tn501 into RP1, Pemberton have been able to demonstrate that this plasmid promoted a high frequency of chromosomal transfer in *R. sphaeroides.* As a result, they have published the first map of auxotrophic markers in the photosynthetic bacteria, which opens the way for additional mapping and strain construction. A single m/ mutation has subsequentlybeen mapped to a position very near the genes for the photosynthetic apparatus.A derivative of RP1, pBLM2, which has enhanced chromosome mobilization ability in *R. capsulata*was isolated by Marrs by screening rare exconjugant clones for donor ability. A frequency of chromosome transfer of 6 x10^{-4}/donor was obtained for some markers and R-primes bearing the genes for photosynthesis were found. Similar constructs should be possible for the m/gens of the organism.This preceding of RP1, pBLM2, which has enhanced chromosome mobilization ability in *R. capsulata* was isolated by Marrs by screening rare exconjugant clones for donor ability, A frequency of chromosome transfer of 6 x 10^4 per donor was obtained for some markers and R-primes bearing the genes for photosynthesis were found. Similar constructs should be possible for the *nif* genes of the organism. These preceding genetic tools have been adaptations of system first described in the enteric bacteria. *R. capsulata* enjoys, in addition, an elegant endogenous system of generalized gene transfer discovered by Marrs. The agent (GTA) that serves as the vector for gene transfer appears to be a small phage-like particle: DNA extracted from the GTA particles is linear, double-stranded of ca. 3 x 10^6 daltons and is randomly packaged from the chromosome. Mapping results with the GTA system have yielded map distances that are remarkably close to the physical distances obtained by restriction endonuclease mapping.This transfer agent has been used to show *nif* transfer between *R. capsulata* mutant strains and to begin to construct a linkage map of m/genes. By applying the ratio test, markers of the same phenotype less than 2500bp apart can be reliably demonstrated to be linked with this agent. Although 13 *nif* mutations have been shown to fall into five linkage groups (Wall and Braddock, unpublished), the question of overall clustering must be settled by different exchange techniques employing larger pieces of the chromosome. Preliminary indications from cloning studies using the heterologous pSA 30 plasmid to identify *R. capsulata nif* structural genes indicated that these genes are clustered.

Regulation of the repression of synthesis of nitrogenase in the Rhodospirillaceae is assumed to be similar to that in other diazotrophs. No activity is measurable in cultures grown in ammonia, complex medium or air. In addition, the polypeptides corresponding to the nitrogenase proteins are absent under these conditions.Enzymatic studies with *R. capsulata* and *Rhodopseudomonas palustris* demonstrated that GS and glutamate synthase are the key enzymes of ammonia assimilation. The involvement of a generalized nitrogen regulatory system in the control of nitrogenase was indicated by the derepression of the enzyme complex in the presence of the GS inhibitor, MSX. Support for the involvement of common regulatory elements for *gin* and *nif* has been derived from the isolation of glutamine auxotrophs lacking GS activity which are derepressed for nitrogenase in the presence of ammonia.

4.8.7 ***Rhizobium* Species**

The agricultural economic importance of the symbiotic associations involving the rhizobial species has focused attention on these organisms for many years. The mechanisms of recognition, infection and nodulation have been investigated extensively; however, the innate difficulties of a developmental symbiosis has showed the acquisition of knowledge which might be used to improve the process. Renewed interest has arisen because of the recent advances in understanding nitrogen fixation in the free-living bacterial systems.

4.8.7.1 Fast Growing Species

Because of the faster doubling times, *R. leguminosarum, R. meliloti, R. trifolii* and *R. phaseoli* have been easier to manipulate by standard genetic techniques than the slower growing species. Mutants have been isolated after chemical mutagenesis and penicillin enrichment techniques *via* classical procedures. Although transformation procedures have been available for most species for many years, little linkage data has resulted from its use. In addition, generalized transduction in *R, meliloti* was reported as early as 1967; however, only one instance of cotransduction of markers has been demonstrated with this system. It is expected that these genetic exchange process will begin to be re-investigated since the establishment of circular linkage maps of the chromosomes has been accomplished *via* conjugation. Transductional analysis will be essential for fine structure mapping and the ability to transforms opens the way for additional cloning manipulations. Indeed, the recent literature reflects this interest since transduction has now been reported for *R. leguminosarum* and *R. trifolii* and improved methods for plasmid transformation of *R. meliloti* have been developed.

Systems of conjugation have played the most prominent role in the genetic investigations of the rhizobial species. Early endogenous conjugational systems have either not been rigorously pursued or *Rhizobium* species used has been questioned. More recently, the existence of conjugative plasmids in three strains of *R. leguminosarum* which code for bacteriocin production and have chromosome mobilization ability (Cma) has been demonstrated. In contrast with the endogenous plasmids which are interesting because of their own genetic content, the plasmids of the PI incompatibility group originally from *Pseudomonas aeruginosa* have been the most productive for genetic analysis. As a result of their use, chromosome linkage maps now exist for *R. meliloti* 2011, for *R. meliloti* 41 and for *R. leguminosarum.* Recombination and linkage between markers in crosses of *R. leguminosarum* with *R. trifolii* or *R. phaseoli* were essentially the same as results obtained from crosses within *R. leguminosarum.* Therefore, the map derived for *R. leguminosarum* is believed to represent all three species. A recent comparison of these maps shows the similarities and complementation obtained among the *R. meliloti* strains and *R. leguminosarum.* Use of the promiscuous plasmids has also made the introduction of transposons into *Rhizobium* a fairly straightforward procedure. Although PI plasmids are generally stable in *Rhizobium* when bacteriophage Mu is present in the plasmid, the plasmid is no longer stably maintained after its introduction by conjugation. When a transposon conferring drug resistance is also included on such a plasmid, selection for that drug resistance after transfer to the plasmid to *Rhizobium* selects for those cells in which transposition has occurred. Two such suicide plasmids (i) one containing Tn7 and (ii) one containing Tn5, have been constructed and used. An additional procedure for transposon mutagenesis has been described by Ruvkun. Any cloned gene can be introduced into *E. coil* and there mutagenized with transposons. If it is not in a transmissible vector, it may be inserted into a broad host-range cloning vehicle and conjugated back into the original host. The cloning vehicle can then be 'chased' from the bacterial cytoplasm by the introduction of an incompatible plasmid while continuous drug resistance selection is made. The majority of drug resistant exconjugant will have undergone a homologous recombinational event, or homogenization, such that the transposon now resides in the chromosomal gene. In *R. leguminosarum R. trifolii* and *R. phaseoli,* the genes for nodulation lost specificity and nitrogen fixation have been reported to be present on large plasmids. In contrast, experiments with *R. meliloti* 102F34 designed to examine the location of m/genes, suggested both a chromosomal position and that at least three discrete units necessary for nitrogen fixation were contained in an 11.2kb fragment. Most intriguing was the recent observation of the occurrence of reiterated *nif D* and *nif H* genes of *R. phaseoli* with at least one copy on a larger plasmid.

4.8.7.2 Slow Growing species

Included in the slower growing group of rhizobial strains are *R. lupini, R. japonicum* and *Rhizobium* sp. or cowpea rhizobia. The genetic analysis of these species lags behind that of the faster growers primarily because of the greater, difficulty encountered in the microbiological manipulations. In contrast to the fast growers, these strains express high nitrogenase activities in defined culture; therefore, these bacteria may ultimately be more amenable to the analysis of functions. Although most of the mutants isolated early were drug resistant, more recently *nif* mutants of *R. japonicum* have been obtained after mutagenesis and screening in planta or explanta. Glutamine auxotrophs of cowpea strain 32H-1 showing impaired nitrogenase activity has also been isolated. Few other auxotrophs have been reported in these strains.Although an endogenous conjugation system was reported for a non-nodulating *R. lupini* strain as early as 1968, methods for gene transfer in other strains of slow growers have only now begun to be developed (Kondorosi and Johnston). Using an uncharacterized bacteriophage, Shah has now generated a linkage map of *R. japonicum* by transduction. This map can now be used to locate and manipulate the genes essential for nodulation and nitrogen fixation. In preparation for a more extensive genetic analysis, Kennedy has reassessed the ability of several strains of cowpea rhizobia to transfer, maintain and express P group plasmids. Strains which have a reasonable frequency of transfer and good plasmid stability during nodulation were identified but no *Cma* was reported. Obviously *Cma* is the next step in the analysis of the slow growers.

4.8.8 Regulation

From studies with cloned DNA fragments containing *nif* genes, it has been shown that the control for expression of these genes in the fast-growing species operates at the level of transcription during nodulation. Similar experiments with slow growers have not yet been done.

The involvement of the ammonia assimilatory system in the regulation of nitrogenase of *Rhizobium* remains an open question. In contrast to *K. pneumoniae,* results with bacteroids have suggested that the assimilation system of the bacteria is essentially non-functional during greatest N_a-fixing activity. Studies with *R. japonicum* bacteroids and free-living bacteria supported this finding in that the ammonia produced from $N_2$was found in the medium. In addition, mutants lacking glutamate synthase and strains naturally lacking glutamate dehydrogenase were normal in their abilities to nodulate and fix N_2.On the other hand, studies with glutamine auxotrophs showed that nitrogenase activity was lacking in the mutant bacteroids. These results must be interpreted with care, since it is now known that all *Rhizobium* species so far examined have two distinct glutamine synthetase enzymes. The two enzymes have different physical properties; GSI undergoes adenylylation in response to a nitrogen signal while GSII does not. The description of an auxotrophic revertant of *Rhizobium* sp. 32H1 in which (a) GSI was constitutively adenylylated(b) GSII was still missing and (c) nitrogenase was constitutively synthesized implies common regulatory elements for GSI and nitrogenase. Because the auxotrophy of GSII mutants could be satisfied by either glutamine or purifies, a role for this enzyme is assimilation has been questioned.

4.9 Applications of Nitrogen Fixation

4.9.1 Physiology of Organisms

Application of the biological N_2 fixation process requires knowledge of the physiology of these organisms. The broad range of the ecological adaptations of nitrogen-fixing organisms permits an even broader range of applications. The capacity to fix atmospheric nitrogen occurs in a large number of diverse bacteria but does not naturally occur in eukaryotes except within symbiotic associations.

4.9.1.1 Aerobes

Aerobic nitrogen-fixing bacteria are found mainly in the family Azotobacteriaceae, which includes the taxonomically similar genera *Azotobacter, Azotococcus, Azomonas, Beijerinckia* and *Derxia.* Members of the genera *Azotobacter, Beijerinckia* and *Derxia* have been demonstrated to fix N_2, but

Parejko could find no evidence for fixation in strains of *Azotomonas* that were examined. Most-of these aerobic bacteria actually fix nitrogen only under microaerophilic conditions. It has been proposed that these organisms adapt physiologically by producing copious amounts of polysaccharide which hinders diffusion of oxygen thereby permitting the functioning of nitrogenase. *Azotobacter* possess very high rates of respiration; indeed, the highest respiratory rates ever measured (QO_2) are those of *Azotobacter.* These high respiratory rates have been called respiratory protection since they maintain intracellular partial pressures of oxygen that are compatible with the functioning of nitrogenase. However, even the *Azotobacter* strains are susceptible to oxygen at partial pressures beyond 20% O_2, or when forced into phosphate limitation. Under these conditions, the nitrogenase is 'switched off or protected by anFe-S protein that forms a complex with dinitrogenase reductase. Nitrogenase is switched on when favorable conditions return. Although *Azotobacter* is generally though of as a free-living nitrogen fixing organism, several species have been reported to form associations with the roots of plants. For example, *Azotobacter paspali* forms a loose association with the roots of *Paspalum notatum* and other grasses.

Aerobic nitrogen-fixing organisms also include iron-and methane-oxidizing bacteria. Three strains of *Thiobacillus ferrooxidans* have been shown to incorporate ^{15}N when growing in culture media at pH values of 2. Also nitrogen fixation among the methane-oxidizing bacteria is common. Dalton has shown that the methane mono-oxygenase enzyme of *Methylococcus capsulata* provides respiratory protection for the organism such that cultures can be conditioned to withstand elevated partial pressures of oxygen without increased respiration.

4.9.1.2 Facultative Anaerobes

Facultatively anaerobic bacteriaincludes species of *Bacillus, Klebsiella, Rhodopseudomonas, Rhodomicrobium* and *Rhodospirillum* able to grow aerobically on fixed nitrogen, cannot fix nitrogen in air. Only under strict anaerobic or extremely low oxygen tensions are these organisms capable of nitrogen fixation. The non-sulfur phototrophic bacteria are able to utilize light energy directly for the fixation of atmospheric nitrogen anaerobically and recently Madigan have demonstrated that *Rhodopseudomonas capsulata* fixes nitrogen microaerophilically in the dark. Certain *Cyanobacteria*, most notably the filamentous *Anabaena* species fix atmospheric nitrogen aerobically. This would appear to be a dilemma since the cyanobacteria conduct plant-type photosynthesis and thereby evolve O_2. Anabaena has overcome this incongruity by morphologically altering approximately every tenth cell along its length into a thick walled 'heterocyst' devoid of photosystem II but possessing the nitrogen-fixing apparatus. The heterocyst supplies fixed nitrogen to the vegetative cells, which in turn provide reduced carbon compounds for the fixation process.

4.9.1.3 Anaerobes

Clostridium is one of the most primitive of organisms and thus its physiology has attracted considerable attention. These organisms are strict anaerobes, but will tolerate oxygen to varying degrees in culture without irreversible denaturation of their nitrogenase. By comparison the nitrogenase of the sulfate-reducing bacteria is impaired by any level of oxygen present in the culture.

4.9.1.4 Symbionts

The complexity of many symbioses has hindered progress elucidating the physiology and biochemistry of the prokaryotic symbiont. Although the free-living state of these organisms is well defined, in many cases the symbiotic state is poorly understood.*Rhizobium* species are defined as heterotrophs able to grow both aerobically and anaerobically in culture. Hanus have demonstrated certain *R. japonicum* and cowpea rhizobia are capable of chemolithotrophic growth. If provided with a fixed nitrogen source. *Rhizobium,* the endophytes of leguminous plants have been demonstrated to fix measurable levels of N_2 when grown in culture on well-controlled, low partial pressures of oxygen,whereas nitrogen fixation can be obtained on defined media in the slow-growing rhizobia *(R, japonicum, R, lupin;* cowpea miscellany), the fast growers *(R. leguminosarum, R. melioti, R. trifolii, R. phaseoli* can fix nitrogen *explanta* only in media containing liquid extract from plant callus growth media. The rates of fixation are extremely low compared to those obtained when these organisms are participating in their respective symbioses.The genus *Frankia* has been designated to encompass those endophytes of symbioses between non-leguminous plants and the actinomycorrhizae. Callaham were the first to successfully isolate the endophyte of these associations, obtaining the organism from *Comptonia peregrina* nodules in pure culture. Subsequently, the endophytes from *Alnus* and *Eleagnus* have also been isolated. The cyanobacteria participate as the nitrogen-fixing endophyte in lichens and with the water fern *Azolla.* The lichens are associations between cyanobacteria and a fungus. In the case of *Peltigeraaphthosa,* a complex symbiosis occurs consisting of an ascomycetous fungus, a species of the cyanobacteria *Nostoc* and a green alga *Coccomyxa* sp. The *Nostoc* in the symbiosis contains a much higher number of heterocysts than is normally found in the free-living culture (20% versus 5-10%; Stewart). *Anabaena azollae,* the endophyte found in the leaf cavities of the water fern *Azolla,* also has an increased number of heterocysts (Anonymous, 2012)

4.9.1.5 Agronomic Applications

The most grandiose application will be the integration of the *nif* genes directly into plant cells such that the transformed eukaryotes now are capable of providing their own fixed nitrogen directly from the atmosphere. Whereas the transfer of the genes may certainly be possible, the expression of the genes at a significant level, provision of sufficient energy for nitrogen fixation without severely affecting plant growth or development, protection of the 0_2 labile nitrogenase proteins and so on, represent intractable barriers that will require extraordinary effort and time to surmount. The most logical intermediate step to achieve nitrogen fixing eukaryotes is via *Agrobacterium, Agrobacterium,* taxonomically related to the *Rhizobium,* is able to infect a wide variety of plants, rather than being restricted to the legumes as are the rhizobia, and form crown galls. This infection and host-bacteria relationship require the transfer or a plasmid from the bacteria into the plant genome. Thus, *Agrobacterium* may be a very useful model system to solve the problems that the nitrogen fixation process may impart upon the host plant. This section will address some of the more partial, short term applications.

4.9.1.6 Rhizobium

Rhizobium spp. has attracted considerable attention as the primary approach to increasing nitrogen fixation potential in crop legumes. The features that make *Rhizobium* symbioses attractive are (a) their symbiotic relationship with major economic/agricultural crops and (b) that the *Rhizobium* and many of the important legumes can be manipulated genetically. Compared to the associative systems, the intimate symbiosis between *Rhizobium* and leguminous plants afford a "more definitive approach toward improvement of its efficiency and effectiveness. During the early stages of plant development, a period of nitrogen limitation is endured before the required nodule mass has been synthesized. The high energy demand of nitrogen fixation then imposes a carbon stress upon the metabolic processes of the plant. This and a multitude of other inter-relationships govern the yield potential of the leguminous crops. To increase yields in the short term, the proper *Rhizobium.* Strain plant cultivar combination as well as the necessary disease resistance traits should be utilized. The use of the proper *Rhizobium* strain-plant cultivar can markedly increase yields, but the farmer often does not have access to the desired information. Moreover, new cultivars are introduced more rapidly than *Rhizobia* strain-plant profiles can be completed.

Typically, fields in which leguminous crops are planted contain large indigenous populations of rhizobia. For example, the majority of these indigenous *R. japonicum* in mid-western soils consist of the 123 serotype, a group which is not conducive to optimal yields with most soyabean cultivars. This serogroup is quite competitive *versus* applied strains and thus forms a high percentage of the bacteroids found within the nodules of plants grown in these soils. Applying strains of rhizobia possessing superior biochemical and genetic traits is futile if they eventually form only a .minority of the nodule bacteroids population. Thus, improved strains of rhizobia need to contain the necessary characteristics for competitiveness as well as any beneficial biochemical or genetic characteristics. Thus, strains of rhizobia should be selected for competitiveness before genetic/ biochemical alterations are made or, conversely, the genes for competitiveness should be transferred to improved strains of rhizobia. However, the parameters controlling competitiveness are not understood at this time, although lectins are believed to be a major factor.The problem of competitiveness may be circumvented by sterilization of the soil or by producing rhizobia strains resistant to those agrichemicals deleterious to rhizobia. At present, soil sterilization methods are impractical on a large scale. A short term solution may be reagent-selective inoculants. Legumes seeds can be coated with various agents deleterious to the indigenous rhizobia. A desired *Rhizobium* strain which can be made resistant to this agent also can be incorporated into the seed coating. The nodules resulting from this treatment contain a high proportion of the desired *Rhizobium*. Thus, full expression of the beneficial traits arc obtained.The list of desired traits is endless and with the development of *Rhizobium* genetics the near future promises an abundance of improved strains. Among these traits is the hydrogen recycling mechanism as exemplified by certain strains of *R. japonicum*. During the nitrogenase reaction, a minimum of 1mol of hydrogen is evolved for each mole of atmospheric nitrogen reduced. Under less favorable conditions, more hydrogen is evolved than nitrogen reduced. In the extreme case that is in the absence of nitrogen only hydrogen is evolved. Evans and associates have identified strains that do not evolve hydrogen during the fixation of atmospheric nitrogen. These strains possess an uptake hydrogenase which consumes all of the hydrogen evolved, *via* nitrogenase, in an oxygen-dependent reaction and concomitantly produces ATP. This ATP can then be recycled for the fixation of nitrogen. Use of these strains in field tests has increased the nitrogen content of soybeans by 10-13%. The transfer of this trait to other rhizobia possessing additional beneficial parameters will be possible in the very near future.

At the present time, identification of the biochemical processes that limit symbiotic nitrogen fixation is more difficult than the genetic transfers. The biochemistry and physiology of the symbioses have not been adequately characterized, so that the limitations of the symbiotic fixation process can be defined. During the peak of nitrogen fixation activity, the symbiotic process is thought to be carbon limited. There is considerable physiological information supporting this concept and it has been calculated that if the peak of nitrogen fixation activity could be lengthened by several days the nitrogen content of soyabean seed theoretically could be doubled, The interdependence of carbon metabolism and nitrogen metabolism in Rhizobium leguminous plant symbioses suggest that any increase in the availability of carbon compounds to the root nodules, or alternatively more efficient utilization of available carbon compounds within root nodules, will increase the nitrogen fixation potential. As an index of photosynthates availability many workers have reported differences between cultivars in the rate of CO_2 uptake in soybeans. However, the rate of CO_2 uptake is not a reliable indicator of net photosynthates production orphotosynthates transport to root nodules. More systematic methodologies for screening cultivars and *cultivar-Rhizobium* strain combinations are required.

Mutations affecting the metabolism of the citric acid cycle or the uptake of these metabolites significantly alters the nitrogen fixation capacity of the symbiosis. Compared to other carbon compounds, the citric and cycle intermediates support greater rates of nitrogen fixation in suspensions of anaerobically isolated bacteroids. However, it is not known how the photosynthetic compounds supplied to the root nodule, primarily glucose or sucrose, are metabolized (Embden-Meyerhof *vs.* Entner-Doudoroff*vs.* Pentose shunt pathway) or in which compartment of the symbiotic tissue (infected plant cell *vs.* non-infected cell *vs.* bacteroid) the metabolism occurs.The genetics and biochemistry of the host plant must also be considered and a number of investigators are focusing on this area. For example, in *Pisum sativum,* where two genes have been found that control nodule number, a correlation exists between nodule number and seed yield in peas some described a soyabean plant line that does not undergo senescence like normal plants. Hopefully these plants will continue to fix $N_2$for prolonged periods and thus lead to increased yields.Perhaps the greatest application of *Rhizobium* research will be in the tropics and subtropics where the majority of the 20,000 species of leguminous plants thrive. These regions coincide with the majority of the world's underdeveloped population. Because of the poor soils in many tropical regions, applications of nitrogen fixation technology have the greatest potential for helping mankind. However, agricultural research on legumes has been conducted primarily in the temperature climates of the more developed countries.

4.9.1.7 Azospirillum

A large number of diazotrophs have been reported in association with plants, usually found near, on or within the roots. These associations are not nearly as specialized in their symbiotic morphology as the *Rhizobium,* but their biochemistry may be equally complex. These associations have generated considerable excitement, since their wider range of association indicates they may be adaptable to many non-leguminous, agriculturally important plants. In the foreseeable future, the expected benefits in terms of plant productivity are considerably less than the *Rhizobium*-leguminous plant symbioses, but they may provide significant crop improvement in soil with poor moderate nitrogen fertility. *Azospirillum* has been the most studied associative organism because (a) it can form associations on a large variety of plants often in high numbers (10^7/g of root) and (b) some reported rates of fixation can significantly affect crop yields. *Azospirillum* has been reported associated with maize, sorghum, sugar cane, rice, millet, oats, rye, barley, forage grasses and the water plant *Spartina alternifora*. Dobereiner have reported rates of fixation *of Azospirillum* on *Digitaria decumbens* as high as 1kg N/ha/d. Although some of the early nitrogen fixation measurements were overestimates due to experimental conditions, more careful work by Okon and others has shown that in Israel, *Azospirillum* association can contribute the majority of the nitrogen required by the plant. In the more regions, particularly the midwestern and northern United States, it appears that *Azospirillum* may contribute little to the nitrogen economy of typical crop plants. However, it may be practical to select cell lines (mutants or genetically improved strains) of *Azospirillum* as inoculum for specific cultivars to form productive associations under a particular set of environmental or ecological conditions.

4.9.1.8 Cyanobacteria

In the United States, the cyanobacteria are usually associated with the eutrophication of lakes, rivers and streams and thus are treated as a scourge. However, they are treated as a boon in the rice growing regions of the world since the *Cyanobacteria* are beneficial to the crop. The cyanobacteria are nurtured and subsequently used as a green manure to fertilize the rice field after it is drained. The extent of the contribution of *cyanobacteria* to rice is not yet known, but a typical rice crop will remove about 50 kg N/ha. Wetland rice can be grown continuously with reasonable yield levels without N-fertilizer additions if adequate cyanobacteria populations are maintained. *Cyanobacteria* frequently dominate the phytoplankton population of eutrophic lakes and account for the major source of fixed nitrogen. In non-eutrophic lakes and rivers and in the open ocean, N_2-fixation rates may be imperceptible. Burris has reported that cyanobacteria attached to rocks in the intertidal zone of the Great Barrier Reef could fix 6.8 to 30.6 kg/N/ha of rock surface per year. Stewart reported rates of fixation of 25kg N/ha/year on the supralittoral fringe of temperate shores. Much higher rates of fixation have been reported in association with coral reefs. Stacey has isolated a rapid growing blue-green alga from the warm shallow coastal area of Port Aransas, Texas capable of fixing 40 kg N/ha/yr.The contribution of cyanobacteria to soils has not been estimated. Although contributions to the nitrogen economy may be quite substantial in certain environments, the contribution on a global scale is rather small.

4.9.1.9 Cyanobacterial Associations

Associations of *cyanobacteria* with eukaryotic cells are rare. The two primarily studied cyanobacterial associations are *Azolla* and the lichens. *Azolla* spp. is some of the few vascular plants capable of forming an intimate association with a blue-green alga. *Azolla* spp. is fast-growing water ferns that contain the *cyanobacteria. Anabaena azollae,* as an endophyte within their leaf cavities. In the symbiotic state, the *Anabaena* expresses an elevated number of nitrogen-fixing heterocysts, as high as 20% of the total cell. In Southeast Asia, *Azolla* are nurtured simultaneously with rice or as a green manure crop in fallow paddies. When grown simultaneously with rice, some species of *Azolla* serve only as a future green source of nitrogen, but other *Azolla* species may leak or excrete ammonia into the aquatic environment which may then be available to the rice crop immediately. Thus, the nitrogen input of *Azolla* blooms toward a particular rice crop is difficult to determine; when *Azolla* are considered solely as a green manure crop, values as high as 250 kg/N/ha have been estimated, but values between 30 to 100 may be more common. *Azolla* have provided a readily available inexpensive source of fertilizer nitrogen to under developed regions of the world, where the cost and technology of applying commercial fertilizer would be prohibitive. Lichens have little direct agronomic importance; however, their ability to fix atmospheric nitrogen and occupy habitats not suitable for other plants makes them important long term investments of fixed nitrogen. Lichens can colonize poor and/ or acidic soils; survive prolonged periods of desiccation and fix N, under severe temperature extremes. Lichens are composed of diverse classes of organisms. The fungal partner is usually an Ascomycetes but may also be a Basidiomycetes or in the Fungi imperfecti group. The cyanobacterial partner may have to provide the photo synthetically fixed carbon in addition to the fixed nitrogen although this depends upon the corn position of the lichen. The reported range for fixation by lichens is 0.2 to 12.0 kg N/ha/yr.

4.9.1.10 Photosynthetic Bacteria

The photosynthetic bacteria, although ubiquitous, probably does not provide meaningful agronomic levels of mixed nitrogen. The photosynthetic bacteria contribute in anaerobic and microaerophilic environments, especially those rich in H_2S but nutrient-poor and illuminated. They have been reported to contribute fixed nitrogen in rice paddies, salt marshes, estuarine muds and sulfur-spring ditches.

4.9.1.11 New Associations

The discovery of *Azospirillum* association with the roots of grasses initiated a renewed interest in associative systems. Although most of this interest has centered on *Azospirillum,* number of other associative systems have been reported. The best defined new associations are (a) sugarcane-Beijerinckia (b) wheat-*Bacillus* spp. arid/or *Ervinia herbicola* (b) rice-*Achromobaacter*, (c) *Paspalum notatum-Azotobacter paspali;* and (5) *Azospirillum.* Because of the importance of wheat as a major agronomic crop, the reports of associations with *Bacillus* spp. and *E. herbicola* have received particular interest and scepticism. Klucas and Pedersen have used *K. pneumoniae* or *E. herbicola* as inoculum on winter wheat and grain sorghum. They reported significant differences between plant cultivars and inoculants, pointing out that plant genotype, species of microorganisms and environmental conditions all play a major role in determining yield. Depending upon the various treatments, inoculation produced increases of up to 44.4g dry wt. or decreases of as much as 24.7g dry wt. Rennie and Larson have shown that inoculation of wheat with *Bacillus* in sterile leonard jars resulted in increased plant nitrogen.

4.9.2 Industrial Applications

4.9.2.1 Chemical Catalysts

The Haber-Bosch process, the principal industrial process for ammonia production since 1913, requires pressures of 350-1000 atm. temperatures of around 350°C plus elementary hydrogen for the reduction of N_2. The usual catalyst is composed of iron. Ruthenium and osmium catalysts are available that perform at lower temperature and which push the equilibrium further toward ammonia, but their greater costs discourage their use. The high energy costs of the Haber process means an entirely different procedure must be developed to provide cheaper ammonia. Even the biological process requires large inputs of energy, but unlike the industrial process it occurs at room temperature and atmospheric pressure, Hopefully, new catalyst and procedures will not only provide cheaper sources of ammonia but will require a minimum amount of equipment and capital investment so that ammonia can be produced where it is needed and the costs of transportation can be reduced. Recently, the emphasis has switched to molybdenum-containing compounds modeled after the FeMo cofactor of dinitrogenase. Shah has demonstrated that isolated FeMoCo is capable of catalytically reducing acetylene to ethylene in the presence of sodium borohydride. Also, Thorneley have demonstrated a hydrazine-like intermediate which they believe is bound end-on to the molybdenum of dinitrogenase. A number of cubane iron-sulfur clusters of the sum formula $[Mo_2Fe_6S_9(SC_2H_s)_8]^{3-}$ or $[Mo_2Fe_6S_8(SC_3H_5)_9]^{3-}$ have been synthesized in which one Mo atom replaces one Fe atom in a corner of a typical Fe_4S_4 cluster. Analysis of these clusters by EXAFS (Extended X-Ray Absorption Fine Structurue) show that they closely resemble native dinitrogenase. A large number of other structures which bear resemblance to the FeMo cofactor of dinitrogenase have also been reported. At present there have been no reports that these compounds function catalytically.Nitrogenase, in addition to reducing nitrogen to ammonia, can reduce acetylene to ethylene, evolve H_2 from protons and reduce a number of other triple-bonded compounds.

4.9.2.2 Ammonia Production

Ammonia production from continuously cultured organisms or organisms bound to solid supports has been widely discussed, but application has been difficult. The requirements for microbial production of ammonia are (a) derepression of the genes for nitrogen fixation, (b) repression of ammonia assimilation enzymes, (c) availability of large quantities of energy for nitrogen fixation and cell viability, (d) export of the ammonia produced by nitrogenase into the surrounding media, and (e) extraction of ammonia from undesirable components in the effluent. There has been considerable variability and instability of genetic alterations described in points (i) and (ii) above during continuous culture experiments. The best strains in terms of longevity of ammonia export are those capable of maintaining low rates of protein synthesis.The energy sources utilized most frequent for culturing nitrogen fixing organisms are expensive reduced carbon sources such as sucrose, glucose, mannitol and/or organic acids. Thus, provision of an inexpensive energy source has been a major limitation. The transfer, of cellulase genes into a nitrogen-fixing organism has not yet been reported, but utilization of cellulose or any other cheap energy source (industrial or commercial wastes) is a necessary prerequisite for microbial production of ammonia to be cost-effective.The overall cost-effectiveness depends upon the rate of ammonia production per culture, the density at which the organisms can be cultured or attached to a matrix and the extent or type of extraction procedure needed. The actual ammonia production per culture or per microbe depends upon the genetic alteration achieved; assuming the rate of ammonia production from genetically altered cells can be equated with the doubling time of the faster growing nitrogen-fixing organisms. The rapid advances of gene cloning and gene regulation coupled with the rising costs and dwindling supplies of fossil fuels indicate biological ammonia production may be cost-effective within the next few decades.

4.9.2.3 Timber Production

Nitrogen is the most common limiting nutrient for timber production as well as agronomic crops. The contribution of nitrogen fixation to forests can occur *via* lichens-cyanobacteria, leguminous *plants-Rhizobium* symbioses and woody plant-actinomycetes symbioses. The most often used biological method of forest fertilization has been the use of actinomycete-nodulated plants, such as *Myrica, Alnus, Pursfiia* and others Actinomycetous root nodules have been reported in more than 140 species of plants, most of which are woody shrubs or trees. Many of these trees provide excellent lumber for building materials, furniture, pulp wood, or fuel.In the Pacific Northwest, red alder has the highest rate of nitrogen fixation (up to 480kg/ha) followed by snowbush (up to 100kg/ha), Scotch boron (160kg/ha) and lupines (100kg/ha). Most reports show mixed stands of alder and Douglas fir increase total timber production, in some cases up to 100%, compared to pure stands of Douglas fir. However, management of these mixed forests can require additional labor and capital. Alder and other native woody nitrogen-fixing plants are considered nuisances since they will outgrow the Douglas fir during the first several years after planting and consequently shade and reduce the growth of the fir. Prohibiting the growth of these nitrogen-fixing plants is difficult, since many are pioneer species and rapidly colonize selective or the clean-cut areas. Proper management requires intercropping of alder several years after the fir have been planted to avoid shading. This demands cultivation of the Douglas fir for the first few years, perhaps additions of chemical nitrogen fertilizers and then planting of the alder. The uses of alder as a second lumber product from this intercropping system are pulp, furniture and fuel. With regard to the application of biological nitrogen fixation to forests, DeBell has stated in many instances the most limiting factors are *not* the need for more scientific research or break through regarding N_a. fixation *per se,* but rather some fundamental biological information as well as appropriate demonstration of benefits.

4.9.2.4 Hydrogen Production

Like biological ammonia production, hydrogen production *via* nitrogen-fixing organisms requires depressed strains capable of utilizing cheap, available energy sources. Unlike ammonia production, cellular export and extraction of H_2 from the media is not a problem. Considerable volumes of H_2 are generated under certain culture conditions. For example, the build-up of hydrogen from *Rhodopseudomonas capsulata* can rupture sealed gas culture tubes. H_2 evolution from *Rhodospirillum rubrum* has been reported at rates 20 ml/h/g/dry wt. of culture.The major technological difference would be product extraction. Since the hydrogen gas will be mixed with whatever gas (or gas mixture) has been used to support cellular metabolism, the extraneous gases will have to be removed. However, the methodologies for purification by liquefaction are available. The extraction requirement would limit small site applications and add the cost of transportation for the consumer. Again the cost-effectiveness depends upon the genetically altered rate of production that can be achieved as well as capital purchases, maintenance and other production and transportation costs.

4.9.2.5 Phytochemical Production

Many of the legumes and non-leguminous plants produce valuable biochemical such as dyes, fibers, flavorings, odors, substitutes for chocolate, citrus, coffee, garlic, licorice, tea, tobacco and vanilla, medicines, pharmaceuticals, oils for fragrances, cooking and machinery and a host of other chemicals. Cultivation of these plants on a commercial scale would require large initial expenditures. However, since the symbionts for many of these plants are known or are obtainable, biological nitrogen fixation is the logical alternative to chemical fertilizer to reduce the cost of applying nitrogen, Bio-alionctics has recently announced that it will use an algal strain to produce commercially 3.8 to 7.6 million of ethanol per year. The algae residue after processing will be sold as fertilizer.

4.9.2.6 Biomass Conversion

The nitrogen fixation process has often been linked with biomass conversion. Utilization of neutralized pulp mill wastes, spoiled grains and wastes from food processing could serve as energy and carbon sources for nitrogen-fixing organisms. Coculturing of diazotrophs and other -microbes has been suggested as a means of providing additional vitamins or minerals not necessarily found in diazotrophs. The resulting biomass could then be processed as food for humans and/or animals. Biomass production rates of BGA in waste water treatment ponds have been reported as high as 100tons/ha/yr with the average centering around 50tons/ha/yr. Major drawbacks have been production costs, nutritional quality, palatability and, when dealing with less developed countries, transportation. The technology required would limit its availability in those areas where these products would be most needed. Alternatively, the biomass produced could be converted into nitrogen fertilizers, either by utilizing the biomass directly or by extraction of the nitrogen-containing components.

4.10 Factors affecting Biological Nitrogen Fixation

4.10.1 Salinity

The symbiotic nitrogen fixation in legume nodules has been shown to be extremely sensitive to water deficits and salinity stress, which could result in decreasing N accumulation and yield of legume crops under these conditions. The effects of salts have been investigated on several legume-rhizobium symbioses. Nitrogenase was substantially inhibited by sodium chloride (NaCl), and this inhibition was associated with a significant decrease in plant growth and N content. The effect of on nitrogenase activity has been associated with changes of the nodule permeability to oxygen diffusion. A large genetic variation has been found among species and cultivars in the nitrogen fixation sensitive to salinity.

4.10.2 Drought

Symbiotic nitrogen fixation is highly sensitive to drought, which results in decreased N accumulation and yield of legume crops. The effects of drought stress on N_2 fixation usually nave been perceived as a consequence of straight forward physiological response acting on activity and involving exclusively one of three mechanism shortage limitation regulation by nitrogen accumulation. The sensitivity of the nodule water economy to the volumetric flow rate of the phloem into the nodule offers a common framework to understand each of these mechanisms. As these processes are sensitive to volumetric phloem flow into the nodules, variations in phloem flow as results of changes pressure in the leaves are likely to cause rapid changes in nodule activity. It could explain the special sensitivity of N fixation to drying soils. Due to the high sensitivity of nodules to phloem volumetric flow into the nodules are a number of possible consequences resulting in a high sensitivity of N_2 fixation to soil drying. Decreased phloem flow would probably decrease the import of carbon to the nodule, although photosynthate supply during water deficits has not been shown to be a major limitation to nodule activity. A lower rate of water flow in the nodule has the potential directly to influence nodule P because of alterations in the configuration of the cortical cells and airspace dependence of nodule export on water flow into the nodule although the phloem also has the potential to have major into on the accumulation of N export products in the nodule.

4.10.3 Temperature

High root zone temperature has been shown to strongly affect bacterial infection and nitrogen fixation in several legume species including soybean. Elevated temperatures may delay nodule initiation and development and interfere with nodule stricture and functioning in temperate legumes, whereas in tropical legumes N efficiency is mainly affected. Low temperaturedelay root hair infection and decrease nodulation and nitrogenase activity.

4.10.4 Soil Acidity

Soil acidity affects all aspects of nodulation and nitrogen fixation from survival and multiplication of the rhizobia in the soil, through infection and nodulation, to nitrogen fixation. The failure of legumes to nodulate under acid soil condition is common especially in soils with pH less than 5.0.

4.10.5 Elevated Atmospheric CO_2

Elevated CO_2 likely to affect carbon cycling and crop productivity by stimulate synthesis. Low rhizosphere CO_2 concentration (below 100ppm) result in significant decline in nitrogenase activity of legume nodules (Bethenod *et. al.,* 1984). However, increasing the partial pressure of CO_2 (pCO_2) in the rhizosphere above ambient did not stimulate activity of soybean. Therefore, direct effects of altered rhizosphere pCO_2 nitrogenase activity are unlikely. Nitrogenase activity is rather oxygen than carbon limited.

4.10.6 Agrochemicals

The use of pesticides has become an essential part of agriculture. Pesticides have different effects on symbiotic systems. For example, the fungicides carbondazim has been found to have no significant effect on number of nodules in cowpea but increased the fresh nodule weight and total N content (Singh *et al.,* 1986), Yet, the same compound decreased he number of *Bradyrhizobium japonicum* surviving on soybean seeds and decreased nodulation and yield in the field, but when associated with oxine copper.

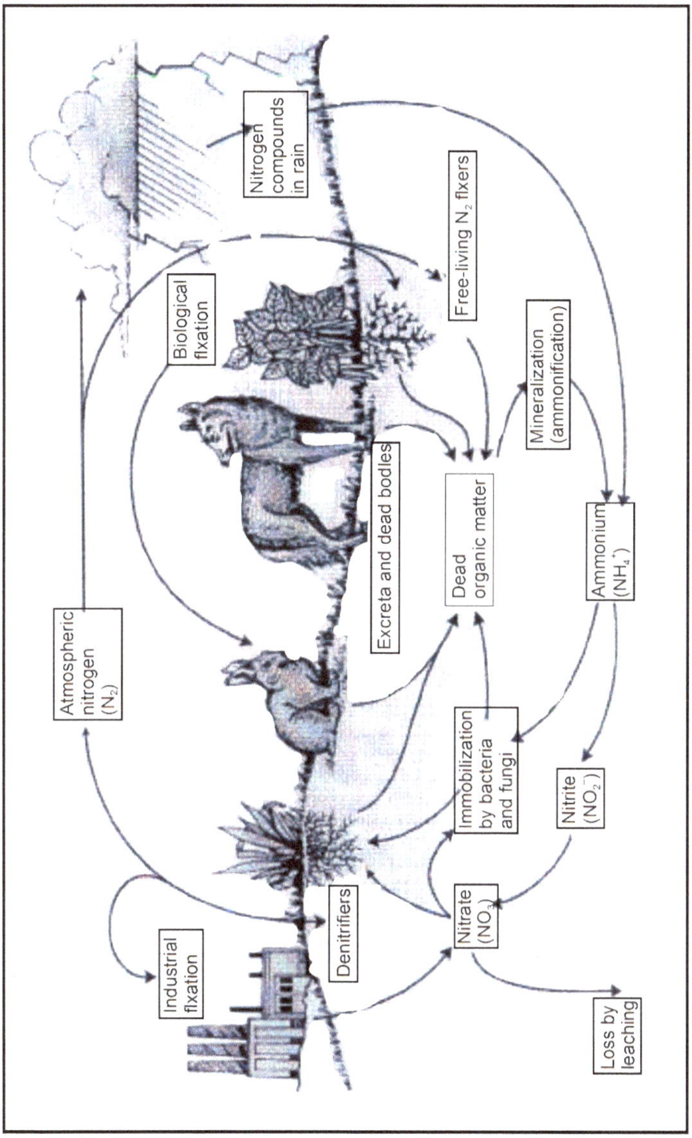

Fig. 4.1 : Nitrogen cycles through the atmosphere as it changes from a gaseous form to reduced ions before being incorporated into organic compounds in living organisms.

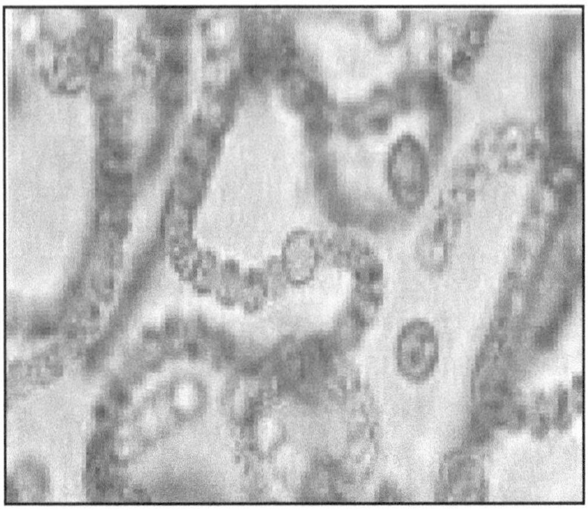

Fig. 4.2 : Heterocysts in Nostoc

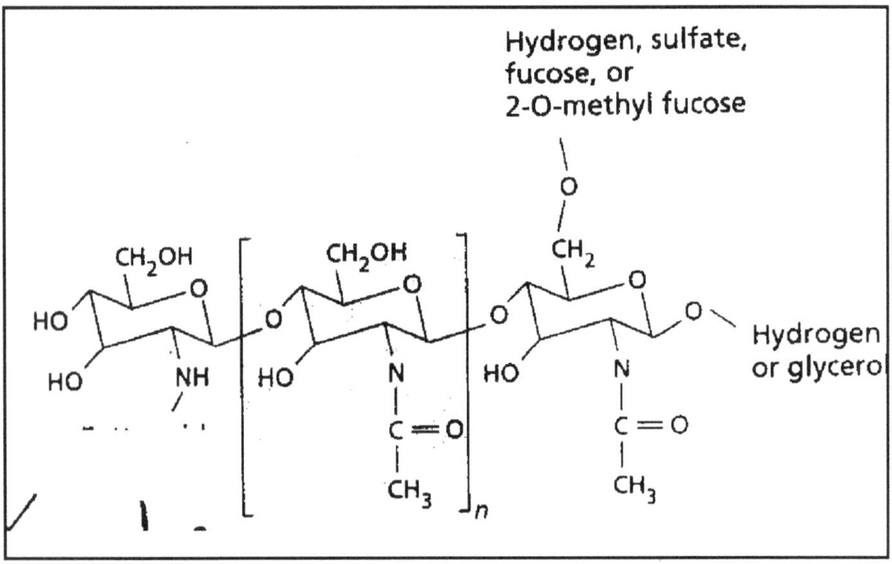

Fig. 4.3 : Nod factors are lipochitin oligosaccharides.

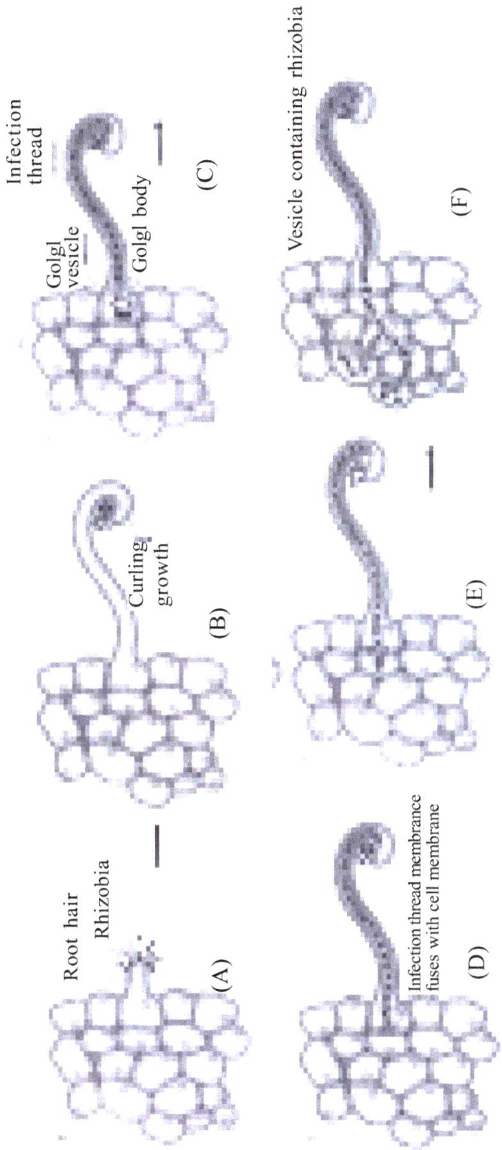

Fig. 4.4 : Formation of Root Nodule

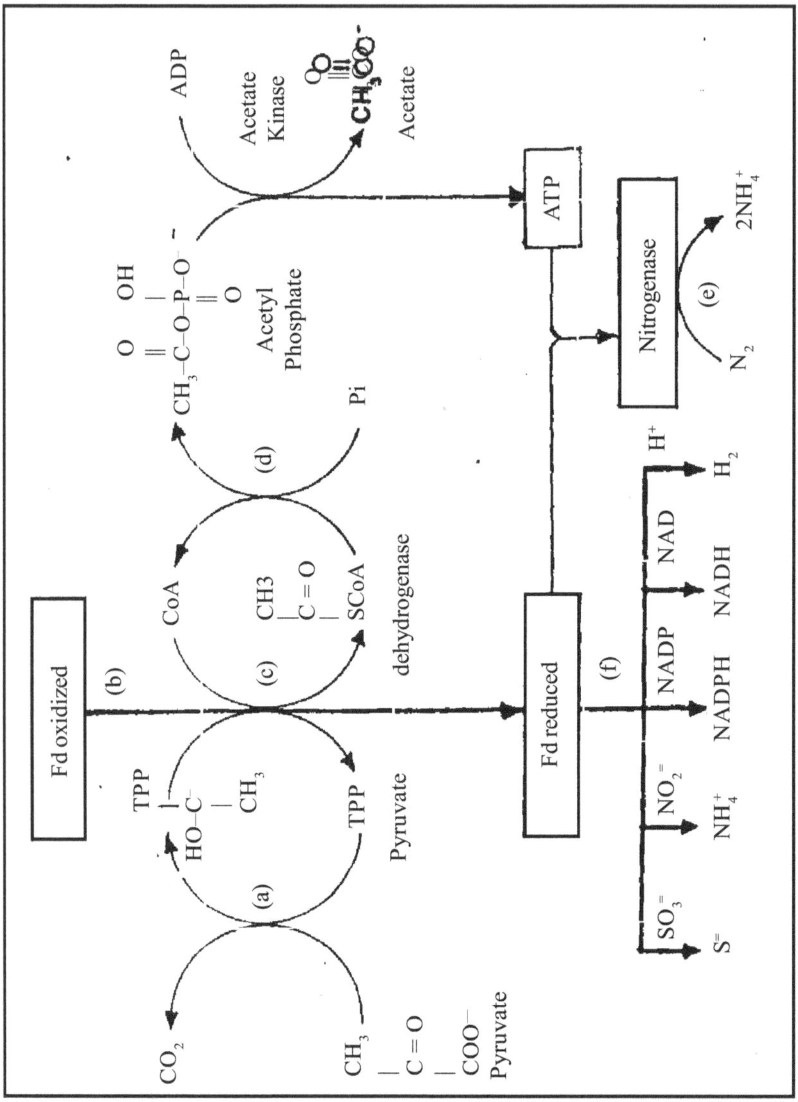

Fig. 4.5 : The ferredoxin linked phosphoroclastic reaction of Clostridium pasteurianun.

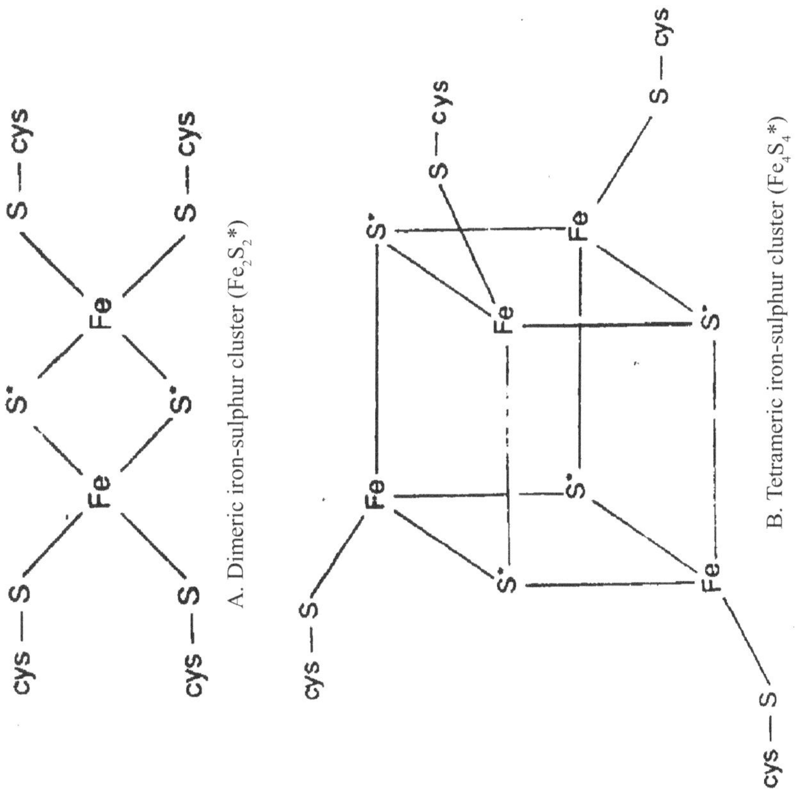

A. Dimeric iron-sulphur cluster (Fe_2S_2*)

B. Tetrameric iron-sulphur cluster (Fe_4S_4*)

Fig. 4.6 : Clusters of iron-sulphur proteins

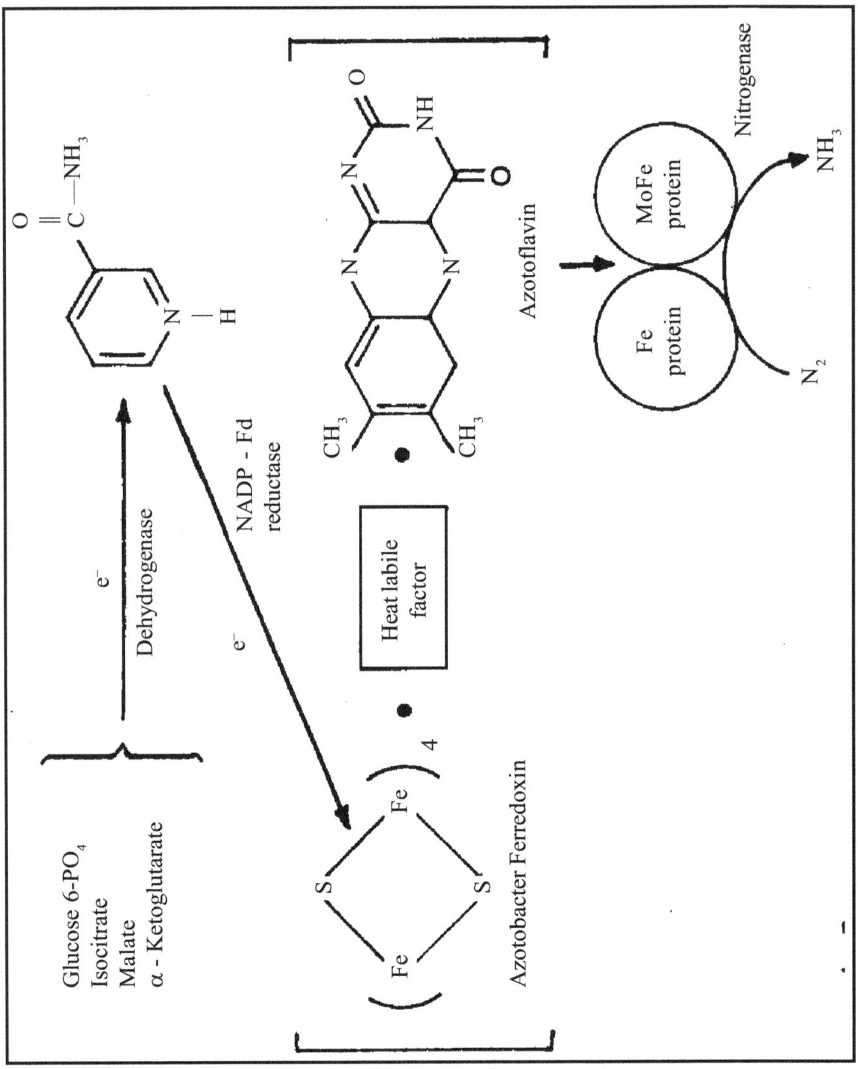

Fig. 4.7 : Electron transport chain (ETC) linking cellular reducing power to nitrogenase in *Azotobacter.*

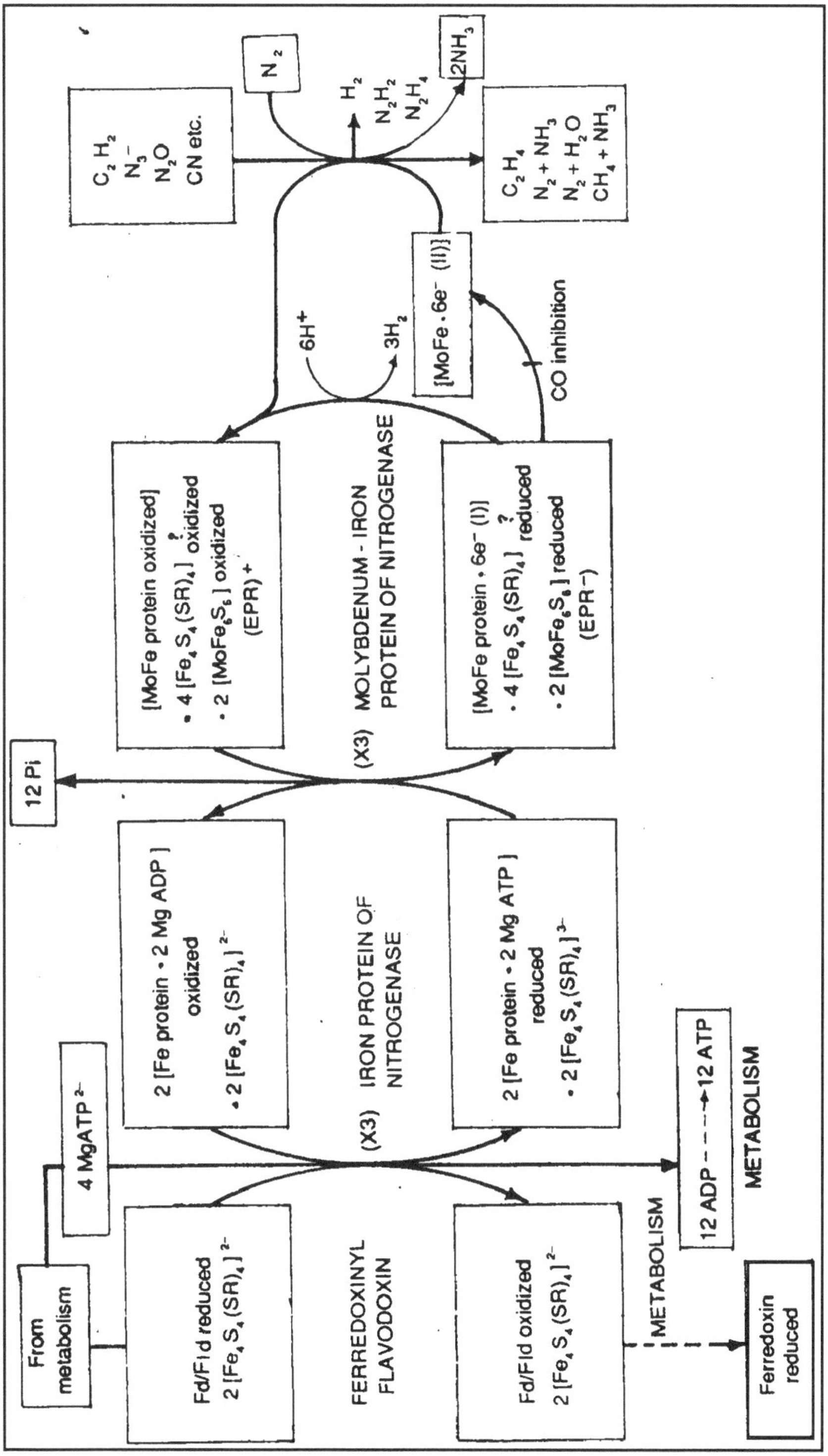

Fig. 4.8 : Mechanism of Nitrogenase Enzyme action (during reduction of NH_2 to NH_4^+).

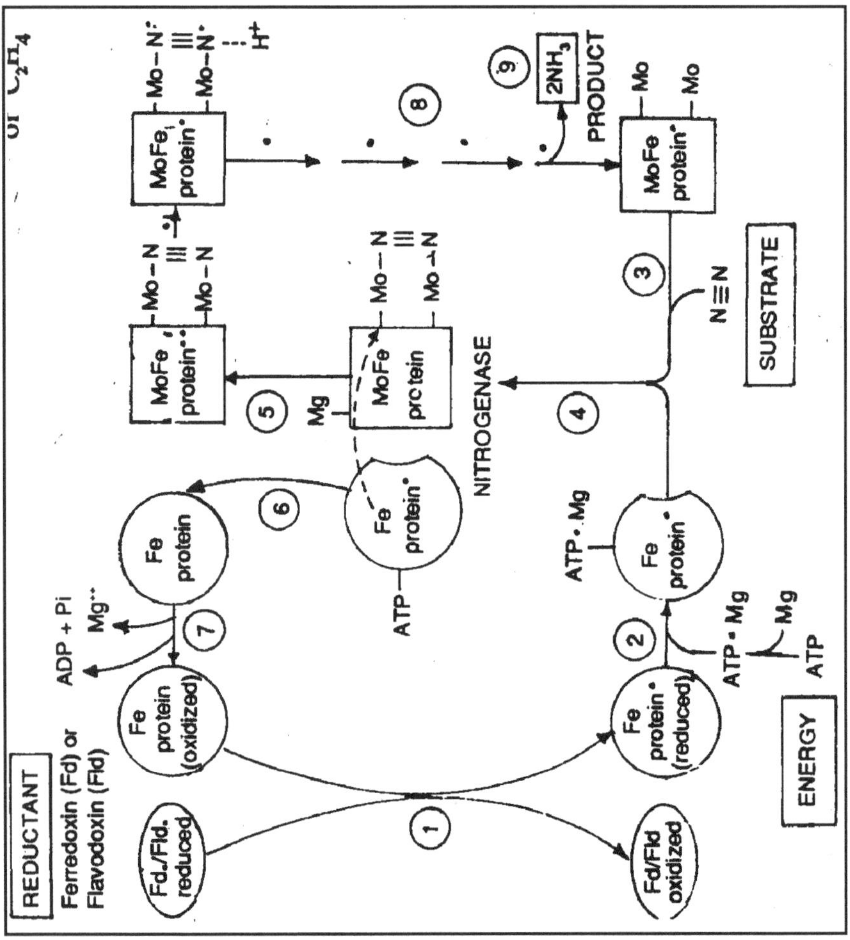

Fig. 4.9 : Action of Enzyme Nitrogenase

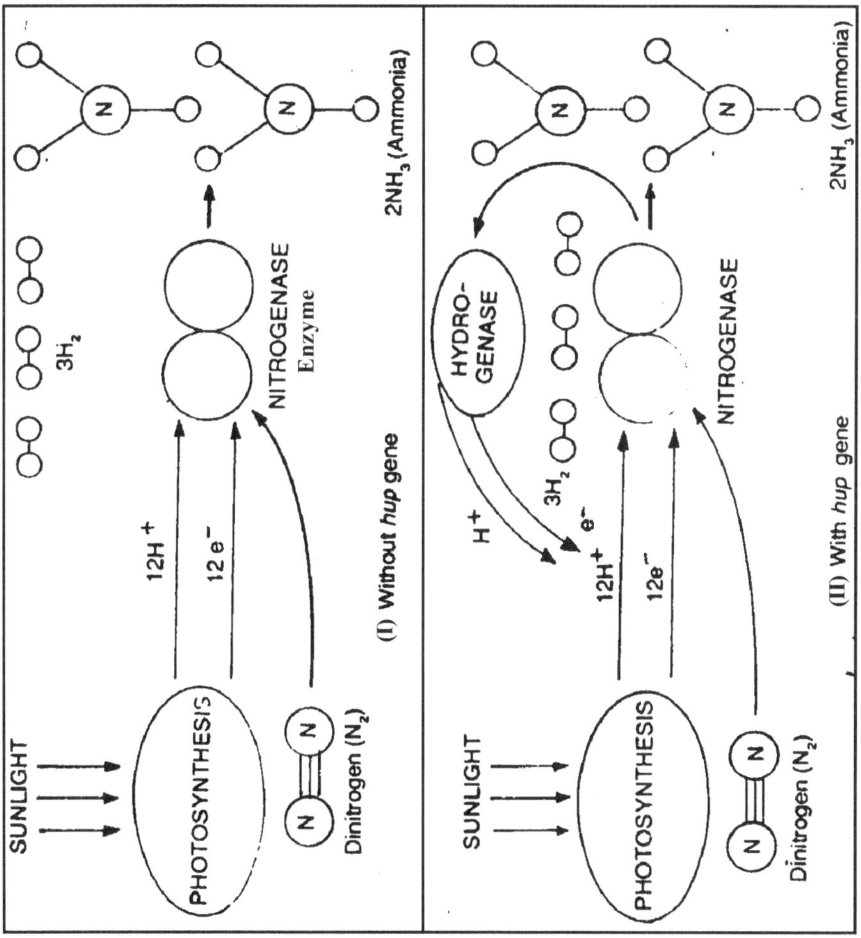

Fig. 4.10 : Hydrogen evolution in *Rhizobium*

CHAPTER - 5

Rhizobium

The role of leguminous plants in improving the soil fertility is known for centuries. The nitrogen uptake in legumes by some unknown process was demonstrated through agronomic experiments by J.B Boussingault in 1838, which was opposed by Liebig and others. The presence of bacteria in the roots nodules of legumes was observed by P.G Lachmann (1858). M. Woronin (1866) further examined and observed them to be rod-shaped and motile. Frank (1879) showed that root nodules are formed by some bacteria which he named *Rhizobium.* W.O Atwater (1884) concluded from his studies on peas that leguminous plants could fix atmospheric nitrogen. Hellriegel and Wilfarth's experiments (1866-88) showed that the growth of non-leguminous plants, barley and oats, etc., was directly proportional to the amount of nitrate supplied while in case of leguminous no such relationship existed and growth of the leguminous plants picked up in some case after the seedling stage. *Rhizobium* was isolated by M.W. Beijerinck (In 1888),a Dutch Botanist from the root nodules in pure culture which lead to study the morphological characteristics and its physiology. He named the bacteria bacillus radicicola. Later, it was renamed as *Rhizobium* by Frank in 1889 *as Rhizobium leguminosarum* (Greek noun *Rhiza* means a root and *Bios* mean life i.e. which lives in root). The efficiency of nitrogen fixed by different plants and strains including cross inoculation groups was examined by A.I. Virtanen and others.

5.1 Classification of *Rhizobia*

Rhizobium belongs to the family Rhizobiaceae, Generally the name rhizobia reveals root and stem nodulating bacteria that live in N_2 fixing symbiosis mainly with legumes plants. It includes six genera. There are four types of *Rhizobium* e.g. *Rhizobium* (fast growers), *Bradyrhizobium* (slow growers), *Azorhizobium* (stem nodulating bacteria), *Sinrorhizobium* (wild strain of soybean etc.). The major species of *Rhizobium* are: *Rhizobium cicero, R.etli., R. japonicum, R. leguminosarum, R. lupine, R. meliloti, R. phaseoli, R trifoli.*, etc. Besides, some names like *Allorhizobium* and *Mesorhizobium* have been considered in the classification.

a) ***Rhizobium japonicum:***It is also known as *Bradyrhizobium japonicum*, the bacterium produces nodules in soybean, *Lupinus* spp and cowpea miscellany. Bacteroids in nodules are longer than normal cells, slender and with only occasionally branched and swollen forms. They are slow growing bacteria (Soybean grain).

b) ***Rhizobium leguminosarum:***This is fast growing species of the pea group which mainly forms nodule in pea, broad beans, lentils and vetch. Bacteroids in nodules are commonly irregular with X, Y, star and club shaped form (Pea grain).

c) ***Rhizobium lupini:***This species cause nodule formation on *Lupinus* sp. and *Ornithopus* sp. Bacteroids are vacuolated rods and seldom branched (Lupin grain).

d) ***Rhizobium meliloti:***The bacterium causes nodule formation in sweet clover, alfalfa and fenugreek; they are rod shaped, branched and grows fast on yeast extract medium (Alfalfa grains).

e) ***Rhizobium phaseoli:***The bacterium produces nodules on temperate species of *Phaseolus*. Bacteroids are rod shaped and often vacuolated with few branched forms (Bean grain).

f) ***Rhizobium trifolium:****Trifolium* spp. nodules are produced by this bacterium. Bacteroids are pear swollen, vacuolated, rarely X and Y shaped (Clover grain).

g) ***Rhizobium* sp.:**They nodulates cow pea group. Bacteroids are irregular shaped (Cowpea grain).

It is worth mentioning that all rhizobia cannot form nodules/fix N on all legumes. There is certain host specificity. In general, *Rhizobium*has been decided into two groups based on their growth habits-

a) **Fast Growers** which produce acid on yeast mannitol agar medium (*R. loti, R. leguminosarum* and *R. meliloti*).

b) **Slow Growers** which produce alkali on yeast mannitol agar medium (*R. japonicum* and *Bradyrhizobium* species infecting tropical legumes). Sometimes, *Rhizobium* is classified in to three groups on the basis of their symbiotic responses as:

- **PE** (Promiscuous and effective e.g. species infecting *Arachis*, *Cajanus* and *Vigna*.
- **PI** (Promiscuous and Infective e.g. species in symbiosis with *Vicia, Lens* and *Pisum*.
- **S** (Specific as in association with *Cicer*).

Rhizobia are capable of nitrogen-fixing symbiosis with more than 1, 130 species of leguminous plants and parasponia among Ulmaceae (Trinick, 1973 and Vincent, 1974). Most rhizobia are specific in their association with legumes. Rhizobium have been taxonomically grouped and designated to a particular species based on crop inoculation group concept proposed by Fred, Baldwin and McCoy (1932). Cross inoculation group of Rhizobium is given in Table 5.1.

Table 5.1 : Cross Inoculation Groups of *Rhizobium*

Types	Bacterium Biovars	Host
Rhizobium (R), fast growing, sub-polar, flagellated strain	*Rhizobium meliloti*	Alfalfa (*Medicago sativa*), Sweet clover (*Melilotus alba*), fenugreek/methi (*Trogonella foenumgraecum*)
	Rhizobium leguminosarum bv. viciae	Pea (*Pisum sativum*) Vetch (*Vicia sativa*), Lentil (*Lens culinaris*), Faba beans (*Vicia faba*)
	Rhizobium leguminosarum bv. phaseoli	Beans (*Phaseolus vulgaris*)
	Rhizobium leguminosarum bv. trifolii	Clovers/Berseem (*Trifolium* sp).
	Rhizobium loti	Trefoil (*Lotus* sp).
	Rhizobium fredi	Soybean (*Glycine max, Glucine soja*)
Bradyrhizobium (B), slow grower, polar or sub polar, flagellated strain	*Bradyrhizobium japonicum*	Soybean
	B. elakanii	soybean
	B. spp *Cowpea*	Cowpea (*Vigna unguiculata*), Green gram (*Vigna radiata*), Pigeonpea (*Cajanus cajan*), Chickpea (*Cajanus cajan*), Peanut (*Arachis hypogea*)
Azorhizobium (A) fast grower	*A. Caulinodans*	Root and stem nodules on Sesbania (*S. rostrata*)

Source : Bhattacharya and Tandon (2012)

The classification of nitrogen-fixing bacteria (formerly and colloquially known as rhizobia– defined as N_2 bacteria that form root nodules on legumes and possess so-called nod (nodulation) and nif (nitrogenase genes) is undergoing rapid and major modifications (Giller, 2001). It has been accepted for a long time that these rhizobia were all closely related and belong to the alpha-proteobacteria. A simple subdivision split those bacteria in fast-growing and slow-growing bacteria. It is now clear that there are several other bacteria, also belonging to the beta-proteobacteria, that possess the same ability to nodulate and to fix nitrogen. The genes that are responsible for these processes are often located on plasmids or symbiotic islands that can be transferred laterally to other bacteria, giving rise to new N_2fixing bacteria. The consequence of horizontal gene transfer is that a bacterial phylogeny (based on ribosomal genes) is not congruent with a phylogeny based on the genes for nodulation or with genes for nitrogen fixation. A comparison of phylogenies based on rDNA, nod genes and nif genes has shown that the nodulating *Methylobacterium* species have derived the nif genes through horizontal gene transfer from *Gluconacetobacter diazotrophicus*, an endophyte that can fix N in association with sugarcane. Species of *Burkholderia* (Beta-proteobacteria) with N_2 fixation capacity have derived these genes from rhizobia from the alpha-proteobacteria (Rivas *et al.,* 2009). Some rhizobia are therefore more closely related to free-living or associative nitrogen-fixers than to true rhizobia. Phylogenetic studies have also shown that certain bacteria that cannot fix nitrogen are still closely related to rhizobia. A famous example is *Rhizobium radiobacter* (formerly known as *Agrobacterium tumefaciens*), a bacterium that forms gall-like swellings on roots of many plant species.

Current taxonomies list 88 species of rhizobia (Rivas *et al.,* 2009) or 98 species (Weir, 2012) in two classes, three orders, seven families and 13 genera (Table 5.2). The number of rhizobia is likely to grow in the coming years. Some other 17 bacteria have been isolated from legume-nodules but are more likely opportunistic colonizers of nodules as they do not possess nod genes. At present taxonomy of rhizobium is undergoing frequent change with recent technological advances like DNA-DNA and DNA-tRNA hybridization, t-RNA catalogue, rDNA sequencing etc. The diversity of rhizobia with respect to genus, species, subspecies, biovars etc. are being studied around the cell protein analysis, multilocal enzyme electrophoresis (MLEE), analysis of cellular fatty acids (FAME), DNA base composition, Restricted Fragment Length Polymorphism (RFLP) etc. Based on 16s rRNA sequence analysis the phylogenic (tree like) structure of *Rhizobium* against nod A (Raychaudhari *et al.,* 2007). The structure varies depending on different nod gene (nod B, nod C and nod J etc.) (Fig. 5.1). Such information is also being generated for other micro-organisms and is primarily of interest to students and researchers in molecular biology or biotechnology.

Table 5.2: Classification of *Rhizobia* after Weir (2012).

Class	Genus	Species Number
Alpha-proteobacteria	*Rhizobium*	30 (+11)
	Mesorhizobium	21(+1)
	*Ensifer (*Formerly *Sinorhizobium)*	17
	Bradyrhizobium	9(+1)
	Phyllobacterium	3
	Microvirga	3
	Azorhizobium	2
	Ochrobacterium	2
	Methylobacterium	1
	Devosia	1 (+1?)
	Shinella	1
Beta-proteobacteria	*Burkholderia*	7 (+?)
	Cupriavidus	1 (+?)

Species number indicates number of rhizobia, between brackets the number of non-rhizobial species of that genus. Source: Koele *et al.,* (2014)

5.2 Characteristics of *Rhizobium*

Rhizobium bacteria are pleomorphic can exist in different forms within a species or strain, generally rod shaped (0.5-0.9x1.2-3.0μm), gram negative, non-spore former, and contain granules of poly β hydroxyl butyric acid (PHB). These are motile by one polar or sub polar flagellum and are able to grow well under oxygen tension (Fig 5.2). For satisfactory growth, the optimum temperature is 25-30^{0}C and optimum pH is 6.0-7.0. Its colonies are circular, convex, semi-transculent, raised and mucilaginous. *Rhizobia* can fix 20-200kg N/ha/yr depending on the crop and growth conditions. Its inoculation is recommended for legumes (Pulses, leguminous, vegetables such as various beans, groundnut, soybean forage, legumes and clovers etc.)

5.3 *Rhizobium* Symbiosis

Legume-*Rhizobium* symbiosis is an important facet of symbiotic nitrogen fixation which is exploited to benefit agriculture and its sustainability. Over a century ago German scientists Hellriegel and Wilfarth experimentally demonstrated the nitrogen fixationin legume nodule by nodule inducing ferment (*Rhizobium*): the stage was set for the popularity of the *Rhizobium* inoculation technology world over. The legume-*Rhizobium* symbiosis is estimated to account for 40% of the worlds combinednitrogen. Pulses are the basic source of protein in Indian diet. They play an important role in farming systems and sustainable agriculture due to their contribution in meeting about 80-90% of the nitrogen requirement of leguminous crops and showed the residual effect on the following crop adding nitrogen to soil organic pool. Such effects are observed either by *in situ* incorporation of legume residues or via cattle dung and urine.In this symbiosis macro-symbiont is the legume plant and micro-symbiont is the prokaryotic bacteria (*Rhizobium*). The Macro-symbiont legume belongs to Leguminaceae, divided into three subfamilies comprising of 700 genera and 14,000 species. 6 Only about 200 of these are cultivated by man. Important legumes cultivated in India are pigeonpea, chickpea, soybean, lentil, lathyrus, rajmash, alfalfa, clover, beans and peas. India accounts for 25% of world pulse production. Atmospheric nitrogen fixation is carried out by the enzyme nitrogenase of the bacterium with the assistance of nodulins (legume plant proteins) and transferred to plant in a true spirit of symbiosis. Estimates of N_2 fixation in different Indian pulse crops are presented in Table 5.3.

Table 5.3: Estimation of N_2 Fixation (kg/ha/yr) by *Rhizobium* in Different Crops.

Crop	N_2 fixed	Crop	N_2 fixed
Chickpea	26–63	Pigeonpea	68–200
Cluster bean	37–196	Soybean	49–130
Cowpea	53–85	Peas	46
Groundnut	112–152	Alfalfa	100–300
Lentil	35–100	Clover	100–150
Mungbean	50–55	Fenugreek	44

Source : Brahmaprakash and Sahu (2012)

Inoculation response of pulses is far from desirable, at best inconsistent and dependent on many variables. Most cultivated soils to legumes are known to harbor cowpea group of rhizobia and nodulation surveys indicate a need for inoculation every season for majority of the pulses cultivated in India. The competition of (inefficient) native strains to (efficient) inoculant strains appears to be a bottleneck in realizing symbiosis: An interaction between two organisms wherein both are benefited higher yields from *Rhizobium* inoculation. The yield increase due to inoculation in pigeonpea varied from 1.2–20.3%, 8–47.8%, and 1.8–26.4% in 1992, 1993 and 1994 respectively, in different locations in India. The grain yield increase may also appear to be an interaction of varieties and strains of *Rhizobium* (Brahmaprakash and Hegde, 2005). A complementary coordinated effort on the part of plant breeders and microbiologists is now necessary to successfully select a high yielding variety with elevated nitrogen fixing abilities for sustainable agriculture.

5.4 Associated Plant Species

Symbiosis with rhizobia and N_2 fixation has been demonstrated in around 20% of the around 13,000 leguminous species, belonging to around 60% of the legume genera (Corby, 1988; de Faria *et al.,* 1989; Sprent, 2001; Sprent, 2007). However, a large number of legumes remain unresearched and the capacity to form symbiosis with rhizobia has not been demonstrated. It is therefore likely that the final number of N_2fixing legumes is much higher, especially in the papilinoid groups. However, from the number of non-fixing genera, especially in the more primitive legume groups, but also in more advanced groups where the ability to fix N has been lost; it cannot be assumed that any legume can fix atmospheric N (Giller, 2001). This may be another difference with AM, where it can almost be assumed that an unknown plant is very likely able to form AM. Some important crops as groundnut (*Arachnis hypogaea*), chickpea (*Cicer arietinum*), soybean (*Glycine max*), pigeonpea (*Cajanus cajan*), New World pulses (*Phaseolus* sp.), Old World pulses (*Vigna* sp.), pea (*Pisum sativum*), lentil (*Lens culinaris*), and faba bean (*Vicia faba*) are associated with rhizobia. Other important legumes occur in pastures such as clover (*Trifolium* sp.) or are trees in agroforestry systems (*Leucaena leucocephala, Faidherbia albida*, and *Acacia* sp.).

5.5 Uptake of Other Nutrients than N and P

Due to increased uptake of N through N_2 fixation, other nutrients are usually limiting. Several elements are in higher demand by legumes than by other plants, and these include phosphorus (P), calcium (Ca), boron (B), molybdenum (Mo), cobalt (Co), and iron (Fe). These macronutrient and micronutrients play important roles in energy generation, nodulation and N_2 fixation (Redondo-Nieto *et al.,* 2003).

P is needed for generation of ATP whereas B is needed for the maintenance of nodule cell wall and membrane structure.(Bolaños *et al.,* 2001 reported the requirement for B has been reported for rhizobial infection and the nodule invasion process, and is also essential for symbiosome development and bacteroids maturation and for early events in plant-bacteria signaling (Redondo-Nieto *et al.,* 2001). Ca in plants is involved in the structure and function of cell wall and membrane, and cytosolic free Ca^{2+} is important as a second messenger in the signaling mechanism of many plant responses (Bush, 1995). A high Ca^{2+} supply increased the number of nodules (Lowther and Loneragan, 1968), and Munns (1970) reported that Ca^{2+} is especially required for early infection events. Furthermore Mo, Co and Fe are crucial for N-fixation as constituents of several enzymes such as nitrate reductase and nitrogenase (Giller, 2001). Some, but not all, legumes have a large demand for K. It is therefore possible that legumes are more prone to micronutrient co-limitation, the more so as availability of these micronutrients is pH dependent. At low pH, Mo deficiency is more likely to occur, whereas Fe deficiency is more likely in alkaline soils. However, the importance of micronutrient deficiency has likely been under estimated.

5.6 Host Relationships

The symbiosis between legume host plant and rhizobia shows several degrees of selectivity, implying that only certain strains of rhizobia can associate with (or nodulate) a given legume species, and there must be recognition between the host legume and the rhizobia (Giller, 2001). Most soils contain indigenous bacteria that are able to form nodules in specific legume plants; however in agriculture inoculation of the soil with specific strains occurs as well (see below). The legume host plant exudes flavonoid and isoflavonoid molecules that stimulate the *nod* genes of the rhizobia. In turn the rhizobia secrete a nod factor into the rhizosphere to confer specificity of recognition 18 between rhizobia and host plant. Upon recognition the plant will start to form nodules that are then invaded by the rhizobia who will start fixing N (Giller, 2001). Maximum N-fixation is only reached when the plant is sufficiently nodulated, which only occurs under continuous N limitation of the soil (Bonilla and Bolanos, 2010). The initial molecular dialogue between rhizobia and legumes is based on the molecular dialogue between AMF and plants (Bucher, 2007).

5.7 Genetic Variations

Genetic variations among species, genetic variation among and between plant species that allows optimizing use through targeted plant breeding.Similar to the AM symbiosis, genetic variation exists among plant varieties, cultivars and landraces for their ability to associate with rhizobia and fix nitrogen. This genetic variation has been extensively researched in cowpea and soybean (Belane and Dakora, 2010). Especially the latter species has been subject to breeding efforts that took the ability to nodulate and fix nitrogen into account. Gwata *et al.,* (2005) have suggested that promiscuity is the result of the presence of both recessive alleles at two loci. For other species efforts to breed for increased N-fixation are uncommon, because many plant characteristics contribute to N-fixation and breeding for increased fixation would need to be environment specific (Giller, 2001). Soybean, however, is being bred according to two alternatives: promiscuity and specificity. In Brazil and North America, soybean was bred for an even more restricted host selectivity such that the plant cannot or hardly any longer associate with the naturally occurring ineffective strains (Herridge and Rose, 2000; Hungria *et al.,* 2006; Peoples *et al.,* 2009). Such breeding inevitably implies the use of inoculation (often through seed coating) that needs to be regularly repeated in order to maintain a sufficient inoculum density. The practice entails higher costs, and these may not be in reach of resource-poor farmers. An alternative soybean breeding strategy was therefore applied at the International Institute of Tropical Agriculture (IITA) in Nigeria, where they developed soybean genotypes with reduced nodulation specificity that can nodulate effectively with indigenous *Bradyrhizobium* strains populations in Nigeria. While these indigenous strains fix lower amounts of N, they guarantee N fixation under a wide range of agro-ecologies and hence avoid the problems of scarcity or absence of highly effective inoculum (Mpepereki *et al.,* 2000; Abaidoo *et al.,* 2007). Similarly,nodulation of a different promiscuous soybean in Zimbabwe were observed byMusiyiwa *et al.* (2005). Promiscuous soybean can still fix higher amounts of N and become more productive if inoculated with more specific strains of *Bradyrhizobium* (Thuita *et al.,* 2012). Also in the case of soybean, formation of nodules is no measure for effectiveness of nitrogen fixation and increased growth, as shown by Zengeni and Giller (2007).

However, we know much more of soybean in this regard than of most other legumes.) Other legumes are less selective, and this has allowed successful introduction of South American bean species (*Phaseolus* sp.) into Africa (Ojiem *et al.,* 2006). Plant introductions have in some cases incidentally introduced rhizobial inoculum into new continents. These exotic inocula may have subsequently switched to indigenous vegetation where they outcompeted native inoculum. At the same time they were less effective in N-fixation with these native plants, thereby shifting the competitive balance between indigenous and exotic legumes allowing the exotic legumes to cause what has been described as invasion meltdown (Rodriguez-Echevarria, 2010; Rodrigeuz-Echevarria *et al.,* 2012). This switching has contributed to the invasions of Australian *Acacia* species around the globe.

Because the indigenous soil rhizobia are probably also more effective in old world than in New World soils, due to a longer history of cropping of legumes, developing promiscuous legume breeds is preferred over breeding for high specificity (Herridge, 2008). Thuita *et al.* (2012) show that promiscuous soybean varieties respond to inoculation, and that nodulation, N_2fixation and biomass yield are improved if the strain is infective and effective. Thus commercial products produced elsewhere can be an important source of effective strains for use in areas where soybean is being introduced or where low populations of indigenous rhizobia hinder biological N-fixation. Nevertheless, They also found no benefits of the commercial products Twin-N and Leguspirflo, which contain endophytic bacteria targeted to increased nodulation of legumes as well as enhance N-fixation in non-legumes. Kiers *et al.* (2007) have suggested that breeding programs for soybean, which took place in field that were well fertilized, may inadvertently have been selected for reduced rhizobial benefit, in a process that is analogous to the decreased mycorrhizal benefit due to evolutionary changes in AMF. For legumes other than soybean, much less research has been carried out, and it remains uncertain under what conditions inoculation with rhizobia may enhance N-fixation and crop yield compared to management of locally existing rhizobial inoculum through good agronomic practice (Giller, 2001).

5.8 Optimal Conditions

The infection rate and effectiveness of rhizobia on legumes are influenced primarily by the N status of the soil, and N-poor conditions are needed for full nodulation and N-fixation (Bonilla and Bolanos, 2010). However, other nutrients should not be limiting, especially P, which can be best prevented by associations with AMF (Bucher, 2007). Under highly acidic conditions (tropical oxisols and ultisols) both P-limitation and aluminum toxicity may hamper successful N-fixation (Hungria and Vargas, 2000). Saint Macary *et al.* (1992) furthermore noted that the right bacterial strains need to be present for the cropped legume and that enough of inoculum or indigenous population is applied or present. Peoples and Herridge (2002) remarked that basic improvements in crop agronomy are probably most promising to enhance crop yield, nutrition (both a balanced availability of N and P, and sufficient K and Ca; but also including trace elements such as Mo, Fe, Co, B), weed control, diseases and pests, and cropping sequence and intensity need to be managed.

5.9 Quantitative Estimates of Nutrients

Kuyper and Giller (2011) tried to quantify in monetary terms the ecosystem services delivered by rhizobia though N savings. That calculation suggested a monetary value of around US $90 billion. Carryover effects of increased N inputs in soils and higher productivity of subsequent crops and over-yielding of cereals in intercropping systems were not included. The non-N benefits of intercropping or rotations with legumes as discussed above (increased organic matter and hence potential mitigation of climate change (Jensen *et al.,* 2012), decreased pathogens, increased soil microbial activity and diversity) are important ecosystem services that have also not been quantified, as they cannot easily be attributed to rhizobia. Finally, as pointed out by Figueiredo *et al.,* (2013), there is close to zero fossil fuel use linked to the use of rhizobia, in contrast with the use of fertilizers, and also lower NO_2 emissions. However, quantification of the amount of N fixed by rhizobia remains difficult (Herridge, 2008). An estimate of amounts of N fixed by different legumes under optimal conditions is given in Table 5.4. Figueiredo *et al.* (2013) divided the difficulty in estimating N-fixation in three classes such as:

a) Methodological problems in field-scale N_2 fixation estimation.

b) Highly variable N_2 fixation rates, which are strongly affected by environmental and agricultural concerns.

c) Difficulty in estimating individual cropping systems, worldwide distribution, and cultivated areas. Yet, as there are global estimates of the amount of reactive N that is added annually to the earth system through legumes that have been planted and managed by humans (Vitousek *et al.,* 2013), a global estimate can still be attempted.

Table 5.4 : Fractions and Amounts of N Fixed (kg N/ha/Yr) by Grain Legumes

Legume species	Fraction of N fixed	Amount of N fixed
Peanut	16-92	21-206
Pigeonpea	0-100	0-166
Chickpea	0-96	0-124
Soybean	12-100	14-188
Lentil	9-91	4-83
Bean	0-73	0-125
Cowpea	32-76	9-201

Source : Koele *et al.,* (2014)

5.10 Legumes, *Rhizobia* and *Arbuscular Mycorrhizae*– A Case of Synergy:

The fact that legumes associate both with rhizobia (for enhanced N) and AMF (for enhanced P) has given rise to the question of in what way these organisms interact. Experiments under controlled conditions have shown that combinations of rhizobia and AM fungi are usually more productive than plants that are only infected by rhizobia or AM fungi. This higher yielding when both microbial symbionts are the interactions between both groups of mutualistic symbionts has synergistic effects, a term that, however, is poorly defined at best. It is probably best to try to define synergy in such a way that a difference can be made between additive effects and synergistic effects. We define synergy if, in an experimental setting where both rhizobia and AM fungi are separate factors, the interaction terms rhizobia × AM fungi are significant. Synergy can then be both positive (when the yield is higher than the additive effect) or negative (when the yield is higher than would be predicted under an additive model). In case there is no significant interaction term, the yield is additive. Based on that criterion, Larimer *et al.* (2010) tested the effect of combinations of rhizobia and AM fungi, and concluded that effects were not synergistic but additive.

5.11. *Rhizobia* and Cereals

Indigenous rhizobia can colonize the rhizosphere of cereals and in some cases provide benefits to growth (Chaintreuil *et al.,* 2000; Yanni *et al.,* 2011). Thus inoculation of non-legumes with rhizobia has attracted interest as biofertilizer (Mia and Shamsuddin, 2010). Matiru and Dakora (2004) showed that roots of sorghum and millet landraces from Africa were easily infected by rhizobial isolates from five unrelated legume genera, and that with sorghum in particular, plant growth and phosphorus uptake were significantly increased by rhizobial inoculation, although it is not clear whether the bacteria increased P uptake or stimulated the plant to take up more Phosphorus. Matiru and Dakora (2004) suggested that field selection of suitable rhizobia/cereal combinations could increase cereal yields through direct benefits by the cereals from the legumes-rhizobia mutualism. Similarly, Yanni *et al.* (2011) pursued the use of rhizobia associated with clover for increased productivity of rice in Egypt. Biswas *et al.,* (2000) found increased uptake of N, P, K and Fe in rice inoculated with rhizobia and other growth-promoting bacteria, even though rice does not commonly associate with rhizobia and it does not form nodules. They argue that the rhizobia and other rhizosphere bacteria altered root morphology (through hormonal effects) including more root hairs, increasing the volume of soil for nutrient uptake. However, Yanni *et al.* (2001) and Vargas *et al.* (2010) warn that the effects of rhizobia on cereal growth and yield are not consistent or even unambiguously demonstrated, and that direct effects (Increased N uptake through N-fixation) and indirect effects (Increased N uptake from mineralized organic N in soil, protection against pathogens) are hard to assess separately.

5.12 Testing of Nodulation Ability by *Rhizobia*

Leguminous plants (soybeans, beans, peas, chickpea, alfalfa) form root nodules that contain populations of *Rhizobium*, a nitrogen fixing bacterium, The symbiotic association the plant and *Rhizobium* is initiated when bacteria in the soil attach to root hairs. This highly specific attachment process is mediated by plant proteins, the lectins, that bind the bacteria to the surface of the root hair, that is then penetrated by the microbes. The bacteria from the root hairs travel to inside and infect the root cells. The infected root cells divide and form a nitrogen fixing nodule. The nodule provides the anaerobic environment necessary for nitrogen fixation. The ability of *Rhizobium* to infect small-seeded legumes can be tested on nitrogen- free media containing agar, vermiculite, sand, etc. There are several methods available for testing *Rhizobium* isolates for nodulation ability (Fig.5.3).

Requirements

Rhizobium trifolii (2-4 days old slants), Berseem (*Trifolium alexandrinum*) seeds, Modified Jensen's agar medium, Culture tubes (150 x 20 mm), H_2SO_4, Dessicator with $CaCl_2$, Black paper, Sterile water, Cotton, Specimen tubes, Sterile forceps, Aluminium foil, Growth cabinet (fitted with air-conditioners or water coolers and warm-white fluorescent tubes supplemented with one or two in can descent lamps for each row of a pair of fluorescent lights), Inoculating loop and Bunsen burner.

Procedure

a) Preparation of nitrogen-free agar medium of the following composition:- $CaHPO_4$(1g), K_2HPO_4(0.2g), $M_gSO_4 7H_2O$(0.2g), NaCl(0.2g), $FeCl_2$(0.1g), Agar(15g), Trace elements, (1.0ml (solution containing Bo-0.05%, Mn-0.05%, Zn-0.005%, Mo-0.005%, Cu-0.002%) and Distilled water (1000ml).

b) Weigh the constituents and dissolve (except K_2HPO_4 and agar) in 200ml distilled water, dissolve K_2HPO_4 in 100ml of distilled water separately. Dissolve agar in 400ml of warm distilled water, Mix all the constituents and raise volume to 10litre by the addition of distilled water. Adjust pH to 6.5-7.0.

c) Place the test tubes and seeds in a desiccator over $CaCl_2$ for 24hrs.

d) Surface sterilizes the seeds with conc. H_2SO_4 for 10-20minutes.

e) Wash the seeds in quick changes of sterile water four to five times to avoid burning of the seed coat.

f) Pour 13-15ml of agar medium into each culture tube.

g) Plug the tubes with loose cotton plugs.

h) Sterilize the plugged tubes at 15 lb/in^2 pressure (121°C) for 20 minutes.

i) Keep the tubes in a slanting position in such a way that the slopes of the agar reach the top of the tube and allow the medium to solidify.

j) Cover the top of each tube by thin circles of aluminium foil held in place with a rubber band.

k) Aseptically make a small aperture on the aluminium foil with a sterile knife and fit the aperture with cotton wool for aeration and for replenishing nutrient solution in tubes.

l) Transfer aseptically, using a pair of sterile forceps, grown seedlings in a second tube (opposite to the hole carrying the cotton plug) of the test tube in such a way that the root system lies on the agar slope and test tube in such a way that the root system comes out of the tube.

m) Cover the tubes carrying seedlings with large specimen tubes containing moist sterile cotton to prevent desiccation of the seedlings.

n) Allow the seedlings to establish in the agar slopes.

o) Remove the cover tube after the establishment of the seedlings.

p) Cover the lower portion of the tube with black paper to avoid light from falling on the root system.

q) Make a suspension of *Rhizobium* from scrapings of growth from 2-4 days old fresh slants.

r) Mix the bacterial suspension with ¼ strength nitrogen free nutrient solution (medium without agar).

s) Pour 5-10ml of the *Rhizobium*-nutrient solution mixture into each tube from the second hole.

t) Incubate the inoculated tubes in growth cabinets for 3 to 4 weeks.

Observation and Results

Remove the seedlings from the inoculated tubes periodically for the development of nodules on the roots. Record the time taken for the development of nodules post-inoculation and number of nodules formed per plant. Cut off a nodule from a legume, stain the bacterium with methylene blue for 30 seconds and observe under oil-immersion for the cells of *Rhizobium*. A loopful of the crushed nodule may be streaked on mannitol-yeast agar and incubated at room temperature for 7 days. Compare the growth and stained preparation of the isolate with the original culture used for inoculation.

5.13 Screening Test for Efficient *Rhizobia*

The conventional method to screen or select the most efficient strains of *Rhizobium* from a large number of strains/isolates is to examine its potential for nitrogen fixing ability on its homologous legume. This is mainly done by growing the plants under aseptic conditions, inoculating the *Rhizobium* strain to test plant followed by examining nodulation and measurement of total N content of the dry matter yield of the plant after 4-6 weeks of plant's growth. The total nitrogen is proportional to dry matter yield of plant, nitrogen per cent remaining almost constant.

5.13.1 For Small Seeded Legumes

Thorton (1930) and Jensen (1942) used different media for seedling agar in nodulation studies various sizes of tubes plugged with cotton based on the need have been used. The volume of agar to be poured into the tube depends on the requirement of study. Gibson (1963) modified the seedling agar by adding several trace elements. Nitrogen controls are usually maintained by adding KNO_3 solution. This method allows the shoot system to grow freely in air while root system is still enclosed in the tube and completed in following steps:

a) Nitrogen free agar is poured (13-15ml) into tubes (150 mm×20 mm), slanted and allowed to solidify such a way that the slope of the agar reaches the top of the tube.

b) The top of each tube is covered with by thin circle of aluminium foil and held in place with a rubber band.

c) A small aperture is made on the foil and fitted with cotton wool which not only affords aeration but also acts as ad inlet for adding nutrient solution in tubes. Sterile conditions have to be maintained to prevent contaminants.

d) Aseptically grown seedlings are placed in a second hole in such a way that root system touches the agar slope and the shoot system comes out of the tube. This is done by using a pair of sterile forceps. This method of growing seedlings is found better than completely enclosed seedling agar method.

5.13.2 For Large-Seeded Legumes

For the screening of large seeds, following three methods are used as (i) Leonard Jar Test. (ii) Pot Culture Test and (iii) Field Test

Leonard Jar method

Leonard (1944) and Vincent (1970) used a wide bottle of 750 ml capacity with its bottom cut off so as to provide a flat ground finish is inverted in a glass jar of suitable dimensions as the neck of the bottle properly fits. A suitable wick (cotton wool or ordinary cotton lamp wick) is passed through the inverted bottle neck in such a manner that it is not only dipped in the nitrate-free nutrient solution in the jar, but also touches upper most region of sand medium in which the seed/seedlings are growing.Seeds are sterilized with 95% ethyl alcohol, immersed for 3 min. in 0.2% $HgCl_2$ acidified with 5 mg/litre HCI (Conc.). Seeds are washed with sterile water(5 times). The seeds then are germinated in the folds of moist filter paper on plain agar (with water) in petri plates/dishes which are inverted to provide seedlings with uniform straight roots. The seedlings are then carefully placed in the Gibson's tube so that the emerging root just touches the surface of the agar slope in the tube. The lower portion of the tubes is put in the racks for convenient handling. One-fourth strength of sterile seedlings solution is added through the hole in the cap to within 12mm of the top of each tube.

Inoculum of specific *Rhizobium* grown for 2-4 days old is added through the hole on the aluminium cap. Upto 2-3 weeks, no watering is needed. After 4-5 weeks, quarter strength seedling solution should be added at 2-3 days intervals. After 7-10 days nodulation in the roots of the plants may be examined. After 4-5 weeks, dry matter yield of plants may be recorded and the nitrogen fixing efficiency is calculated in terms of increase in dry weight of plants over control.

The cotton wick is meant for steady supply of plant nutrients to the growing plant by the capillary action of the sand. The sand substrate is wetted with one-fourth strength of the nutrient solution until the solution begins to drip through the wick at the bottom. The jar at the bottom is filled with same nutrient solution and the top portion of the assembly is covered with a petri dish. The assembly may be covered with water proof paper bag with a help of rubber bands so that the roots of the plants during growth are protected from light as well as contamination by other microorganisms through the interface between two glass assemblies. The assembly is sterilized in an autoclave at 20lbs (9.08kg) pressure at 126°C for 2hrs and used after it has cooled.Surface sterilized seeds and washed with sterile water are sown on the sand, on germination, the petri dish is removed, seedlings inoculated with the culture suspension and covered with sterilized gravel of safeguard from aerial contamination. The proper controls are kept for evaluation.

Pot Culture tests

Pot testing of the efficiency of a few efficient rhizobial strains were carried out by growing host legumes in sterilized soil and/or in unsterilized soil. In these cases the efficiency of inoculated strains of *Rhizobium* was measured from the nodulation taking into account number of nodules and/or dry weight of nodules and dry weight or vegetative part viz. shoots. By following the above plants test, efficient nitrogen fixing strains of *Rhizobia* for each legume crop can be selected. Selection at this stage is done depending on the performance under pot culture conditions using unsterile soil which have to be tested under field conditions.Pot culture testing of the efficiency of *Rhizobial* strains are carried out by growing the specific host legumes acid for 2 days. The acid was removed and sand was washed several times with tap and distilled water to get rid of the last traces of acid. Neutral sand allowed to dry, autoclaved at 20lbs pressure for 20 min. and put in pots. *Rhizobium* inoculated seeds are sown and irrigated with nitrogen free nutrient solution at required intervals through container placed in bottom of the pot.

Field tests

Finally, the rhizobial strains selected as above are further subjected to the field inoculation test. In this, the few efficient strains of rhizobia (obtained as above) were used to inoculate seeds of legumes growing in the field in randomized plots. Plots may serve as another control. Each rhizobial strain used for inoculation serves as one treatment. Sufficient number of replications for each treatment are kept so that after 4-6 weeks growth (or at the pre-flowering stage) plants from 2-3 plots from each treatment are uprooted yield or grains obtained due to inoculation with different rhizobial strains with those of uninoculated control and fertilizer nitrogen control. From the results of such field experiments the rhizobial strains with inherent nitrogen-fixing efficiency (as tested under laboratory aseptic conditions or pot culture conditions) are further screened for their efficiency, leading to increased yield of grain of host legume under field conditions, where the inoculated rhizobial strains are to compete with the indigenous *Rhizobium* of the same species. *Rhizobium* strains, which proved to be efficient by the aseptic culture or pot culture conditions, did not always perform well under field conditions. This is mainly because efficient rhizobia strains (used are inoculum) may not be always more competitive than the indigenous rhizobia (which are inefficient) and therefore, are not able to occupy more indigenous sites on the roots of the host legume. Hence larger number of nodules formed in the roots of the host legume will be due of the inefficient indigenous rhizobia and only a few will be due to the efficient inoculated strain. Under such conditions inoculation with efficient strains of rhizobia will now be result in significant nodulation and increase in nitrogen fixation and grain yield. This means that rhizobial strains selected should not only be efficient nitrogen fixers but also be more competitive than the indigenous rhizobia of the same species.

5.14 Determination of Competitiveness of Inoculated *Rhizobial* Strains:

Determination of Competitiveness viz. percentage of nodules formed in a host legume due to the specific, efficient, inoculated strains and those formed due to the indigenous strains can be determined by two methods such as (i) Serological method and (ii) Antibiotic resistance method.

5.14.1 Serological Method

Heavy suspension rhizobial strains used as inoculum in physiological saline solutions (Antigen or ag) is injected in rabbits by standard procedures (Dadarwal and Sen, 1974) and ultimately the blood of the injected rabbit is collected and the red blood cells are separated by centrifugation. The liquid portion of the blood (generally called serum but when the serum is collected after injection of antigens it is called antiserum due to the formation of antibodies, is separately preserved. This is antiserum (AsI) prepared against the antigen (Ag1) that is the *Rhizobium* used for inoculation of the *Rhizobium* used as inoculum or with the nodule (suspension) formed due to the inoculated *Rhizobium*. No other *Rhizobium* strains or nodules formed by other rhizobia will act positively with the above antiserum (As). Hence, to determine the percentage of nodules formed due to inoculated *Rhizohium* (R_1), all the nodules (or a representative numbers of nodules) formed on the legume host are collected, suspension prepared and tested against the antiserum (As_1) prepared against the inoculated *Rhizobium* (Gaur and Sen, 1974). The nodules that will react positively will be considered to have been formed due to the inoculated *Rhizobium* strain. Thus, from the total number of nodules examined and number of the nodules giving positive reactions, percentage of nodules formed due to inoculated strain could be calculated. In the same manner the antiserum or different rhizobial strain (used as inoculum) could be prepared and percentage of nodules formed due to the inoculated strain could be determined. The strain or *Rhizobium* biofertilizer which produces maximum number of nodules can be considered as the most competitive.

5.14.2 Antibiotic Resistance Method

Rhizobium strains of the same species differ in their antibiotic resistance pattern. Different antibiotics like penicillin, streptomycin, terramycin and chloramphenicol etc. can be used as a marker for the determination of the resistance of different strain so *Rhizobium* used as inoculum. Nodules formed on the legume due to the inoculation of different strains are detached and their suspensions (in saline) are examined for their antibiotic resistance and compared with those of the inoculated strain. Nodules suspension showing similar antibiotic resistance pattern as that of the inoculated strain can be considered to be formed by the latter. In this method, percentage of nodules formed by the inoculated strain can be determined (Stein *et al.,* 1982). All the efficient strains (as selected by aseptic cultures and pot cultures test) are, therefore, examined for their ability to form nodules even in presence of indigenous rhizobia. Of all the efficient (with high inherent nitrogen-fixing ability) strains of *Rhizobium*, the one which forms maximum numbers of nodules (60-80% nodules) are then finally selected as efficient as well as competitive strains of rhizobia suitable for being used as an inoculum.

5.15. Carrier-based Mass Production Technology of *Rhizobium*

Mass culture production of rhizobia had its beginning in 1895 when Nobbe and Hiltner introduced a laboratory grown culture of *Rhizobium* spp. under the name 'Nitragin'. Prior to 1923 cultures of *rhizobia* were mostly prepared in small cans or bottles. Matchette (1927) patented a method for large scale production of rhizobia by submerged fermentation with the development of a device for artificial aeration. There are several forms of inoculants viz. agar-based, broth, cultures, frozen concentrate, carrier based-granular soil inoculum, porous gypsum granules, natural peat granules and freeze-dried, which can be used for inoculating seed or soil. The choice of the type of inoculants to be made available depends on several factors such as volume of demand, organizational considerations (production for immediate use in nearby locations or R&D purposes), economics and it'seffectively for distribution over long distance and storage. In India commercial production of legume inoculants started in late 1960's when soybean was introduced for large-scale production in the country. However, legume inoculants for fodder crops like berseem/lucerne were prepared in laboratories of some Agricultural Colleges for use at Agricultural Farms and supply to some progressive farmers.The growth of bacterial cells when mixed with carriers (moist solid medium) is known as carrier-based inoculants.The carrier based inoculant is the most acceptable form due to better survival of rhizobia on seed than do agar, broth or reconstituted lyophilized cultures, and longer shelf life, ease of manufacture and distribution. Inoculant production can be divided into four phases such as

a. Selection and testing of strains of *Rhizobium.*

b. Preparation of broth culture.

c. Preparation of carrier based inoculants, and

d. Packing, labelling and storage.

5.15.1 Selection and testing of Rhizobium strains

The basic criteria for strain selection are the ability of the rhizobia to form effective nodules with all the hosts for which they are recommended and their capacity to do their best under a wide range of field conditions (Vincent, 1956).

Maintenance of selected cultures: Maintenance of selected rhizobial strains requires methods that provides a maximum storage period (years) with minimal chance for variations (mutations) and contamination. Ideally this should be combined with tests for checks on serological type, host specificity and effectiveness. Maintenance of the selected cultures is done by (a) freeze drying with additives and (b) without freeze drying equipment.

- **Freeze-drying with additives** such as glucose, sucrose, peptone, dextran or mixture of these were successful with other organisms. Drying with 10% glucose, 10% sucrose or preferably 10% sucrose + 5% peptone solution is generally used.
- **Without freeze-drying equipment,** satisfactory storage may be achieved by drying cultures on procelain beads over a desiccant (silica

gel) in a screw-cap bottle. This technique can be used with minimal facilities and bottle may be sampled as required (one bead at a time). However, the disadvantage is that the stock bead culture must be renewed every 2-3 years to maintain a maximum number of viable cultures.

5.15.2 Preparation of Broth Culture

Rhizobia are heterotrophic and generally not highly demanding in their nutrient requirements. Cheap sources of raw materials are usedfor mass production of inoculants. Most of the rhizobia readily utilize mono- and di-saccharides. Fast growing ones utilize both mono-and di-saccharide while slow growing strains prefer pentoses (arabinose). Sucrose is economical, therefore preferred for mass production. Efficient utilization of these carbon sources, however, depends on aeration and the method of sterilization. Small quantities of yeast extract are used as source of organic nitrogen and growth factors. *Rhizobia* grow best in the temperature range 30-32^{0}C and pH of the broth is adjusted to 6.5-7.0 with KOH or H_2SO_4.

Broth cultures with high population of rhizobia are desired to prepare high quality carrier based inoculant. The old culture (3-7 days) is examined microscopically and gram-staining test carried out to make sure it is free from contamination. About 3-6ml of sterile water is transferred aseptically into the tube. The bacterial growth is scrapped with the help of an inoculation needle and the resulting bacterial suspension is transferred into a 250ml conical flask containing 100ml yeast extract mannitol broth which was plugged with non-absorbent cotton and sterilized at 15lbs pressure for 20 min and subsequently cooled. The flasks are incubated at 28-30°C on a rotary shaker for 2 to 7 days depending on the nature of *Rhizobium*.

A short fermentation time is desired to reduce contamination. Increased number of viable cells may be achieved by varying the source and concentration of the yeast extract and mannitol taking into consideration the type of strain used. Submerged culture technique is generally used for mass production of rhizobia.

5.15.3 Carrier Materials and Preparation of Carriers

The term 'carrier' refers to moist solid medium viz. peat, lignite, charcoal, soil, compost/ farmyard manure, pressmud, vermiculite, coirdust, polymers and their mixture which is impregnated with viable *Rhizobium* culture for preparation of specific inoculants.

- **Peat soil** is rich in organic matter content and widely used as carriers for preparation of legume inoculants. Good quality pet is not available in India although medium quality peat souls amounting to 5.5 MT are found in Niligeri valley in Ootacamund which are being used a carriers for biofertilizers in small quantities. It contains 21-56% organic matter was procured by IARI from Ooty during 1969- 70 and was found to be a good carrier for *Rhizobium* production. A count of 30×10^8 peat, was usually recorded after one week quality peat, other alternate carriers were tested. (Brockwell *et al.,* 1995).

- **Lignite** contains about 28% organic matter and 92% water holding capacity. It can also be obtained from Palana (Rajasthan) and Singrauli (Madhya Pradesh) and 3MT annually is available at Neyeli (Tamil Nadu)
- **Charcoal** is another carrier available in plenty. It contains 77% organic matter and 90% water-holding capacity.
- **Compost/Farmyard manure** is easily available and usually contains about 30% organic matter and 50% water-holding capacity.
- **Press mud**, a byproduct of sugar factory is generated @ 5.45MT a year and readily available in almost all States on India (Jauhri and Koshi, 1984).
- **Polymer** crystals are water extending crystals proved to be very effective synthetic carrier for *Rhizobium* inoculants. It is non-toxic and holds moisture 30 times of its own weight. According to availability and cost at production site, choice is either lignite or charcoal.
- **Fly ash,** a residue of burnt coal (100MT/year) is being generated by 82 thermal power plants in the country which is not only a pollution hazard but is also a cause of concern for its disposal. A study undertaken on fly ash as a carrier of biofertilizer production shows that further investigations should be taken up to promote its use (Sudhakar *et al.,* 2004).

Mostly all carriers are used. Although charcoal is used by production units, quality criteria like fineness, extent of pyrolysis, water-holding capacity, organic carbon, active surface area etc. are not properly followed resulting in production of inoculant containing inadequate number of *Rhizobia* and low shelf life.

The carrier materials are sundried to 5% moisture and pulverized to fineness of 100-200 mesh. Coarser materials form balls on mixing with broth and seeds are not properly coated with the inoculant. Lignite powder being acidic (pH5.5) usually one kg of lime ($CaCo_3$) is mixed with 9 kg of lignite to bring the pH to about 7. Charcoal powder being neutral in pH does require lime. to enhance its effectively for multiplication of rhizobia, 1% $CaCO_3$, 0.5% KH_2PO_4 and 25% soil are thoroughly mixed with charcoal. Finally the carrier is mixed with 10% water before sterilization. The carrier material must be sterilized before blending with broth culture to suppress the proliferation of contaminants. Better growth of *Rhizobium* is obtained in sterile carrier. Best method is irradiation for carrier sterilization. It is capable of penetration the carrier material killing microorganisms. For bulk preparation, using autoclave, lime mixed lignite or charcoal powder is filled up to two third capacity of steel trays (size depending on the size of the horizontal autoclave) and sterilized at 15 PSI for 4hrs continuously of 1hr each for 3 successive days. The carrier materials may be packaged and sealed in autoclavable polypropylene bags which can stand a temperature of 121°C and sterilized.For preparation of carrier based inoculants, the rhizobial number of broth must attain a minimal population of 10^9 cells/ml, which is mixed with sterilized carrier. The broth is added to one-third of the water-holding capacity of carrier. Blending is done by through mixing of broth culture with sterilized carrier in trays manually. In bulk operation, mechanical mixtures may be used.

5.15.4 Packing, Labelling and Storage

The blended carrier is kept for 2-4hrs for curing. After curing, the inoculant is packed in polypyropylene bags, made of flexible sheet of pouch type of 50-75μ milky opaque to prevent excess of light, low density, pouch type and 15 cm ×25cm in size. It provides an excellent barrier to water loss. Dispensing of 200g inoculant in each packet is done either manually of by using automatic dispenser. The bags are sealed by an electric sealer leaving about 2/3rd vacant space to give proper aeration to the inoculant. Low density polypropylene (LDPE) packets are used universally, as they can be readily sealed; allow high gas exchange with relative low moisture loss (Thompson, 1991). If production technology is based on sterilization of carrier material through autoclave, polypropylene of high density polypropylene (HDPE) pouches should be used. LDPE cannot withstand the sterilization temperature of 121°C. In sterilization through irradiation, LDPE bags should be used as polypropylene is unstable when exposed to irradiation. The inoculant packet is labelled with relevant information regarding culture viz. name of the culture, crop for which intended, expiry date, directions for storage/use and precautions certification (ISI) mark if any. Packets (Inoculants) must be incubated for a week in the room at 30°C so that the bacteria multiplies and attains the required count. The packets may then be stored in a cold room at 4°-15°C.

5.16. Mode of *Rhizobium* application

Seed Application

The coating of seeds with the *Rhizobium* inoculant of drenching inoculant suspension on crop rows are based on the principle to place the organisms on seed coat or at closest proximity of roots in rhizosphere. Cultures are mixed and coated on seeds so as to load each seed with 10^3-10^4 viable cells. The adhesives like gum arabic 5-10% solutions are prepared to serves as a sticker of *Rhizobium* cells on seeds. Usually 200g *Rhizobium* inoculant are used for treating 10-12kg pulse seeds. On an average 500g carrier based inoculant is sufficient to treat seeds required to be sown in one hectare under normal soil conditions 10^3-10^6 when seeds are sown under problem soil adverse conditions. Treated seeds must be sown immediately and should not be mixed with acid forming chemical fertilizers.

Method of Seed Pelleting

The technique of coating of inoculated seeds with inert materials to protect them from problem soils and stresses is known as Pelleting. This provides prolong survival to rhizobia on seed. The coating materials are calcium carbonate (lime) for acidic soils and gypsum for saline and alkali soils. Use of rock phosphate showed good results. The technique is followed as:

Pure gum arabic (100g, lightcolured, non-toxic or preservative free) is dissolved in 250ml hot water with mixing. The solution is neutralized with calcium carbonate. The carrier based inoculum (100g) is thoroughly mixed with the gum solution which swells to about 350ml. This total quantity of slurry is sufficient to inoculated 7kg of small, 14kg of medium or 28kg of large seeds (Roughley, 1970).The adhesive and seed should be mixed together in a mixer so that the seeds are coated uniformly. Three and a half kg of calcium carbonate is required to pellet the inoculated quantity of different sizes of seeds mentioned above. Calcium carbonate and *Rhizobium* inoculated seeds are quickly mixed usually within 1-2 minutes until all seeds are evenly pelleted.

Precautions: Precautions should be taken to avoid soft or powdery pellets which may be either due to addition of too much of lime or uneven mixing. Pasty-looking pellets with some exposed seed surfaces indicate use of too much of adhesive when can be rectified by adding more lime. Clumping of small seeds may be caused due to addition of too much of adhesive which cannot be rectified even by more of lime.

5.17 Isolation and Identification of *Rhizobium*

Rhizobium strains can be isolated from the root nodules of legumes as follows .

Requirements

- The apparatus required are such as an incubator; sterilized water; petri dish, glass rod, test-tube (sterilized)and inoculation needle.
- The reagents required are such as Ethyl alcohol (95%), Mercuric chloride (0.1% weight/volume), and Congo red yeast extract mannitol agar (CRYEMA) plate.

The procedure consists of two parts such as (i) nodule selection and (ii) surface sterilization:

Part 1-Nodule Selection

a) Uproot a healthy legume plant from which a *Rhizobium* strain is required.

b) Wash the roots of the plant gently under tap-water.

c) Select pink-coloured healthy nodules, and separate them from the root with the help of a sharp clean blade.

d) Put the nodules in a test-tube, and wash them thoroughly 4–5 times with tap-water so that the surface of the nodules becomes soil-free.

e) Immerse undamaged nodules in 95% ethyl alcohol for 5-10 seconds, and rinse them with sterilized water.

f) Transfer the nodules to a sterile test-tube, and keep the nodules immersed in 0.1% mercuric chloride ($HgCl_2$) solution for 2-3 minutes.

g) Shake the test-tube from time to time in order to remove air bubbles from the surface of the nodules and to bring fresh sterilant in contact with the surface of the nodules.

h) Decant off the mercuric chloride solution, and flood the nodules with sterilized water 5-8 times in order to remove all the sterilant.

Part -2: Surface Sterilization:Plating of Nodule Suspension

a) Add a few drops of sterilized water to the test-tube, and crush the nodules in the tube with the help of a sterile glass rod.

b) The suspension so formed has a very high count of bacteria. Dilute this suspension by adding 2-4 ml of sterilized water to the tube.

c) Take one drop or 0.1ml of the suspension, and put it on a plate containing CRYEMA medium.

d) Spread the suspension over the plate gently with the help of a sterile glass spreader. Repeat this spreading for three different plates containing the same medium.

e) Invert the plates, and keep them for incubation at 28–30°C until small, elevated, round-shaped colonies develop. If the colonies are white, translucent, glistening and elevated, they may be expected to be colonies of *Rhizobium*. Any colony absorbing the red colour of the Congo red indicator may be assumed to be contaminated.

f) After proper identification, *Rhizobium* from the colony so developed can be taken and multiplied in order to obtain a pure culture through the inoculation and fermentation process.

Identification: The following culture tests need to be carried out in order to check whether the *Rhizobium* colony as developed in the above incubation is pure or contaminated with common contaminants such as *Agrobacterium*.

a) Growth in alkaline medium: *Agrobacterium radiobacter* can be detected by drawing streaks on Hoffer's alkaline medium (pH 11) where *Rhizobium* does not grow, while *A. radiobacter* does.

b) Growth in glucose peptone agar: *Rhizobium* shows little or no growth on glucose.

c) Gram test: It should be test negative.

5.18 Commercial Production of *Rhizobium* Biofertilizer

Various media with different composition are available for growing *Rhizobium*species. For commercial production, a liquid medium(without agar) such as Norris and Date (1976)is used. There are two as (i) Non-sterile and (ii) Sterile production system is used for the commercial production of *Rhizobia*. Quantities of chemicals as required for the production of 25, 50 and 100 tonnes of *Rhizobium* biofertilizer are given in Table 5.5.

5.18.1 Non-Sterile Production system

This system uses an unsterilized carrier and procedure is complete in following steps as-

a) Prepare liquid media in conical flasks.

b) Sterilize the media in an autoclave.

c) Add pure culture at 10 ml/litre of sterilized medium in each flask. This is called broth.

d) Incubate the flasks on a rotary shaker at 28-30°C for 36-48 hrs. Thereshould be a population of 10^8-10^9 cells/ml at the end of the incubation. Broth can also be incubated in a fermenter of suitable size.

e) Neutralize the carrier (lignite) with a suitable chemical (lime), and add gumacacia at 4 g per 100g of carrier.

f) Mix the broth culture with the carrier to attain a moisture content of about 30-40 %, and incubate at 28°C for 24hrs. Leave for curing for 2-3 days.

g) Dispense the cured broth-mixed carrier in polyethylene pouches. Pouches of 250g, 500g or 1kg are prepared to meet the requirement of different quantities of seed to be treated for sowing in a given area.

h) Seal the pouches by heat sealer.

i) Put the pouches in printed polyethylene packets.

j) Store the packets at less than 20°C until dispatch.

Table 5.5 : Chemicals required for the production of *Rhizobium* biofertilizer

Chemicals required	Biofertilizer (30–40% moisture)		
(In kg)	**25 tonnes**	**50 tonnes**	**100 tonnes**
Mannitol	87.50	175.00	350.00
Yeast Extract	8.50	17.50	34.80
K_2HPO_4	4.50	8.50	17.50
$MgSO_4.7H_2O$	7.00	14.00	28.00
NaCl	1.75	3.50	7.00
$FeCl_3.6H_2O$	0.10	0.20	0.50

5.18.2 Sterile Production System

This system uses a sterile carrier and the procedure is completed in following steps as-

a) Prepare liquid media in conical flasks.

b) Sterilize the medium in an autoclave.

c) Add pure culture at 10ml/litre of sterilized media in each flask. This is called broth.

d) Incubate the flasks on a rotary shaker at 28-30°C for 36–48hrs. There should be a population of 10^8–10^9 cells/ml at the end of incubation. Brothcan also be incubated in a fermenter of suitable size.

e) Neutralize the carrier with a suitable chemical (lime), and add gum acaciaat 4g/100g of carrier.

f) Put the carrier in polypropylene bags, and autoclave for 4-6hrs.

g) Seal the pouches after autoclaving and keep them for 2 days before inoculation.

h) Wipe the pouches with alcohol.

i) Dispense the desired broth culture aseptically with an automatic dispenserby making holes in the polypropylene pouches or packets.

j) Seal the hole immediately with tape.

k) Blend the packets only manually to ensure that broth mixes with the carrier adequately (the sealed pouches/packets have to be manually manipulated toensure mixing).

l) Incubate the pouches for further multiplication of microbes by keeping the pouches in the incubator at 28°C for 24hrs. Leave for 2-3 days for curing.

m) Put the pouches in printed polyethylene packets.

n) Store the packets at less than 20°C until dispatch.

5.19 Stem Nodules

Certain aquatic and water tolerant legumes are known to bear nitrogen fixing nodules on stems. These stem nodules are caused by *Rhizobium*. Only some species of *Aeschnomene* and *Sesbania* are known to bear stem nodules and have been investigated in this regard (Fig 5.4). Recently, renewed interest on these stem nodulating plants has been evidenced in view of the potential benefit that could be achieved if similar stem nodulation of legumes. N fixing stem nodules have been recorded in warty lecellate barks of various trees inhabited by Enterobacteria. Stem nodules offer a challenging and convenient system to investigate the inter-relationship between photosynthesis and BNF. The cortex of stem nodules is green due to chlorophyll and the bacteroids tissue is pink possibly due to legHb tissue for the process of the nitrogen fixation can easily be traced and measured with accuracy by the use of labeled carbon. The stem nodules can easily be peeled off at the base for laboratory experiments. The genetic basis for the stem nodulation habit may have to be understood in great so as to induce stem nodules in cultivated crops to secure better BNF of atmospheric nitrogen. The bacterium, *Aeschynomene indica*, a gram (-) coccoid prokaryotic with an elaborate internal membrane system reminiscent of certain photosynthetic bacteria has been discovered. Ultra structurally, this organism bears a striking resemblance to the purple photosynthetic bacteria of the genus *Rhodopseudomonas* such as *R. capsuoata* and *R. spharoides.*

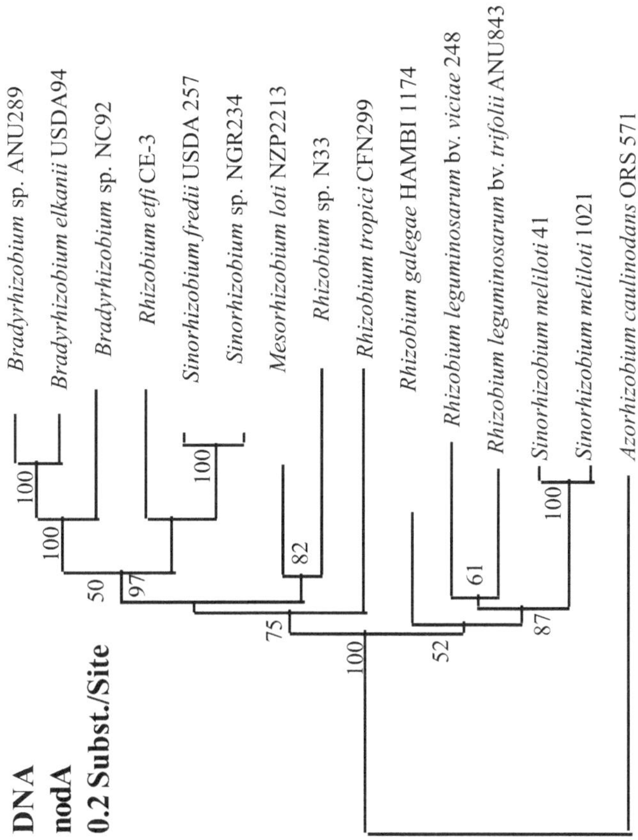

Fig. 5.1 : The structure varies depending on different nod gene

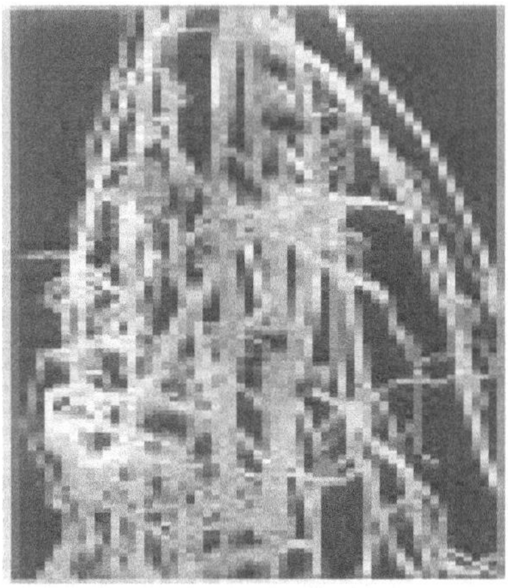

(A)

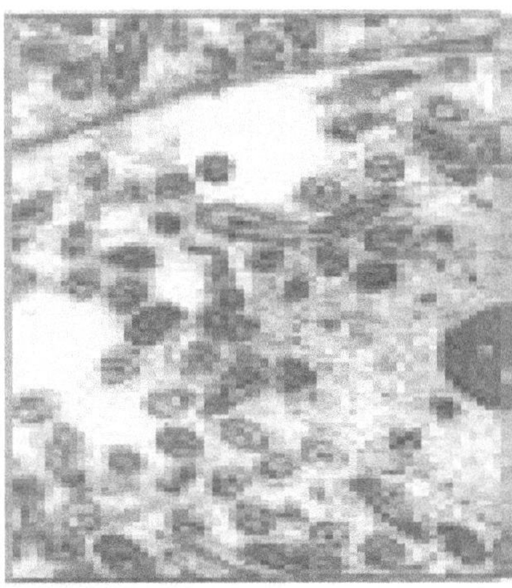

(B)

Fig. 5.2 : (A) Root Nodules

(B) Rhizobium bacterium

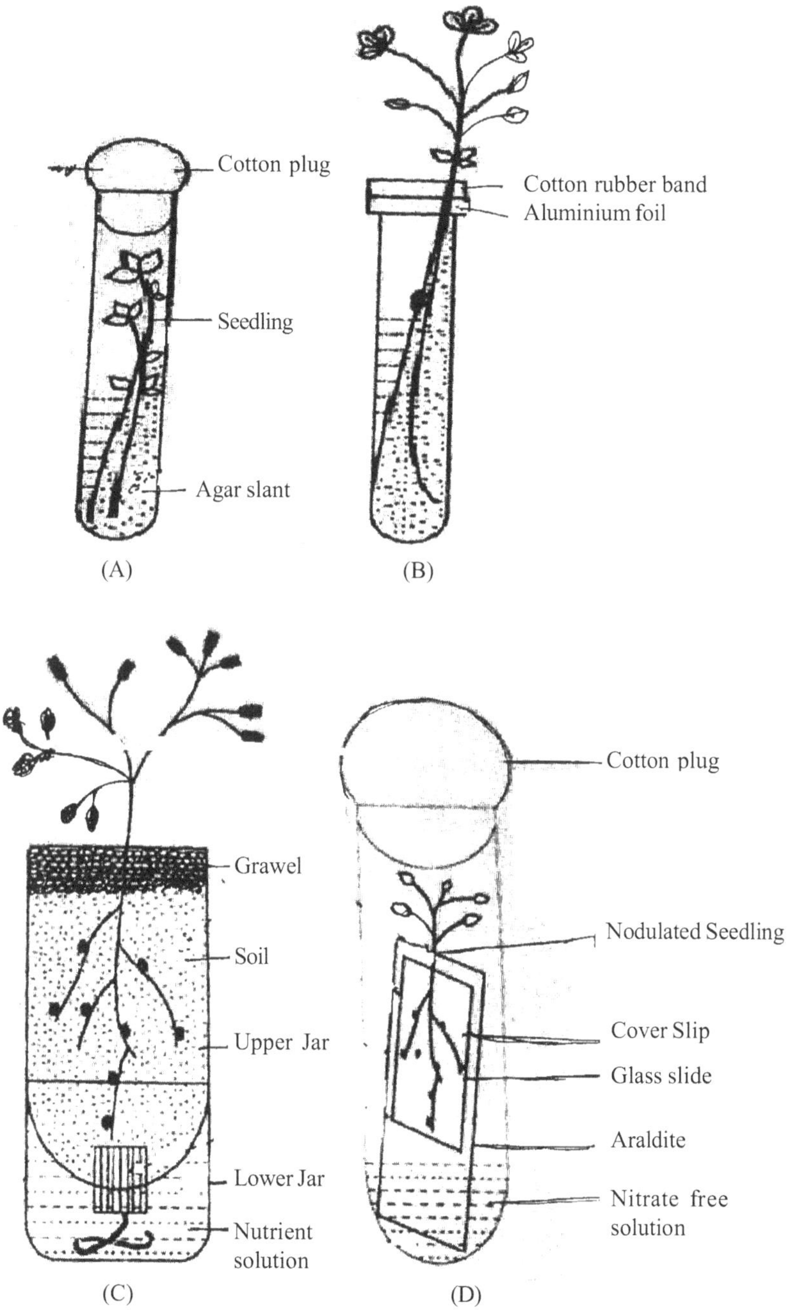

Fig. 5.3 : Four commonly used techniques for testing nodulation ability by rhizobia; (A) Jensen's seeding agar method. (B) Agar seeding method modified by Gibson. (C) Leonard jar assembly for growing large seeded legumes. (D) Fahraeus seeding tube for studying infections of legume.

Fig. 5.4 : Stem Nodules

CHAPTER - 6

Azolla

Azolla is a free-floating water fern that floats in water and fixes atmospheric nitrogen in association with nitrogen fixing blue green alga *Anabaena azollae*. *Azolla* fronds consist of sporophyte with a floating rhizome and small overlapping bi-lobed leaves and roots. Rice growing areas in South East Asia and other third World countries have recently been evincing increased interest in the use of the symbiotic N_2 fixing water fern *Azolla* either as an alternate nitrogen sources or as a supplement to commercial nitrogen fertilizers. *Azolla* is used as biofertilizer for wetland rice and it is known to contribute 40-60kg N/ha per rice crop. The agronomic potential of *Azolla* is quite significant particularly for rice crop and it is widely used as biofertilizer for increasing rice yields. Rice crop response studies with *Azolla* biofertilizer in the People's Republic in China and in Vietnam have provided good evidence that *Azolla* incorporation into the soil as a green manure crop is one of the most effective ways of providing nitrogen source for rice. The utilization of *Azolla* as dual crop with wetland rice is gaining importance in Philippines, Thailand, Srilanka and India. The important factor in using *Azolla* as a biofertilizer for rice crop is its quick decomposition in soil and efficient availability of its nitrogen to rice. In tropical rice soils the applied *Azolla* mineralizes rapidly and its nitrogen is available to the rice crop in very short period. The common species of *Azolla* are *A. microphylla, A. filiculoides, A. pinnata, A. caroliniana, A. nilotica, A. rubra* and *A. mexicana* (Fig. 6.1).

Importance of *Azolla* as Feed Supplement

The Milk Producers Union also involved in the training and marketing of *Azolla*. They are purchasing *Azolla* fronds from the village level Azolla growers both under wet and dry conditions. Around 400 rural women and 370 tribal people have been trained on the cultivation of *Azolla* through this project. In a review on *Azolla*, Wagner (1997) calls *Azolla* "a green gold mine, this fact can be ignored with regard to substituting it for inorganic fertilizer and other important uses mentioned above to conserve the environment as well as to maintain the sustainable agriculture. The biochemical constitution along with the rapid multiplication rates makes *Azolla* ideal organic feed substitutes for livestock.

A. microphylla

A. filiculoides

A. pinnata

A. nilotica

A. mexicana

Fig. 6.1 : Different sps. of *Azolla*

They can easily digest the floating fern on the water due to high protein content and low lignin content. Experience of the farmers suggested that cows and buffaloes which consume*Azolla* recorded an increase of milk production by 15-20%. The quality of milk was also found to be good and the scientists had found animals feeding on *Azolla* to be healthy. Similarity chickens feeding on the fern were said to be registering an increase in egg production. Sheep, goat, pig and rabbit have found a new fodder in *Azolla* (Anonymous, 2012).

6.1 *Azolla* acts as Green manure and Production Technology

In tropical regions, maintaining the soil fertility cannot be solved only by paying attention to chemical fertilizers and organic manures alone but also to biological inputs like *Azolla* and blue green algae particularly in rice cropping systems. *Azolla* is endowed with microsymbiont *Anaebaena azollae,* a nitrogen-fixing cyanobacteria (blue-green algae). Ultra structural studies identified several morphologically distinct eubacteria in the leaf cavities of *Azolla* species. Thus, it is now well established that there is a tripartite symbiosis among *Azolla,* cyanobionts and eubacteria. However, the type of bacteria that are present in the cavity is controversial. It results in high biomass production alongwith appreciable content of protein. The cultivation of *Azolla* was practiced in China and Vietnam over several centuries but in a very limited area until the end of 1950s having humid sub-tropical climate with wide variations in seasonable temperature. Research, conducted in these countries during 1960s, helped to overcome the problems of fluctuating temperature (low in winter and high in summer) as well as against various pests. The efficiency of *Azolla* culture was also improved by cultural practices. With the oil crisis during 1970s, other countries outside China and Vietnam showed interest of adopt *Azolla* as green manure. Research programmes with view to develop *Azolla* culture were introduced with the help of Food and Agriculture Organizations (FAO) and International Rice Research Institute (IRRI) the results showed varying success in other Asian countries as well as in Africa and America.

6.2 Morphology of *Azolla*

The name *Azolla* is derived from Greek word azo (to dry) and allyo (to kill) meaning that plant dies when it dries. The genus *Azolla* established by J. B. Lamark as early as 1783 (Svenson, 1944) was placed in the family Salvinaceae of the order Salviniales. The small fast growing free floating fern has global distribution. *Azolla* is a dichotomously branched free floating aquatic fern naturally available on moist soils, ditches and marshy ponds. The shape of Indian species is typically triangular measuring about 1.5 to 3.0cm in length, 1 to 2cm in breadth. Fronds has tiny roots usually associated with rich microphylla (Roger and Renaud,

1979) short branched stem called rhizome covered with small alternate overlapping leaves the sporophyte has dorsiventral organization (Peters and Calvert, 1983) and each leaf is divided into dorsal and ventral lobe., the ventral lobe is thin almost colourless and distal half is only one celled thick. The aerial dorsal leaf lobe has multilayered mesophyll adaxial and abaxial epidermal tissues, numerous stomata and single celled papillae. In the dorsal leaf lobe there is an ellipsoidal cavity which is formed by the enfolding of the adaxial epidermis. The cavity largely filled with gases is lined with mucilage (Lumpkin and Pluckmett, 1980) which contains the cyanobiont *Anabaena azollae* (Peters *et al.,* 1980) and a gram positive non-nitrogen fixing bacteria (Hates *et al,* 1980) identified as Arthrobacter species (Grilli *et al.,* 1988).

An *Azolla* plant, often called a frond consists of a main rhizome rarely exceeding 3 or 4 cm in length. It floats on water surface and is covered by small alternate leaves closely overlapping the rhizome. From the main rhizome emerges at regular intervals, secondary rhizomes having the same general characteristics as the first one, followed by third which subsequently produces the fourth. This growth pattern gives the plant a triangular or circular shape depending on the species. Unbranched, adventitious roots are formed on the lower part of the rhizome towards the water where they grow vertically.Each leaf is made of two lobes i.e. dorsal and ventral lobe. The dorsal lobe is thicker, chlorophyllous develops on the upper side of the stem and a central cavity shelters *Cyanobacteria* (blue-green algae). *Anabaena azollae* a prokaryotic organism occurs as unbranched filaments formed by two types of cells. The more numerous, measuring 6 mm x 10 mm is called vegetative cells and the other called heterocysts, are slightly larger and have thicker cell walls. The heterocyst of the *Anabaena* is the site of nitrogen fixation.

In very young leaves, filaments are made up of vegetative cells only and the number of heterocysts progressively increases during the leaf development until it accounts 30 to 40% of all cells. Finally in some cases, some of the cells change to akinites, and the presence of these conservation forms in the leafy cavity is not well understood.

6.3 Taxonomy of *Azolla*

Azolla can be easily distinguished from various floating aquatic plants like *lemna* or *salvinia,* but specific differentiation within *Azolla* may be difficult. On the basis of the morphology of sporocarp, the genus of *Azolla* has been divided into twosections with a total of seven species.

1. ***Rhizosperma*:** The megasporocarps have 9 floats and the glochidia are absent or unbarbed.

A. nilotica: It is found from Sudan to Mozambigue and is a native of in east Africa. In favourable conditions, it grows profusely reaching several decimeters. *Glochidia* are absent.

A. pinnata: It is the only species which is divided into 2 cultivars viz. A. *pinnata* var. *pinnata,* and *A. pinnata* var. *imbricata.*

Glochidia unbarked: A. pinnata var. pinnata. It is found throughout Africa South of Sehel, Medagascar and Australia. It is fairly easy to identify by the morphologyof its vegetative structure, shaped like christmas tree showing main axis and regularly arranged secondary branches.

A. pinnata var. *imbricata:* It is native of subtropical and tropical Asia. It is widely distributed in India and this variety was used in China and Vietnam for many centuries. It resembles the var. *pinnata* but its shape is not clearly triangular.

2. ***Azolla*:** The megasporocarps have 3 floats and the glochidia are barbed.

A. ucaroliniana, A. mexicana* and *A. umicrophylla: All the 3 originate from temperate, subtropical and tropical regions of North and South America. These species differ from each other mainly by buildup of their megasporal wall and thus are often difficult to identify at the vegetative phase.

A. filiculoides* and *A. rubra: *Azolla filiculoides* and *rubra* are native of America and the Korea, Japan and New Zealand, respectively. The two species differ from the caroliniana group by the presence of unicellular trichomes on the abaxial surface of the upper lobe of their leaves.

6.4 Reproduction in *Azolla*

6.4.1 Sexual Reproduction

In certain environmental conditions, *Azolla* undergoes a complex nature of a sexual reproduction. Instead of lower leaf lobe, the plant produces a group of 2 or 4 structures called sporocarps which are protected by an envelope (involucre). Each sporocarp in turn is made of a biassisial envelop, the indusium, which harbors either around hundred sporangia or a single one.

Each sporangium produce 32 or 64 spores which because of their small size are called microspores, consequently sporangia and sporocarps which protect them are microsporangia and microsporocarps. The latter are spherical, diameter about 2 mm, yellowish and can be seen at the lower side of the plant with the naked eye. In the second case, a lone sporangium produces a single larger spore, hence the names megaspore, megasporangium and megasporocarp even though the latter is smaller ($\pm$ 0.5 mm) than the microsporocarps. At maturity stage, the spherical megaspore at this stage is topped by 3 of 9 spongy structures called floats which in turn are covered by conical hood made of upper part of the indusium which is highly lignified and protects at its summit a small colony of *Anabaena*.

At maturity, the microsporocarps liberate their microsporangia into water. Each of these will disintegrate into 3 or 4 small spongy masses called massulae which contain the microspores. The massulae are attached to the megaspore through glochidia at the bottom of the water. The microspores in the massulae germinate and develop into a small prothalus producing male gametes, (antherozoids) ciliated and mobile. The megaspore (megasporocarp) germinates and becomes a prothalus producing female gametes (oosphere). The antherozioids cross with the oosphere through an unknown process and fertilization is achieved. The embryo starts growth and as it grows upwards its apex pushes away lignified indusium and freeing itself from the sporocarp, forms a new plantlet which floats on the surface of water. Megaspores and microspores are resting stages of *Azolla*. By harvesting them by special methods, one can maintain the plant for several months. However, *Azolla* rarely produces sporocarps and the environmental conditions favourable for sexual reproduction are poorly understood. Moreover, the harvest of large quantities of sporocarp is difficult and mass production of *Azolla* biomass from sporocarp is a slow process.

6.4.2 Vegetative Multiplication

The vegetative multiplication of *Azolla* is through fragmentation of the plant. The most developed secondary rhizome breaks off from the main plant by the formation of abscission layer at their base, giving rise to new plants.

6.4.3 Relationship between Symbiotic Partners

The main relation between *Anabaena* and *Azolla* are the supply to the fern of nitrogenous compounds fixed by the bacteria from the atmosphere which in turn are using carbon compounds photosynthesized by the fern.

The symbiotic association can grow and develop without any source of nitrogenous compounds. This is possible by the fact that *Anabaena azollae* has enzymatic complex, nitrogenase, capable of converting N_2 to NH_3. The fixation of nitrogen into ammonia occurs in the heterocysts which also supports vegetative cells and growth of *Azolla*. The heterocysts excrete a significant amount of fixed ammonia into the leaf cavity from where it is taken up by the fern. *Azolla* and vegetative cells of *Anabaena* are photosynthetic. With non-symbiotic *Anabaena* species such as *A. cylindrica,* the organism should be self-sufficient in carbon compounds and cannot under these conditions transform more than 3 to 5% of its cells into heterocysts. In mature *Azolla* leaves the proportion of heterocysts exceeds 30%. This is obvious that the part of the carbon requirements of the microsymbiont is met from *Azolla.*Some recent observations showed that the bacteria of the genus *Arthrobacter* may also be involved in the symbiotic process of nitrogen-fixation which needs to be examined fully.

6.5 Mass Multiplication of *Azolla* under Field Conditions

A simple *Azolla* nursery method for large scale multiplication of *Azolla* in the field has been evolved for easy adoption by the farmers.

Materials: 40sq.m area plot, Cattle dung, Super phosphate, Furadan and Fresh *Azolla* inoculum.

Procedure

- Select a wetland field and prepare thoroughly and level uniformly.
- Mark the field into one cent plots (20×2m) by providing suitable bunds and irrigation channels.
- Maintain water level to a height of 10cm.
- Mix 10kg of cattle dung in 20 litres of water and sprinkle in the field.
- Apply 100g super phosphate as basal dose.
- Inoculate fresh *Azolla* biomass @ 8 kg to each pot.
- Apply super phosphate @ 100g as top dressing fertilizer on 4th and 8th day after *Azolla* inoculation.
- Apply carbofuran (furadan) granules @ 100 g/plot on 7th day after *Azolla* inoculation.
- Maintain the water level at 10cm height throughout the growth period of two or three weeks.
- Observations.
- Note the *Azolla* mat floating on the plot. Harvest the *Azolla*, drain the water and record the biomass.

Method of Inoculation of Azolla to Rice Crop

The *Azolla* biofertilizer may be applied in two ways for the wetland paddy. In the first method, fresh *Azolla* biomass is inoculated in the paddy field before transplanting and incorporated as green manure. This method requires huge quantity of fresh *Azolla*. In the other method, *Azolla* may be inoculated after transplanting rice and grown as dual culture with rice and incorporated subsequently.

A. Azolla Biomass Incorporation as Green Manure for Rice Crop

- Collect the fresh *Azolla* biomass from the *Azolla* nursery plot.
- Prepare the wetland well and maintain water just enough for easy incorporation.
- Apply fresh *Azolla* biomass (15t/ha) to the main field and incorporate the *Azolla* by using implements or tractor.

B. Azolla Inoculation as Dual Crop for Rice

- Select a transplanted rice field.
- Collect fresh *Azolla* inoculum from *Azolla* nursery.
- Broadcast the fresh *Azolla* in the transplanted rice field on 7th day after planting (500 kg/ha).
- Maintain water level at 5-7.5cm.
- Note the growth of *Azolla* mat four weeks after transplanting and incorporate the *Azolla* biomass by using implements or tranctor or during inter-cultivation practices.
- A second bloom of *Azolla* will develop 8 weeks after transplanting which may be incorporated again.
- By the two incorporations, 20-25t of *Azolla* can be incorporated in one hectare rice field.

6.6 Productivity of *Azolla*

The value of *Azolla* depends on its biomass which is determined by several environmental factors. The growth of *Azolla* at the water surface under optimum conditions described earlier could be represented by a sigmoid (S Shaped) curve, slightly asymmetrical, with the inflexion point as generally situated before the biomass reaches half the maximum value.

The exponential phase (Phase I) depends on the species until the biomass covers the available water surface which represents fresh matter yield of about 1 kg/m^2. Later, the growth becomes practically linear (Phase II) and continues until the biomass yields about 2 kg/m^2 in some favourable cases, and even upto

8 kg/m^2 in some exceptional cases. This is followed by stationary (Phase III). The maximum biomass is reached (Phase IV). Afterwards the degeneration of the biomass starts (Phase V).

The growth curve depends on the environment conditions and *Azolla* ecotypes. It is generally useful to plant a mixture of 2 or 3 different ecotypes, each of them characterized by high productivity during different seasons of the year.

6.6.1 Biomass Produced

The fresh matter (100-200 kg/day) is procured daily from area of 1,000 m^2 or 1,000 to 2,000 kg/ha/day, is produced. The high nitrogen-fixing ability, rapid growth, high biomass yields and nitrogen content indicate its potential as a biofertilizer for increasing rice yields with low cost and ecofriendly inputs. Watanabe *et al.*, (1980) established nitrogen-fixation by *Azolla* @ about 1.1 kg.N/ha/day. Growth rates were summarized by Becking (1979) which showed that doubling time varies between 2 and 10 days for most species. Maximum biomass ranged from 0.8 to 5.2 ton dry matter/ ha and averaged 2.1 ton dry matter/ha (Kikuche*et al.,* 1984) Nitrogen content ranged from 20-146 kg/ha with an average of 70 kg N/ha.

6.6.2 Chemical Composition

The chemical composition of*Azolla* includes Carbon 43.0%, Ash 10-20%, Nitrogen 2-7%, C/N ratio (3.5 Av) 6-20 (12 common), Essential Amino Acids 50% of total, Soluble Carbohydrates 3.5%, Starch 4-10%, Lipids 3-6%, Cellulose 10-20%, Hemicellulose 10-20%, Lignin in young growth, $\pm$ 5%, Lignin in mature growth 30%, Toxins ND.

6.7 Effect of Environmental Factors

6.7.1 Water

Azolla prefers to grow in a free floating state. It requires some amount of standing water and its multiplication is drastically reduced, if there is not enough water and soil is just moist. This is probably due to the peculiar structure of its stomata that do not control the transpiration, without water the plant dies within a few hours. It can also survive rooted in the soil, if the soil is permanently humid but in that case, its productivity is lower, over population occurs indeed rapidly in such conditions, since the fronds are unable to fragment and disperse. *Azolla pinnata* generally is the least able to support this situation not because it is less efficient in holding water but because its root system is less efficient and, therefore, it is less able to extract water in those conditions. A water depth of at least 5-10 cm is recommended for its multiplication but higher water depth does not have any adverse effect until the availability of plant nutrients become limiting (Singh, 1989).

6.7.2 Temperature

A temperature of 25-30^{0}C (Peters *et al.,* 1980) is considered optimum for most species but temperature responses varies among different species. Some strains grow, though slowly at temperature below 10^{0}C and can survive frost conditions for a few days. *Azolla filiculoides* and *A. rubra* are tolerant to low temperatures and cannot withstand temperature over 30^{0}C whereas *A. caroliniana, A. microphylla, A. nilotica* and some varieties of *A. pinnata* are relatively tolerant to high temperature (Tung and Watanab, 1983; Watanab and Berja, 1983). High temperature also encourages the building up of insects and pest population. *Azolla* culture in China since 1960 has been made possible by introduction of exotic species with different temperature requirements and also development of methods to protect *Azolla* against extreme winter or summer temperatures.

6.7.3 Light

The growth rate of *Azolla* increases with light intensity up to 1000m Einstein $m^{-2}s^{1}$which corresponds to about 50 k lux or 50% of maximum natural light intensity. The growth limiting effect on *Azolla* is more pronounced when it is grown alongwith rice, because stage (Singh and Singh, 1997). The light saturation for nitrogenase activity and photosynthesis under laboratory is reported between 5 and 20k lux but the growth under field conditions is maximum at 50 k lux (Singh and Singh, 1990). There also appears to be synergistic interaction between tolerance to high temperature and light intensity. At higher light intensities the optimum temperature for growth of *Azolla* moves to higher side. The extreme temperature results in reddish brown colouration of the fern but this change has not been reported to affect the nitrogen fixation.

6.7.4 Soil pH

Azolla grows satisfactorily in a pH range from 4 to 10. Indirect effects may cause unavailability of certain essential elements and at pH levels the growth of microsymbiont (*Cyanobacteria)* may be reduced resulting in poor growth of *Azolla.* The average soil pH for successful growth of *Azolla* was reported between 4.5 and 8.0 but the optimum pH appears to between 5 to 7 (Peters *et al.,* 1980; Singh, 1979). Highly acidic and alkaline soils are not suited for multiplication of *Azolla.*

6.7.5 Salinity

The information on tolerance of *Azolla* to salinity is inconclusive but electrical conductivity beyond 3 mhos cm is probably always unfavourable to growth of *Anabaena azollae* which is more sensitive to salinity than *Azolla* and this results in production of *Anabaena-free Azolla,* inefficient in nitrogen-fixation. Saline soils usually having pH above 7 are deficient in phosphorus, zinc and copper which is not conducive for the growth of *Azolla.*

6.7.6 Mineral Nutrients

6.7.6.1 Nitrogen

Azolla is capable of fixing atmospheric nitrogen even if the field is fertilized with mineral nitrogen and inhibiting effect of NH_4^+, NO^-_3 on nitrogenase is not reported. It appears that with *Azolla-Anabaena* symbiosis the nitrogenase seems to be protected in one way or the other depending on the species and the ecotypes. Application of urea or ammonium sulphate improves the yield of both the rice crop and *A. pinnata* (Manna and Singh 1988, Singh *et al.,* 1992). These findings suggest that the combined use of nitrogenous fertilizer and *Azolla* increased the yield of rice.

6.7.6.2 Phosphorus

Phosphorus is a key element and its deficiency results in poor growth, red colouration, curling of root hairs and reduced N content. A deficiency in phosphorus ($<$0.22% of the dry matter) is accompanied by reduction of nitrogenase activity and nitrogen content in *Azolla. Azolla* can accumulate phosphorus in the range of 1 to 2% of the drymatter depending on the P availability. The average Phosphorus content of the soils in which *Azolla* grew satisfactorily was 25ppm, however, it varied depending on nature of species (Kushari and Watanabe, 1991). There are conflicting reports regarding the levels of P necessary for sustained growth. In laboratory studies, a concentration of 0.06 ppm is reported as being sufficient whereas in field conditions 0.3 to 1 ppm is required. Among the *Azolla* spp. examined, *A. pinnate* appears to have lowest P requirement. Sah *et. al.,* (1989) reported from studies under conditions of continuous flow of the growth medium that minimum P concentration was about 15 ppm. Additionally their findings have showed that the internal tissue – P necessary for sustained growth is about 0.1% of the dry weight. The research study involving different strains of *A. pinnata, A. mexicana* and *A. microphylla* showed that *A. pinnata* was more productive (as measured by total biomass) than any of other species. However, when P concentrations were higher ($<$ 6 ppm). *A. microphylla* (IRRI 4018) and to a lesser extent *A. mexicana* (IRRI 2026) were more productive than *A. pinnata* species (Kushari and Watanabe, 1991). Kushari and Watanabe (1992) reported that *A. pinnate var. pinnata* (IRRI 7001) grew best at low phosphorus levels (0.02 and 0.03 ppm), however at 0.06 ppm, *A. microphylla* (IRRI 4018) outperformed all other species tested. Based on their studies, it appears that *A. pinnata* is better performer under field conditions with low P content.

6.7.6.3 Potassium

Positive effects of potassic fertilizer on the growth of *Azolla* and also a trend of luxury consumption of the element by the fern has been observed.

6.7.6.4 Iron and Molybdenum

The cases of iron and molybdenum deficiency as well as cases of iron toxicity were observed.

6.7.7 Insect Pests and Diseases

Many insects especially larvae of *Pyralidae (Lepidoptera)* and *Chironomidae (Diptera)* feed on *Azolla* especially in tropics and at high atmospheric humidity. Gasteropods reported as pests on *Azolla* but less damaging when compared with other insects. Insecticides commonly recommended to control rice insects, are equally effective against major insects of *Azolla.* The use of insecticides increases the growth and nitrogen fixation by fern. Insecticides such as carbofuran, methomyl, monocrotophos, diazinon can be sprayed for control of insect pests which should not cause environmental problems. If the *Azolla* is to be recycled through fish ponds or is to be fed to livestock or poultry the toxic and non-biodegradable pesticides should not be used. The selection of *Azolla* species which are resistant to insect pest is essential to avoid use of pesticides. The biological control and use of biopesticides neem extract or other techniques should be employed. Some natural products viz. Neem (*Azadirachta indica*) origin or *Albizza lebbek* (Sirish) are reported to be effective against insects feeding on *Azolla* (Reynaud and Franche, 1986; Kushari and Teheruzzaman, 1991). The high temperature, humid environment and the population densities along with the damage to fern by insects induce secondary infections by fungi viz. *Rhizoctonia* and *Sclerotium* and some bacteria. This usually occurs when the culture becomes old or its mat is left harvested for a longer period particularly in hot summer. Fungal infections can be controlled by use of fungicides. Algal growth should be removed manually so that growth of *Azolla* is not hampered.

6.8 Different Methods for Propagation and Use

Azolla is to be kept in vegetative form to make it available whenever needed for use. It is almost impossible to obtain spores for the conservation of the material in dry state. Moreover, even when this is possible, the time required to produce sufficient biomass from spores is usually too long.

- **Nurseries:** Standing water can normally be used for maintenance of an *Azolla* population. *Azolla* develops in the paddy fields as soon as they are flooded, either spores of previous seasons, or vegetative parts which have survived in soil. The spontaneous presence of *Azolla* all the year round is doubtful and the vegetative forms have to be kept active in nurseries especially designed for that purpose. The maintenance of healthy *Azolla* culture should be made the responsibility of village communities. The farmers should get the effective healthy inoculant at appropriate time to multiply them. A few hundred square meters having depth of water not exceeding 5 to 10 cm is desirable but may require frequent adjustments of water level depending on the climatic conditions viz. rainfall etc. Slight shading of *Azolla* beds to avoid scorching sun under certain climatic conditions may be desirable. The use of creepers is recommended for protection from wind, if necessary.
- **Weed Removal:** In the growth phase the biomass production of *Azolla* is maximum at 1 kg/m^2.This prevents the growth of other plants viz. *Lemna, salvinia, Eichhornia, Spirodela* or different alga. Overcrowding of *Azolla* plants should be avoided because it reduces the productivity. Excess biomass should be removed depending on the daily need or at least once a week.
- **Fertilization:** Some biomass are regularly harvested, nutrient removal is taking place. Hence the plant nutrients except nitrogen should be applied. Fertility of nursery can be maintained by recycling of crop wastes, animal residues, house-hold ash and waste material. Split phosphate fertilization is required depending on the harvest and yield of biomass.

6.9 Advantages of *Azolla*

- It easily grows in wild and can grow under controlled condition also.
- It can easily be produced in large quantity required as green manure in both the season's viz. kharif and rabi seasons.
- It can fix atmospheric CO_2 and nitrogen to form carbohydrates and

ammonia respectively and after decomposition it adds available nitrogen for crop uptake and organic carbon content to the soil.

- The oxygen released due to oxygenic photosynthesis, helps the respiration of root system of the crops as well as other soil micro-organisms.
- It solubilizes Zn, Fe and Mn and makes them available to the rice.
- *Azolla* suppresses tender weeds such as *Chara* and *Nitella*in a paddy field.
- *Azolla* releases plant growth regulators and vitamins which enhance the growth of the rice plant.
- *Azolla* can be a substitute for chemical nitrogenous fertilizers to a certain extent (20 kg/ha) and it increases the crop yield and quality.
- It increases the utilization efficiency of chemical fertilizers.
- It reduces evaporation rate from the irrigated rice field.
- *Azolla* fixes nitrogen at substantial rates.
- *Azolla* has rapid growth.
- Since *Azolla* floats at the water surface, it cannot complete with rice for light and space.
- In most climates, *Azolla* grows best under a partial shade of vegetation which a rice canopy, in its early and intermediate stages of growth can easily provide.
- When rice approaches maturity, due to low light intensities under the canopy and depletion of nutrients, *Azolla* begins to die and decompose, thus releasing nutrients into the medium.
- *Azolla* decomposes rapidly and therefore the nitrogen it has fixed and the phosphorus and other nutrients it may have observed from the water, perhaps in competition with the rice are rapidly released back in to the medium and made available for uptake by rice during grain development.
- *Azolla* has great ability than rice to accumulate potassium in its tissues in low potassium environment; thus, after decomposition, it makes this nutrient available to rice.
- A thick *Azolla* mat in a rice field has the side benefit of suppressing weeds.

6.10 Functions of *Azolla*

- **Green manure in Rice Field:***Azolla* is quickly decomposed being rich in nitrogen content and a low C: N ratio organic material. It is used in areas where water is available before transplanting. Fresh *Azolla* is inoculated @ 1 to 2 tonnes/ha in a well prepared and leveled field and the water level is maintained at 5-10cm depth. Single superphosphate (25-50 kg/ha) in split doses and insecticide (2-3 kg/ ha) are also applied for better growth of *Azolla*. The duration of monoculture to form a thick mat over the surface of water takes about 10-20 days. Water is drained from the field and *Azolla* growth is ploughed and incorporated into the soil. Depending on the resources, incorporation is done by hand, by hoes, by rotary or conical weeders by ploughing or harrowing.The extent of preliminary draining is a determining factor for a successful incorporation. If rice transplantation follows the *Azolla* monoculture which may be carried out in such way that about 20% of the *Azolla* will float again upon flooding and form a new carpet between the rice plants while the bulk 80% of its biomass shall remain underground and will be mineralized into available form of plant nutrients. One crop of *Azolla* normally gives 10-15 tonnes green manure.
- **Advantages of Monoculture:** In tropical soils, the fertilizing effect of the monoculture is apparent during the first few weeks of growth and development of rice. It is generally comparable to 30 units of nitrogen. While functioning as green manure *Azolla* monoculture also inhibits the growth of weeds.
- **Dual Cropping:** Growing *Azolla* simultaneously with rice seedlings after transplantation is called dual cropping/ culture. About a week after transplanting rice seedlings, the field is inoculated with fresh *Azolla* @ 0.5-1.0 tonn/ha. *A. microphylla* inoculated @ 200 kg/ha could multiply faster having 15-25 tonnes of biomass within 2-3 weeks period. Lower loads of inoculation shall take longer duration to establish in rice fields. *Azolla* technology is very efficient in nitrogen-fixation and biomass production during wet season due to better environmental conditions prevailing during the second season rice in India particularly. Cloudy days coupled with low temperatures (18^0-$25^0 \pm 1^0$C) favours its vegetative multiplication. Intercropped *Azolla* is not fertilized but if soil is poor in phosphorus, it can be applied @ 20 kg/ha including the recommended dose of phosphorus for the rice crop. Insecticide viz. furadan at 2.5 kg/ha are applied to control insect pests. *Azolla* forms

a thick mat within 15-20 days and it starts decaying. Field is drained off and *Azolla* is incorporated in soil. It is possible to raise 2 dual crops during a single rice crop period. The growth of *Azolla is* significantly influenced by rate and time of inoculation, method of planting, density of rice crop, application of fertilizers and pesticides (Singh, 1989; Singh and Singh 1990). On an average one crop of *Azolla* grown as green manure provided 30 kg N/ha and response of rice is equivalent to the application of 30 kg/N/ha as chemical nitrogen. *Azolla* green manuring plus one dual cropping gives about 60 kg N/ha and *Azolla* green manuring plus two dual cropping gives about 90 kg/ha. Azolla improve the physical and chemical properties of the soil and further improve the crop yield reported by Bhuvaneshwari and Kumar (2013).

- **Biofertilizer in Rice Crop:** *Azolla* has been used as a bio-fertilizer for rice crop in China, Vietnam, Philippines, Thailand, Korea, India, Bangladesh, Sri Lanka, Nepal, Burma, Indonesia, Brazil and West Africa but its full potential has not been harnessed in agriculture. The symbiotic association draws the energy from photosynthesis carried out by fern to fix atmospheric nitrogen which is of the order of 150-200 kg/ha. *Azolla* in rice system contributes about 40-60 kg N/ha crop (Kannaiyan, 1989). The advantage of using *Azolla* as biofertilizer for rice crop is because of its quick decomposition and mineralization of its N in wet soil and rendered available to rice crop (Kannaiyan, 1990). The growth rate of *Azolla* is very rapid so much so that it doubles it weights within 2-3 days. Due to fast relative growth rate the biomass production is heavy in a short time which can be used as biofertilizer for rice. *Azolla* has the potential of maintaining an exponential growth rate under optimum conditions. The growth of *Azolla* is initially slow both in the growth medium and in rice fields followed by fast growth (Kannaiyan, 1993). Phosphorus fertilization is necessary for its growth and N_2 fixation by *Azolla* (Kannaiyan *et. al.,* 1981 and Singh and Singh, 1989). *Azolla* grown in phosphorus deficient solution had a reddish brown discolouration that spread from the centre of the frond to the tip of the body with reduction in size of the frond. Singh (1977) recommended 4-8 kg P_2O_{5-}/ha per week for large-scale multiplication of *A. pinnata* under field conditions. Kannaiyan *et al.,*(1982) demonstrated the effectiveness of split application of phosphorus for *Azolla* growth and multiplication. Animal dung @ 1 to 1.5 tonn/ha or cattle slurry 2,000-3,000 litre/ha can be used as an alternative of superphosphate (Singh *et. al.*, 1993).

- **Nitrogen Recuperation:** *Azolla* fixes atmospheric nitrogen in wet lands. The nitrogen is made available to cultivated plants after the decay of the fern. The incorporation or burying of an *Azolla* carpet of good quality has similar effect to those obtained by applying 40-50 kg/ha of nitrogen/ hectare.
- **Organic Matter Addition:** High productivity of *Azolla* allows large quantities of organic matter additions which also affects physical conditions of soil viz. soil structure, water holding capacity and cation exchange capacity.
- **Potassium:** *Azolla* can accumulate this element in soils low in potassium than does rice. The potassium becomes available to the crop as soon as *Azolla* starts decomposing. Therefore, *Azolla* can act indirectly as potassic fertilizer for the rice crop.
- **Control on Weeds:** The photosynthetic activity under *Azolla* mat is highly reduced for want of light. Therefore it checks growth and development of weeds. This herbicidal effect depends on the type of dominant weeds but is often significant and appreciated by farmers whose weeding labour is largely minimized. *Azolla* also prevents the proliferation of algae harmful to the development of the crop.
- **Water Saving and Buffering Effect on Temperature:** The evapotranspiration of water surface covered with *Azolla* is much less when compared with from a non-covered water surface. Therefore it saves irrigation and is economical. The presence of *Azolla* mat has a buffering effect on daily temperature fluctuations under it.
- **Source of Energy/Food Source:** It is a good raw material for biogas production and can be blended with rice straw to produce both fuel for cooking and spent slurry after biomethnation as organic fertilizers for use in sustainable agriculture. In temporary over-production, *Azolla* can be converted into compost. *Azolla* should be mixed 5% soil and 5% superphosphate to reduce the loss of nitrogen. Blending *Azolla* with rice straw will give good results and losses of nitrogen can be reduced.
- **Fresh *Azolla*:** Its very high water content renders its transportation over long distances inconvenient and expensive, besides it does not suit all animals. Drying by spreading for a few hours will normally reduce the water content from 95% to less than 50%.
- **Dried *Azolla*:** Sun drying can be advised as any dehydration process may not be economical due to very high content of water in it. This

requires adequate drying area near the harvesting site, appropriate climatic condition and enough manpower. The operation requires initial spreading of the biomass a layer of not more than 10 cm thicknesses which is turned two times a day for 3 to 4 days. It facilitates storage and transportation over long distances as well as production of pellets.

- **Food for Fishes, Pigs and Chickens:** *Azolla* is probably preferred than other aquatic plants and consumed by many species of fish. It is, however, important to maintain an equilibrium between the population of fish and that of *Azolla* either by introducing when necessary, biomass of *Azolla* collected from elsewhere or by harvesting the excess biomass to keep the *Azolla* population in a linear phase. The *Azolla* may be interesting for *pisciculture* but cannot alone ensure high productivities. In China, *Azolla* is most often supplied in fresh conditions at a daily rate of 1-4kg depending on the age of the pigs either as such or mixed with other food. According to some estimates, 100 kg of fresh *Azolla* would allow the production of 1 kg meat. Fresh or dried *Azolla* is fed to chickens and can be an effective additive in chicken feed (Ali and Leeson, 1995). It is estimated that 20% of the commercial feed ration can be supplemented by using 100-300 g/day. *Azolla* can be used as an alternate source of protein for animals, poultry and ruminants (Tamang and Samanta, 1993; Ali and Leeson, 1994, 1995; Jayaraman *et al.,* 1995). This is getting popular as food for many of animals. *Azolla* can be incorporated into concentrated feeds upto about 20% for goats (Tamang and Samanta, 1993).

6.11 *Azolla* – The Best Feed for Cattle and Poultry

Azolla is a free floating water fern that floats in water and fixes nitrogen in association with the nitrogen fixing blue green algae, *Anabaena azollae*. *Azolla* is considered to be a potential biofertilizer in terms of nitrogen contribution to rice. Long before its cultivation as a green manure, Azolla has been used as a fodder for domesticated animals such as pigs and ducks. In recent days, *Azolla* is very much used as a sustainable feed substitute for livestock especially dairy cattle, poultry, piggery and fish. *Azolla* contains 25-35% protein on dry weight basis and rich in essential amino acids, minerals, vitamins and carotenoids including the antioxidant b carotene. Chlorophyll a, chlorophyll b and carotenoids are also present in *Azolla*, while the cyanobiont *Anabaena azollae* contains chlorophyll a, phycobiliproteins and carotenoids. The rare combination of high nutritive value and rapid biomass production make *Azolla* a potential and effective feed substitute for live stocks.

6.11.1 Inputs Required

Azolla fronds, Polythene sheet, Super phosphate and Cow dung.

6.11.2 Methodology

The area selected for *Azolla* nursery should be partially shaded. The convenient size for *Azolla* is 10 feet length, 2 feet breadth and 1 feet depth. The nursery plot is spread with a polythene sheet at the bottom to prevent water loss. Soil is applied to a depth of 2cm and a gram of super phosphate is applied along with 2kg of vermicompost or cow dung in the nursery for quick growth. *Azolla* mother inoculum is introduced @ 5 kg/plot.

The contents in the plot are stirred daily so that the nutrients in the soil dissolve in water for easy uptake by *Azolla*. *Azolla* is harvested fifteen days after inoculation at the rate of 50-80 kg/plot. One third of *Azolla* should be left in the plot for further multiplication. Five kg cow dung slurry should be sprinkled in the *Azolla* nursery at ten days intervals. Neem oil can be sprayed over the *Azolla* at 0.5 levels to avoid pest incidence.

Animal	**Dosage/day**
Adult cow , Buffalo, Bullock	1.5-2 kg
Layer, Broiler birds	20 – 30 grams
Goat	300 – 500 grams
Pig	1.5 – 2.0 kg
Rabbit	100 gram

6.11.3 Value of the Technology

The egg yield is increased in layer birds due to *Azolla* feeding. The *Azolla* fed birds register an overall egg productivity of 89.0% as against 83.7% recorded by the birds fed with only concentrated feed. The average daily intake of concentrated feed is considerably low (106.0g) for birds due to *Azolla* substitution as against 122g in the control birds. More importantly *Azolla* feeding shows considerable amount of savings in the consumption of concentrated feed (13.0%) leading to reduced operational cost. By considering the average cost of the concentrated feed as Rs. 17/ Kg, a 13.0% saving in the consumption ultimately leads to a feed cost savings of 10.0 paise/day/bird and hence a layer unit maintaining 10,000 birds could cut down its expense towards feed to a tune of Rs.1000/day.

6.11.4 Benefits

The *Azolla* feeding to layer birds increase egg weight, albumin, globulin and carotene contents. The total protein content of the eggs laid by the *Azolla* fed birds is high and the total carotene content of *Azolla* eggs (440g/100g of edible portion) is also higher than the control. The rapid biomass production due to the high relative growth rate, increased protein and carotene contents and good digestibility of the *Azolla* hybrid Rong ping favour its use as an effective feed supplement to poultry birds.

6.10.5 Effect of Azolla hybrid Rong Ping on the nutritional value of Egg:

Parameters	***Azolla* egg**	**Control**	**% Increase Over Control**
Egg Weight (g)	61.20	57.40	6.62
Albumin (g /100g of edible portion)	3.9	3.4	14.70
Globulin (g /100g of edible portion)	10.1	9.5	6.31
Total protein (g/100g of edible portion)	14.0	12.9	8.52
Carotenes (μg/100g of edible portion)	440	405	8.64

6.11.6 Application

In Indian conditions, agriculture is very much coupled with poultry farming. *Azolla* is an important low cost input, which plays a vital role in improving soil quantity in sustainable rice farming. The twin potentials as biofertilizer and animal feed make the water fern *Azolla* as an effective input to both the vital components of integrated farming, agricultural and animal husbandry.

6.11.7 Limitation

Azolla is a water fern and requires a growth temperature of 35-38°C. The multiplication of *Azolla* is affected under elevated temperature. Hence adopting this technology in dry zones where the temperature exceeds 40°C is difficult.

6.11.8 Achievements

Azolla hybrid Rong Ping had been selected to supply to the tribal population. *Azolla* mother inoculum nursery was laid out in villages with the help of Krishi Vigyan Kendra, TNAU, Coimbatore. Karamadai, women entrepreneurs were selected and one day training was imparted to them on the cultivation of *Azolla*. Wet biomass (Starter inoculum) enabled them to initiate commercial *Azolla* cultivation in their backyards.

CHAPTER - 7

Azotobacter

The first species of the genus *Azotobacter*, named *Azotobacter chroococcum* family Azotobacteriaceae, was isolated from the soil in Holland in 1901. *Azotobacter* represents the main group of heterotrophic free living nitrogen-fixing bacteria principally inhabiting neutral or alkaline soils. *Azotobacter chroococcum* and *Azotobacter agilis* were studied by Beijerinck (1901). In subsequent years several other types of *Azotobacter* group have been found in the soil and rhizosphere such as *Azotobacter vinelandii*, Lipman (1903); *Azotobacter beijernckii*, Lipman (1904); *Azotobacter paspali*, Döbereiner (1966), *Azotobacter nigricans*, Krasil'nikov (1949); *Azotobacter armenicus*, Thompson and Skerman (1981); *Azotobacter salinestris*, Page and Shivprasad (1991). *A. paspali* while *A. agilis* was categorized under the genus *Azomonas* (Table 7.1).

7.1 Morphology and Taxonomy

The bacterium is heterotrophy, aerobic organism, fix dinitrogen as non-symbionts and besides occurs in the rhizoshpere of many plants. The bacterium is rod shaped gram negative, polymorphic, size ranging from 2-10 x 1-2.5 μm. Young cells have peritrichous flagella and are used as locomotory organs, The old bacteria produces encapsulated forms and are heat resistant (Fig. 7.1). Like *Azospirillum, Azotobacter* also contains poly β-hydroxy butyrate and produces polysaccharides and melanin. The ideal pH is 6.8-7.8 and temperature 35 to 40°C. *Azotobacter* is found in rhizosphere of many plants and fixes nitrogen and produces hormones like growth substance. Cell occurs singly, in pairs and sometimes in chains with large capsules. As cultures ages the cells become coccoid, form cysts and turn yellow or cinnamon with water insoluble pigment. The cell size and shape vary considerable with spieces, strains, age of culture and growth conditions; *A. chroococcumm* cells have acuminated ends, while *A. beijerinckii* are oval with rounded ends. However, the cells of *A. beijerinckii* are larger than that of *A. chroococcum*. Young cells of *A. vinelandii* closely resemble to *A. chroococcum* type, being comparatively smaller. Cells of the family Azotobacteriaceae are usually variable in morphology. The vegetative

Table 7.1 : Important characteristics of few Azotobacter spp.

Parameters	Azotobacter spp.					
	A. chroococcum	***A. beijerinckii***	***A. vinelandii***	***Azomonas agilis***	***A. indignii***	***A. paspali***
Habit	Soil and water	Soil and water	Soil and water	Water	Water	Soil
Size of the Cell (μm)	2.0-3.0 × 3.0-6.0	4.6 × 2.4	3.4 × 1.5	3.3 × 2.8	3.2 × 0.5	
Cyst Formation	Present	Present	Present	Absent	Absent	Absent
Flagellation	Peritrichous	-	Peritrichous	Peritrichous	Polar	Peritrichous
Motility	Motile especially in young culture or if grown in ethanol	non-motile	Motile (in old cultures motility may be unnotices)	Motile	Motile	motile
Characters of pigment	Bark brown to black (water insoluble)	Yellowish to pale brown does not diffuse into water	Yellowish gree fluorescent diffuses into water		-	-
Pigment formation	With ageing	With ageing	In young cultures	Formed in young culture	-	
Utilizes sodium Benzoate	Yes in some cases	Yes (grown in a conc. of even 5%	Yes (grow in a conc. of even 1%)	No	No	Yes
Utilizes mannitol Benzoate	Yes	Yes	Yes	No	No	No
Utilizes Rhamnose Benzoate	Yes	Yes	Yes	No	No	Yes
Utilizes starch	Yes	No	No	No	NO	No

Sources : Gair (2010)

cells give rise to specialized spherical resting cells known as cysts. Each cyst is produced from a single vegetative cell. There is a distinct morphological difference exists between vegetative cells and cysts. As compared to the rod-shaped structure of vegetative cells, the cysts under a light microscope appear as spherical cells with contracted cytoplasm and a double contoured thick wall. During cyst germination, among vegetative cell emerges from the cyst structure leaving behind a residual portion of the cyst coat. Encysted cells have the ability to tolerate adverse conditions of desiccation, sonic treatment and ultraviolet radiations which are hardly tolerated by the vegetative cells. Motility in most of the cells is carried out by means of petrichous flagella. Among the three different typical species, *A. beijerinckii* are less motile when compared with *A. vinelandii* and *A. chroococcum* has the highest motility of the three. This serves to differentiate among the main species of *Azotobacter*. The pigment is commonly attributed to the presence of melanin. The melanin is formed as a result of oxidation of tyrosinase, a copper-containing enzyme.The characteristic growth of the colonies together with the development of pigments form the basis of distinguishing strains characters. But all strains show utilization of citrate, acidification of litmus milk, negative MR and VP tests. However, they do not utilize rhamnose.

The strains of *A. chroococcum*, however, display variation iwanin morphological, cultural and physiological attributes. The main physiological attributes of practical importance are ability to fix molecular nitrogen, synthesis of growth promoting substances and production of antibiotic metabolites.Its distribution and abundance in soil is primarily determined by soil pH and organic matter content. *Azotobacter* has a full range of enzymes needed to perform the nitrogen fixation: ferredoxin, hydrogenase and an important enzyme nitrogenase. (Karunakaran *et al.*, 2014). Nitrogen fixation is achieved by the enzyme nitrogenase, which reduces N_2 to ammonia. However, this enzyme is extremely sensitive to oxygen in *Azotobacter* species. High respiration rates and conformational protection of the enzyme are suggested as two factors which make nitrogen fixation possible in an aerobic environment.

7.2 Distribution

Azotobacters are found in rhizosphere of many plants that fix nitrogen produce growth promoting hormones. The population in soil ranges from negligible to 10,0000/gm of soil depending on organic content, microbial interactions and temperatures. *Azotobacter* is active N_2 fixers under lab conditions, ubiquitous in geographic distribution and can use a wide variety of carbon and energy sources for their growth. A survey of sorghum and maize growing areas showed that the most probable number (MPN) of N_2 fixers varied 1000 fold, from 100 to 100,000/g soil. Isolates obtained from these soils, 42% showed nitrogenase activity *in vitro*. In another study, out of 546 isolates obtained from the rhizosphere of pearl millet grown at ICRISAT Center only 17% isolates showed nitrogenase activity, *in vitro*. Pearl millet rhizosphere was dominated by *Azospirilla* (72% of total N_2 fixing isolates) followed by *Enterobacter* (12%), *Azotobacter* (11%) and 5% *Pseudomonads*. Similar information is lacking for other crops (Gaur, 2010). *Azotobacter chroococcum* strains are isolated by subramoney from red loamy soils in India. Studies regarding the distribution of *Azotobacter* species indicated that they are widely spread and in top layers and the population decreases with depth. *Azotobacter chroococcum* and *A. beijerinckii* are deemed to be the most commonly occurring species. The later appeared in acid soils while *A. chroococcum* in neutral and alkaline soils.Several *Azotobacters* have the unique characteristic of producing extracellular slime. Although its formation is related to nutritional conditions such as type of carbon and nitrogen sources, it is above all generically controlled character which is specific for different species. Rhizosphere of the crop plants is considered to be congenial habitat where it exists in comparatively high number when compared with soil. Bacterisation of soil or seed material with artificial inoculation helps to raise the population 2 to 10 fold. However, some plants exert inhibitory effect during first 2-3 weeks where population becomes less (R: S ratio lesser than 1). It is pertinent to select right type of strains of *A. chroococcum* to have proper establishment of the bacterium on chosen hosts. Such preferential establishment of strains may be ascribed to the competitive advantage conferred on the genotype of culture to utilize efficiently root exudates of the given crop plant.The multiplication of *Azotobacter* in soil showed that its population in soil rarely goes beyond $10^3$to 10^4/g soil (Darzneik, 1961). In Indian soils the population of *Azotobacter chroococcum* ranges between 10^3 and 10^4/g soil and rarely exceeds 10^4/g (Gaur and Misra, 1978).

7.3 Screening of *A. chroococcum* strains

The nitrogen fixing ability of the strain cannot be a sole criterion to select it for efficiency. In fact, few low nitrogen-fixing strains were excellent in field performance when compared with the high nitrogen fixing strains. Therefore, nutritional, stimulatory and therapeutic capabilities of different strains of *A. chroococcum* have to be taken into consideration. The close scrutiny of the characters of these strains and their influence on plant growth and yield has helped to evolve a kind of procedure to select an appropriate efficient strain of *A. chroococcum* for a particular crop plant. By and large, the homologous isolates (crop-wise) which improve the germination of seeds are preferable for overall efficiency to increase the yield. The stepwise selection of *A. chroococcum* as follows:-

a) Strains should be homologous (crop-wise selection).

b) Strains should be large-sized, chromogenic forming pigment of brown to black colour.

c) Strains should have a moderate capacity to fix nitrogen.

d) Strains should be able to improve seed germination.

e) Strains should properly establish on the roots of crop plants.

After screening on the above selection criteria they should be tested on crop productivity in field conditions.

7.4 Maintenance and Culturing

Azotobacter isolates are maintained on nitrogen free media and periodically sub-culture. Lyophilization of cultures helps in preservation of different isolates. The pure and efficient cultures are subjected to vigorous testing in pot and field conditions. If the culture proves beneficial for crops is then released for mass production of the Inoculant.

7.5 Isolation of *Azotobacter*

Requirements

The apparatus required consists of incubator, test-tubes, Petri dishes, glass rods, inoculation needles, pipettes, and conical flasks&the reagents required aresterile water, Jensen's medium.The microbes are isolated from the soil sample (collected from the rhizosphere) by the soil dilution method and plating on an N-free agar medium.

Procedure

The procedure is

1) Collect moist soil from the rhizosphere, and make a suspension of 10g of this soil in 90ml of sterile water in a conical flask.

2) Shake the suspension for about 5 minutes.

3) Dilute this suspension serially as follows: (i) arrange at least 6 test-tubes, each containing 9ml of sterile water, in a test-tube stand (ii) Take 1ml of the suspension from the conical flask aseptically and add to the first test-tube containing 9 ml of sterile water. Total volume will be 10ml. Shake the test-tube vigorously (iii) Pipette out 1ml of the suspension from this test-tube and add it to the second test-tube and shake (iv) Repeat this process serially until the last test-tube.

4) Pipette out 1ml of liquid aseptically from the third, fourth, fifth and sixth test-tubes (to start with a minimum dilution of 1000 times), and pour onto separate N-free agar medium (Jensen's medium) plates. Spread the aliquot on the plates, and mark the plates.

5) Invert the plates, and incubate them at about 28 ± 2°C for at least three days.

6) After incubation, soft, flat, transparent or milky, mucoid colonies of *Azotobacter* will develop on the plates.

7) Pick up a single colony aseptically with the help of a sterilized inoculation needle and streak it on an N-free agar media plate.

8) Invert and incubate the plate at 28 ± 2°C. In this way, pure colonies of *Azotobacter* will be formed.

After proper identification, *Azotobacter* from the colony so developed can be taken and multiplied in order to obtain a pure culture through the inoculation and fermentation process.

Identification

Azotobacter has the ability to produce pigment. Different species of *Azotobacter* produce different pigments, thus facilitating their identification. The pigments produced by important species are:

a) *Azotobacter chroococcum* – brown

b) *Azotobacter beijerinckii* – light brown

c) *Azotobacter vinelandii* – greenish yellow

d) *Azotobacter insignis* – light brown

e) *Azotobacter agilis* – green

f) *Azotobacter macrocytogenes* – purple

Although pigment production does not give a confirmation of the species, it isa good indication of their presence.The secretion of gum or polysaccharide is another important characteristic, as the formation of cysts by all the species (for withstanding adverse conditions).

7.6 Commercial production of *Azotobacter*

The production process is similar to that for *Rhizobium* production. However, *Azotobacter*-specific broth is prepared. This is the same medium as that for laboratory use (Jensen's or Ashby's media) except that agar is not added in the preparation of broth to be used for carrier-based commercial production. Quantities of chemicals required for the production of 25, 50 and100 tonnes of *Azotobacter* biofertilizer are given in Table 7.2.

Table 7.2: Chemicals required for the production of *Azotobacter* biofertilizer

Chemicals Biofertilizer (30–40% moisture)	**25 tonnes**	**50 tonnes**	**100 tonnes**
Sucrose	150.00	300.00	600.00
K_2HPO_4	7.50	15.00	30.00
$MgSO_4.7H_2O$	3.75	7.50	15.00
NaCl	3.75	7.50	15.00
$FeSO_4$	0.75	1.50	3.00
$CaCO_3$	15.00	30.00	60.00

Note: Calculated on the basis of Jensen's medium.

Source : In: Guide to laboratory establishment for plant nutrient analysis, Food and Agriculture Organization of the United Nations Viale delle Terme di Caracalla, 00153 Rome, Italy

7.7 Crop Responses

A low N_2-fixing *A. chroococcum* (M-4 strain)were found to have broad spectrum utility to enhance growth and yield of crops viz. sorghum, cotton and potato (Shende and Apte, 1982). While probing the effect of inoculation with *Azotobacter* on the growth of the plant, it became clear that the manifestation of such effect begins spontaneously even during the germination of inoculated seeds. The improvement in germination of seeds of cotton, rice, maize, wheat, soybean and bengal gram was confirmed from time to time (Shende *et al.*, 1977). It can be attributed to the production of growth promoting substances. Relevance of the improved seed germination is of the paramount importance as the inoculation with *Azotobacter* can assure the maintenance of proper plant stand of field crops per unit area. The stimulatory effect of *Azotobacter* on tomato and wheat plants was reported by different workers (Jackson *et al.*,1964; and Brown and Burlingham, 1968). The multiplication of *Azotobacter* in soil is intensified with the addition of straw, manure and also various composts containing cellulose reported by Kundu and Gaur (1980). Gaur and Rai (1982) investigated the effect of paddy-straw, partially degraded paddy-straw or farmyard manure on nitrogen fixation by *Azotobacter* or *Azospirillum*. Farm yard manure was found a better carbon source for nitrogen-fixation by *Azotobacter* whereas partially degraded paddy-straw for *Azospirillum*.

Azotobacteria genus synthesizes auxins, cytokinins, and GA–like substances, and these growth materials are the primary substances controlling the enhanced growth. These hormonal substances, which originate from the rhizosphere or root surface, affect the growth of the closely associated higher plants. *A. chroococcum* can promote the growth of various crops by some mechanisms such as production of IAA, GA and cytokinin (Paul and Verma, 1999). Seed inoculation of faba bean with *Rhizobium leguminosarum* bv.*viceae* and five different *A. chroococcum* under antibiotic conditions of culture resulted in significant effects on nodulation and plant growth at the flowering (Rodelas, 1999). Crops receiving *Azotobacter* inoculation alongwith moderate levels of nitrogen gives similar grain yield as the crops receiving higher doses of mineral fertilizers but uninoculated. This indicated the possibility of reducing application by inoculating the crop with appropriate strains. This inoculum is useful for cereals and non-leguminous crop plants. The benefits accrue from their inoculation in the form of increased grain yield, biomass and N uptake. These may be ascribed to N addition through biological nitrogen fixation, development and branching of roots, production of plant growth hormones like IAA, GA etc. enhancement in the uptake of NO_3^-, NH_4^-, $H_2PO_4^-$, K and Fe etc. and improved water status of plants, increased nitrate reductase activity and production of antifungal compounds. *Azotobacter* has ability to produce antifungal antibiotics and fungi decompound against pathogens like *Fusarium, Alternaria* and *Trichoderma.*The strains of *Azotobacter* effective for the plant growth were rarely useful to control the plant disease. It was felt that therapeutic strains were needed to be used for controlling a particular specific disease. The population of *Azotobacter* is generally low in the rhizosphere of the crop plants and in uncultivated soils. Dutta and Singh (2002) reported a significant in seed yield (7.86q/ha) in rapeseed and mustard due to inoculation with *Azotobacter*. Similarly, positive reports on application of *Azotobacter* and *Azospirillum* on the yield of mustard (*Brassica juncea*) are available (Tilak and Sharma, 2007).

Das and Saha (2007) concluded that the combined inoculation of *Azotobacter, Azospirillum* along with diazotrophs increased grain and straw yield of rice by 4.5 and 8.5kg/ha, respectively. Undergreenhouse conditions plant height, leaf number/plant, number of primary and secondary branches/ plant, fresh and dry weight of whole plant, number of siliqua/plant, seeds/siliqua of brown sarson increased significantly with *Azotobacter* inoculation than no inoculation with seed and stover yield of 10.107g/pot and 22.400g/pot respectively (Wani *et al.*,2013). The challenge to the research community will be to develop systems to optimize beneficial plant-endophyte bacterial relationships (Sturz *et al.,* 2000). In order to guarantee the high effectiveness of inoculants and microbiological fertilizers it is necessary to find compatible partners, i.e. a particular plant genotype and a particular *Azotobacteria.* Co-inoculation of *R. leguminosarum* and *A. chroococcum* can improved some of growth indices under drought conditions reported by Dashadi *et al.,* (2011). *Azotobacteria* genus synthesizes auxins, cytokinins, and GA–like substances, and these growth materials are the primary substances controlling the enhanced growth reported by Wani *et al.,* (2013).

A. chroococcum and *Azospirillum lipoferum* were found to have not only the ability to fix nitrogen but also the ability to release phytohormones similar to gibberellic acid and indole acetic acid, which could stimulate plant growth, absorption of nutrients, and photosynthesis (Essam *et al.,*2013).The results of the trial conducted on mustard yield increased by about one quintal over application of 20kg N alone. Thus, by spending Rs. 12.50 only per hectare on biofertilizers, there is net saving of Rs. 800/ha. Application of 20kg N plus *Azotobacter* seed inoculation resulted in 56.6 and 10.8% higher gross income over control and application of 20kg N alone respectively (Table 7.3). Thus cost benefit ration by using low cost benefit input (*Azotobacter* biofertilizers) in mustard is approximately 1: 65. Although application of 40kgN/ha (recommended dose in mustard) results in higher net income but the FYM treatment (organic manure) also enriches the soil with other nutrients, thereby, improving the overall fertility of the soil. In comparison with 20 kgN plus *Azotobacter* seed inoculation also, the results are less beneficial but it reduces the dependence on inorganic fertilizer which may be benefited specially to the farmers with limiting resources. Because both the 20kg plus *Azotobacter* as well as FYM plus *Azotobacter* treatments are definitely superior over control as they results in net benefit of nearly Rs. 3000/ha over control. Therefore, farmers are advised to inoculate the seeds of mustard with *Azotobacter* at the time of sowing. For direct application in the field, it has been observed that if *Azotobacter* is applied along with FYM, the response is more as compared to its application alone.

Table 7.3: Effect of Organic, Inorganic and Biofertilizer on Grain Yield (q/ha) of Mustard and Net Income

Treatment	Grain yield	Gross Income (Rs/ha)	Expenditure on nutrient (Rs/ha)	Net Income (Rs/ha)
Control	5.8	5220	-	5220
20kg N/ha	8.3	7370	118.7	7351.3
Azotobacter seed inoculation	9.3	8280	131.2	8148.8
4 tonnes of FYM and Azotobacter seed inoculation	9.5	8550	492.5	8057.5
40kg N/ha	9.9	8910	237.4	8672.6

Source : In: The Complete Technology Book on Biofertilizers and Organic Farming, 2012

7.8 Gmax Nitromax

Gmax Nitromax is combined product of *Azotobacter* and *Azospirillum*. *Azospirillum* works well with all crops whereas *Azotobacter* gives excellent results when soils have rich organic matter content (more than 2%). As a combined product of *Azospirillum* and *Azotobacter*, Gmax Nitromax combines the benefits of both inoculants in a single formulation. The product is available as liquid formulation and lignite carrier based formulation.

Composition

Azospirillum & *Azotobacter* 1% (w/w), Sticking agent–CMC–1%., Inactive Ingredients 98.0% (w/w), (Moisture 35%, lignite 63%)

Packing

Carrier based formulation is available in attractive one kg laminated poly pouches. The packing is moisture proof and well tolerates transportation and handling. The product is also supplied in bulk packing of 50kg/25 kg sizes in HDPE bags. Liquid formulation is available in one litre HDPE containers and in 200 lr barrels.

Method of application

The mixture of *Azotobacter, Azospirillum* (Gmax Nitromax) and phosphate solublizing bacteria (Gmax Phosphomax) can be used for non-leguminous crops-like cereals, oilseeds cotton, vegetable and all other crops which are usually planted as seeds.

Seed treatment

For 1kg of seeds about 40g of Gmax Nitromax product is required. Mix required quantity of Gmax Nitromax with equal quantity of rice gruel. With this mixture, mix the required amount of seeds shade dry and after 24hrs, use the seeds of sowing.

Seedling Root dip

Mix 2kg of Gmax Nitromax with 50 lits of water. In this mixture, keep the seedling roots in immersed condition for 10 minutes and use the seedlings for transplantation in the field.

Soil application

Mix about 5kg Gmax Nitromax with 100kg of organic manure, keep the mixture under shade for one week and apply to the soil.

Drip irrigation

Dilute 5kg of Nitromax in 100 litrs of chemicals free, good quality water. Filter the mixture with a pure cloth; use the filtered solution in drip irrigation for one acre.

Recommended crops

Gmax Nitromax is recommended for all crops except all kinds of legumes. Gmax Nitromax can be used along with Gmax Phosphomax, Gmax Tricon and Gmax Fyton to provide nitrogen, phosphatic nutrition and to provide plant protection from diseases along with plant growth promotion.

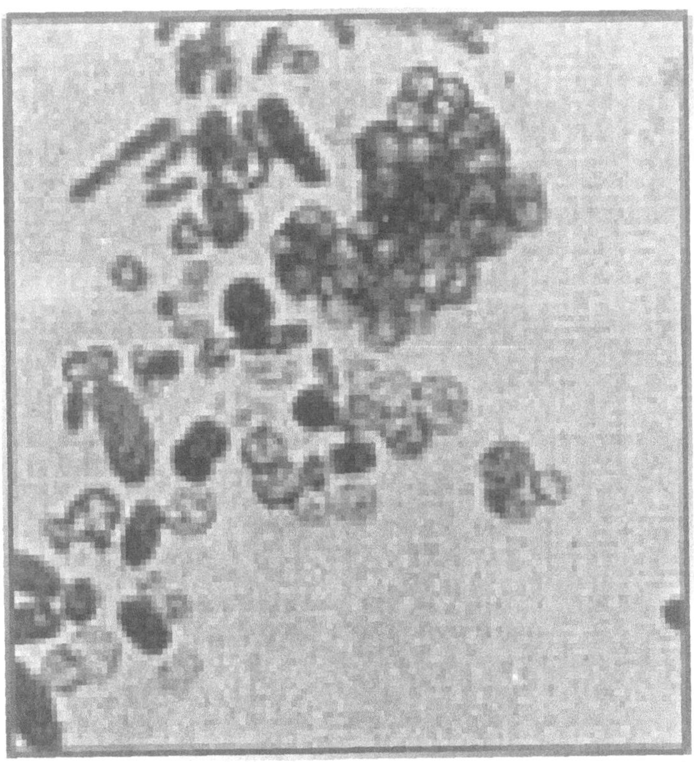

Fig. 7.1 : *Azotobacter* sps.

Chapter - 8

Azospirillum

Azospirillum is a spiral-shaped nitrogen fixing bacterium. It also produces hormones and vitamins. Important species are *Azospirillum brasilense* and *Azospirillum lipoferum*. It is widely distributed in soils and grass roots.The genus *Azospirillum* belongs to family Spirillaceae with four commonly available species viz. *A. brasilence, A. lipoferum, A. amazonense* and *A. halopraeferens or A. seropedicae* (Fig. 8.1). Genus *Spirillum* was created in 1832 by Ehrenberg distinguishing it from phototrophic spirilla which were placed in a separate genus *Thiospirillum* (Ehrenberg, 1838). They are Gram negative, motile, vibroid and contain polyhdroxy butyrate granules. In the presence of ammonium salts it grows aerobically and in microaerophilic conditions, fixes atmospheric nitrogen.The members of *Azospirillum* are aerobic and utilize carbon sources from organic acids and sugars such as L-arabinose, D-fructose, D-glucose, sucrose and pectin and nitrogen sources through the fixation of nitrogen, ammonia, amino acids and nitrate.

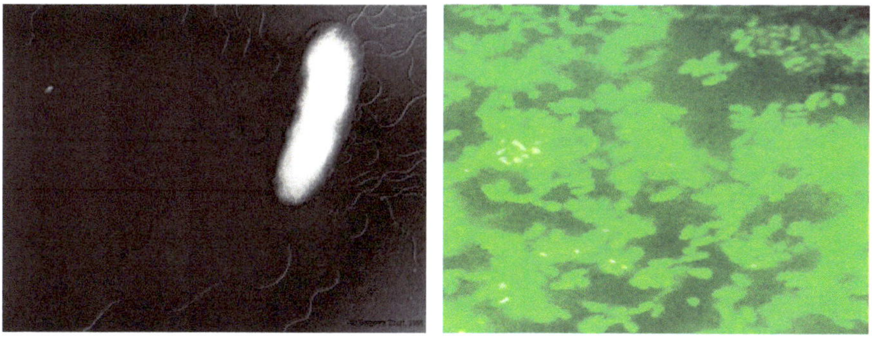

A. brasilense *A. lipoferum*

Fig 8.1 : Different sps. of *Azospirillum*

8.1 Occurrence

Azospirillum spp. is of ubiquitous distribution in many parts of the world consisting of tropical, sub-tropical and temperate climatic conditions and also with various standing crops grown in a variety of soil types. *Azospirillum* spp. Fixes atmospheric nitrogen and has been isolated from the rhizosphere of a variety of tropical and sub-tropical non-leguminous plants. Lakshmikumari *et al.,* (1976) reported the occurrence of nitrogen fixing spirillum in the roots of rice, sorghum, and maize. Various cereals have responded differently to filled inoculation with this organism (Barber *et al.,* 1976). Okon and Gonzalez (1994) evaluated world wide data accumulated over the past 20 yrs on field inoculation with *Azospirillum* and concluded that these bacteria are capable of promoting the yield of agriculturally important crops in different soil and climatic regions.

Azospirillum is confined to the root system of those tropical grasses where a C_4pathway (Hatch and Slack Pathway) is operative because they grow and fix nitrogen on salt of organic acids such as malic and aspartic acid (Dobereiner and Day,1974;Arun, 2007). Besides C_4 plants, several C_3 plants including weeds were investigated and an abundant distribution of the organism was recorded in their roots. Dobereiner *et al.,* (1976) reported than less than 10% of soils and roots from temperate regions contain *Azospirillum* whereas more than 50% of tropical samples were positive. Alluvial soils were more favourable for them than eroded hilly soils. Although pH of the soil had a strong influence on the distribution of *Azospirillum,* sporadic occurrences could be demonstrated in soils having pH as low as 4. The optimum pH is around seven *Azospirillum* were isolated from tropical grasses and xerophytes growing in arid zone of India. Most isolates *of Azospirillum brasilense* were tolerant to chloride and sulphates, but are sensitive to bicarbonates (Rao and Venkateswarlu, 1985).A new nitrogen fixing bacterium, *Herbaspirillum,* taxonomically related to *Azospirillum* was isolated from Brazil (Baldani *et al.,*1986). It was found to be associated with grasses.

8.2 Growth Media

Azospirillum is gram negative, motile generally vibroid in shape and contains ploy â hydroxybutyrate granules. *Azospirillum* produces white dense and undulating fine pellicles (a thin skin or film specially one that reflects a part of the light falling upon it and limits the rest of the light through it and that is used for dividing a beam of light) is a very characteristics on the semi-solid malate medium. Selective medium and techniques is now available for *A. brasilense.* The technique essentially consists of growing on successive liquid enrichments in N free semi- solid medium supplemented with cycloheximide (250mg/L), followed by the most probable number counting method and N free supplemented with streptomycin sulphate (200mg/L), sodium deoxycholate (200mg/L), 2,3,5 tri phenyl tetrazolium chloride (15mg/L) and congored (1000mg/L). The sizes of the cells are variable even among isolates from the same roots. Microscopic examination reveals polymorphism but the dominant forms on solid medium are characteristic curved rods of varying sizes with prominent refractive fat droplets.

Electron micrographs show lateral and single polar flagellum. When grown in nitrogen free media it behaves as microaerophilic and when supplied with fixed nitrogen it grows as an aerobe. The salts of organic acids such as malate, succinate, lactate and pyruvate have been found to be satisfactory oxidizable carbon and energy sources. Certain strains were found to require low levels of yeast extract for growth in mineral media and to grow in glucose as carbon source, whereas strains that did not require yeast extract failed to grow on glucose. On these and several other observations the existence of two or possibly three groups of organisms within the species *S. lipoferum* was established.

8.3 Taxonomy

A taxonomy study of *S. lipoferum* group was conducted based on the DNA homology and the other testes.DNA bases compositions of 39 strains of *Spirillum* were determined by Hylemon *et al.* (1973). A uniform methodology provided a basis for comparison of the strains. The results together with previous serological data and with previous results for two strains of *Spirillum volutans* indicated that the present genus *Spirillum* should be divided into three genera, with the original name being restricted to obligatory micro-aerophilic fresh water forms having a large cell diameter and a DNA base composition of 36 to 38 mol %G+C. This genus contained only one species, *Spirillum volutans.* The generic name *Aquaspirillum* was proposed for the aerobic, fresh water forms having a DNA base composition of 49 to 65 mol% G+C. Thirteen species are included in this genus. The third generic name was proposed for marine forms which did not attack carbohydrates and which possess a DNA base composition of 42 to 48 mol% G+C. The new genus *Azospirillum* was further divided into two distinct groups such as (a) *A. brasilense* and (b) *A. lipoferum.* Group II was distinguished from group I by their ability to use glucose as a sole carbon source for growth in nitrogen free medium, by their production of an acid reaction in a peptone based glucose medium, and by their requirement of biotin and by their formation of wide, longer-S-shaped or helical cells in semi-solid nitrogen free malate medium, Morphological, physiological and nutritional characteristics of different strains of *Spirillum* and *S. lipoferum* were studied by Hylemon *et al.* (1973); Okon *et al.* (1976a) and Tarrand *et al.* (1978); but immunofluoresence and immunodifusion studies have indicated antigenic difference (DePolli*et al.*,1980). *A. brasilense* contains two subgroups *nir*$^+$ and *nir* based on ability to denitrify nitrate (Neyra *et al.*,1977). Both groups possess assimilatory and dissimilatory nitrate reductase, but *A. brasilense* nir$^+$ does dissimilate nitrate (Magalhaes *et al.* 1978). Hylemon *et al.* (1973) reported that all strains hadgiven catalase-positive test,whereas negative tests for catalase in five strains of *Azospirillum* *lipoferum* was noted by Tarrand *et al.,*(1978). The catalase reaction was divided by Tarrand and his coworkers (1978) into three categories i.e., strong, weak and negative.All *Spirillum* strains were negative for indole and H_2S production, pectinase activity, cellulose hydrolysis of starch and gelatin, methyl red test and Voges Proskauer test. Some of the strains turned litmus milk alkaline (Tarrand *et al.,* 1978). Sizes of all colonies ranged from 0.5-1.5mm in diameter except for six colonies which formed pin-point. All colonies were white, circular and convex with smooth edges, except for two strains, which formed white colonies with fimbriated edges. The fimbriated colonies were formed partially below surface (Hylemon *et al.,* 1973). Okon *et al.,*(1976a) noted that colonies of *S. lipoferum* tend to develop a light pink pigment on nutrient or potato agar. However, certain strains formed very deep pink colonies on MPSS agar medium.

8.4 Physiological Effects

A. lipoferum (AL) and *A. brasilense* (AB) can grow vigoursly in salts of malic, succinic, lactic and pyruvic acid. AL is able to utilize glucose and sucrose as sole carbon sources for growth and N_2 whereas these sugars are poor substrate for AB. The correlation between organic acids as potentially good substrate for this bacterium and their accumulation in the roots of grasses has been emphasized in the support of the suggestions of a symbiotic association between these bacteria and grasses. Further these bacteria require low oxygen for the expression of nitrogenase activity. Experiments indicated that tricarboixlic acid pathway is operative these organism while glycolysis and pentose pathway is of minor importance. Micro-aerophilic studies have been reported that there is another mechanism by which *Azospirillum* fixes N_2. In contrast nitrate metabolism in *S. lipoferum* has received little attention. The process of assimilatory and dissimilatory nitrate reduction in other bacteria is well known (Burns and Hardy, 1975), but Neyra *et al.* (1976) for the first time reported that some of the strains of nitrogen fixing *S. lipoferum* are able to bring about demtrification. Neyra and Burkum (1976) demonstrated that dissimilatory nitrate reductase occurs at membrane level. Better understanding of these transformations is necessary to assess the economic importance of these organisms. *S. lipoferum* grows vigorously on malate, succinate, lactate or pyruvate, moderately on galactose or acetate and poorly on glucose and citrate (Okon *et al.,*1976). It reduces $^{15}N_2$ acetylene reduction rates decrease rapidly when the pH of the culture rises above 7.8. These organisms are highly aerobic when grown in ammonia solutions. However, *S. lipoferum* reduces N_2 well only under micro-aerophilic condition. The optimum pressure of oxygen for acetylene reduction by a stagnant culture is 0.006-0.02 atmospheres depending upon the cell density. Shaking *S. lipoferum* with air temporarily inactivates its nitrogenase and reactivation is inhibited by chloramphenicol. The organisms assimilated 20-24mg of N/g of organic acid oxidized during growth*Azospirillum* spp. belong to the carbon compounds and adequately low level of facultative endophytic diazotrophs groups which colonize combined nitrogen (Andrew *et al.*, 2007). *Azospirillum* directly benefits plants by improving the fixing activity of bacteria in rhizosphere of plants and helps in shoot and root development (Gonzalez *et al.*, 2005).

8.5 Maintenance of Cultures

Azospirillum cultures can be maintained on ammonium chloride containing agar medium formulated by Oken and coworkers. On this medium, bacterium does fix nitrogen but proliferates profusely under aerobic conditions. The number of cells in the both can be counted by planning dilutions on ammonium chloride agar medium. In the large scale multiplication, the bacterium can be grown in ammonium chloride medium in flasks on the rotary shaker at temperature of 35^0C. The best carriers for this bacterium is powered sterilized farm yard manure and soil of FYM alone and charcoal and on this carrier the bacterium survives for a period of 6 months.

8.6 Isolation of *Azospirillum*

Azospirillum bacteria occur inside as well as outside plant roots, plant roots are taken in order to isolate *Azospirillum*.

8.6.1 Requirements

The apparatus required consists of an incubator, sterilized water, petri dish, glass rod, test-tube, inoculation needle. The reagents required areEthyl alcohol, 0.1% mercuric chloride, Phosphate buffer.

8.6.2 Procedure

1) Take roots of any field crop plants.
2) Wash the roots first under tap-water, and then surface sterilize with 95% ethyl alcohol for about 5 seconds.
3) Cut the roots into small pieces (0.5cm long), and wash them with 0.1% mercuric chloride for 1 minute, followed by washing with sterile water and then with phosphate buffer.
4) Place the pieces of roots in screw-capped tubes containing semi-solid sodium or calcium malate medium, and incubate at 28–30 °C for 3–4 days.

5) After proper incubation, a white pellicle of *Azospirillum* will develop 1–2cm below the upper surface of the medium.

6) Transfer the isolate (white pellicle) 3–4 times to semi-solid calcium malate medium contained in screw-capped tubes.

7) Thereafter, make serial dilutions up to 10–10, and then incubate 0.1ml from the last dilution to a fresh tube/plate containing semi-solid calcium malate medium for growth.

8) Incubate the tube/plate for 3–4 days. Observe the growth of *Azospirillum*.

9) After proper identification, *Azospirillum* from the colony so developed can be taken and multiplied in order to obtain a pure culture through the inoculation and fermentation process.

8.6.3 Identification

Azospirillum organisms are Gram-negative, curved, and rod-shaped of varying size. They contain poly-β-hydroxybutyrate granules. *Azospirillum* shows spiral movements. The formation of white pellicles on semi-solid calcium malate medium is a characteristic of *Azospirillum*. *Azospirillum* forms round-shaped colonies on the solid malate medium. *Azospirillum* micro-organisms are producers of strong bases. Hence, when they grow in a medium containing Bromothymol blue indicator, they change the colour of the medium to blue.

8.7 Commercial Production of *Azospirillum*

The procedure is same as for *Rhizobium* and *Azotobacter*. The brothis prepared in N-free bromothymol blue medium (excluding the use of agar)and/or Okon's modified liquid medium (excluding bromothymol blue because bromothymol blue is only for identification/confirmation –once it is confirmed, there is no need to add it again). Lists the quantities of chemicals/reagents required for the production of 25,50 and 100 tonnes of *Azospirillum* are given in Table 8.1.

Table 8.1: Chemicals required for the production of *Azospirillum* biofertilizer

Chemicals Biofertilizer (30–40% moisture)	25 tonnes	50 tonnes	100 tonnes
K_2HPO_4	45.00	90.00	180.00
KH_2PO_4	30.00	60.00	120.00
$MgSO_4$	1.50	3.00	6.00
NaCl	0.75	1.50	3.00
$CaCl_2$	0.15	0.30	0.60

NH_4Cl	7.50	15.00	30.00
Malic acid	37.50	75.00	150.00
NaOH	22.50	45.00	90.00
Yeast extract	0.40	0.80	1.50
Na_2MoO_4	0.015	0.03	0.06
$MnSO_4$	0.01	0.02	0.04
H_3BO_4	0.01	0.02	0.04
$Cu(NO_3)_2$	0.005	0.01	0.02
$ZnSO_4$	0.015	0.03	0.06
$FeCl_3$	0.015	0.03	0.06
Bromothymol blue (0.5% alcoholic)	15 litres	30 litres	60 litres

Note: Calculated on the basis of Okon's medium.

Source : In: Guide to laboratory establishment for plant nutrient analysis, Food and Agriculture Organization of the United Nations VialedelleTerme di Caracalla, 00153 Rome, Italy

8.8 Production of Growth Promoting Substances

Tien*et al.,*(1979) found increased yield of pearl millet and attributed this increase due to Indole acetic acid (IAA), Gibberellic acid (GA) and Cytokinin-like substances. IAA was formed from tryptophan by *A. brasilense.* Umaligarcia *et al.,*(1980) found that young inoculated roots of pearl millet produced more mucilaginous sheath root hairs and lateral roots than control. At early stages of growth, inoculation increased the numbers of roots and their branching. Root segments of inoculated corn and wheat showed significant increase 30-50% over control in the uptake rate of NO_3, K and H_2PO_4. *Azospirillum brasilense* (strains 201, 251, 8, 15 and 304) and *A. lipoferum* (strains 20, 254, 12, 353 and 102) were tested for production of growth promoting substances in the culture medium (Saxena, 1983). *Azospirillum* strains were grown for 5 days at 37°C in 250ml flasks containing 100ml Okon's medium (Okon *et al.,*1977). Bacterial cultures after their growth were centrifuged at 20000 rpm for 20 minutes. Each centrifugate was acidified to pH 3.0 with N-HCl. The acidified centrifugates was divided into 2 equal fractions (A and B). Fraction A was extracted for auxinlike substances and fraction B for gibberellin like substances.

- ***Auxin-like substances:*** 50ml of supernatant fluid (Fraction A) was shaken 3 times with ethyl acetate (10ml) at 30 minutes interval. Combined ethyl acetate extracts were evaporated to dryness in a current of air and residue dissolved in 3ml methanol. Extracts were examined by paper partition chromatography using freshly prepared solvent mixture, isopropanol: ammonia solution: water (10:1:1).
- ***Gibberellin-like substances:*** 50ml (Fraction B) was shaken with 0.5g activated charcoal for two hrs. The charcoal was separated by

centrifugation, extracted with aqueous acetone (16ml 95%). The acetone was evaporated in a current of air at room temperature and the moist residue shaken twice with 1.5 times its volume of ethyl acetate. The combined ethyl acetate extract was evaporated to dryness in cold air. The residue was dissolved in 3ml methanol and examined by paper partition chromatography using freshly prepared solvent mixture as mentioned above.

Chromatography

0.1ml of the extract was spotted in bands on narrow strips (45cm x 6cm) of chromatography paper (Whatman No.l). Authentic IAA, GA_3 and control were also run. With freshly prepared solvent mixture (Isopropanol:Ammonia solution: water, 10:1:1) spotted chromatograms were equilibrated in solvent-saturated air for 2hrs and then developed with solvent by descending chromatography. After the solvent phase reached to solvent front, marked at 30cm from the base line, chromatograms were removed, marked with lead pencil to the level of maximum flow and then were air dried in dark. One set of strips for IAA and the other set for GA were dipped in 5% (v/v) cone, sulphuric acid in methanol, dried in hot air and exposed for 15 minutes to ultra violet radiation (wave length 350µm.). IAA was identified by yellow fluorescence at Rf0.3 and 0.5 and GA_3 by green fluorescence at Rf0.5 and 0.7. Chromatograms strips not treated with chromogenic reagents were dried for 7 days to remove solvents which were cut into 10 equal parts representing the sequence of Rf values 0.1 to 1.0 and were eluted separately for bioassays.

8.9 Bioassay for Auxins and Gibberellins

8.9.1 Auxin-like-substances

Nitsch and Nitsch (1956) test had given First internode test and it is used to measure the auxin activity of the elutes with following steps:

a) Seeds of Oat (*Avena sativa*) var. Kent were soaked in tap water for 4hrs at room temperature.

b) These were washed thoroughly with distilled water and planted on wet filter paper in aluminium trays.

c) Trays were kept at an angle of 45° to facilitate straight growth of internode and covered with black polythene paper. Seeds were allowed to grow at 25°C for 4 days in complete darkness. Afterwards all manipulations were performed under green light.

d) When the first internode reached about 2.5cm in length and coleoptile of 0.5cm, 4mm sections were cut with a conventional cutter, 2mm below the coleoptile node.

e) The sections were washed in glass distilled water for one hour.

f) 5 sections were taken in a test tube containing 1ml of the phosphate buffer (1.794g K_2HPO_4, 1.019g citric acid monohydrate and 20g sucrose dissolved in 1 litre).

g) Chromatogram pieces of different R_f values were eluted separately in 3ml distilled water. One ml of this elute was added to respective tube with 5 internode sections.

h) Tubes were allowed to rotate around a horizontal axis at about 1 rpm in the dark for about 24hrs at 25°C.

i) The lengths of sections were then measured using enlarger (×5).

j) The per cent increment over control in the length of internode sections was calculated.

k) Likewise different concentrations of IAA ranging from 0.001/ug to 20µg were also tested with the internode sections and a standard curve was drawn for determination of IAA in the culture supernatant.

The effect of elutes from chromatograms of each of the microbial extract were tested on the length of first internode sections of oats (var. Kent) and the results showed the production of auxin-like substances and the nature and level of auxin produced varied among themselves (Table 8.2). None of the cultures showed auxin activity throughout the chrornatogram strip.

Table 8.2 : Auxin activities in the culture filtrate of *Azospirillum*

Cultures of Azospirillum Spp.	Isolate number	Percent increase in length of oat coleoptile over control									
		R_f values									
		0.1	0.2	0.3	0.4	0.5	0.6	0.7	0.8	0.9	1.0
A. brasilense	(201)	7		16	20	20	16				
A. lipoferum	(20)		13	13	20	25			29		
A. lipoferum	(254)			20	29	26			15		
A. lipoferum	(12)			20	20	18	15				
A. brasilense	(251)		26	25	30	25	30				
A. brasilense	(8)			30	31	26	30				
A. brasilense	(15)			17	12	16			19		
A. brasilense	(304)			28	20	25	25				
A. lipoferum	(353)			17	25	17	7				
A. lipoferum	(102)			32	17	16				5	

Source: Gaur (2010) In: Biofertilizers in Sustainable Agriculture, ICAR, New Delhi

The maximum activity was observed between Rf 0.3 to 0.5 in all the cases.The total activity of auxin was higher in *A. brasilense* (251) followed by *A. brasilense* (8), *A. lipoferum* (20) and *A. brasilense* (304). The other organisms occupied intermediate positions.The substances at Rf 0.3 to 0.5 corresponded to the same position of authentic IAA and behaved similarly in the bioassaytest.The maximum extension of 32% in oat coleoptile was observed in culture extracts of *A. lipoferum* at R_f 0.3 which corresponded to 2.5ug IAA/ml of original culture.

8.9.2 Gibberellin-like-substances

Ogawa (1963) has given modified technique of seeds of rice (var. Indira (a mutant of Tainan-3) and it is used for rice second leaf sheath growth in response to chrornatogram pieces prepared from different cultural extracts. The method was as follows:

a) Seeds of Rice (*Oryza sativa*) var. Indira were kept in running water for 4hrs.

b) These were allowed to sprout under immersed condition in a germination chamber at 25°C.

c) Uniformly sprouted seeds were collected and 5 seeds were placed in each tube containing chrornatogram pieces of each Rf and 4ml of distilled water.

d) The tubes were kept under continuous illumination in a temperature controlled chamber at 26°C and light intensity of 160 foot candles.

e) Two ml of double distilled water was added on the 4th day.

f) The second leaf sheath had attained the maximum expansion in about 7 days. It was then measured and the data were compared with dosage response curve of different GA_3 concentrations ranging from 0.001 to 20μg.

The effects of elutes of each Rf zone from the chromatograms were tested on the growth of second leaf sheath of rice (variety Tainan-3 mutant Indira) and the per cent increases in length were recorded (Table 8.3).

Table 8.3 : Gibberellin activity in the culture filtrate of *Azospirillum*

Cultures of *Azospirillum* Sp.	**Isolate number**	**Percent increase in length of oat coleoptile over control**									
		R_f values									
		0.1	**0.2**	**0.3**	**0.4**	**0.5**	**0.6**	**0.7**	**0.8**	**0.9**	**1.0**
		0.1	0.2	0.3	0.4	0.5	0.6	0.7	0.8	0.9	1.0
A. brasilense	201					10.2	20.5	33.5	25.5	15.0	10.0

A. lipoferum	20					11.5	25.5	30.7	21.0	18.5	9.5
A. lipoferum	254		15.5		10.5	25.8	30.9	28.7	26.5		5.8
A. lipoferum	12				17.5	20.5	26.8	29.7	30.8	10.5	9.7
A. brasilense	251				16.5	19.8	31.8	36.8	24.5		8.5
A. brasilense	8						29.7	35.8	27.8		12.6
A. brasilense	15						30.5	39.7	29.2	11.5	
A. brasilense	304						25.6	26.8	29.7		10.8
A. lipoferum	353					14.5	50.8	55.7	30.5	14.8	12.5
A. lipoferum	102						30.5	42.5	40.6	25.6	10.5

Source : Gaur (2010) In: biofertilizers in Sustainable Agriculture, ICAR, New Delhi

A. brasilense (strain 251) showed increase at R_f 0.3 which corresponded to 0.25μg lAA/ml of R_f the original of the culture. *A. lipoferum* (20) at R_f 0.4 showed 20% increase which was equivalent to 0.067μg IAA ml of original culture. *A. brasilense* (15) at R_f 0.4 showed 12% in growth producing 0.007μg IAA/ml of the original culture. Other R_f values were also observed which could not be identified.The results showed that the percent increase in length of second leaf sheath over control was noted at 0.5-1.0 R_f values in most of the cultures. Some cultures showed the activity at R_f 0.4 also. Most of the cultures showed the highest activity at and around R_f 0.7.Highest level of total gibberellins was observed in *A. lipoferum* (353) followed by *A. lipoferum* strains 102, 12 and 254. *A. lipoferum* strains were found superior to *A. brasilense* (201, 251, 8, 15 and 304). The activity at R_f 0.5 to 0.7 corresponded to the same position as that of authentic GA_3.

A. lipoferum (353) showed 55.7% increase in second leaf sheath of rice at Rf 0.7 followed by *A. lipoferum* (102) 42.5%, *A brasilense* (1 5) 39.7% and *A. brasilense* (251) 36.8% increase which corresponded to 0.08, 0.05, 0.047 and 0.042μg GA_3/ml of original culture respectively. The studies clearly showed that *Azospirillum* spp. besides nitrogen fixation is also good producers of plant growth promoting substances.

8.10 Response to *Azospirillum* Inoculation in different Crops

As *Azospirillum* sp. fixes appreciable amount of atmospheric nitrogen and remains in loose symbiosis with tropical plant roots, experiments were conducted to use *Azospirillum* as inoculant to supplement the nitrogen need of cereal and leguminous plants (Umaligarcia *et al.,* 1980; Schank *et al.,* 1981). Smith and coworkers (1978) were the first to use nitrogen fixing *Azospirillum* sp. as inoculant on commercial scale to test whether these organisms supplement the nitrogen need of field crops or not. Nur *et. al.,* (1980b) conducted trials under greenhouse condition on *Zea mays* and *S. italica* with *Azospirillum* isolated from *Cynodondectylon*and found increase in dry matter yields and total nitrogen content, whereas no increase in total nitrogen content in the millet (*Panicum glaucum*) reported by Barber *et al.,* (1976).

Azotobacter colonizing the roots not only remains on the root surface but also a sizable proportion of them penetrates into the root tissues and lives in harmony with the plants. They do not, however, produce any visible nodules or out growth on root tissue. Fulchieri and Frioni (1994) observed that maize inoculated with *Azospirillum* had enhanced dry weight of seed by 59% and also the yield which was similar to 60kg urea N/ha. Increased yield in pear millet obtained by inoculation of *Azospirillum* was due to production of indole acetic acid (IAA), gibberellins, and cytokine like substances by the bacterium and their subsequent effect on the plant. The results of the various experiments conducted throughout India have clearly shown that *Azospirillum* can be used as a potential biofertilizer in both expensive and intensive agriculture. In developing countries like India, the use of *Azospirillum* as a biofertilizer would not only to the nitrogen supplementation to crops but also help in improving the fertility of soil in the long run.

Hartmann *et al.* (1988) reported the influence of aminoacids on nitrogen fixation ability and growth of *Azospirillum* spp. The utilization of amino acids forgrowth and their effects on nitrogen fixation differgreatly among the several strains of each species of *Azospirillum* spp. The different utilization of variousamino acids by *Azospirillum* spp. may be important fortheir establishment in the rhizosphere and for their associative nitrogen fixation with plants. Kucey *et. al.* (1993) indicated that upto 18% of the plant nitrogenwas derived from nitrogen fixation. All wild type *Azospirillum* strains were found to fix N efficiently eitheras free-living or in association with plants (Hurek *et. al.,*1988). Heulin *et. al.* (1989) reported that association with rice, *Azospirillum lipoferum* N-4 contributed about 66% ofthe total N in plants as was demonstrated with N isotope studies. The mesophilic lac-z-marked *Azospirillum lipoferum* ALP-3 did not grow and fix nitrogen at 45°C outof 40 thermo-tolerant marked mutants, developed from ALP-3 and screened for N-fixing ability at 30°C and 45°C,only 14 mutants could grow and fix nitrogen at 45°C (Anand *et al.,*1999). Panwar *et al.,* (2000) reported the response of *Azospirillum* and *Bacillus* on growth and yield of wheat under field conditions. Gadagi *et al.* (2004) examined the effect of combined *Azospirillum* inoculation and nitrogen fertilizer on plant growth promotion and yield response of the blanket flower *Gaillardia pulchella*. *Azospirillum* strain OAD-2 inoculation significantly increased plant height, number of leaves per plant, branches per plant and total dry mass accumulation in *G. pulchella* than other inoculations or un-inoculated controls. *Azospirillum* strains OAD-2 and OAD-11 can play an important role in the N nutrition of *G.pulchella.*

Omay *et al.,* (1993) reported *Azospirillum* spp. inoculation in wheat, barley and oats seeds in green house experiments. For wheat significant increases were obtained when the inoculation was associated to 100% of the recommended nitrogen. For barley the presence of the inoculants substituted 20%of the recommended nitrogen fertilization. For oats the inoculation with *Azospirillum* spp. RAM-7 did not provide a significant increase in grain yields. Gadagi Ravi *et al.* (2003) reported that the higher increase in IAA in *Azospirillum amazonense* inoculation on growth, yield and N_2 fixation of rice. Bacteria of the genus *Azospirillum* stimulate plant growth directly either by synthesizing phyto-hormones or by promoting nutrition by the process of biological nitrogen fixation. All *A. amazonense* strains tested produced indole, but only 10% of them showed high production. The nitrogenase activity also was variable and only 9% of isolates showed high nitrogenase activity and the majority (54%) exhibited a low potential. Saikia *et al.,*(2007) described that dinitrogen fixation activity of *Azospirillum brasilense* in maize. The presence of nitrogenase activity was noticed in plants treated with *Azospirillum.* Plant root- rhizosphere N-fixing bacterial interactions are required before any consistent, significant and beneficial N-fixing association can be developed. Shelf life of the *Azospirillum* may be increased; this will be helpful to enhance the crop productivity. Fillinger*et al.,* (2001) reported that trehalose is required for the acquisition of tolerance to a variety of stresses in the filamentous fungus. Baubu *et al.,* (2002) reported that shelf life of *Azospirillum* inoculants was improved by the addition of polymer, chemicals and amendments in the lignite carrier reported by Saranraj *et al.,*(2013) and Sivasakthi *et al.,*(2013). Application of BGA and *Azospirillum* increased the all yield attributing aspects in the rice reported by Mishra *et al.,*(2013). Table 8.4 shows the response of *Azospirillum* inoculation in different crops.

Table 8.4: Response to *Azospirillum* Inoculation in different Crops

Crop (%)	**N applied (Kg/ha)**	**No. of Isolations**	**Increases in Yield over control**
Rice	040	510.57	21.40
Wheat	040	313.80	4.41
Barley	040	612.95	16.70
Oats	040	611.44	42.23
Pearl millet	0	2	6.31

Source : In: The Complete Technology Book on Biofertilizers and Organic Farming, 2012

8.11 Gmax Nitromax

Azospirillum is nitrogen fixing bio inoculant suitable for all crops except legumes. GMax AgroTech is capable to produce ten tonnes of technical grade *Azospirillum brasilense/Azospirillum lipoferum* per day. This product is also available in combination with Azotobacter as combined formulation. Gmax Nitromax is combined product of *Azospirillum* and *Azotobacter*. *Azospirillum* works well with all crops whereas *Azotobacter* gives excellent results when soils have rich organic matter content (more than 2%). As a combined product of *Azospirillum* and *Azotobacter*, Gmax Nitromax combines the benefits of both inoculants in a single formulation.

8.11.1 Formulation and Composition

The product is available as liquid formulation and lignite carrier based formulation. *Azospirillum* and *Azotobacter* 1%(w/w), sticking agent–CMC (1%) and inactive ingredients 98.0% (w/w).

8.11.2 Packing

Carrier based formulation is available in attractive one kg laminated poly pouches. The packing is moisture proof and well tolerates transportation and handling. The product is also supplied in bulk packing of 50kg/25kg sizes in HDPE bags. Liquid formulation is available in one litre HDPE containers and in 200litre barrels.

8.11.3. Description

Azospirillum is an associative type of bacteria, living in close proximity with the root zone. During unfavorable conditions it forms cyst which helps to tide over unfavorable conditions. Thus it thrives and maintains its population during favorable conditions. It fixes atmospheric nitrogen, provides 30-50% of nitrogen requirement of the plant. It produces plant growth hormones such as auxin and cytokinin.

8.11.4 Method of application

The mixture of *Azospirillum*, *Azotobacter* (Gmax Nitromax) and Phosphate solubilizing bacteria (Gmax Phosphomax) can be used for non-leguminous cropslike cereals, oilseeds cotton, vegetable and all other crops which are usually planted as seeds.

8.11.4.1 Seed treatment

For one kg of seeds about 40gm of Gmax Nitromax product is required. Mix required quantity of Gmax Nitromax with equal quantity of rice gruel. With this mixture, mix the required amount of seeds shade dry and after 24hrs, use the seeds of sowing.

8.11.4.2 Seedling Root dip

Mix 2Kg of Gmax Nitromax with 50 litres of water. In this mixture, keep the seedling roots in immersed condition for 10 minutes and use the seedlings for transplantation in the field.

8.11.4.3 Soil application

Mix about 5kg Gmax Nitromax with 100kg of organic manure, keep the mixture under shade for one week and apply to the soil.

8.11.4.4 Drip irrigation

Dilute 5kg of Nitromax in 100 litrs of chemicals free, good quality water. Filter the mixture with a pure cloth; use the filtered solution in drip irrigation for one acre.

8.11.4.5 For liquid inoculants

Soil application Mix 3 litres of Gmax Nitromax liquid formulation with 100kg of organic manure, keep the mixture under shade for one week and apply to the soil.

8.11.5 Recommended Crops

Gmax Nitromax is recommended for all crops except all kinds of legumes. Gmax Nitromax can be used along with Gmax Phosphomax, Gmax Triconand Gmax Fyton to provide nitrogen, phosphatic nutrition and to provide plant protection from diseases along with plant growth promotion.

CHAPTER - 9

Blue Green Algae (BGA)

BGA belong to a class of prokaryotic photosynthetic microorganisms also known, as cyanobacteria are capable of fixing atmospheric nitrogen aerobically. Most abundantly found in tropical zone, never form fibrous growth and are enveloped by a sheath. The trophic independence of algae for C and N makes them responsible for a natural fertility buildup of rice field soils. The nitrogen fixing ability of BGA was first recognized by Frank and Prantil in 1889. The occurrence of the BGA in the rice field was reported by FE Fritsch in 1907. Its role in rice filed as biological n fixer was demonstrated by P.K. De in 1939. BGA comprise about 2,500 species under 150 genera. More than 125 strains of nitrogen fixing living BGA are common in flooded rice ecosystems. The predominant nitrogen fixing genera are *Nostoc, Anabaena, Aulosira, Tolypothrix, Calothrix, Gloeotrichia, Apanothece* etc. The order Nostocales includes *Anabaena, Calothrix, Tolypothrix* and *Nostoc* etc. Some good strains used for BGA biofertilizers are: *Anabaena variabillis*, *Nostoc muscorum*, *Aulosira fertilissima* and *Tolypothrix tenius* (available from the National Centre for BGA at the IARI, New Delhi). Detailed classification of BGA has been carried out by Desikachary (1959) and Fogg *et al.* (1973) and the information is still being updated. The key to the oxygenic photosynthetic bacteria is as below:

a) Contain chlorophyll a and have Phycobillins. Group I: Cyanobacteria

b) Contain both chlorophyll a and b, lack Phycobillins. Group II: Prochorales

9.1 Characteristics

BGA is prokaryotic unicellular, colonial filamentous, photosynthetic, nitrogen fixing aerobic organism. It contains three distinct cell types such as (a) vegetative cells, spores (akinites) and heterocysts. The heterocysts are believed to be the site of nitrogen fixation. Cells are spherical or cylindrical in shape and doubling time is 20 hrs. It contains pigments like chlorophylla, β carotene, flavacin, blue

C- phycocyanin and red C- phycoerythrin. It multiplies by binary or multiple fission of cells, called as bacocysts.Some of the cyanobacteria besides fixing nitrogen are also known to produce auxin, indole-3-acetic acid, indole-3-propionicacid or 3-methyl indole and vitamin B_{12} (Misra and Kaushik, 1989). Tripathi *et al.* (2008) studied the role of blue green algae biofertilizer in ameliorating the nitrogen demand and fly-ash stress tothe growth and yield of rice and concluded that use of BGA along with other amendments improved the growth, yield and mineral composition of the rice plants besides reducing the highdemand of nitrogen fertilizers. The plant growth promoting effect of BGA biofertilizers havebeen proven by many workers in rice, wheat and other crops (Dhar *et al.,* 2007; Nain *et al.,* 2010).

9.2 Nitrogen Fixation

Nitrogen fixing cyanobacteria are most wide spread as Nitrogen fixers on earth. Cyanobacteria or BGA are the diverse group of prokaryotes (Fig. 9.1). The activities of nitrogen fixing organisms provide an important source of nitrogen to the marine ecosystem (Gonzalez *et al.,* 2005). They also grow and fix nitrogen in terrestrial environment, from rain forest to desert. BGA are able to survive in the extreme environment and have ability to fix nitrogen because of the capacity to fix nitrogen and used as biofertilizer. In addition to contributing N, BGA add organic matter, secrete growth promoting substances like auxins, vitamins, mobilize insoluble phosphate and improve physical and chemical nature of the soil. Nitrogen fixation by BGA takes place with the help of nitrogenase enzyme at the site of heterocysts (specialized cells on the filament). The fixed N is available to rice plants only after it is released into the surroundings, either as extra cellular products or by mineralization of their intracellular contents through microbial decomposition after the death of algae. At the vegetative stage of rice crop, the nitrogen fixed by BGA is more useful. BGA can fix N independently in the paddy fields or in symbiosis with *Azolla*. Fixed nitrogen may become available to rice plants only after its release into the surroundings either as extra products or by mineralization of their cellular contents through microbial decomposition after the death of algae.Inoculation of the flooded paddy field with BGA can fix 25-30 KgN/ha crop.

9.3 Mass Cultivation of BGA

Several methods are used for the production of BGA, whereas open air cement tank method is considered to be easy and cost effective. The cement tanks are permanent structures and can be cleaned easily. On the other hand, the galvanized iron sheet trays are very expensive, prone to rusting and difficult to clean. Similarly, the polythene lined pits do not last for more than 3-4 harvests and thus become costly. Production of BGA inoculum can be done in the following steps.

1) In an open space,cemented tanks of size 5x1.5x0.3m are constructed. The four corners of the tank should be round. The inner walls and floor of the tank should be glazed smooth. Water tap is set up at about 25cm height at one of the broad sides and a drainage pipe fitted with a stop cock on the opposite wall at the bottom. An overflow outlet is provided at about 20cm height above the drainage outlet. The length of the tanks can be increased. A space of about 1.0m between the tanks should be left for operational convenience.

2) Spread 10kg soil and add 200g of single super-phosphate per tank. The soil should be loam or sandy loam and not heavy type like black cotton or clay soil. Heavy soil can bemix with appropriate quantity or river bad sand. Light soil facilitates formation of algal flakes, which easily separate our from the soil. The soil should be preferably taken a fallow land since it is low in nitrogen and microbial flora.

3) Fill the tanks with upto height of about 15cm and add insecticide (50% EC or 30% EC, 10-15ml malathion) to prevent breeding or mosquitoes and other insects. Mix the contents thoroughly and allow standing till the supernatant becomes clear. Sprinkle 200g of BGA culture on the water Surface.

4) Under favourable conditions as temperature 30°C and above, the growth of BGA will be rapid and a thick algal mat is formed on the surface of the water in about 10-15 days. At this stage, formation of another BGA layer can be seen on the surface of the soil. During this period, add water periodically to maintain the water level around 10cm.

5) Closely monitor the BGA coming up in the tanks by periodically examining the algal growth using the microscope. One can alternatively use the iodine test of differentiate between green and blue-green algae. The green algae turn dark violet or black in colour with iodine.

6) Alkaline conditions with pH around 8, appreciably prevent contamination with green algae.

7) Stop adding water to the tanks only after a thick BGA mat is formed and allow the contents of dry without draining the water.

8) When completely dry, the algal mat will form flakes which will separate out from rest of the soil. These flakes are collected, sun dried and packed in the polythene bags.

9) Fill the tanks again with water, with water, put fresh soil, starter culture

and super-phosphate and repeat the process. Single harvest from a tank yields about 7 to 10kg of soil based algal flakes.

10) Instead of using the soil based starter culture, one can use laboratory grown liquid culture of rice field's BGA to inoculate the tanks. Even the dominating native BGA, directly collected from the rice fields can also be used.

Precautions

- In northern plains, BGA multiplication can be practiced in the month of march to october.
- The inoculum packets should be kept away from pesticide and fertilizers in a dry place.
- The algae material should be completely sundried before packing.
- The quantity of water to be added to the tank should be worked out depending on the local climatic conditions.
- Application of low doses of phosphatic fertilizers after BGA inoculation accelerates the algal establishment.
- Recommended pest control measures and other crop management practices do not interfere in the establishment of BGA in the rice fields.
- Sun dried BGA in the polythene packs can be stored at room temperature for along 3 years without loss in viability.

Other Method

Producing algal biomass in aseptic conditions and them mixing with a suitable carrier material is being done in some countries. The method ensures desired titer value of intended cultures and longer shelf like of the inocula.

Method for Field Application

Broadcast the dried algal flakes on standing water at the rate of 10-20kg/ha, one week after transplantation. Addition of excess quantity of algal flakes accelerates the multiplication. The filed should be kept water-logged for about 10 days after inoculation to allow good growth of BGA. Establishment of the algal inoculum in the field can be seen in the form of floating algal biomass on the surface of water or as numerous, small glistening air bubbles adhering to the soil surface. These indications are best observed in the afternoon.

9.4 Algae in Agriculture

Ninety per cent of the total paddy area is situated in tropical south and south-east Asia. In India, it is most important and extensively grown food crop occupying 38 million hectares, about 37% of total area under cereals. Eighty seven per cent of the rice fields in the country accounts for holdings of 1-4ha (43% irrigated and 47% unirrigated) and 13% of holdings are less than one hectare in size (Gaur, 2010). Thus, the Indian rice farmers largely belong to small and medium level farmers. The most promising way is to tap the biological nitrogen fixation by blue greens in rice fields.Unlike bacteria which need exogenous energy to fix atmospheric nitrogen, these algae convert solar energy to energy required for nitrogen fixation. These algae can thus be used as biological input in rice fields. An all India survey of rice field soils showed that out of 2,213 soil samples examined only 33% were found to harbour useful nitrogen fixing BGA and their relative abundance showed wide variations (Venkatarama, 1987). The algae that are generally used as biofertilizer in rice are species of *Aulosira, Tolypothrix, Scytonema, Nostoc, Anabaena, Plectonema,* etc. as composite culture. BGA act as a supplement to the N fertilizers contributing up to 30KgN/ha. It increases the crop yield between 5-25% (Bhattacharjee and Dey, 2014).

9.5 Nitrogen-Fixation and Growth Promoting Substances

N is the key input required in large quantities for low land rice production. Soil N and BNF by associated organisms are major sources of N for low land rice. The 50-60% N requirement is met through the combination of mineralization of soil organic N and BNF by free living and rice plant associated bacteria. To achieve food security through sustainable agriculture, the requirement for fixed nitrogen must be increasingly met by BNF rather than by industrial nitrogen fixation. BGA forms symbiotic association capable of fixing nitrogen with fungi, liverworts, ferns and flowering plants, but the most common symbiotic association has been found between a free floating aquatic fern, the *Azolla* and *Anabaena azollae* reported by Mishra *et al.,* 2013. Nitrogen fixed by blue-green algae becomes available to the associated plant either by exudation or their mineralization after the death of the blue-greens. In paddy fields, the death of algal biomass is associated with drying cycles and their mineralization is rapid due to drying and wetting cycles.The transfer of algal nitrogen to higher plant other than rice has been demonstrated qualitatively in natural ecosystem by several workers using N^{15} tracer techniques. Recovery of nitrogen by the rice crop varies from 14 to 52% depending on the fresh or dried method of application etc. The agronomic trials conducted with rice crop commonly show that nitrogen contributed by algae is of the order of 20 to 30kg/ha. Besides the contribution of nitrogen, growth-promoting substance liberated by these algae play an important role in sustaining the crop yield. Production of auxin-like substances and vitamins by *Cylindrospermum musicola* increased the root growth and yield of rice crop plants. A number of growth promoting substances such as amino acids, sugars polysaccharides, Vitamin (B_{12}, Nicotinic, Pantothenic, Folic acid) growth hormones (IAA, 3-methyl Indole) have been documented (Misra and Kaushik 1989; Kaushik 1998).

The combined use of chemical fertilizers and algae input always shows higher yields over chemical fertilizer application alone. Such gains are higher with lower levels of fertilizers and get reduced but effects are obtained upto 100-120 kg N/ha. Venkatraman (1992) reported an average increase of 7.29% in the yield of rice in Tamil Nadu when the fields received 100 kg N/ha plus BGA inoculation over 100 kg/ha alone. Kaushik (1998) reported that the trails conducted with BGA at farmers' field in Haryana state increased the yield of paddy by 10-15% whereas a net saving of 25-30 kg N/ha/season was obtained. The results in general show the following benefits due to BGA inoculation of paddy fields (Gaur, 2010).

- Application of BGA alone can contribute 20-30kg nitrogen per hectare due to nitrogen fixation while supplementing the nitrogen requirement with biological resources, a farmer could save to the extent of 20 to 30kg N/ha without compromising with the normal yield.
- The integrated use of chemical fertilizers and BGA has been found to be ecologically sound and economically viable.

9.6 Soil Properties

The growth of algae improves the physical, chemical and biological properties of soils. The formation of soil aggregates was significantly enhanced due to algal growth. The infiltration rates, hydraulic conductivity and permeability of soils due to algae inoculation were improved. Soil compaction was reduced particularly on the upper soil surface due to inoculation of algae in corn (Schaefer, 1986). The effect on nutrients such as phosphorus, micronutrients, and exchange cations like sodium. In salt affected soils, exchangeable sodium gets reduced and the algae inoculation effects are comparable to application of gypsum in such soils.

9.7 Soil reclamation

Cyanobacteria being phototrophic, continuously add organic compounds during their growth on the upper soil surface. Due to repeated BGA inoculations organic carbon content of soil was augmented. The synthesis and production of polysaccharides improved the soil aggregation and formed stable soil aggregates. The species of cyanobacteria such as *Aphanocapsa, Aphanothece, Gloeocapsa, Schizothrix, Lyngybya, Scytonema, Aulosira* and *Tolypothrix* have been reported to have muciligenous sheath.

9.8 BGA under field conditions

Optimum pH for the growth are generally reported in the range of pH 6.5 to 8.5 However, some strains of cyanobacteria can grow can grow under extreme acidic and alkaline conditions such as pH 3.8 and 10.5. The establishment of algae is also influenced by the nature of nitrogenous fertilizer and pesticides applied to paddy fields. Fertilizer nitrogen has deferential effect on the processes of photosynthesis and nitrogen fixation. The nitrogen level which limits the process of nitrogen fixation appreciably augments the biomass production upto 75μg N/ml, most the cyanobacteria show concentration dependent increase in the growth and uptake of nitrogen. The uptake of nitrogen is reduced, the activation of nitrogenase activity leads to enhanced nitrogen fixation by the algal growth. Thus use of BGA in presence of nitrogenous fertilizer is beneficial for the productivity of the rice crop. Herbicides are commonly used in rice cultivation which has been reported not to interfere in the growth and nitrogen fixation of cyanobacteria when these are applied at the recommended levels of application.

9.9 Blue Green Algae as Food and Medicine

Fresh water blue-green algae produce exotoxins. When water bloom of these toxic species are present in reservoir, lake or pond, the cells and toxins can become concentrated enough to cause illness or death in almost any mammal, bird or fish which ingests enough of toxic cells or extracellular toxins. It can also cause gastro-enteritis and contact irritations in users of certain recreational and municipal water supplies. However, some of the beneficial ones are being used as poultry and cattle feed supplements.

Most of the sea weeds contain 16-30% protein on dry basis which is somewhat higher than in cereals, eggs and fish. Some species of green and red algae are used as salad, curry soups and vegetable. *Gracilaria edulis*, a red algae is used for making gruel in the costal areas of Tamil Nadu and known as 'Kanji pasi'.

Sea weeds were harvested for centuries particularly in China and Japan where it formed as part of their staple diet. These algae contain as many as 60 trace elements besides proteins, vitamins, bromine and substances of antibiotic and stimulatory nature. The red sea weed *Porphyra* which is rich in protein and essential amino acids and used as culinary items known as 'Laver' in Britain and 'Nori' in Japan. The major seaweeds growing areas are coastal areas of Gujarat state and Tamil Nadu. The salt and water lakes in the east coast of India such as Chilka and Pulicat lakes have good growth of *Gracilaria* spp. commonly used for preparation of Agar.

A number of species have been found to have antibiotic and anticoagulant properties. Carrageenan is reported to have curative property in ulcer therapy and alginates are found to prolong the rate of activity of certain drugs. Agar and algin are being extensively used in pharmaceutical preparations. The use of *spirulina* in medicine is a new development in India. The presence of active principles of biomedical importance has created a great awareness for advanced research in many marine and fresh water algae.

9.10. Algae as Early Colonizers

They appear early in barren or denuded areas, which is note-worthy. Algae are primary colonizers of areas destroyed by volcanic eruptions. As the algae, grow, die and decay, the environment becomes congenial for higher plants. In both volcanic and eroded areas the *Cyanophyceae* seem to be primary colonizers.Distinct growth/ bloom (10^6 cells/cm^2) appears at the surface of land if the moisture level in high. In some desert soils receiving occasional precipitation their growth may be seen on the soil surface following a period of rainy season. Algae have a great agronomic significance in paddy fields. Nitrogen fixing BGA such as *Aulosira fertilissima, Anabaena, Calothrix, Nostoc, Oscillatoria* and *Tolypothrix* are used as biofertilizers.

9.11 Algal Blooms

9.11.1 Blue-Green Harmful Algal Blooms

Most algae are harmless and are an important part of the food web. Algae are naturally present in slow moving streams, lakes, marine waters and ponds in low numbers. Certain types can become abundant and form blooms under the right conditions. Some algae can produce toxins that can be harmful to people and animals. These are collectively called harmful algal blooms (HABs). Algae blooms most frequently occur in nutrient-rich waters, particularly during hot, calm weather. Because it is hard to tell a harmful algae bloom from other algae blooms, we recommend avoiding contact with any floating rafts, scums, and discolored water. Find out what waterbodies have a blue-green algal bloom notice.

9.11.2 Freshwater Blue-green Algal Blooms

Blue-green algae can form HABs that discolor the water or produce floating rafts or scums on the surface of the water. These can cause health risks to people and animals when they are exposed to them. Blue-green algae blooms can occur in freshwater lakes and ponds and can reduce the recreational value of a water body, due to unpleasant appearances and odors, and can cause a variety of ecological problems, such as reduced oxygen levels. They also have the potential to form harmful (toxic) blue-green algal blooms, although the factors that cause blue-green algae to produce toxins are not well understood. Harmful blue-green algae blooms can cause health effects when people and animals come in contact with them. Symptoms can include nausea, vomiting, diarrhea, skin or throat irritation, allergic reactions or breathing difficulties. Blue-green algae can also produce toxins that affect the liver and nervous systems when water is consumed in sufficient quantities.

9.11.3 Marine Algal Blooms

Different algae can form harmful algal blooms in marine waters, producing marine blooms and a variety of biotoxin. However, cyanobacteria are related more closely to bacteria than to algae. Freshwater cyanobacterial harmful algal blooms (CyanoHABs) can use up the oxygen and block the sunlight that other organisms need to live. They also can produce powerful toxins that affect the brain and liver of animals and humans.

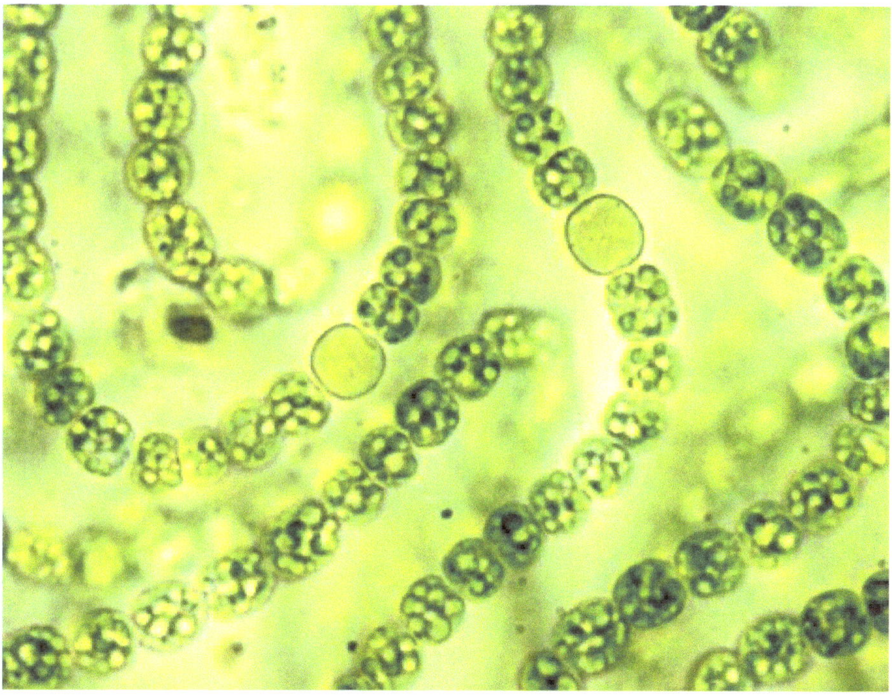

Fig. 9.1 : Blue Green Algae

Chapter - 10

Phosphorus Solubilizing Micro-Organisms (PSMs)

Phosphorus is the second important key element after nitrogen as a mineral nutrient in terms of quantitative plant requirement (Donahue *et al.,* 1990), which is required in optimum amount for proper growth of plants and soil micro-organisms. Being a constituent of ATP, it is involved in various processes such as cell division, energy transduction through photosynthesis and biological oxidations and uptakeof nutrients. The average soil contains 0.05% phosphorus, but only one tenth (1/10) of this is available to plants due to its poor solubility and chemical fixation in soil (Barber, 1984). About 98% of Indian soils have inadequate supply of available phosphorus (Ghosh and Hassan, 1979). A survey of Indian soils showed that out of 363 districts, only 2.2% were high, 51.5% medium and 46.3% were low in phosphorus content (Ghosh, 1982).An adequate supply of phosphorus during early phases of plant development is important for laying down the primordia of plant reproductive parts. It plays significant role in increasing root ramification and strength thereby imparting vitality and disease resistance capacity to plant. It also helps in seed formation and in early maturation of crops like cereals and legumes. The deficiency of phosphorus reduces the size of plants and imparts it deep green colour.Poor availability or deficiency of phosphorus markedly reduces plant size and growth. To satisfy crop nutritional requirements, P is usually added to soil as chemical phosphorus fertilizer; however synthesis of chemical P fertilizer is highly energy intensive processes, and has long term impacts on the environment in terms of eutrophication, soil fertility depletion, and carbon footprint. Moreover, plants can use only a small amount of this P since 75 to 90% of added P is precipitated by metal–cation complexes, and rapidly becomes fixed in soils. Such environmental concerns have led to the search for sustainable way of phosphorus nutrition of crops.Phosphorus plays an indispensable biochemical role in photosynthesis, respiration, energy storage and transfer, cell division, cell enlargement and several other processes in the living plant. It helps plants to survive winter rigors and also contributes to disease resistance in some plants (Sagervanshi *et al.,* 2012). Phosphate fertilizers can also be used to immobilize heavy metals in soil. Insoluble phosphate compounds can be solubilized by organic acids and phosphatase enzymes produced by plants and micro-organisms (Park *et al.,* 2010). Application of biological fertilizers such as biological phosphate fertilizers improves soil fertility (Yousefi *et al.,* 2011).

The problem of P fertilization may become serious in coming years because of the fact that manufacture of phosphatic fertilizers requires the use of non-renewable resources such as high grade rock phosphate and sulphur, which are getting depleted progressively and becoming higher. The situation is further aggravated by the fact that P is readily fixed in the soil and the average utilization efficiency of added fertilizer P by plants ranges from 15-25% (Gaur, 1982). India has about 100 to 150 MT of phosphatic rock deposits and most of them are low grade with impurities and not suited for superphosphate manufacture (Gaur, 1985). Therefore, their utilization as direct fertilizer is of economic importance. Although rock phosphate is recommended in acidic soils, there are large areas of arable land world over which are non-acidic. In this regards, phosphate-solubilizing micro-organism (PSM) have been seen as best eco-friendly means for phosphorus nutrition of crop. Although, several bacterial (*Pseudomonas* and *Bacilli*) and fungal strains (*Aspergilli* and *Penicillium*) have been identified as PSM their performance under *in situ* conditions is not reliable and therefore needs to be improved by using either genetically modified strains or co-inoculation techniques.

10.1 Phosphatic Compounds in Soil and Other Sources

Phosphorus plays an important role in virtually all major metabolic processes in plant including photosynthesis, energy transfer, signal transduction, macromolecular biosynthesis and respiration (Khan *et al.,* 2010) and nitrogen fixation in legumes (Saber *et al.,* 2005). Although P is abundant in soils in both inorganic and organic forms, it is a major limiting factor for plant growth as it is in an unavailable form for root uptake. Inorganic P occurs in soil, mostly in insoluble mineral complexes, some of them appearing after frequent application of chemical fertilizers. These insoluble, precipitated forms cannot be absorbed by plants (Rengel and Marschner, 2005). Organic matter is also an important reservoir of immobilized phosphorus that accounts for 20–80% of phosphorus in soils (Richardson, 1994). Only 0.1% of the total phosphorus exists in a soluble form available for plant uptake (Zhou *et al.,* 1992) because of its fixation into an unavailable form due to P fixation. The term P fixation is used to describe reactions that remove available phosphate from the soil solution into the soil solid phase (Barber, 1995). There are two types of reactions such as (a) Phosphate sorption on the surface of soil minerals and (b)Phosphate precipitation by free Al^{3+} and Fe^{3+} in the soil solution (Havlin *et al.,* 1999). Soils that exhibit highest phosphorus fixation capacity occupy 1,018Mha in the tropics (Sanchez and Logan, 1992). It is for this reason that soil phosphorus becomes fixed and available P levels have to be supplemented on most agricultural soils by adding chemical P fertilizers, which not only represent a major cost of agricultural production but also impose adverse environmental impacts on overall soil health and degradation of terrestrial, freshwater and marine resources (Tilman *et al.,* 2001). The repeated and injudicious applications of chemical P fertilizers, leads to the loss of soil fertility (Gyaneshwar *et al.,* 2002) by disturbing microbial diversity, and consequently reducing crop yield. The long-term effect of different sources of phosphate fertilizers on microbial activities includes inhibition of substrate-induced respiration by streptomycin sulphate (fungal activity) and actidione (bacterial activity) and microbial biomass carbon (Bolan *et al.,* 1996). Similarly, the application of triple superphosphate (94 kg/ha) has shown a substantial reduction in microbial respiration and metabolic quotient (Chandini and Dennis, 2002).Moreover the efficiency of applied P fertilizers in chemical form rarely exceeds 30% due to its fixation, either in the form of iron/aluminium phosphate in acidic soils (Norrish and Rosser, 1983) or in the form of calcium phosphate in neutral to alkaline soils (Lindsay *et al.,* 1989). It has been suggested that the accumulated P in agricultural soils would be sufficient to sustain maximum crop yields worldwide for about 100 years if it were available (Khan *et al.,* 2009a, b). A major characteristic of phosphorus biogeochemistry is that only 1% of the total soil P (400 to 4,000 kg P/ha in the top 30cm) is incorporated into living plant biomass during each growing season (10 to 30kg P/ha), reflecting its low availability for plant uptake (Blake *et al.,* 2000; Quiquampoix and Mousain, 2005).

Furthermore,phosphorus is a finite resource and based on its current rate of use, it has been estimated that the worlds known reserves of high quality rock P may be depleted within the current century (Cordell *et al.,* 2009). Beyond this time the production of P based fertilizers will require the processing of lower grade rock at significantly higher cost (Isherwood, 2000). The realization of all these potential problems associated with chemical P fertilizers together with the enormous cost involved in their manufacture, has led to the search for environmental compatible and economically feasible alternative strategies for improving crop production in low or P-deficient soils (Zaidi *et al.,* 2009). The use of microbial inoculants possessing P-solubilizing activities in agricultural soils is considered as an environmental-friendly alternative to further applications of chemical based P fertilizers.Total P_2O_5 contents of rock phosphates from different sources such as India, North Africa, USA and USSR etc., varies considerably. The water-soluble phosphorus in rock phosphates is negligible. Although, rock phosphates are recommended in acidic soils, but their use in neutral to alkaline soils may not be economical unless efficient phosphate solubilizing micro-organism are introduced in the rhizosphere of crops.

10.2 Constraints in using Phosphate Fertilizers

There is global concern about the energy and costs involved in mining the rock phosphate and its transport from manufacturing sites to farm crop fields. Mining phosphate minerals and spreading phosphorus fertilizers over the landscape is neither eco-friendly, economically feasible nor it is sustainable and it poses following constraints as-

a) Emission of the fluorine as the highly volatile and poisonous HF gas,

b) Disposal of gypsum, and

c) Accumulation of cadmium (Cd) and other heavy metals in soil and possibly crops as a result of repetitive use of phosphorus fertilizers.

At present mining rate (about 7,100 MT/annum), reserve will be depleted in about 500 to 600 years. In India, deposits of sufficiently enriched phosphatic rocks are limited and hence it imports 2MT of rock phosphate annually. About 98% of cropland in India is deficient in available forms of soil phosphorus and only 1-9% has high phosphorus status (Sharma *et al.*, 2013). Intensive cropping pattern during this green and white revolution has also resulted in widespread deficiency of phosphorus. Although various amendments are available for management of P in different soil, all are costlier and practically difficult. Thus, even if the total soil P is high and also if P fertilizers are applied regularly, pH dependent chemical fixation determines the quantity of available phosphorus.

The holistic phosphorus management involves a series of strategies involving manipulation of soil and rhizosphere processes, development of phosphorus efficient crops and improving P recycling efficiency. Microbial mediated phosphorus management is an ecofriendly and cost effective approach for sustainable development of agricultural crops. Micro-organisms are an integral component of the soil P cycle and are important for the transfer of phosphorus between different pools of soil phosphorus. PSMs through various mechanisms of solubilization and mineralization are able to convert inorganic and organic soil P, respectively (Khan *et al.*, 2009a) into the bioavailable form facilitating uptake by plant roots. It is important to determine the actual mechanism of P solubilization by PSMs for optimal utilization of these micro-organisms in varied field conditions. Hence, it is imperative to better understand the plant-soil-microbial P cycle with the aim of reducing reliance on chemical P fertilizers. This has led to increased interest in the harnessing of micro-organism to support phosphorus cycling in agro ecosystems.

10.3 Phosphate Solubilizing Micro-organisms (PSMs)

PSMs include different groups of micro-organisms particularly bacteria and fungi, which have been reported to render inorganic phosphatic compounds soluble. Such bacteria and fungi grow in medium with insoluble phosphates such as tricalcium, ferric, aluminium and rock phosphates and bone meal. PSMs besides assimilating phosphorus for their own requirement, release sufficient quantities in excess of their needs. The solubilization is not restricted to calcium salts only but salts of other cations as well are solubilized. The genera of bacteria such as *Pseudomonas, Bacillus, Aspergillus* and *Penicillium* have been reported to be active in the solubilization process (Gaur, 1990), and the counts of PSB may range between $10^4$to 10^6/g soil (Sharma *et al.,* 2013).

10.3.1 Occurrence and Isolation of PSM

Pikovskaya (1948) recommended a medium containing tricalcium phosphate as an insoluble phosphate sources together with glucose, yeast extract and other salts. This medium is still very widely used for isolation, enumeration and maintenance of phosphate dissolving micro-organism. Addition of insoluble phosphate compounds like di-or tricalcium phosphate as fine powder in the medium has also been suggested by some workers (Bunt and Rovira, 1955; Louw and Webley, 1958). Both the methods such as fresh precipitation of dicalicum phosphate or hydroxyapatite and incorporation of finely divided powder of insoluble phosphate compounds in the medium are followed for isolation of phosphate dissolving micro-organism, enrichment culture technique appears to be most suitable.

During the last two decades, knowledge on PSMs increased significantly (Rodriguez and Fraga 1999; Richardson, 2001). Several strains of bacterial and fungal species have been described and investigated in detail for their phosphate-solubilizing capabilities (Glick, 1995; He *et al.,* 1997). Typically such micro-organisms have been isolated using cultural procedures with species of *Pseudomonas* and *Bacillus* bacteria (Illmer and Schinner, 1992) and *Aspergillus* and *Penicillium* fungi being predominant (Wakelin *et al.,* 2004). These organisms are ubiquitous but vary in density and mineral phosphate solubilizing ability from soil to soil or from one production system to another. In soil, P solubilizing bacteria constitute 1-50% and fungi 0.1-0.5% of the total respective population. They are generally isolated from rhizosphere and non-rhizosphere soils, rhizoplane, phyllosphere, and rock P deposit area soil and even from stressed soils using serial plate dilution method or by enrichment culture technique (Zaidi *et al.,* 2009).

The concentration of iron ore, temperature, and C &N sources greatly influence the P-solubilizing potentials of these microbes. Among the various nutrients used by these micro-organism, ammonium salts has been found to be the best nitrogen source followed by asparagine, sodium nitrate, potassium nitrate, urea and calcium nitrate (Ahuja *et al.,* 2007). Since 1948, when Pikovskaya suggested that microbes could dissolve non-readily available forms of soil P and play an important role in providing P to plants, numerous methods and media, such as Pikovskaya (Pikovskaya, 1948), bromophenol blue dye method (Gupta *et al.,* 1994) and National Botanical Research Institute P (NBRIP) medium (Nautiyal, 1999) have been proposed. The source of insoluble phosphate in the culture media to isolate PSM is a major issue of controversy regarding the isolation of PSM in true sense. Commonly used selection factor for this trait, tricalcium phosphate (TCP), is relatively weak and unreliable as a universal selection factor for isolating and testing PSMs for enhancing plant growth. The use of TCP usually yields many (up to several thousand per study) isolates of "supposed" PSM. When these isolates are further tested for direct contribution of phosphorus to the plants, only a very few are true PSM. Other compounds are also tested, but on a very small scale. These phosphates, mainly iron/aluminium phosphate and several calcium phosphates are even less soluble than TCP in water, because soils greatly vary in pH and several chemical properties; it appears that there is no metal-Phosphate compound that can serve as the universal selection factor for PSM. Here multiple sources of insoluble phosphate are recommended.

The selection of the metal-phosphate candidates for potential PSM will depend on the type of soil (alkaline, acidic, or organic-rich) where the PSM will be used. Adding calcium phosphate compounds (including rock phosphates) for alkaline soils, iron/aluminium phosphate compounds for acidic soils, and phytates for soils rich in organic phosphorus (Bashan *et al.,* 2013a, b). Both bacterial and fungal strains exhibiting P solubilizing activity are detected by the formation of clear halo (a sign of solubilization) around their colonies. Production of a halo on a solid agar medium should not be considered the sole test for phosphorus solubilization. When colonies grow without a halo after several replacements of the medium, an additional test in liquid media to assay phosphorus dissolution should be performed and the few isolates that are obtained after such rigorous selection should be further tested for abundant production of organic acids and the isolates complying with these criteria should be tested on a model plant as the ultimate test for potential P solubilization (Bashan *et al.,* 2013a). The viable microbial preparations possessing P-solubilizing activity are generally termed as microphos (Zaidi *et al.,* 2009). The phosphate-solubilizing microbes showing greater solubilization (both qualitatively and quantitatively) of insoluble P under *in vitro* conditions are selected for field trials prior to production in bulk for ultimate transmission as a biofertilizer. Once a potential isolate is identified, it must be further tested for direct contribution to P plant nutrition and not necessarily to general growth promotion, as commonly done because promotion of growth, even by PSB, can be the outcome of other mechanisms. (Bashan *et al.,* 2013a) and ability to solubilize phosphorus is not necessarily correlated with the ability to promote plant growth (Collavino *et al.,* 2010). The production of biofertilizer and its acceptance by farming communities are closely linked. For uptake by farmers, quality management is essential and must be performed consistently in order to supply reliable and contaminant-free bio products. As far as *in vitro* field trials are concerned the establishment and performance of these PSM inoculate developed in laboratory is largely hampered by environmental variables including salinity, pH, moisture, temperature and climatic conditions of the soil. Moreover, it is also known that inocula developed from a particular soil fail to function as effectively in soils having different properties (Rodriguez and Fraga, 1999). Hence there is a need to study PSM activity in correlation with these factors before PSM application as a biofertilizer. Protocol for isolation and effective inoculants development of PSM based biofertilizer has been shown in Fig.10.1.

10.3.2 Biodiversity of Phosphate Solubilizing Micro-organisms (PSMs)

A substantial number of microbial species exhibit phosphorus solubilization capacity; these include bacteria, fungi, actinomycetes and even algae (Table 10.1). In addition to *Pseudomonas* and *Bacillus*, other bacteria reported as P-solubilizers include *Rhodococcus, Arthrobacter, Serratia, Chryseobacterium, Gordonia, Phyllobacterium, Delftia*sp. (Wani *et al.*, 2005; Chen *et al.*, 2006), *Azotobacter* (Kumar *et al.*, 2001), *Xanthomonas*(De Freitas *et. al.*, 1997), *Enterobacter, Pantoea*, and *Klebsiella* (Chung *et al.*, 2005), *Vibrio proteolyticus*, *Xanthobacter agilis* (Vazquez *et al.*, 2000). Furthermore, symbiotic nitrogenous rhizobia, which fix atmospheric nitrogen into ammonia and export the fixed nitrogen to the host plants, have also shown PS activity (Zaidi*et al.*,2009) For instance, *Rhizobium leguminosarum* bv. *Trifolii* (Abril *et al.,* 2007), and Rhizobium species nodulating Crotalaria species (Sridevi *et. al.,* 2007) improved plant P-nutrition by mobilizing inorganic and organic P. Various Phosphorus solubilizing (PS) bacteria have also been isolated from stressed environments for example the halophilic bacteria *Kushneria sinocarni* isolated from the sediment of Daqiaosaltern on the eastern coast of China, which may be useful in salt affected agricultural soils (Zhu *et al.,* 2011).

In soil, P-solubilizing fungi constitute about 0.1–0.5% of total fungal populations (Kucey, 1983). Moreover, P-solubilizing fungi do not lose the P dissolving activity upon repeated sub culturing under laboratory conditions as occurs with the P-solubilizing bacteria (Sperber, 1958a, b; Kucey 1983). Moreover, fungi in soils are able to traverse long distances more easily than bacteria and hence, may be more important to P solubilization in soils (Kucey, 1983). Generally, the P-solubilizing fungi produce more acids than bacteria and consequently exhibit greater P-solubilizing activity (Venkateswarlu *et al.,* 1984).

Among filamentous fungi that solubilize phosphate, the genera *Aspergillus* and *Penicillium* (Fenice *et al.,* 2000; Khan and Khan, 2002; Reyes *et al.,* 1999, 2002) are the most representative although strains of *Trichoderma* (Altomare *et al.,* 1999) and *Rhizoctonia solani* (Jacobs *et al.,* 2002) have also been reported as P solubilizers.

Arthrobotrysoligo spora (Nematofungus) also has the ability to solubilize phosphate *in vivo* as well as *in vitro* (Duponnois *et al.,* 2006). Among the yeasts, only a few studies have been conducted to assess their ability to solubilize phosphate these include *Yarrowialipolytica* (Vassilev *et al.,* 2001), *Schizosaccharomyces pombe* and *Pichia fermentans*. As more studies are conducted, a wider diversity of phosphate-solubilizing filamentous fungi is expected to be described. Of those identified, many are commonly found in agricultural soils such as *Penicillium* sp., *Mucor* sp. and *Aspergillus* sp. which has been shown to increase plant growth by 5–20% after inoculation (Gunes *et al.,* 2009). The P-solubilizing ability of actinomycetes has attracted interest in recent years because this group of soil organisms is not only capable of surviving in extreme environments (drought and fire.) but also possess other potential benefits (production of antibiotics and phytohormone-like compounds) that could simultaneously benefit plant growth (Fabre *et al.,* 1988; Hamdali *et al.,* 2008a, b). Approximately 20% of actinomycetes can solubilize P, including those in the common genera Streptomyces and Micromonosporareported by Hamdali *et al.,* (2008a).

In addition to bacteria, fungi and actinomycetes, algae such as cyanobacteria and mycorrhiza have also been reported to show P solubilization activity. The interactive effects of AMF and rhizobacteria on the growth and nutrients uptake of *Sorghum bicolor* were studied in acid and low availability phosphate soil. The microbial inocula consisted of the AMFs (*Glomus manihotis* and *Entrophospora colombiana)*, PSB (*Pseudomonas* sp.), results indicated that the interaction of AMF and the selected rhizobacteria has a potential to be developed as biofertilizers in acid soil. The potential of dual inoculation with AMF and rhizobacteria needs to be further evaluated under different crop and agro climatic conditions, particularly in the field (Widada *et al.,* 2007).

Table 10.1 : Biodiversity of Phosphate Solubilizing Micro-organisms (PSMs)

Sr No.	Micro-organisms
1.	**Bacteria:** *Alcaligenes* sp., *Actinomadura oligospora, Agrobacterium* sp., *Aerobactor aerogenes, Achromobacter* sp., *Azospirillum brasilense, Bacillus* sp., *Bacillus circulans, B. cereus, B. mycoides, B. polymyxa, B. coagulans, B.chitinolyticus, B. fusiformis, B. pumils, B. megaterium, B. subtilis, Bradyrhizobium* sp., *Brevibacterium* sp., *Citrobacter* sp., *Escherichia intermedia, Erwinia* sp., *Enterobacter asburiae Flavobacterium sp., Micrococcus sp., Nitrobacter* sp., *Pseudomonas* sp., *P. putida,, P. striata, P. fluorescens, P. calcis, Nitrosomonas* sp., *Rhizobium meliloti, Serratia phosphoticum, Thiobacillus ferroxidans, T. thioxidans* and *Xanthomonas* sp.
2.	**Fungi :** *Aspergillus awamori, A. niger, A. tereus, A. flavus, A. nidulans, A. foetidus, A. wentii., Alternaria teneius, Achrothcium* sp. *Cephalosporium* sp., *Cladosprium* sp., *Curvularia lunata, Cunnighamella, Candida* sp., *Chaetomium globosum, Fusarium oxysporum, Humicola inslens, Humicola lanuginosa, Helminthosporium* sp., *Myrothecium roridum, Morteirella sp., Micromonospora* sp., *Mucor* sp., *Oideodendron* sp., *Penicillium digitatum, P. funicolosum, P lilacinium, P balaji, Paecilomyces fusisporous, Pythium* sp., *Phoma* sp., *Populospora mytilina, Rhizoctonia solani, Rhizopus* sp., *Schwanniomyces occidentalis, Sclerotium rolfsii. Trichoderma viridae, Torula thermophila,*
3.	**Actinomycetes :** *Actinomyces, Streptomyces.*
4.	**Cyanobacteria:** *Anabena* sp., *Calothrix braunii, Nostoc* sp., *Scytonema* sp.,
5.	**VAM:** *Glomus fasciculatum.*

(*Source*: Sharma *et al.*, 2013)

10.3.3 Solubilization of Insoluble Phosphates in Liquid Medium

Goswami and Sen (1962) studied solubilization of tricalcium phosphate in liquid medium by *Bacillus megaterium var. phosphalicum,* Fosfo-24 and the organisms isolated from cassia gland. Maximum solubilization of phosphate under cultural conditions occurred within a week. *In vitro* solubilization of insoluble phosphates by soil fungi isolated from forest tree seed beds was shown by Agnihotri (1970). Of the 18 fungi tested, 14 solubilized tricalcium phosphate, 11 fluorapatite and 7 hydroxyapatite and the rest failed to solubilize them. *Aspergillus niger, A. flavus, Fusarium oxysporum, Sclerotium rolfsii, Cylindrocaldium* sp. and *Penicillium* sp. showed significant solubilization of all the three phosphates tested. Damping off fungi viz., species of *Pythuim* and *Rhizoctonia* failed to dissolve any of these phosphatic compounds. Bardiya and Gaur (1972) tested the solubilization of tricalcium phosphate by *Bacillus megaterium* var. *Phosphaticum* (PB), *B. megaterium* (R_3, 28), *B. polymyxa* (H3, H5, H10), *B. circulans, Escherichia freundii* (1_3) and Czechoslovakian culture (Fosfo 24). *B. pulvifaciens* (R5) solubilized 76 mg P_2O_5 out of 100mg P_2O_5 followed by *B. pulvifaciens HI* (57.0mg), *B. megaterium* (PB) 36.5mg, *B. pumilus* (31mg),

B. megaterium 28 (26mg) *B. polymyxa,* H-5 (22mg). The others solubilized varying amounts ranging between traces to small amounts.Bajpai and SundraRao (1971) selected four bacteria *Bacillus megaterium* var.*phosphaticum* (USSR), *Bacillus megaterium* (I_1,) *Bacillus circulans* (1_2) and *Escherichia freundii* (1_3) for solubilization of tricalcium phosphate in the liquid medium at different pH containing a known quantity of tricalcium phosphate and incubated for two weeks. The results showed that the solubilization of tricalcium phosphate at a wide range of pH. *B. megaterium* (Indian strain) solubilized good amounts of phosphate even at pH 9. The optimum pH for maximum solubilization of phosphate for each of the organisms was different. *Pseudomonas* spp. isolated from Maharashtra soils could solubilize 13-58% of tricalcium phosphate in a liquid medium. *P. Putida* gave maximum dissolution (Ostwal and Bhide, 1972). The speceis of *Aspergillus, Penicillium, Bacillus* and *Streptomyces,* were isolated from a laterite soil by Banik and Day (1981). Tricalcium phosphate was solubilized to a maximum extent followed by aluminium and iron phosphate. Bacteria were found most active followed by *Penicillium* and *Aspergillus*. Streptomyces were least effective. Similar, observation in respect to the activity of *Streptomyces* against rock phosphate solubilization was made by Bardiya and Gaur (1974).

Pseudomonas and *Aspergillus* were isolated from rhizosphere of coconut and cocoa (Nair and Rao, 1977). Ortuno *et al.,* (1978) reported that on a tricalcium phosphate substrate, solubilization phosphorus started after 24 hrs of incubation and reached a maximum on the 7th and 15th day with *Aspergillus niger* and *Pseudomonas fluorescens*, respectively. The periodic solubilization of ferric and aluminium phosphate was investigated in liquid medium (Gaur and Gaind, 1983). The maximum solubilization of aluminium and iron phosphate by *Aspergillus awamori* was achieved on 10th and 11th day. The maximum amounts of P_2O_5solubilized by *P. striata* from dicalcium, tricalcium, aluminium and iron phosphate were 30.8, 58.4, 4.24 and 26.6% respectively (Table 10.2). *Bacillus cereus, Pseudomonas fluorescens, Aspergillus niger* and *Penicillium pinophillum* were isolated from the desert soils of Rajasthan (India) Venkateshwarlu *et al.* (1984) reported the solubilization of rock phosphate and hydroxyapatite by Gram-positive and Gram-negative rods and cocci-shaped bacteria and genera of fungi such as *Aspergillus, Penicillium* and *Rhizopus*. Gaur *et al.* (1973) reported the presence of gram-positive and sporeforming bacteria and fungi. Among the isolates from rock phosphate, *Aspergillus carbonum* proved the best as they utilized three different types of rock phosphate as well as organic phosphates such as lecithin and calcium phytate. The cultures isolated from alluvial soil were identified as *Bacillus* sp. and *Aspergillus* spp. These cultures which were isolated as organic phosphate mineralizer were found incapable of solubilizing rock phosphate. However, they solubilized tri-calcium phosphate to an appreciable extent. Solubilization of dicalcium and tricalcium phosphate by thermophilic bacteria, actinomycetes and fungi was examined. *Bacillus stearothermophilus*, *Thermoactinomycetes* sp. etc. proved to be good phosphate solubilizers at high temperatures (Sujata *et al.*, 2004).

Table 10.2 : Solubilization of different insoluble phosphorus by *Aspergillus awamori*

Sl. No.	Treatments	Days of incubationnet P_2O_5 solubilized (mg) average 3	6	10	11	Average
1.	Dicalcium phosphate	13.3	18.2	18.9	17.0	17.0
2.	Tricalcium Phosphate	12.3	17.4	20.0	19.7	17.3
3.	Aluminium phosphate	3.4	8.5	15.5	12.1	9.9
4.	Ferric phosphate	3.4	12.5	12.1	15.0	10.7

Source : Gaur (2010)

Biofertilizers in Sustainable Agriculture, Directorate of Information and Publications of Agriculture, Indian Council of Agricultural Research, New Delhi.

10.3.4 Solubilization of Rock Phosphate

Earlier attempts on isolation of PSM were confined to solubilization of tri-calcium phosphate which was more easily solubilized than rock phosphate. Bardiya and Gaur (1972) showed that out of 14 tricalcium phosphate solubilizing bacteria tested; only one culture could solubilize 5 to 10% of rock phosphate whereas others were not effective. Bardiya and Gaur (1974) isolated by enrichment technique, rock phosphate dissolving bacteria, fungi and yeasts from the rhizosphere of leguminous crop and soils samples obtained from rock phosphate deposits area (Table 10.3 and 10.4).

Table 10.3 : Efficiency of Rhizosphere Isolates in rock phosphate solubilization

Rhizosphere	Organism	No. of isolates	pH change	P_2O_5solubilized (mg/100 ml)
Control	-	-	6.5	0.9
Black gram, Groundnut	*Aspergillus* sp.	11	3.3-5.5	3.1-25.6
Green gram, Soybean	*Penicillium* sp.	6	3.6-5.2	5.6-17.6
Black gram, Groundnut	Yeast	6	3.9-4.7	1.3-10.1
Green garm, Soybean	*Aspergillus* sp.			
Cowpea, Green gram	*Curvularia* sp.	3	4.5-5.7	4.1-11.0
Black gram	*Trichoderma* sp.	2	5.0-5.4	5.9-10.3
Cowpea, soybean	-	-	-	

Source : Gaur (2010)

Table 10.4 : Efficiency of Isolates of rock phosphate deposit soils for rock phosphate solubilization

Location	No. of Isolates	Characterization	pH change	P_2O_5 solubilized (mg/100 ml)
Control	Uninoculated	-	6.2	1.31
Kanpura 1	1	Gram (+) sporulated motile bacterium	5.9	1.5
Kanpura 1 Matoon 1 and III	3	Yeasts	4.0-4.9	1.4-2.4
Kanpura 1 Matoon II	2	Actinomycetes	3.6-4.7	1.5-3,2
Kanpura 1 Matoon 1, II and III	8	Gram (-) rods, non-capsulated non-sporulated non-motile bacteria	3.2-3.5	5.3-9.0
Kanpura 1 and II Matoon 1, II and III	20	Gram (-) rods non-capsulated non-sporulated motile bacteria	2.8-3.2	9.5-15.2

Source : Gaur (2010)

Six isolates of yeast were obtained from rhizosphere soils of cowpea, soybean, black gram and groundnut. One of the isolates solubilized the least amount of 1.3mg P_2O_5 whereas the other five cultures solubilized between 9 to 10 mg P_2O_5. The most effective culture was indentified as *Schwanniomyces occidentalis.* Eleven isolates belonging to *Aspergillus* group solubilized from 3.1 to 26.6mg P_2O_5. The pH of the medium became acidic and was found in the range of 3.3-5.5. *Aspergillus awamori* isolated from the rhizosphere of soybean crop showed the maximum dissolution of rock phosphate. Six *of Penicillium* sp. showed varying degree of efficiency in rock phosphate solubilization. *Penicillium digitatum* obtained from the rhizosphere of soybean crop showed maximum solubilization (17.6mg P_2O_5). Three isolates of curvularia sp. solubilized between 4.1 to 11.0mgP_2O_5. Two isolates of *Trichoderma* sp. solubilized between 5.9 to 10.3mg P_2O_5. The results indicated that rhizosphere of cowpea and soybean crops harbourded efficient rock phosphate solubilizing fungi and yeasts.The effective strains of bacteria were found in the soil samples of rock phosphate deposit area obtained from Udaipur. The soil samples from Kanpur and Matoonhar boured the most efficient rock phosphate dissolving bacteria as 20 isolates solubilizing P_2O_5 in the range of 9.4 to 15.2 mg were obtained. The medium was found acidic with a pH range between 2.8 to 3.2. The isolates were Gram-negative, non-spore former, non-capsulating and motile rod shaped bacteria. They were identified as *Pseudomonas striata* and *P. rathonis.*

Eight cultures of Gram-negative, rod shaped, non-capsulating and non-motile bacteria solubilizing comparatively less rock phosphate between 5.3 to 9.0 mg P_2O_5 were also isolated from the soils. They also produced less acidity.The solubilization of ^{32}P tagged hydroxyapatite by *Pseudomonas striata, Bacillus polymyxa* and *Aspergillus awamori* was studied in liquid medium (Arora and Gaur, 1978). The extent of phosphorus solubilization was increased progressively upto 5th week of incubation, *A. awamori* was the best followed by *P. striata* and *B. polymyxa.* At the sixth week of incubation, there was a substantial increase in the solubilization. This can be explained by the fact that during the process of phosphate solubilization, a part of the phosphate is assimilated by the concerned organisms which will be released after the autolysis of the aged cells.The dissolution of different kinds of low grade powdered rock phosphate using efficient cultures of bacteria *Pseudomonas striata, Bacillus polymyxa, B. megaterium, B. pulvifaciens, B. circulans* and *Citrobacter fraundii* and fungi *Aspergillus awamori, Penicillium digitatum, Aspergillus niger* (ANI) and *A. niger.* (AN_{30}) were examined in liquid medium (Arora and Gaur, 1979). *P. striata, B. Polymyxa, B. megaterium* and *B. pulvifaciens, A. niger* and *P. digitatum* effectively solubilized tricalcium phosphate. Rock phosphates were also solubilized but relatively to a lesser extent than tricalcium phosphate. Among the fungi, *Aspergillus awamori* and *A. niger,*while among bacteria, *Pseudomonas striata* and *Bacillus polymyxa* were found to be the best.The phosphorus solubilizing ability of the selected efficient strains of fungi,*Aspergillus awamori* and *Penicillium digitatum* and bacteria, *Pseudomonas striata* and *bacillus polymyxa* were confirmed on Pikoskaya's solid and liquid medium by other workers (Wani and Patil 1979, and Wani *et al.,* 1979). *Pseudomonas aeruginosa* and *Bacillus megaterium* were isolated from rhizosphere of rice, Jute and *dhaincha* (Sattar and Gaur 1985). Rao *et al.,*(1989) reported about the isolation of *Pseudomonas fluorescens, P. putida* and *P. syringae* from Antarctica Lake.Gaind and Gaur (1991) isolated thermotolerant, *Bacillus circulans* and *B. subtilis* from rhizosphere soils from oat and pigeon pea. *Pseudomonas* spp. viz., *P. aeruginosa, P. cepacia, P. fluorescent* and *P. putida* from wheat rhizosphere (Fretias and Germida, 1990) and *Bacillus* spp. viz. *B. licheniformis, B. mycordes,* and *B. megaterium* from paddy rhizosphere (Watanabe and Hayano, 1993) were isolated. Several strains of *Aspergillus* and *Penicillium* and three bacterial strains were isolated from rhizosphere of soybean, maize and chilli crops. *Pseudomonas* and *Penicillium* were isolated from the forest soils of Austria (Illemer and Schinner, 1992). Out of 30 cultures of bacteria and fungi, 26 isolates solubilized varying amounts of phosphorus. Some of the isolates showed positive antifungal activities as well and were identified as *Pseudomonas* sp., *Pseudomonas striata* (IARI culture) continued to perform best in phosphate solubilization (Srivastava *et al.*, 2004).

10.3.5 Preparation of PSM Inoculants

For production of PSM inoculants, a suitable and economical earner material has to be located. Peat in not available in India and therefore, alternate carrier such as wood charcoal and soil mixture after due studies were selected for the purpose (Gaur and Gaind, 1984). Two parts of ground charcoal (100mesh) and one part of soil should be mixed along with 0.5% KH_2PO_4 by weight and the moisture were maintained at 10%. Trays containing the mixture (8kg/tray) were sterilized at 15lb pressure for 3 hrs alternately for 3 days. One ml of homogenized suspension of different cultures of *Pseudomonas striata, Bacillus polymxa, Aspergillus awamori* or any other efficient selected culture was inoculated to the flasks containing 400ml of sterilized above-mentioned broth and the culture were allowed to grow for 4-5 days. To 8kg of the carrier 1,200-1,500 ml of specific broth culture were added and thoroughly mixed so as to maintain it at approx. 40% moisture and pH nearly neutral. Two hundred grams of its inoculantwere filled in polythene packets, incubated at room temperature of 26±2°C in duplicate. The shelf life should be monitored at monthly intervals by counting the presence of PSM in the carrier by the techniques elaborated elsewhere (Gaur, 1990).

10.3.6 Directions for Use of PSM Inoculants

Dissolve 100g of sugar or gur in 0.5 to 1 litre boiled and cooled water. Empty the contents of packet into the solution and slurry made. Pour the quantity of seeds intended to be sown in one acre or half a hectare in a suitable container and add the slurry. Continuously mix with a clean rod or hand so that the seeds are pelleted with culture properly. Spread the inoculated seeds on a clean cloth or gunny bag or newspaper in the shade to dry and avoid exposure to sunlight. After drying, sow the seeds without delay. In case of transplantation crops like paddy or vegetables, the roots of the seedling should be dipped in the culture suspension for 2-3hrs and then transplanted. While inoculating plantation trees, flower beds or roses, it should be mixed with good quality farmyard manure or compost and then should be applied in the pit near the roots of trees or applied to flowers near plants roots and mixed well so that it is not exposed to direct sun.Free-living heterotrophic nitrogen fixing bacteria are very widely distributed taxonomically, geographically and ecologically. But only a few species are known to possess the potential for high amount of nitrogen fixation. *Azospirillum* has shown high nitrogen fixation attributes.

10.3.7 Mechanism of Phosphorus Solubilization by PSM

In a review of phosphorus chemistry in soils, Sims and Pierzynski (2005) identified the major processes of the soil phosphorus cycle that affect soil solution phosphorus concentrations as (a) dissolution– precipitation (mineral equilibria), (b) sorption– desorption (interactions between P in solution and soil solid surfaces), and (c) mineralization–immobilization (biologically mediated conversions of P between inorganic and organic forms).The main P solubilization mechanisms employed by soil micro-organism include such as:

(1) Release of complexing or mineral dissolving compounds e.g. organic acid anions, siderophores, protons, hydroxyl ions, CO_2.

(2) Liberation of extracellular enzymes (biochemical P mineralization), and

(3) Release of P during substrate degradation (biological P mineralization) (McGill and Cole 1981). Therefore, micro-organism play an important role in all three major components of the soil P cycle (i.e. dissolution–Precipitation, sorption-desorption, and mineralization–immobilization). Additionally these micro-organisms in the presence of labile C serve as a sink for P, by rapidly immobilizing it even in low P soils; therefore PSM become a source of P to plants upon its release from their cells. Release of P immobilized by PSM primarily occurs when cells die due to changes in environmental conditions, starvation or predation. Environmental changes, such as drying– rewetting or freezing–thawing, can result in so-called flush-events, a sudden increase in available P in the solution due to an unusually high proportion of microbial cell lysis (Butterly *et al.*, 2009). (Grierson *et al.,* 1998) found that about 30–45% of microbial P (0.8–1 mg/kg) was released in a sandy spodosol in an initial flush after drying–rewetting cycles within the first 24 hrs.

10.3.7.1 Inorganic P solubilization by P-solubilizing micro-organism

It occurs mainly by organic acid production either by(a) lowering the pH, or (b) by enhancing chelation of the cations bound to P (c) by competing with P for adsorption sites on the soil (d) by forming soluble complexes with metal ions associated with insoluble P (Ca, Al, Fe) and thus P is released(Table 10.5). The lowering in pH of the medium suggests the release of organic acids by the P-solubilizing micro-organism (Whitelaw 2000; Maliha *et al.*, 2004) via the direct oxidation pathway that occurs on the outer face of the cytoplasmic membrane (Zaidi *et al.,* 2009). These acids are the product of the microbial metabolism, mostly by oxidative respiration or by fermentation of organic carbon sources (glucose) (Atlas and Bartha 1997; Trolove *et al,*. 2003) or such organic acids can either directly dissolve the mineral P as a result of anion exchange of phosphate by acid anion or can chelate Fe, Al and Ca ions associated with P

(Omar, 1998). The monovalent anion phosphate $H_2PO_4^-$ is a major soluble form of inorganic phosphate, which usually occurs at lower pH. However, as the pH of the soil environment increases the divalent and trivalent forms of Pi (HPO_4^{-2} and HPO_4^{-3} respectively) occur. Thus, the synthesis and discharge of organic acid by the PSM strains into the surrounding environment acidify the cells and their surrounding environment that ultimately lead to the release of P ions from the P mineral by H^+ substitution for the cation bound to phosphate (Goldstein, 1994). The prominent acids released by PSM in the solubilization of insoluble P are gluconic acid (Di-Simine *et al.,* 1998; Bar-Yosef *et al.,* 1999), oxalic acid, citric acid (Kim *et al.,* 1997), lactic acid, tartaric acid, aspartic acid (Venkateswarlu *et al.,* 1984).

Evidence from an abiotic study using HCl and gluconic acid to solubilize P also indicated that chelation of Al^{3+} by gluconic acid may have been a factor in the solubilization of colloidal Al phosphate (Whitelaw *et al.,* 1999). Organic acids produced by P-solubilizing micro-organism can be detected by high performance liquid chromatography and enzymatic methods (Parks *et al.,* 1990; Whitelaw, 2000). However, acidification does not seem to be the only mechanism of solubilization, as the ability to reduce the pH in some cases did not correlate with the ability to solubilize mineral P (Subba Rao, 1982). Altomare *et al.* (1999) investigated the capability of the plant-growth promoting and bio-control fungus *T. harzianum* T-22 to solubilize in vitro insoluble minerals including rock phosphate. Organic acids were not detected in the culture filtrates and hence, the authors concluded that acidification was probably not the major mechanism of solubilization as the pH never fell below 5. The phosphate solubilizing activity was attributed both to chelation and reduction processes.

Although, organic acid has been suggested as the principal mechanism of P solubilization, the solubilization of insoluble P by inorganic acid (HCl) has also been reported, although HCl was able to solubilize less P from hydroxyapatite than citric acid or oxalic acid at same pH (Kim *et al.,* 1997). Bacteria of the genera *Nitrosomonas* and *Thiobacillus* sps. can also dissolve phosphate compounds by producing nitric and sulphuric acids (Azam and Memon, 1996). According to the sink theory, P-solubilizing organisms remove and assimilate P from the liquid and hence, activate the indirect dissolution of calcium phosphate compounds by consistent removal of P from liquid culture medium. For instance, the P content in the biomass of *Pseudomonas* sp. and *P. aurantiogriseum* were similar to those observed in non-P-solubilizing micro-organism (Illmer *et al.,* 1995) which can be explained by the fact that the P content in biomass of organisms is consistently correlated with the decomposition of P containing organic substrates (Dighton and Boddy, 1989).

Table 10.5 : Important PSM, their ecological niches and organic acids produced

	Organism	Ecological niche	Predominant acids produced	References
1.	PSB	Soil and phosphate bearing rocks	ND(not determined)	Kovskaya (1948)
2.	PSB	Bulk and rhizospheric soil	ND(not determined)	Gerretson (1948)
3.	*Aspergillus niger, Penicillium* sp.	Soil	Citric, Glycolic, Succinic, Gluconic, Oxalic and Lactic acid	Sperber (1958a, b)
4.	*Arthrobacter* sp. *Bascillus* sp., *Bacillus firmus B-7650*	Wheat and cowpea	Lactic andCitric acid	Bajpai and Sundara Rao (1971)
5.	*Aspergillus* sp., *Penicillium* sp., *Chaetomiumnigri-color*	Lateritic soil	Oxalic, Succinic, Citric and 2-ketogluconic acid	Banik and Dey (1983)
6.	*A. japonicus, A. foetidus*	Indian Rock Phosphate	Oxalic, Citric, Gluconic, Succinic and Tartaric acid	Singal *et al.*,(1994)
7.	*Aspergillus flavus, A. niger, Penicillium canescens*	Stored wheat grains	Oxalic, Citric, Gluconic and Succinic acid	Maliha *et al.*,(2004)
8.	*Aspergillus niger*	Tropical and subtropical soil	Gluconic and Oxalic acid	Chuang *et al.*,(2007)
9.	*Bacillus amylolique faciens, B. licheniformis, B. atrophaeus, Penibacillus macerans, Vibrio proteolyticus, xanthobacter agilis, Enterobacter aerogenes, E. taylorae,*	Mangrove ecosystem	Lactic, Itaconic, Isovaleric, Isobutyric and Acetic acid	Vazquez *et al.*, (2000)

Contd...

	E. asburiae, Kluyvera cryocrescens, Pseudomonas aerogenes, Chryseomonas Luteola, Penicillium rugulosum			
10.	*B. pumilus var.2; B. subtilis var.2; Actinomaduraoli-gospora; Citrobacter* sp.	Giant Cardon cactus (*P. pringlei*) growing in ancient lava	Gluconic, Propionic, Isovaleric, Heptonic, Caproic, Isocaproic, Formic, Valeric, Succinic, Oxalic, Oxalacetic and Malonic acid	Puente *et al.,* (2004a)
11.	*B. pumilus CHOO8A; B. fusiformis*	Cholla cactus (Opuntia Cholla)		Puente *et al.,* (2004a)
12.	*Bacillus* sp. SENDO 6	Giant Cardon cactus (*P. pringlei*)	Gluconicacid Propionic, Isovaleric, Formic, Succini and, Lactic acid	Puente *et al.,* (2009a, b)
13.	*Bacillus megaterium, Pseudomonas* sp., *Bacillus subtilus*	Rhizospheric soil	Lactic, Malic acid	Taha *et al.,* (1969)
14.	*Enterobacter agglomerans*	Wheat rhizosphere	Oxalic and Citric acid	Kim *et al.,*(1997)
15.	*Escherichia freundii*	Soil	Lactic acid	Sperber (1958a, b)
16.	*Enterobacter intermedium*	Grass rhizosphere	2-ketogluconic acid	Hwangbo *et al.,* (2003)
17.	*Pseudomonas putida* M5TSA,	Wild cactus *Mammillaria*		Lopez *et al.,*

	Enterobacter sakazakii M2PFe, and *Bacillus megaterium* M1PCa	*fraileana*		(2011)
18.	*P. fluorescens*	Root fragments and rhizosphereof oil palm trees	Citric, Malic, Tartaric and Gluconic acid	Fankem *et al.*, (2006)
19.	*P. trivialis*	Rhizosphere of *Hippophaer hamnoides* growing in the cold deserts of Lahaul and Spiti in the trans-Himalayas	Lactic and Formic acid	Vyas and Gulati (2009)
20.	*P. radicum*	Rhizosphere of wheat roots	Gluconic acid	Whitelaw *et al.*, (1999)

Source : Sharma *et al.* (2013)

The other mechanism is the production of H_2S, which react with ferric phosphate to yield ferrous sulphate with concomitant release of phosphate (Swaby and Sperber, 1958). It has been suggested that MPS activity occurs as a consequence of microbial sulphur oxidation (Rudolph, 1922), nitrate production and CO_2 formation. These processes result in the formation of inorganic acids like sulphuric acid (Sperber 1958a). However, their effectiveness has been less accepted than the concept of involvement of organic acids in solubilization (Kim *et al.,* 1997).

H+ excretion originating from NH4+ assimilation as proposed by Parks *et al.* (1990):

It could be the alternative mechanisms of phosphorus solubilization. An HPLC (High Performance Liquid Chromatography) analysis of the culture solution of *Pseudomonas* sp., in contrast to the expectation, did not detect any organic acid while solubilization occurred (Illmer and Schinner, 1995). They also reported that the most probable reason for solubilization without acid production is the release of protons accompanying respiration or NH_4+ assimilation. Krishnaraj *et al.* (1998) have proposed a model highlighting the importance of protons that are pumped out of the cell to be the major factor responsible for P solubilization. Here direct role of organic or inorganic acids has been ruled out. For some micro-organism, NH_4+ driven proton release seems to be the sole

4tmechanism to promote P solubilization. Asea *et al.*, (1988) tested two fungi as *Penicillium bilaii* and *Penicillium fuscum*, for their ability to solubilize phosphate rock in the presence of NH_4+ or without N addition, and showed that only *P. bilaii* maintained the ability to decrease the pH and mobilize P when no N was supplied. In a study of *Pseudomonas fluorescens*, the form of C supply (e.g. glucose versus fructose) rather than N supply (e.g. NH_4^+vs. NO_3^-) had the greatest effect on proton release (Park *et al.*, 2009). Further, the involvement of the H^+ pump mechanism in the solubilization of small amounts of P in *Penicillium rugulosum* is reported (Reyes *et al.*, 1999). Acidification of the rhizosphere of cactus seedlings (giant cardon, *Pachycereus pringlei*) after inoculation with the plant growth-promoting bacterium *Azospirillum brasilense*, in the presence or absence of ammonium and nitrate, was studied and it was assumed that the effect of inoculation with this PGPB on plant growth, combined with nitrogen nutrition, might be affecting one or more of the metabolic pathways of the plant which increases proton efflux from roots and liberation of organic acid, leading to rhizosphere acidification (Carrillo *et al.*, 2002). This indicates that for different species, different mechanisms are responsible for proton release, only partly depending on the presence of NH_4^+. Goldstein (1995) suggested that extracellular oxidation via direct oxidation pathway may play an essential role in soils where calcium phosphates provide a significant pool of unavailable mineral phosphorus. This has been confirmed by some researchers (Song *et al.*, 2008) by biochemical analysis of lowering of pH in insoluble P solubilization by *Burkholderia cepacia* DA23.

10.3.7.2 Organic P solubilization

It is also called mineralization of organic phosphorus. Mineralization of soil organic P (Po) plays an imperative role in phosphorus cycling of a farming system. Organic P may constitute 4 to 90% of the total soil P (Khan *et al.*, 2009b). Such P can be released from organic compounds in soil by enzymes:

10.3.7.2.1 Non-specific acid phosphatases (NSAPs) which dephosphorylate phospho-ester or phosphoanhydride bonds of organic matter. Among the variety of phosphatase enzyme classes released by PSM, phosphomonoesterases (often just called phosphatases) are the most abundant and best studied (Nannipieri *et al.*, 2011). Depending on their pH optima, these enzymes are divided into acid and alkaline phosphomonoesterases and both can be produced by PSM depending upon the external conditions (Kim *et al.*, 1998; Jorquera *et al.*, 2008). Typically, acid phosphatases predominate in acid soils, whereas alkaline phosphatases are more abundant in neutral and alkaline soils (Eivazi and Tabatabai 1977; Juma and Tabatabai 1998; Renella *et al.*, 2006). Although plant roots can produce acid phosphatases they rarely produce large quantities of alkaline phosphatases, suggesting that this is a potential niche for PSM (Juma and Tabatabai 1998; Criquet *et al.*, 2004). It is also difficult to differentiate between root- and PSM-produced phosphatases (Richardson *et al.*, 2009a, b) but some evidence suggests that phosphatases of microbial origin possess a greater affinity for Po compounds than those derived from plant roots (Tarafdar *et al.*, 2001). The relationship between PSM introduced into soil, phosphatase activity and the subsequent mineralization of Po still remains poorly understood (Chen *et al.*, 2003)

10.3.7.2.2 Phytases, which specifically cause release of phosphorus from phytate degradation. In its basic form, phytate is the primary source of inositol and the major stored form of phosphorus in plant seeds and pollen, and is a major component of organic phosphorus in soil (Richardson, 1994). Although the ability of plants to obtain phosphorus directly from phytate is very limited, yet the growth and P-nutrition of *Arabidopsis* plants supplied with genetically transformed with the phytase gene (phyA) derived from *Aspergillus niger* (Richardson *et al.*, 2001). This led to an increase in P-nutrition to such an extent that the growth and phosphorus content of the plant was equivalent to control plants supplied with inorganic phosphorus. Hence micro-organisms are in fact a key driver in regulating the mineralization of phytate in soil and their presence within the rhizosphere maycompensate for a plants inability to otherwise acquire P directly from phytate (Richardson and Simpson, 2011).

10.3.7.2.3 Phosphonatases and C–P lyases, that cleave the C–P bond of organophosphonates (Rodriguez *et al.*, 2006). It is therefore clear that phosphorus solubilization by PSMs has been a subject of analysis and research for a long time and still the research seems to be in its infancy. It occurs through different mechanisms and there is considerable variation amongst the organisms in this respect. Each organism can act in one or more than one way to bring about the solubilization of insoluble P. Though it is difficult to pin point a single mechanism, production of organic acids and consequent pH reduction appears to be of great importance. Different mechanisms involved in the solubilization and mineralization of insoluble phosphorus by naturally-occurring microbial communities of soils are briefly illustrated in Fig.10.2.

10.3.8 Role of Siderophores in P Solubilization

Siderophores are complexing agents that have a high affinity for iron and are produced by almost all micro-organism in response to iron deficiency. Thus siderophores act as solubilizing agents for iron from minerals or organic compounds under conditions of iron limitation. There are approximately 500 known siderophores, with the majority of them being used by a wide range of micro-organism and plants and some of them being exclusively used by the microbial species and strains that produce them (Crowley, 2007). Studies have reported the release of siderophores from PSM (Vassilev *et al.,* 2006; Hamdali *et al.,* 2008a); however, siderophore production has not been widely implicated as a P-solubilization mechanism. Considering the dominance of mineral dissolution over ligand exchange by organic acid anions as a P-solubilizing mechanism (Parker *et al.,* 2005), the potential role of siderophores in enhancing P availability should be obvious.

10.3.9 Role of Exopolysaccharides (EPS) in P Solubilization

The role of polysaccharides in the microbial mediated solubilization of P was assessed by Yi *et al.,* (2008). Microbial exopolysaccharides (EPSs) are polymers consisting mainly of carbohydrates excreted by some bacteria and fungi onto the outside of their cell walls. Their composition and structures varies; they may be homo or heteropolysaccharides and may also contain a number of different organic and inorganic substituents (Sutherland, 2001). Four bacterial strains of *Enterobacter* sp. (EnHy-401), *Arthrobacter* sp. (ArHy-505), *Azotobacter* sp. (AzHy-510) and *Enterobacter* sp. (EnHy-402), possessing the ability to solubilize TCP (tri calcium phosphate), were used to assess the role of exopolysaccharides (EPS) in the solubilization of P (Yi *et al.,* 2008). These phosphate solubilizing bacteria produced a significant amount of EPS and demonstrated a strong ability for P-solubilization. However, further studies are necessary to understand the relationship between EPS production and phosphate solubilization.

10.3.10 Role of Sulphur in Phosphate Solubilization

The process of manufacturing of phosphatic fertilizers involves the acidulation of high grade rock phosphates with mineral acids to convert insoluble P into a soluble form. Sulphuric acid has been the most commonly used acid for this purpose. Lipman *et al.* (1916) advocated the use of rock phosphate and sulphur for increasing P availability in soil. This idea was 'floated on the premise that thiobacilli oxidize sulphur in the soil to sulphuric acid which, would then solubilize P. They incubated rock phosphate with soil and sulphur and found that there was increase in water soluble and ammonium citrate soluble P. In a greenhouse study Neller (1956) found that addition of sulphur had a marked effect on P solubilization from rock phosphate and the availability of residual P from rock phosphate increased upto 13% when sulphur was added. A granular fertilizer prepared by pelleting rock phosphate with sulphur and inoculating with sulphur oxidizing bacteria has been reported by workers in New Zealand and Australia and such preparations are called "Biosuper" (Rajan, 1982, 1987). The term "Biosuper" (mixture of rock phosphate and sulphur inoculated with sulphur oxidizing *Thiobacillus*) was first used by Swaby (1975). Biosuper pellets (Rajan and Edge, 1980) have studied the pattern of dissolution of rock phosphates and biosuper to find out the extent of release of P during soil incubation. The oxidation of sulphur and dissolution of rock phosphate reached a maximum after 3 to 4 weeks after their addition. Biosuper was found to be a good P source for perennial ryegrass which increased the ryegrass yield by 13 to 25% (Rajan and Edge, 1980). Generally rock phosphate to sulphur ratio of 5:1 has been used (Rajan and Gillingham, 1986) for the preparation of Biosuper. Rajan (1983) evaluated the effect of sulphur content on the effectiveness of Biosuper and observed that ryegrass yield decreased slightly when ratio of rock phosphate to sulphur was widened to 6:1. Further widening of the ratio to 7:1 was not satisfactory as it reduced the yield drastically.

Sulphur is an expensive input and has to be imported in many developing countries. Moreover, the use of this treatment (rock phosphate + Pyrite + *Thiobacillus thiooxidans)* was not found superior to rock phosphate inoculated with phosphate dissolving fungi *(Aspergillus awamori)* where addition of either sulphur or pyrite was not required (Patil *et al.*, 1974-78). The result showed that phosphate solubilizing fungi increased the available phosphorus status of the soil, phosphorus uptake by the plant, its root development and dry matter yield of the crop was improved. The treatment involving rock phosphate + pyrite + *Thiobacillus thiooxidans* was only next in performance to the above treatment.

10.3.11 Effect of Pyrite in Rock Solubilization

The microbial leaching of low grade ores to extract valuable metals like copper, zinc and uranium largely depends on the oxidizing action of *Thiobacillus ferrooxidans* and *T. thiooxidans* (Torma and Binhegy, 1984). Pyrite, a natural sulphur mineral is oxidized to produce .sulphuric acid. Similar approach involving addition of pyrite to rock phosphate offers an attractive proposition for increasing the availability of P from rock phosphates. Experiments conducted to determine if low grade pyrite could be used in combination with rock phosphate to improve the availability of P in neutral and alkaline soils has given contradictory results. Rastogi *et al.* (1976) observed mobilization of P from rock phosphate by pyrite in an alkaline soil but there was no effect in neutral and acidic soils. Khanna *et al.*, (1979) observed that pyrite did not help in the solubilization of rock phosphate in a calcareous soil. On the other hand, Singh *et al.*,(1979) have shown the utility of narrowing the ratio of rock phosphate to pyrite from 1:0.16 to 1:1 in improving the availability of P from rock phosphate.In glasshouse experiments with Mussoorie rock phosphate with Maize (*Zea mays*) and berseem (*Trifolium alexandrinum*) as test crops in a neutral soil; the addition of low grade pyrite brought about only a modest improvement in availability of phosphorus. Rock phosphate alone or rock phosphate with pyrite (1:5) were less efficient than superphosphate (Gupta and Mishra 1978). Mishra *et al.* (1981) have reported that Mussoorie rock phosphate mixed with pyrite (1:2.5) was a good source of P to potato in a neutral sandy loam soil and gave yields similar to single superphosphate. Sharma *et al.* (1983) found that Mussorie rock phosphate was 40.5% as effective as superphosphate in lentil (*Lens culinaris*). Its efficiency was improved to 50.4% when it was mixed with 25% by weight of pyrite.

10.3.12 Production of Growth Promoting Substances

Evidence continues to gather that microbial growth factors are present in the soil and the soil microflora includes many micro-organism for which growth factors are essential as they cannot synthesize where as other are able to synthesize them. The quantitative data on the incidence of soil bacteria requiring a definite growth factor was reported by Lockhead and Thexton (1952). Out of 534 strains isolated from a field soil on a non-selective basis 41 (7.6%) required Vitamin B_{12} as essential nutrilite. Lockhead (1957) examined the capacity of 316 cultures for the synthesis of thiamin, biothin, riboflavin, Vitamin B_{12} and the terregens factor. It was observed that riboflavin was produced by a large number of isolates followed by thiamin and Vitamin B_{12}.

Sankram (1960) reported that a large number of bacterial isolates from wheat, berseemand cowpea were able to synthesize vitamin $B_{12.}$ Mishustin and Naumova (1962) reported that azotobacterin and phosphobacterin acted by suppressing fungi in the root zone and also produced biotic substances that stimulated plant growth. Naumova *et al.* (1962) reported that *Bacillus megaterium* var. *phosphaticum* and *B. fluoroscens* accumulated in the medium biologically active compounds such as auxins, gibberellins, vitamins *etc.*, which can stimulate plant growth and inhibit the growth of fungi like *Fusarium* and *Alternaria.*Brown (1972) reported that bacteria producing one or two growth promoting substances mostly belong to the genera of *Nocardia* and *Arthrobacter.* Species of *Flavobacterium* and *Brevibacterium,* although not stimulated by roots, isolated from rhizosphere produced plant growth promoting substances. *Pseudomonas, Flavobacterium* and *Brevibacterium* spp. Isolated from root-free soil also produced growth hormones. Barea *et al.* (1976) reported that out of fifty PSB, twenty synthesized all the 3 type of plant hormones, 43 produced IAA, 29 formed gibberellins and 45 cultures produced cytokinin-like substances. Plant growth inhibitors were also detected in cultures of some isolates. Azcon *et al.* (1978) described the cell free supernatants and in some cases the whole bacterial cultures of phosphobacter in behaved as pure hormone in improving dry weight of plants and infection levels compared with mycorrhizal control plants.

Ana *et al.* (1981) compared the relationship of vitamin production and ability to dissolve insoluble calcium phosphate by bacteria isolated from rhizosphere, rhizoplane and control soil. Among isolates producing one or more vitamins 76.4, 37.5 and 57.5% of control, rhizosphere and rhizoplane isolates, respectively. However, phosphate solubilizers from rhizosphere and rhizoplane were found more active vitamin producers than solubilizers from control soil and non-solubilizers from any of the 3 regions. Production of Vitamin B12, riboflavin and niacin by rhizosphere isolates and of riboflavin by rhizoplane isolates was also found to be correlated with P-solubilizers but similar relationships were also observed for control isolates. The tested phosphate dissolving bacteria, *Bacillus polymyxa* (H_5) *B. polymyxa* (P) *B. pulvifaciens*(R,) *B. pulvifaciens* (R_1) and *Pseudomonas striata* (27) and fungi, *Aspergillus awamori* (F_{18}), *A. niger* (An) and *Penicillium digitatum* (F_{18}) produced auxins and gibberellins to an appreciable extent. The variation in nature and quantity of plant hormones was observed among the strains. The study suggests that beneficial effects of these organisms may also be due to the production of growth promoting substances besides phosphate solubilization (Sattar and Gaur, 1987).

10.4 Plant Growth Promoting Attributes of PSM

Besides making soluble P accessible for uptake by plants, there have been a number of reports on plant growth promotion by these micro-organism (Gaur and Ostwal, 1972). This is achieved by production of plant beneficial metabolites, such as phytohormones, antibiotics, or siderophores. Various PSM preparations have been shown to promote the growth of many crops. Endophytic bacteria isolated from rhizoplane of cacti growing in bare lava rocks,not only significantly mobilized Phosphate and other minerals (Puente *et al.*, 2004a, 2009a) but also promoted growth of wild cactus species (Puente *et al.*, 2004b). The mechanisms involved in plant growth promotion by PSM are outlined in Fig. 10.3.

Table 10.6 : Plant Growth Promotion by PSM

PSM Bioinoculant	Crop benefited
B. firlmus NCIM 2636	Paddy in acid soils
G. fasciculatum	Banana
B. megaterium+G. fasciculatum	Banana
Phosphobacterium	Sword bean variety SBS 1
P. striata	Soybean in sandy alluvial soil
P. striata	Chickpea

Source : Sharma *et al.*, (2013)

10.5 Genetic Engineering of PSM

Although knowledge of the genetics of phosphate solubilization is still scanty, and the studies at the molecular level in order to understand how precisely the PSM brings out the solubilization of insoluble P are inconclusive (Rodriguez *et al.,* 2006). However, some genes involved in mineral and organic phosphate solubilization have been isolated and characterized. Initial achievements in the manipulation of these genes through genetic engineering and molecular biotechnology followed by their expression in selected rhizobacterial strains open a promising perspective for obtaining PSM strains with enhanced phosphate solubilizing capacity, and thus, a more effective use of these microbes as agricultural inoculants. The initial achievement in cloning of gene involved in P solubilization from the Gram negative bacteria *Erwinia herbicola* was achieved by Goldstein and Liu (1987). Similarly the napA phosphatase gene from the soil bacterium *Morganella morganii* was transferred to *Burkholderia cepacia* IS-16, a strain used as a biofertilizer, using the broad-host range vector pRK293 (Fraga *et al.,* 2001). An increase in extracellular phosphatase activity of the recombinant strain was achieved.Introduction or over-expression of genes involved in soil phosphate solubilization (both organic and inorganic) in natural rhizosphere bacteria is a very attractive approach for improving the capacity of micro-organism to work as inoculants. Insertion of phosphate-solubilizing genes into micro-organism that do not have this capability may avoid the current need of mixing two populations of bacteria, when used as inoculants (nitrogen fixers and phosphate-solubilizers (Bashan *et al.,* 2000). There are several advantages of developing genetically-modified PSM over transgenic plants for improving plant performance:

a) With current technologies, it is far easier to modify a bacterium than complex higher organisms,

b) Several plant growth-promoting traits can be combined in a single organism, and

c) Instead of engineering crop by crop, a single, engineered inoculant can be used for several crops, especially when using a non-specific genus like *Azospirillum* (Rodriguez *et al.,* 2006).

Some barriers should be overcome first to achieve successful gene insertions using this approach, such as the dissimilarity of metabolic machinery and different regulating mechanism between the donor and recipient strains. Despite the difficulties, significant progress has been made in obtaining genetically engineered micro-organism for agricultural use (Armarger, 2002). Overall, further studies on this aspect of PSM will provide crucial information in future for better use of these PSM in varied environmental conditions.

10.6 Phosphate-Solubilizing Fungi and Bacteria, Controlled Vs. Field

Phosphorus fertilizers are a finite resource that is calculated to run out within the next few decades (Cordell *et al.,* 2009). Current fertilizer use efficiency is often quite low (estimated 10-20% in many soils) because the strength of the soil sink (through the presence of metal oxides and metal hydroxides) very largely surpasses the capacity of the plant to quickly take up this phosphorus. The phosphorus not taken up by plants enters in pools that are quite stable and where the exchange rate with soil solution phosphorus is far too slow to impact plant growth. However, phosphate accumulation in these stable pools could ultimately build up a phosphate store in soil that would gradually reduce the need for phosphorus fertilizer (Sattari *et al.,* 2012) although the contribution that this would make towards solving the problem of peak phosphorus is still very difficult to estimate. Also estimating the rate at which phosphorus runs out is beset with several difficulties because of variable estimates of P stocks that can be economically mined, the quality of these additional P reserves (as sources of rock phosphate also contain cadmium and uranium) and future changes in phosphorus recycling in industrialized countries (Van Kauwenbergh, 2010). But irrespective of the exact estimate, inefficient use of phosphorus fertilizers is a problem for sustainable agriculture and for the challenge to feed the world and raising use efficiency will be needed.

Raising fertilizer phosphorus use efficiency can be achieved through two mechanisms- increasing scavenging soils for available phosphorus through mycorrhizal associations or through plant breeding that results in improved root systems and increasing mining where sparingly soluble phosphorus sources are made available. Phosphate-solubilizing fungi and bacteria could enhance P nutrition of plants through P mining and hence sustainable P recycling within soils. Certain micro-organism have the ability to solubilize P from these metal (hydr-)oxides, making P plant available. Richardson and Simpson (2011) summarized three mechanisms by which micro-organism can increase P for plant uptake:

1) Increased root growth (as a general property of PGPR, often due to hormonal effects; this being part of the scavenging mechanism);
2) Alteration of sorption equilibria in the soil, rendering soil P plant available; and
3) Induction of metabolic processes that are effective in directly solubilizing and mineralizing P from sparingly available forms of soil inorganic and organic P, such as exudation of organic acids or anions, siderophores or enzymes (phosphatases such as acid or alkaline phosphate mono-esterase and phytase).

Research on PSB and fungi with similar capabilities has often remained in the stage of experimental investigations under controlled conditions, and scaling up to field level has turned out to be difficult. Khan *et al.* (2007) noted that solubilization of phosphate compounds by naturally abundant phosphate-solubilizing micro-organisms (PSMs) is very common under *in vitro* conditions, but that the performance of PSM *in situ* has been contradictory. The widespread application of P-solubilizing bacteria and fungi therefore remains limited by a poor understanding of microbial ecology and population dynamics in soil, and by inconsistent performance over a range of environments (Whitelaw, 2000; Khan *et al.,* 2007). Problems encountered in this respect are that the production of organic anions is carbon demanding. In the lab this problem is solved by adding easily available C sources to a microbial culture and then assessing dissolution of P from sparingly soluble minerals. However, adding such compounds in the field may result in competition with other (rhizosphere) micro-organism that may be competitively superior (they convert that carbon into microbial biomass rather than into organic acids or anions), and the organic anions themselves may also be degraded by other rhizosphere organisms. Addition of carbon sources and subsequent microbial growth could also result in microbial nitrogen immobilization. While phosphatases could release phosphate molecules from organic compounds, it is likely that the rate-limiting step in soils (from the perspective of increasing P availability) is not the quantity of phosphatases but the availability of organic phosphate, as these compounds are to a very large extent also strongly adsorbed to metal oxides and hydroxides (Tinker and Nye, 2000). One could try to achieve P dissolution through PSM by adding these organisms to sources of sparingly soluble P (rock P), adding organic sources (e.g., organic waste) and then adding the soluble P as fertilizer. However, the then easily available P added to soil will still be subject to the same kind of (strong) adsorption and occlusion reactions as easily soluble mineral fertilizer P (Koele *et al.,* 2014).

10.7 Effect of PSMs on Growth, Yield and Phosphorus Economy

Inoculation of plants with PSMs generally results in improved plant growth and yield, in particular, under glasshouse conditions (Zaidi *et al.,* 2009; Khan *et al.,* 2010). More importantly, investigations conducted under field level using wheat and maize plants have revealed that PSMs could drastically reduce the usage of chemical or organic fertilizers (Singh and Reddy, 2011). As reported by Vessey and Heisinger (2001), enhancement of plant phosphorus nutrition might be due to stimulation of root growth or elongation of root hairs by specific micro-organism, thus no direct increase in the availability of soil phosphorus is always expected.PSMs have been isolated from soil of various plants such as maize (Alam *et al.,*2002), oil palm (Fankem *et al.,* 2006), soybean (Son *et al.,* 2006), aubergine and chili (Ponmurugan and Gopi, 2006), mustard (Chandra *et al.*, 2007), rice (Chaiharn and Lumyong, 2009), walnut (XuanYu, 2011) and in different vegetables as pea, spinach, onion, potato (Aliai *et al.,* 2013). Better crop performance was reported to be achieved from several horticultural plants and vegetables, which were successfully inoculated with PSB (Young *et al.*, 2003).Phosphorus use efficiency in agricultural lands could effectively be improved through the inoculation of relevant PSMs, which is in fact, an integrated and sustainable mean of nutrient management of crop production systems. Apart from phosphate solubilizing abilities, some of these micro-organism can benefit plant growth by several different mechanisms such as enhancing nitrogen fixation, plant hormone production etc. As an eco-friendly and economical alternative to provide substantial amount of soluble P to plants for growth promotion is the exploitation of P solubilization and mineralization traits of PSB. Additionally, PSB not only protect plants from phytopathogens through the production of antibiotics, HCN, phenazines and antifungal metabolites, etc. (Upadhayay and Srivastava 2012; Singh *et al.*, 2013), but also promote plant growth through N_2 fixation (He *et al.,* 2010). Although phosphorus PSMs are abundant in many of the soils, isolation, identification and selection of PSMs have not as yet been successfully commercialized, thus application is still found to be limited. Investigations on the subject are often designed to confirm a specific response of PSMs to a particular environment, thus large scale application in field level is still limited.

10.8 Factors affecting Phosphate Solubilization

Several factors affect the phosphorus solubilization such as-

10.8.1 Rock Phosphate Concentration

There was a significant improvement in the solubilization of rock phosphate with increasing concentration of P_2O_5 up to 100mg added to 100ml medium by bacteria or fungi and higher amounts than this did not show any further improvement. Gaur and Sachar (1980) showed that lower the quantity of phosphate added to the medium, greater was the solubilization percentage by *Aspergillus awamori*.

10.8.2 Particle Size of Rock Phosphate

Solubilization of Mussorie rock phosphate of different mesh sizes by *Pseudomonas striata* and *Aspergillus awamori* in the liquid medium reported by Gaur (1986). The findings showed that the maximum phosphate was rendered soluble when rock phosphate particle sizes ranged between 30-59 and 60-99 mesh. The efficiency of microbial solubilization was reduced in presence of finer rock phosphate particle of 150-250 mesh as compared to coarser particles. These result showed that grinding or rock phosphate beyong 60-99 mesh is not advantageous for effective solubilization by fungi or bacteria.

10.8.3 Carbon Sources

Rose (1957) reported that glucose or xylose were the best sources of energy for fungi in liquid media where as other workers reported that either yeast extract or soil extract was essential for proper growth of PSM under culture conditions (Katznelson and Bose, 1959; Katznelson *et al.,*1962). The solubilization of rock phosphate using different types of carbon sources was studied by Gaur (1990). Glucose, sucrose, galactose and arabinose were good sources of energy and carbon for *Pseudonomas striata* and these sources were effective in that order. The yeast, (*Schwanniomyces occidentalis*) utilised a few carbon sources such as sucrose, glucose, arabinose, xylose and glactose, whereas *Penicillium digitatum* solubilized maximum P_2O_5 in presence of glucose followed by sucrose, mannitol, arabinose and xylose. This organism except galactose could use the rest carbon sources quite efficiently in relation to phosphate solubilization. Gaur and Gaind (1983) reported that during 15th day of incubation in liquid medium maximum rock phosphate solubilization by *Aspergillus awamori* occurred with sucrose or mannitol as compared to glucose, it is evident that sucrose is a cheaper source than glucose or mannitol. It was observed that by increasing the concentration of glucose as carbon source from 1 to 3% in broth the quantity of phosphate solubilized was enhanced. The results also indicated that greater acidity was produced when glucose was used at higher rate (3 and 2%) as compared to 1%.

10.8.4 Nitrogen Sources

Pseudomonas striata strains utilized different sources of nitrogen such as ammonium sulphate, asparagine, urea, ammonium nitrate, potassium nitrate, calcium nitrate except sodium nitrate. *Schwanniomyces occidentalis* utilized ammonium or amino nitrogen only. They could not utilize nitrate form of nitrogen. Ammonium sulphate was the best source both for the bacteria and the yeast. The fungi could solubilize a good amount of rock phosphate in presence of different nitrogen sources. The best sources for *Aspergillus awamori* was ammonium nitrate followed by asparagine, *Penicillium digitatum* was effective in presence of sodium nitrate followed by potassium nitrate.

10.8.5 pH

The results showed that the optimum pH for bacteria was around 6 and the activity of the fungi or the yeasts was the best in the pH range between

5 to 6. The fungal activity was greatly reduced at pH 7 to 8.

10.8.6 Temperature

The optimum temperature for best activity of *Pseudomonas striata* ranged between 20 to 30°C. The temperature range between 25-30°C was most favorable. For the fungi, 30°C was the optimum. The activity of these organisms was drastically reduced at higher temperature such as at 35° and 40°C. There is a need to isolate active phosphate solubilizing organisms which can grow at higher temperature and at least tolerate upto 45°C.

10.9. Current Trends

Phosphorus is an important limiting factor in agriculture production, and considering the negative effects of chemical P fertilizers, microbial intervention of PSM seems to be an effective way to solve the phosphorus availability in soil. However, P-solubilization in soil is much more difficult to study than solubilization of P in broth culture. The crops respond differently to the inoculation of PSMs and are dependent on several factors such as the soil temperature, moisture, pH, salinity, and source of insoluble P, method of inoculation, the energy sources and the strain of microorganism used. Hence study of PSM activity in correlation with these factors has to be done extensively before PSM can be used as a biofertilizer with promising results. The successful implementation of this approach has already been demonstrated in the fields by various workers, to a limited extent. However, the large scale use of this technology would benefit from additional studies, particularly those directed towards understanding how the interaction between soil and microbial system might be facilitated.

The organisms involved in phosphorus (P) cycling in soils are highly varied, and micro-organism probably play the most important role. However, more than 99% of soil micro-organism have not been cultured successfully (Torsvik and Ovreas, 2002). Therefore, culture-independent methods are required to study the function and ecology of microbes involved in P cycling in soils. Molecular approaches for such culture-independent methods have been developed in the recent past. The molecular techniques based on nucleic acid composition like LMW RNA (Low Molecular Weight RNA) profiling and Polymerase Chain Reaction (PCR) based techniques, are excellent tools for this purpose, as they are precise, reproducible and not dependent on culture media composition or growth phase of micro-organism (Peix *et al.*, 2007). Molecular-based techniques also provide new opportunity to detect the presence and abundance of specific micro-organism or to quantify the expression of target genes directly in soil or in the rhizosphere with high levels of sensitivity. For example, specific primers based on conserved regions have been described for various micro-organism associated with P mobilization, including mycorrhizal fungi, *Penicillium* sp., and *Pseudomonas* sp. (Oliveira *et al.,* 2009), as have primers that are directed at traits such as bacterial phytases (Jorquera *et al.,* 2011).

Microarrays composed of suites of functional bacterial genes and arrays for phylogenetic analysis of bacterial diversity based on 16S-RNAgene sequences along with next-generation sequencing and soil microbiome analyses; provide further application for assessment of diversity surrounding particular traits or functional groups of micro-organisms (Richardson and Simpson, 2011). Collectively, these tools provide new opportunities to address key questions in microbial community ecology and to assess the survival and persistence of specific inoculants under different environmental conditions. Looking at the possible avenues which can open up with exploring these environmental friendly micro-organism, it is necessary to study the composition and dynamics of these microbial populations to reach a better understanding of soil PSM diversity and P uptake by plants.

10.10 Future Prospects

Despite their different ecological niches and multiple functional properties, P-solubilizing micro-organism have yet to fulfill their promise as commercial bio-inoculants. Current developments in our understanding of the functional diversity, rhizosphere colonizing ability, mode of actions and judicious application are likely to facilitate their use as reliable components in the management of sustainable agricultural systems. Although significant studies related to PSM and their role in sustainable agriculture have been done over the last few decades, the required technique remains in its infancy. Nevertheless with an awareness of the limitations of existing methods, a reassessment can be expected, so that the use of PSM as potential biofertilizers in different soil conditions becomes a reality. Enhancement in the use of PSM is one of the newly emerging options for meeting agricultural challenges imposed by the still-growing demand for food. Thus, more than ever, obtaining high yields is the main challenge for agriculture. In addition, in recent years both producers and consumers have increasingly focused on the health and quality of foods, as well as on their organoleptic and nutritional properties. Hence, this biotechnology is also likely to ensure conservation of our environments. However, before PSM can contribute to such benefits, scientists must learn more about them and explore ways and means for their better utilization at the farmers' fields. Future research should focus on managing plant–microbe interactions, particularly with respect to their mode of actions and adaptability to conditions under extreme environments for the benefit of plants. Furthermore, scientists need to address certain issues, like how to improve the efficacy of biofertilizers, what should be an ideal and universal delivery system, how to stabilize these microbes in soil systems, and how nutritional and root exudation aspects could be controlled in order to get maximum benefits from PSM application. Biotechnological and molecular approaches could possibly develop more understanding about PSM mode of actions that could lead to more successful plant–microbe interaction. Efforts should also be directed towards the use of PSM to reduce pesticide applications. In brief, PSM biotechnology provides an excellent opportunity to develop environment-friendly phosphorus biofertilizer to be used as supplements and/or alternatives to chemical fertilizers.

10.11 Conclusions

Phosphorus is a vital element in crop nutrition. Adverse environmental effects of chemical based P fertilizers, depleting resources of high grade Phosphatic rocks and their skyrocketing prices have compelled us to find a sustainable approach for efficient P availability in agriculture to meet the ever increasing global demand of food. Soil micro-organism are involved in a range of processes that affect P transformation and thus influence the subsequent availability of P (as phosphate) to plant roots. In particular, micro-organism can solubilize and mineralize P from inorganic and organic pools of total soil P. The use of efficient PSMs opens up a new horizon for bettercrop productivity besides sustaining soil health. However, the viability and sustainability of PSMs technology largely depends on the development and distribution of good quality inoculants to farming communities. Therefore, there is a need for extensive and consistent research efforts to identify and characterize more PSM with greater efficiency for their ultimate application under field conditions. Soil Scientists and Microbiologists have a great accountability to the society to find ways and means as to how soil phosphorus could be improved without applying the chemical P fertilizers under different agro-climatic regions of the world.

The promise of exploiting soil micro-organism to increase mobilization of soil P remains. Whether or not this will be achieved through better management of soil microbial communities, by development of more effective microbial inoculants, through the genetic manipulation of specific organisms, or with a combination of these approaches is not known. What is clear though is that soil micro-organism play an important role in the mobilization of soil P and that detailed understanding of their contribution to the cycling of P in soil-plant systems is required for the development of sustainable agriculture and our movement from a green revolution to an evergreen revolution can be accomplished.

Phosphorus supply through biological means is a viable substitute as phosphate solubilizing bacteria (PSB), phosphate solubilizing fungi (PSF) and actinomycetes have been reported to be active in conversion of insoluble phosphate to soluble primary and secondary orthophosphate ions (Chabot *et al.*, 1993). Soil micro-organism have great potential in providing soil phosphates for plant growth. Phosphorus biofertilizers can help to increase the accessibility of accumulated phosphates for plant growth by solubilization (Goldstein, 1986; Gyaneshwar *et al.*, 2002). Moreover, micro-organism involved in P solubilization as well as better scavenging of soluble P can enhance plant growth through biological nitrogen fixation (BNF), enhancing the availability of other essential elements and by production of plant growth promoting substances (Gyaneshwar *et al.*, 2002). Apart from playing an important role in biogeochemical cycles these soil bacteria have been used for decades in crop production. Due to their colonization in the rhizosphere and active role in promoting plant growth these bacteria are also termed as plant growth promoting rhizobacteria (PGPR) (Hayat *et al.*, 2010, Kang *et al.*, 2012). Application of phosphorus along with PSB improved P uptake by plants and yield representing that PSB solubilize phosphates and activate phosphorus in crop plants (Rogers, 1993). In this respect, biofertilization technology has substantially minimized the production costs and evades the environmental hazards at the same time (Galal *et al.*, 2001).

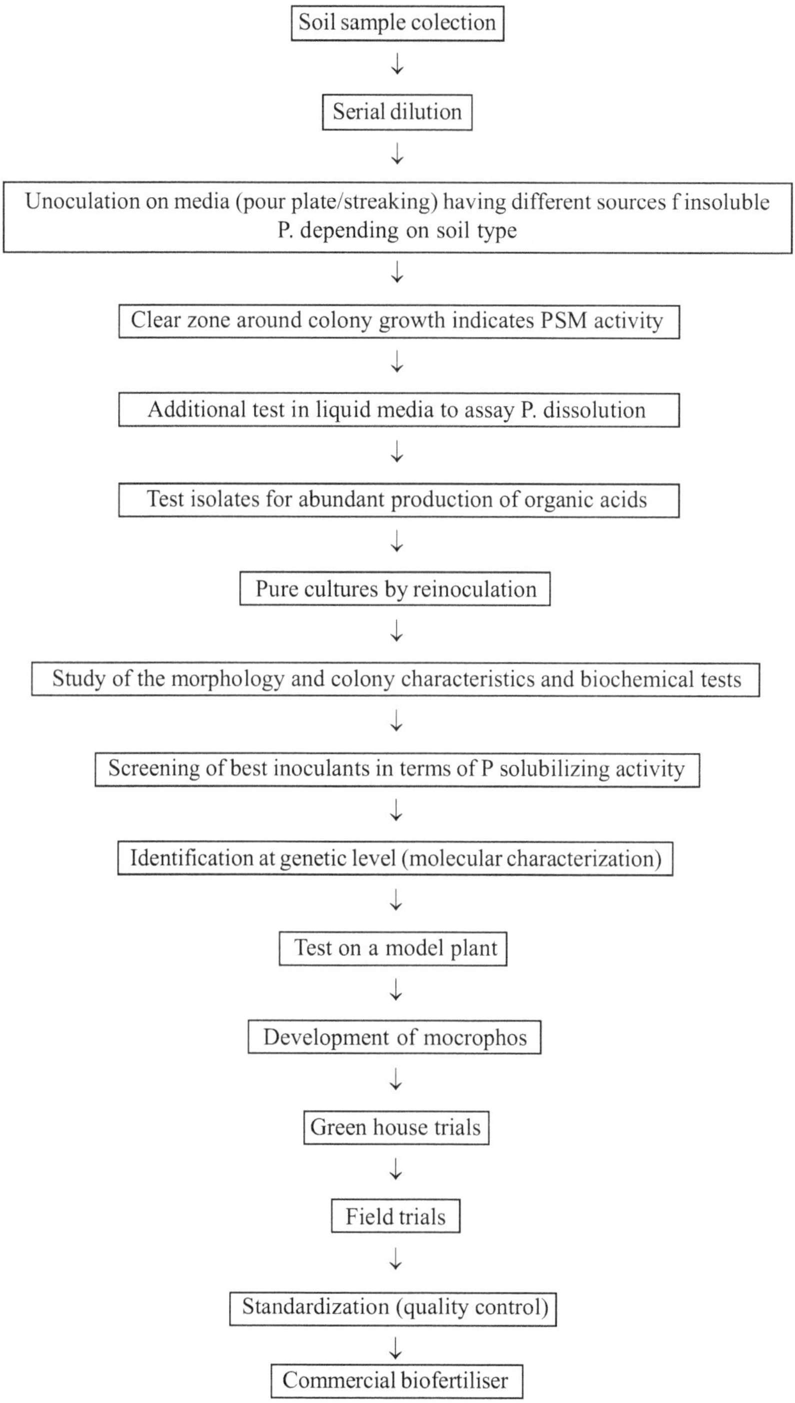

Fig. 10.1 : Protocol for isolation and development of effective inoculants of PSM based biofertilizer

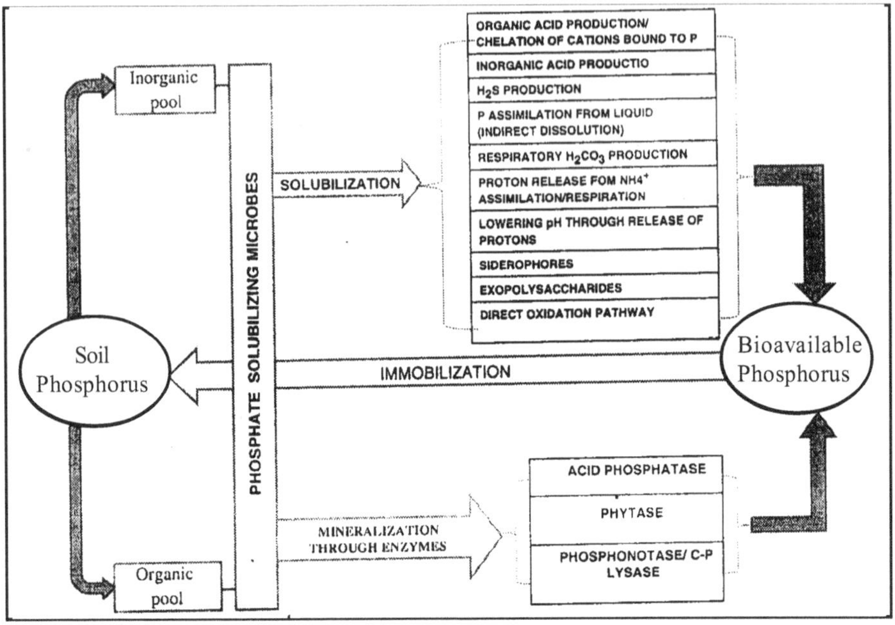

Fig. 10.2 : Schematic representation of mechanism of soil P solubilization/mineralization and immobilization by Phosphate Solubilizing Micro-organisms (PSMs)

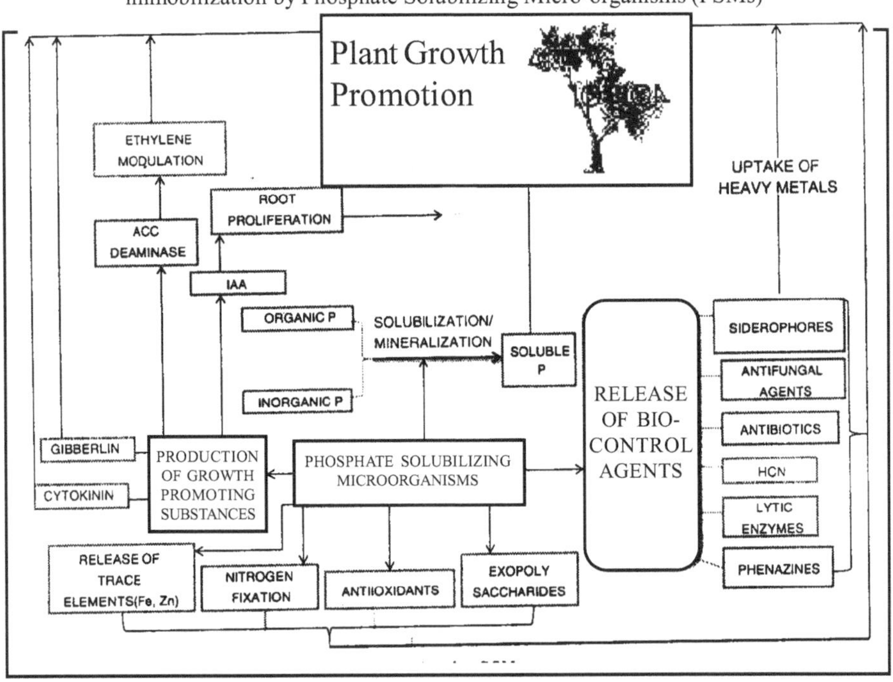

Fig. 10.3 : Possible mechanism involved in plant growth promotion by Phosphate Solubilizing Micro-organisms (PSMs)

CHAPTER - 11

Mycorrhizae

11.1 Introduction

Our Country is a fast developing nation for the overall development and has proved that the country which was known for a begging bowl for food grains has turned up towards self-sufficiency and also has excess production to feed the present level of population. From about 50 MT of food grain production in 1950-51, harvest in 2010-11 reached over 241 MT, highest ever along with high production of fruits, vegetables, milk, egg and fish. This increased food production in the country has mainly been achieved through productivity gains, as the net sown area has remained static for the last four decades. Our burgeoning problems of fast population growth has been a serious concern and by 2020, India would require more than 280MT of food grains, 66MT of oil seeds, 95MT of fruits, 150MT of milk and 11.5MT of fish. The foremost challenges of elevating malnutrition, enhancing productivity and crop diversification can be met through better resource management by developing more productive, more nutritious, but less input demanding crops (Ayyappan, 2011). Recent developments in the field of globalizing market price and enhanced cost of production has drastically affected the crop production through costly resources with the decreasing available land for farming due to industrialization and infrastructure. The poor and marginal farmers are the real culprit with the high escalating cost of fertilizers particularly for phosphatic fertilizers with the withdrawal of subsidy on the fertilizers and has compelled the researcher to work on the less input demanding crops with high fertilizer use efficiency. After the evolution of green revolution in late 1960 onwards, the farmers use the fertilizers as the developed varieties were responsive to fertilizers and other agriculture inputs and it was noticed that the phosphatic fertilizers, which were applied to the soil were not fully utilized in the same crop year and remained adhere to the soil particles which were not readily available to the crops immediately. Such non-available forms of phosphorus can be made available to the clients through the process of solubilizing and or mobilizing of phosphorus using the economical and cheap source such as fungi (AM) and bacteria (Phosphorus solubilizing bacteria).

Mycorrhiza word, first used by a German Researcher Frank (1885), originates from the Greek word myco, meaning (fungus) and rhiz, (root) (Verma *et al.,* 2008). Mycorrhiza is a symbiotic association between a group of soil fungi called arbuscular mycorrhizal fungi (AMF) and plants. Roots of most flowering plants live in mutual symbiosis with mycorrhizae which bio-tropically colonise the root cortex and extra-metricial mycelia help the plants to obtain plant nutrients from soil (Barea, 1991). These fungi are ubiquitous in soil and are found in the roots of many Angiosperms, Gymnosperms, Pteridophytes and Thallophytes (Mosses *et al.,* 1981). The mycorrhizal fungi perform the function of root hairs. The fungus takes carbohydrates from the plant and in turn supplies the plants with nutrients, hormones and protects it from root pathogens. The mycorrhizal plants have greater tolerance to toxic heavy metals, high soil temperature, soil salinity, unfavourable soil pH and to transplantation shocks. They play an important role in increasing plant growth and nutrients uptake (Krishna and Bagyaraj, 1993). These fungi can be used to promote biotechnologically developed plants in soil (Varma and Schuepp, 1994). Of all the AM fungi, endomycorrhizae are more important as bio- fertilizers. Biofertilizers are the promising alternative for the biological management of soil fertility and productivity. Among them mycorrhiza have a great potential in biomass production and soil enrichment on wastelands (Kashyap *et al.,* 2005).

The successful association between plants and AMF constitutes a strategy to improve the nutritional status of both associates, which reduces the use of fertilizers specially phosphorus nutrition (Almagrabi and Abdelmoneim, 2012). The AMF take carbohydrates compounds from their plant host, while the plants benefit from the association by the increased nutrients uptake, which improve tolerance to abiotic stress (drought or salinity), as well as enhanced plant disease control (Linderman, 1994, Song *et al.,* 2011). The water stress is considered the main factor that causing limitations to plant growth. The effects of drought on plant growth depend on several factors such as plant genetic resistance, stage of growth and duration of plant exposed to drought (Panozzo and Eagles, 1999; Echave *et al.,* 2005; Song *et al.,* 2011). The AMF are playing a vital role in sustainable agriculture because they enhance plant water relations, which improve the drought resistance of host plants (Nelsen, 1987). The abilities of specific associations between plants and AMF to tolerate drought are of a great interest. The results in several studies on drought stress conditions indicated that the plant biomass, chlorophyll contents and rate of transpiration were greater in plants inoculated with AMF compared with plants without AMF infection (Ruz-Lozano *et al.,* 1995; Augé, 2001; Beltrano *et al.,* 2003; Asensio *et al.,* 2012). Also AMF have been observed effects on stomatal conductance with similar frequency under amply watered and drought stress (Bethlenfalvay *et al.,* 1988; Henderson and Davies, 1990; Augé *et al.,* 1992; Davies *et al.,* 1993). AMF symbiosis has also affected stomatal sensitivity to atmospheric water status (Huang *et al.,* 1985). AMF effects on plant water relations and metabolism during drought have been associated with morphological and phenological effects. AMF Acacia (Osonubi *et al*., 1992) and rose (Henderson and Davies, 1990) showed more leaf abscission during drought than plants untreated with AMF, while wheat treated with AMF showed less leaf necrosis and leaf drop (Ellis *et al.,* 1985).

Soybeans plant treated with AMF had less drought-induced pod abortion than untreated plants (Busse and Ellis, 1985). Leaf movements were greater in AMF than in Leucaena free AMF inoculation (Huang *et al*., 1985). AMF plant leaves had lower cuticle weight and less epicuticular wax than plant leaves free from AMF infection (Henderson and Davies, 1990). AMF maize had relatively more green leaf area than non-mycorrhizal maize after drought (Subramanian *et al.,* 1995) and AMF symbiosis delayed leaf senescence of alfalfa in drought conditions (Goicoechea *et al.,* 1997).

11.2 Mycorrhiza Helper Bacteria – Three-Way Interactions

Duponnois & Garbaye (1991) and Garbaye (1994) observed the effect of the bacterium *Pseudomonas fluorescens*, significantly stimulated the formation of ectomycorrhiza by the fungus *Laccaria laccata* (later reclassified as *L. bicolor*), and termed these bacteria mycorrhiza helper bacteria (MHB). Frey-Klett *et al.* (2007) provided further information on the effects of MHB during the symbiosis, including improved conductivity of the soil and responsiveness of the roots to fungal recognition and establishment. Moreover, they reported that both the fungus and root select the bacterial population in the rhizosphere soil, promote the growth of fungus, and determine the receptivity of the root to the fungus. Most attention has been devoted to specific helper bacteria of the ectomycorrhizal symbiosis. However, specific bacteria also occur around and on the hyphae of AMF. Artursson *et al.* (2006) reviewed these interactions between AM fungi and rhizosphere bacteria, and concluded that the underlying mechanisms behind such tripartite associations were not well understood and more insight into these interactions is imperative for optimization of the effective use of AM fungi in combination with their bacterial partners as a tool for increasing crop yields. For that reason no commercial inocula of MHB are available on the market. These include interactions between AMF and the nitrogen-fixing rhizobia (Barea *et al.,* 2002) and interactions between AMF and phosphate-solubilizing bacteria (PSB) that can release phosphorus from sparingly soluble P sources (Kucey *et al.,* 1989; Bidondo *et al.* 2012). The release of flavonoids and strigolactones (Steinkellner *et al.,* 2007) and physical contact between bacteria and mycorrhizal fungi are important for the establishment of these three-way interactions (Bonfante and Anca, 2009). This three-way interaction has been studied in terms of increased plant nutrition and growth (Calvaruso *et al.,* 2007, Koele *et al.,* 2009, Frey-Klett *et al.,* 2011) from which it appeared that microbial niches and functioning are closely related, and the interactions are largely dependent on environmental factors. However, field studies need to confirm the symbiotic three-way interactions to have ecological benefits. Several of these bacteria as *Pseudomonas fluorescens* have been described as having beneficial effects on plant growth and are known as plant growth-promoting rhizobacteria (PGPR).

11.3 Classification of Mycorrhizae

Frank (1885) distinguished two main types of mycorrhizae such as 'ecotrophic' (ectomycorrhiza) and 'endotrophic' (endomycorrhiza) on the basis of the fungal vegetative filaments (hyphae) in relation to the host root tissues. According to recent information, seven different types of mycorrhizae are known (Harley and Smith, 1983;Bhandari and Mukerji 1993). These types are as follows (Fig. 11.1):

- Ectomycorrhiza
- Endomycorrhiza
- Ectendomycorrhiza
- Arbutoid mycorrhizae
- Monotropoid mycorrhizae
- Orchidaceous mycorrhizae
- Ericoid mycorrhizae

11.3.1 Ectomycorrhiza

They form a sheath around roots but lack intracellular penetration of the cortical cells, and commonly found on the roots of forest trees like pine, oak, beech and eucalyptus. These form a thick mantle (covering) outside the roots. Ectomycorrhiza is visible to the naked eye as *fuzzy* covering on the root. Inside the root they are present in intercellular spaces and forms a characteristic structure called "Hartig net". At times or as a stage in development, the "Hartig net" may become totally absent. This stage is known as 'perirhizic' mycorrhiza. Basidiomycetes belonging to *Amanitataceae, Boletaceae, Cortinariaceae, Russiclaceae, Tricholomataceae, Rhizopogonaceae* and *Sclerodermataceae* are the fungal symbionts often encountered in these associations (Marx, 1972). Tree species belonging to families viz. *Pinaceae, Salicaceae, Betulaceae, Fagaceae, Tibiaceae, Rosaceae, Leguminosae* and *Juglandaceae* are the hosts (Meyer, 1973). Ectomycorrhizae benefit by improving plant growth of

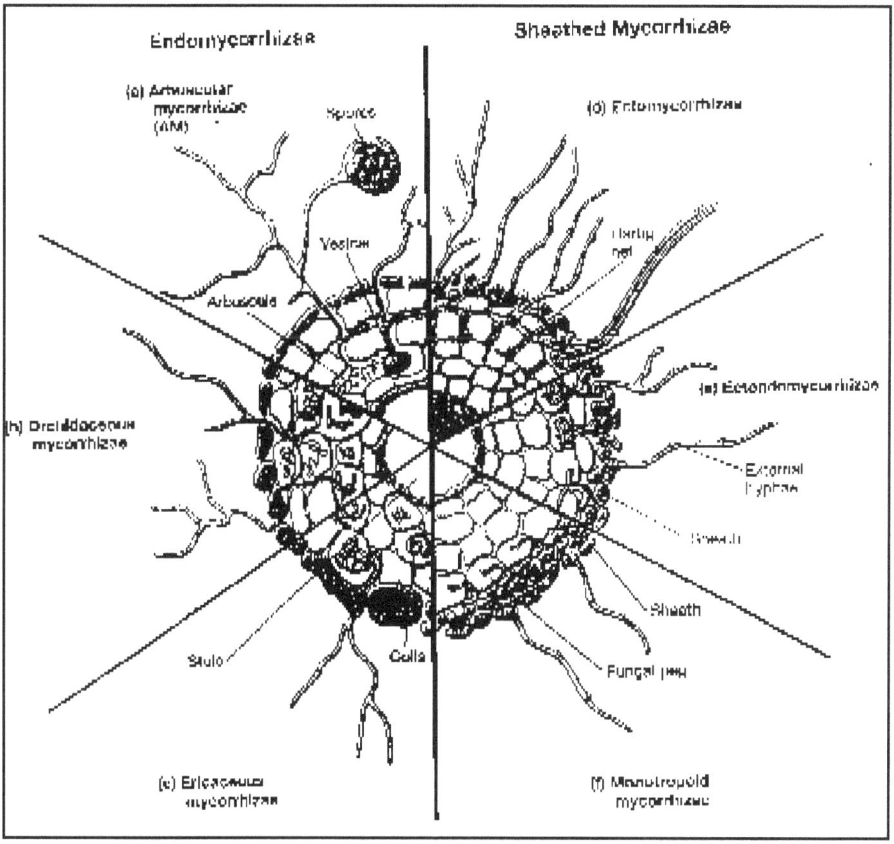

Fig. 11.1 : Mycorrhizae (Root cross sections illustrate different mycorrhizal relationship).

forest seedlings. The fungal partner of ectomycorrhiza can be grown on defined synthetic medium.

11.3.2 Ectendomycorrhiza

In this form, the sheath may be much reduced or even absent. The 'Hartig net' is well developed and the hyphae also penetrate into the host tissues.

11.3.3 Arbutoid Mycorrhiza

This form has a well-developed 'Hartig net'. Sheath is usually much reduced or even absent. The fungal hyphae penetrate into the cells of the host in addition there is intracellular penetration. They form fungal coils in the cells. It belongs to Basidiomycetes and the hosts belong to Ericales.

11.3.4 Monotropoid Mycorrhiza

The monotropoid mycorrhizal roots are found in Monotropaceae, a set of achlorophyllous plants. They have well developed fungal health enclosing the roots. They form 'Hartig net'; the hyphae enter the cells of the epidermis or cortex. Specialized underbranched haustoria penetrate the epidermal cells and go through a complicated development as the host plant grows flowers. The fungus is often associated with the mycorrhizal roots of adjacent plants.

11.3.5 Orchidaceous Mycorrhiza

The host may be wholly or partially achlorophyllous for part of their life cycle. They are associated with Basidiomycetes that are cellulolytic and lignolytic. The fungus penetrates host tissue and after a period of vegetative activity undergoes local or general disintegration. In nature orchid embryos normally become infected with the fungus. The fungus enters the protocorm (germinating seed) and stimulates the growth of protocorms mainly through carbon and phosphorus nutrition. Orchids are generally propagated through tissue culture. It is observed that although there is a lag in buildup of mycorrhizal infection, the fungus spreads readily from plant within the glasshouse making them mycorrhizal. The fungus commonly associated with Orchids is *Tulasnella* spp. which can be cultured in laboratory media.

11.3.6 Ericoid Mycorrhiza

In Ericaceae and related families, the hairs like roots are enmeshed in an extensive external weft of hyphae which also penetrate the cortical cells of the rootlets. These fungal symbiont are members of Ascomycetes. The Basidiomycetes genus *Clavaria* has also been reported to form the symbiotic association but normally do not form the sheath. Ericoid mycorrhizae are associated with healthy plants like blue berries, *Azaleas* spp, *Rhododendron* etc. The fungus involved is *Pezizellaericae* which can be cultured in the laboratory. Inoculation with the fungus has been reported to augment the yield of blue berries by about 15%. Commercial inoculum of fungus prepared in New Zealand is available in the trade name '*Mycoaid*'.

11.3.7 Endomycorrhizae (Vesicular-Arbuscular Mycorrhiza)

Vesicular-arbuscular (VA) mycorrhiza is geographically ubiquitous and occurs in plants growing in different ecological regions (Hayman, 1982). VAM are found on the roots of all plants except on a few vascular plants such as *Cruciferae, Chenopodiaceae, Cyperaceae and Junaceae* (Hirrel *et al.,* 1978, Ocampo *et al.,* 1979)., which do not form (Smith and Gianinazzi-pearson 1988). This is the commonest type of mycorrhizae formed on the roots of an enormously wide variety of host plants by aseptate fungi belonging to Endogonaceae (Gerdemann and Trappe, 1974). These non-septate Zygomycetous fungi belong to genera of *Glomus, Gigaspora, Acaulospora, Enterophospora, Sclerocystis* and *Scutellospora.* There are over 200 species of VAM fungi reported so far. AM replaced the earlier term "Vesicular-arbuscular mycorrhiza" (VAM) because not all endomycorrhizas of this type develop vesicles, but all form arbuscules.The nomenclature refers to the formation of vesicles and arbuscules, typical morphological structures formed by these fungi. AM fungi form symbiotic associations with members of Bryophytes, Pteridophytes, Gymnosperms and Angiosperms throughout the world. The arbuscular mycorrhiza is by far the commonest in agricultural and horticultural plants.The symbiosis with AMF (phylum Glomeromycota) in crop plants and wild plants, including the ancestors of cultivated plants) is widespread (Smith and Read, 2008; Smith and Smith, 2011). AMF are especially known for their major role in P acquisition through improved scavenging (Lambers *et al.,* 2008). Dead material of hyphae also makes a substantial contribution to the soil organic matter pool (Verbruggen *et al.,* 2012a).As has been stated "in agricultural field conditions plants do not, strictly speaking, have roots, they have mycorrhizas." Growth of a host plant can be improved by mycorrhizal colonization provided that soil-available P (or another element where plant uptake is diffusion-limited, such as Zn or Cu is a limiting factor for plant growth. The degree of growth improvement is affected by factors such as host plant species, fungal species, and soil conditions (Tawaraya, 2003). AMF account for 5-50% of the biomass of soil microbes (Olsson *et al.,* 1999), and biomass of hyphae of AMF may amount to 54-900 kg/ha (Zhu and Miller, 2003). In summary, mycorrhizal fungi are the interface between plant roots and the soil matrix, and mediate ecosystem processes at the root-soil interface, including (but not restricted to) nutrient acquisition and uptake (Dickie *et al.,* 2013).

Through the increased soil volume (extension of the depletion zone around roots due to the fact that the diffusion rate of P through soils is slower than uptake rate around roots and hyphae) accessed by the arbuscular mycorrhizal hyphae, the plant has greater access to orthophosphate, inorganic phosphate in the soil solution (Smith and Read, 2008). AMF assist largely in the uptake of P under P-limiting soil conditions. At P-sufficiency or even P-excess in the soil solution, the mycorrhizal contribution to P uptake diminishes both due to plant regulation of the extent of the mycorrhizal symbiosis and direct negative effects of excess P on fungal performance. While attention has been more focused on the regulatory role by the plant on the extent of mycorrhizal root colonization and the extent of extra radical hyphae, separating the impact of a direct (soil-mediated) from an indirect (plant-mediated) effect of nutrient excess is likely important in case of foliar application of fertilizer (as under that condition only the indirect pathway is operative). It is likely that the importance of the direct pathway has been underestimated. Whereas the main mechanism for P uptake by AMF is through the larger soil volume exploited by the fungal hyphae, additional mechanisms for enhanced P acquisition have been suggested. One major cause for the different perspective on a mycorrhizal role in the uptake of P from organic sources is that under many conditions it is still unclear whether P mobilization from organic sources is limited by enzyme availability or by availability of non-absorbed phosphorus (Tinker and Nye, 2000). Similarly, studies by Tawaraya *et al.* (2006) and Arocena *et al.* (2011) have suggested that mycorrhizal plants have access to mineral forms of phosphorus through the excretion of low molecular weight organic acids or anions, especially under acidic conditions, such as citrate, oxalate or malate. However, recent experiments have provided no support for the ability of the fungi to produce these compounds and have indicated that the mycorrhizal symbiosis downregulates exudation of organic anions (Gao *et al.,* 2008; Ryan *et al.,* 2012) implying that other, still poorly characterized mechanisms are responsible for a faster uptake from the sorbed phosphorus pools (Cardoso *et al.,* 2006).

Some studies have also shown uptake of nitrogen by AMF from organic substrates (Cheng *et al.,* 2012; Veresoglou *et al.,* 2012), but the extent to which arbuscular mycorrhizal plants benefit from increased access to these pools is still under debate. Hodge and Fitter (2010) suggested that AMF use organic N predominantly for their own nutrition and do not transport it to the host plant. In general, nutrient immobilization by mycorrhizal fungi under nutrient-poor (especially nitrogen-deficient) conditions may be an important explanation for lack of positive plant responses. AMF can take up nitrogen in the form of ammonium or nitrate (Cardoso and Kuyper, 2006), but according to Smith and Smith (2011), it is unknown whether arbuscular mycorrhizal N uptake fulfills the host's N needs. Hodge and Fitter (2010) argued that N taken up by AMF remains within the fungi and is not further transported to the host plant. However, plant uptake of N is affected indirectly by enhancing N-fixing microorganisms that can co-occur with AMF.

Colonization of plant roots by AMF not only affects plant mineral nutrition, but can also have numerous other beneficial effects under stress conditions (Pozo and Azcon-Aguilar, 2007; Miransari, 2010; Seguel *et al.,* 2013). AMF can work like filters; binding heavy metals at the hyphal surface and/or in a glycoprotein called glomalin, and thereby protect their hosts from toxic metal concentrations in soil at high availability of heavy metals (Audet and Charest, 2007). Because of larger size of mycorrhizal plants, heavy metal concentrations could also be diluted in plant tissue. At low availability of these metals, such as in Zn-deficient soils, the mycorrhizal symbiosis increases access to these essential metals (Gao *et al.,* 2007; Cavagnaro, 2008). Heavy metal stress alleviation by AMF was reviewed by Hildebrandt *et al.* (2007). AMF also can protect plants against conditions of suboptimal water potential (Birhane *et al.,* 2012).

Auge (2001) reported the impact of AMF on plant-water relations and drought. AMF can increase plant water uptake by the increased soil volume that the hyphae scavenge, by improved hyphal-soil contact, by improving the plant nutrient status (P and K), by increasing photosynthesis through enhanced leaf stomatal conductance or by sink stimulation (Auge, 2001; Kaschuk *et al.,* 2009). Worchel *et al.* (2013) and Kivlin *et al.* (2013) performed meta-analyses of fungal symbionts' effects on plant growth under global change and found that all fungal symbionts including AMF reduce negative effects of drought on plants. Protective effects against biotic stress (pathogenic bacteria, fungi or nematodes) have been described consistently for below-ground interactions (Singh and Vyas, 2009; Smith *et al.,* 2010). Protective effects are explained by various mechanisms including priority of occupation of root cells, improved mineral nutrition of AM plants, priming of plant defense reactions by AMF and antagonistic effects on micro-organisms (Singh and Vyas, 2009; Cameron *et al.,* 2013). These latter authors also proposed the term mycorrhiza-induced resistance. The effectiveness of this symbiosis for pathogen control depends on early colonization of the mycorrhizal fungus compared to that of the pathogens. A specific protective effect has also been noted against the parasitic plants *Striga, Orobanche* and *Phelipanche*, parasitic plants that are especially damaging under nutrient-poor (nitrogen-deficient, phosphorus-deficient) conditions (Lendzemo *et al.,* 2006). Negative responses by mycorrhizal fungi on other agricultural weeds have also been reported, but it is still unclear whether and how mycorrhizal management could contribute to weed control (Veiga *et al.,*

2011). Conversely, cases have been described whether mycorrhizal plants, including weeds during the fallow stage can help preventing long-fallow disorders in Australia (Thompson *et al.,* 2013). For almost all agricultural crops, the symbiosis with AMF is the natural, normal situation, and the non-mycorrhizal situation often used in comparative research is non-natural. Pictures of large mycorrhizal plants and small non-mycorrhizal plants should therefore not be interpreted as an argument for inoculation, but rather as a warning how much additional effort and costs are needed to achieve plants of the same size. Such experiments with non-mycorrhizal controls are often executed under controlled conditions in pots or small containers, and this set up may create further problems when it comes to scaling up mycorrhizal experiments to field conditions. While mycorrhizal fungi usually increase phosphorus uptake efficiency by the crop, it has also been observed that the phosphorus concentration of leaves of mycorrhizal plants are higher than those of non-mycorrhizal plants (Treseder, 2013).

Colonization

VAM colonization starts from hyphae arising from soil borne propagules. When it reaches to the cortex, hyphae grow into cells by tree like dichotomous branching to give arbuscules. When colonization is well established, oval structures called vesicles may form which have storage function (Barea, 1991). Mycorrhizal propagules can be severely influenced by damage to vegetation and soil resulting from natural disturbances or human intervention. This destructive process include intense fires (Wicklow-Howard, 1989) soil erosion exposing subsurface soil (Day *et al.,* 1987) by burrowing animal (Koid and Mooney, 1987) and volcanic activities (Allen, 1988). Human activities such as top soil removal/disturbances and stock piling during mining (Stahl *et al.,* 1988) and clear cut logging (Newman, 1985). Agricultural operations such as tillage (Evans and Miller, 1988) long fallow period (Thompson, 1987) soil compaction, growth of non-mycorrhizal crops, biocide application (Hayman, 1986) and flooding of rice (Hall, 1987) can adversely affect mycorrhiza. Reduction in mycorrhizal natural inocula following ecosystem disturbances is the main cause for the reduced symbiotic association. Introduction of mycorrhizal fungi as inocula will improve soil condition and its health (Danielson, 1985; Stahl *et al.,* 1988).

Early studies in India have shown the occurrence of *Rhizophagus* in association with *Litchi chinensis* (Pandey and Misra, 1971) and of VAM in Coconut (Lilly, 1975). Thapar and Khan (1985) during their survey in eight states at 31 locations isolated 15 species of VAM belonging to various genera of which *Glomus macrocarpum* was the most widely prevalent species. Rao *et al.* (1987) observed VAM colonization in twenty aromatic plant species examined. The presence of VAM on different tree species was recorded by

Jagpal and Mukerji (1987), Redhead (1977) reported VAM association in the moist low land rain forests and dry sahel region of Nigeria. Hafeel and Gunatilake (1988) found 95% of tree species occuring in tropical forests of Sri Lanka to be endomycorrhizal. Chalermpongse (1987) found VAM association in most tree species in the dry deciduous and semi evergreen dipterocarp ecosystem in Thailand.

Associated Plant Species

As indicated above, almost all plants are mycorrhizal under conditions of normal soil fertility, and under such conditions, mycorrhizal plants usually show a higher fitness (better growth, higher seed production, enhanced stress resistance, etc.) than non-mycorrhizal plants. Differential performance of mycorrhizal and non-mycorrhizal plants is called mycorrhizal responsiveness. Mycorrhizal responsiveness is due to the interaction of plant species or plant genotype, mycorrhizal fungal species and soil conditions. Historically, responsiveness to mycorrhiza has been called plant dependency on mycorrhiza. However, Janos (2007) argued that dependency is a genetic trait of the plant only; and that dependency and responsiveness are not necessarily correlated. Tawaraya (2003) listed arbuscular mycorrhizal responsiveness (called dependency by him) of different plant species and cultivars. He noted that on average cultivated plants showed lower responsiveness than their wild relatives and progenitors. This observation raises questions about the impact of plant breeding on mycorrhizal responsiveness. Based on his list, an earlier list by Habte and Manjunath (1991), who tried to categorize different crops in terms of responsiveness (dependency), and meta-analyses by Lekberg and Koide (2005) and Hoeksema *et al.* (2010) we can conclude the following:

1) Ten major food crops of the world (maize, wheat, rice, potato, sweet potato, cassava, soybean, sorghum, yam, plantain) are all responsive to AM.

2) The majority of all published studies show a positive response to AM. Only 10-20% of the studies did not show a response or showed a negative response. Cases of negative response (parasitism) have received substantial attention in the agronomic literature and have given rise to the often-expressed opinion that in highly fertilized agricultural systems (especially with P fertilizers) mycorrhiza is not only wasteful but often detrimental because costs (carbon use by the fungus) outweigh the benefits (increased access to nutrients and/or secondary benefits).

3) Mycorrhizal responsiveness can often be predicted from plant traits.

Plants with relatively thick unbranched roots with few root hairs (e.g., species of *Allium* like onion or leek, or cassava) are much more responsive to mycorrhiza than species with thinner, more branched roots with many root hairs (most cereals). N_2fixing legumes, which have a higher demand of P than most other plant species, are also quite responsive to mycorrhiza.

4) Only very few annual crops are non-mycorrhizal as members of the *Brassicaceae* (cabbage and canola) or *Amaranthaceae* (spinach and sugar beet) families. Inclusion of these crops in rotations likely reduces mycorrhizal inoculum potential, having adverse effects on subsequent crops that are dependent on mycorrhiza. Especially with *Brassicaceae*, it is likely that reduction of mycorrhizal inoculum (also due to toxic compounds produced by those plants) can have large carryover effects.

5) Some plants that have the ability to form mycorrhiza do not show them under conditions that are unfavourable for the fungi. Flooded rice for instance is poorly colonized by mycorrhizal fungi, but rice growing under more aerobic conditions (including rice grown under the system of rice intensification is often well-colonized and shows growth response to mycorrhiza (Watanarojanaporn *et al.,* 2013). Agricultural practices as intensive tillage, use of certain fungicides can reduce mycorrhizal inoculum. It is likely that such agricultural practices over ecological and evolutionary time scales have adverse effects on plants and mycorrhizal fungi (Johnson, 1993; Kiers *et al.,* 2002; Verbruggen and Kiers, 2010), especially in countries with highly industrialized agriculture. A study in Kenya, on the other hand, showed a much smaller response in mycorrhizal fungal species composition due to tillage and fertilizer application (Muriithi-Muchane, 2013).

6) As almost all plants are normally mycorrhizal, the presence of mycorrhiza should be assumed for an unknown plant.

7) Tree crops are always mycorrhizal.

11.4 Crop Genetic Variation in Mycorrhizal Responsiveness

Tawaraya (2003) reported that wild ancestors and relatives of cultivated plants often show a higher responsiveness to mycorrhiza than crops. Similar observations have been made for older crop cultivars compared to recently released crop cultivars. Such observations have given rise to the hypothesis that plant breeding under conditions of high soil fertility have inadvertently selected against the ability of plants to associate with mycorrhizal fungi and to benefit from the mycorrhizal symbiosis, and that therefore ecological agriculture needs to reconsider the genetic basis of crops in order to breed them for higher mycorrhizal responsiveness under less nutrient-rich conditions. Crops that show genetic variation in responsiveness (tested by growing different cultivars under uniform soil with the same mycorrhizal fungus) include: chili pepper, papaya, chickpea, oil palm, strawberry, soybean, barley, onion, welsh onion,peanut, oat, peach palm, tomato, plantain, tobacco, olive, common bean, pea, rice, millet, sorghum, wheat, cowpea, maize and breadfruit tree. However, research on possibilities to breed plants for increased responsiveness has indicated several problems.

(i) Assessment of responsiveness (a ratio of performance of mycorrhizal plants relative to non-mycorrhizal plants) is problematical, because high responsiveness of a cultivar can be both due to good performance in the mycorrhizal condition and also to poor performance in the non-mycorrhizal condition. Consequently, breeding for yield stability (including yield stability with variable amounts of mycorrhizal inoculum) can increase performance of non-mycorrhizal plants, which then translates into lower mycorrhizal responsiveness.

(ii) A search for molecular traits as quantitative trait loci (QTLs) that correlate with mycorrhizal responsiveness did indicate that many of those QTLs are also indicative for root traits as specific root length (SRL), root hair length and frequency, root biomass, etc.) and that direct breeding for root traits may be more efficient than breeding for responsiveness.

(iii) It is likely that a comparison of different cultivars would show extensive genotype×soil, genotype×AM fungus, and genotype ×soil×AM fungus interactions, making direct application of that knowledge for field management difficult. Critical reviews on the topic of the possibilities of plant breeding for increased mycorrhizal responsiveness have been provided by Sawers *et al.*, (2010) and Scholten and Kuyper (2012). These papers also provide references to the earlier studies that formed the basis for the hypothesis that modern plant breeding has inadvertently selected against mycorrhizal responsiveness.

11.5 Specificity-Host Preference

It is well known that VAM fungi are not host specific. However, the degree of VA mycorrhizal colonization and its effects can differ with different host-endophyte combinations. Plants like soybean and maize appear to be more mycorrhizal than other crops like potato or bean (Schenck and Kinlock (1974). Woodhouse (1975) reported that phosphorus deficiency in a host cell results in the release of a signal, possibly in the form of phosphates that disrupts the action of protective host lectins. The fungi then invade the cells and the phosphate deficiency is corrected. Thereupon, lectins resume activity preventing pathogen entry. One theory is that phosphate deficiency includes membrane leaked which in turn favour fungal invasion (Ratnayke *et al.,* 1978; Graham *et al.,* 1981). This explains why most plant species are susceptible to VAM infection. However, VAM do not normally exhibit any species or strain specificity to a particular host plant, except, perhaps some host preferences. The extent to which a specie or variety is mycorrhizal is determined by the background of the host, that is the kind of root exudates of the host plant or other barriers to VAM infection. In tree crops very few reports are available in this aspect. Since the early endophyte screening trials, it has become obvious that VAM fungi differ greatly in their symbiotic effectiveness. Though a particular VAM fungus can infect and colonize many host plants, it has a preferred host which exhibit maximum symbiotic response when colonized by a particular VAM fungus (Reena and Bagyaraj, 1990). The tailoring of plant-fungus combinations for maximum effect may lead to developing plants that will form association with certain specific VAM fungi (Gianianazzi *et al.,* 1989). Another form of specificity is ecological specificity. Under controlled conditions, it may be possible to get mycorrhizal establishment in plants. However, when the same symbionts are introduced in the field conditions, mycorrhizal infection may be retarded or prevented in certain areas while it may do better at other places. The ecology specificity does operate in nature which influences the prevalence of indigenous dominant VAM flora. Koske (1987) suggested that selected of fungal flora is not so much dependent on nutrients, organic matter or other variable as it is on the climate.

11.6 Host Relationships

AMF receive carbon from the host plant in return for nutrients, water and protection against toxic metal levels and pathogens. However, despite the perceived mutualistic symbiosis of AMF and plants, negative growth responses of AMF on plants have been repeatedly reported (Johnson *et al.,* 1997). Hoeksema *et al.* (2010) reported that around 20% of all studies did not demonstrate an improvement of plant performance and a part of these even resulted in mycorrhizal plants being smaller than non-mycorrhizal plants. Such studies have given rise to the concept of carbon costs of mycorrhizal symbiosis. The underlying model assumes a carbon for nutrient trade; and with increasing nutrient availability and decreasing costs for the plant to acquire the nutrients, the final balance becomes negative (Grimoldi *et al.,* 2006). This phenomenon, negative mycorrhizal responsiveness, has been called 'parasitism' (Johnson *et al.,* 1997). Causes of negative responsiveness are sometimes used as an argument why mycorrhizal associations are unimportant in intensive agriculture. However, carbon costs have seldom been quantified through measuring photosynthesis and carbon allocation to the fungus. Kaschuk *et al.* (2009) observed that at the same internal nutrient concentrations of nitrogen and phosphorus mycorrhizal plants have higher photosynthesis rates, a process called sink stimulation by microbial root symbionts. The same phenomenon has been demonstrated for rhizobia. The increases in carbon gain were calculated as roughly equivalent to the carbon used by the fungus (around 10% of photosynthesis), suggesting that the mycorrhizal symbiosis might often come 'for free.' The debate of the usefulness of the mutualism-parasitism continuum is still ongoing (Johnson and Graham, 2013).

Negative growth effects may be due to other factors than carbon use by the fungus (Grace *et al.,* 2009). They include nitrogen immobilization by the fungus, suppression of root exudates, changes in hormonal balance, root architecture and root to shoot ratio. In the wheat belt of SE Australia, wheat varieties with low AMF colonization show higher production than varieties with higher colonization, and this may be related to the suppression of exudate production in these P-poor soils (Ryan *et al.,* 2005). Finally, it has been suggested that the carbon invested by plants should not only be compared to the immediate return (of phosphate or other nutrients), but should also be looked at as long-term investment by the plant, ensuring a continuous, stable supply of mineral nutrients (Landis and Fraser, 2008).

11.7 Optimal Conditions

In natural ecosystem soils, pH is the most important determinant of AMF community composition, showing that environmental factors shape the AM composition (Dumbrell *et al.,* 2010). At small spatial scale within the soil (e.g., the rhizosphere), secondary factors regulate AMF community composition such as root exudates (which also depend on the internal nutrient status of the plant). Similar studies in agricultural systems point to large roles for phosphorus availability and degree of disturbance as determinants of the community composition of AMF (Oehl *et al.,* 2011). Rotations may also prevent dominance of fungi that cause a lower benefit for the host plant (Johnson *et al.,* 1992) through a process of negative feedback. Maintaining mycorrhizal diversity is also important because of the suite of other ecosystem services provided by the symbiosis (Gianinazzi *et al.,* 2010; Fester and Sawers, 2011). More important than species composition is the presence of sufficient inoculum at the start of the growing season. Smith and Smith (2011) reported that high density of AMF inoculum in soils is important to achieve crop colonization by AMF and subsequent increases in growth and yield. AMF inoculum of agricultural soils is encouraged by minimum or zero tillage and break crops that associate with AMF, avoidance of long periods of bare fallow, burning of crop residues or frequent cropping with non-host species such as members of the Brassicaceae or Chenopodiaceae (e.g., canola or beet) families (Abbott and Robson, 1991; McGonigle *et al.,* 2011). Negative impacts of long bare fallows on subsequent growth of flax in

Australia were shown to be caused by reduced mycorrhizal potential (Thompson *et al.,* 2013). Negative effects of soil disturbance on mycorrhizal fungi and hence early uptake of nutrients have been shown for maize in Canada (Miller, 2000) in the earliest growth phase, where disturbance resulted in early phosphorus deficiency despite adequate fertilizer application, and for legumes where disturbance resulted in lower colonization and lower amounts of nitrogen fixed by soybean (Goss and De Varennes, 2002).

AMF colonization of the host plant is also impacted by soil fertility and often fertilizer addition (especially phosphorus addition) will reduce AMF colonization (e.g., Abbott and Robson, 1982; Johnson, 1993). Addition of phosphorus fertilizer impacts the mycorrhizal fungus both directly (via the soil) and indirectly (through the plant), while there is plant control over mycorrhizal colonization by the plant through the internal phosphorus status (and such control also regulates competition among different mycorrhizal fungi), it is likely that a major effect of phosphorus is due to the direct negative effect. It is likely that foliar application will have a less negative effect on the symbiosis, although the topic has not often been studied. Not surprisingly therefore, a role for and the benefits of AMF symbiosis in fertilizer-intensive agriculture has been questioned (Ryan and Graham, 2002). These authors pointed out a larger role for mycorrhiza in the forms of agriculture that rely more on organic amendments and less on inputs of mineral fertilizer, such as organic agriculture in the northern hemisphere (Oehl *et al.,* 2004). However, the relevant question is not whether mycorrhiza inoculum addition can increase yields above yield increases due to fertilizer, but whether similar yields can be obtained but with lower fertilizer amounts, because the mycorrhizal symbiosis increases the phosphorus uptake efficiency. From that perspective the negative effects of unbalanced fertilization (which can also cause secondary deficiencies in micronutrients such as zinc), disturbance and use of fungicides need attention. It has been claimed that these human practices select over ecological and evolutionary time scales for mycorrhizal fungi that are on average less beneficial for the plant and/or provide less other ecosystem services (Kiers *et al.,* 2002; Verbruggen and Kiers, 2010). Such effects have indeed been shown in Europe, but far less so in a study in Kenya (Muriithi-Muchane, 2013). However, in European agricultural fields cases have been reported where the higher mycorrhizal inoculum from organic fields reduced nutrient losses, but this ecosystem service was traded off against crop productivity (Verbruggen *et al.,* 2012a).

11.8 Screening and Selection of Efficient VA Mycorrhiza

Some of the commonly used spore types are *Glomus fasciculatum, G. mossae, G. etunicatum, G. tenue* and *Gigaspora margarita.* The fungi selected for inoculation into agriculture soils must be able to increase uptake of plant nutrients and persist in soil. The colonization, efficiency, survival etc. in the rhizosphere and in soil are important. Govinda Rao *et al.* (1983) screened several fungi for best symbiotic response using host tested in pot, microplot and field trial. It is important to add same number of propagules of different fungi based on most probable number (Sieverding, 1991).

11.9 Role of VA-Mycorrhizal Fungi in Crop Growth

VAM fungal hyphae proliferate both outside and inside the root, with internal hyphae differentiating into arbuscules and vesicles. The arbuscules are formed only on the internal hyphae in response to contact with host corticalcells; they are sites of nutrient exchange between the plant and the fungus. Vesicles which contain high concentrations of triglycerides at maturity are thought to serve as storage and survival organs.Benefits from VAM symbiosis accrue because VAM hyphae extend beyond the root hair and increase the absorptive surface area of the root.

11.10 Uptake of Other Nutrients than N and P

Uptake of other nutrients is variable and affected by nutrient limitations (Lambert *et al.*, 1979). Zn uptake is increased by AMF (Jansa *et al.*, 2003; Gao *et al.*, 2007), and also K, S, Ca, and Cuuptake has been shown to be increased through the arbuscular mycorrhizal symbiosis (Rhodes and Gerdemann, 1980; Marschner and Dell, 1994;). The mycorrhizal contribution to enhanced uptake depends on whether the uptake of nutrients is mainly determined by mass flow or by diffusion: uptake of nutrients that react strongly with the solid soil phase and where diffusion is the main pathway for uptake is strongly enhanced by the mycorrhizal symbiosis (next to P, also Zn, Cu, to a smaller extent K and ammonium), whereas for nutrients where mass flow is the main pathway (nitrate, sulfate, calcium) the mycorrhizal contribution is more limited. The importance of mycorrhiza for Zn uptake is also important those cases where phosphorus deficiency is remedied by high P fertilization and where high P doses result in reduced mycorrhizal activity. In such cases, Zn-uptake could be reduced, which together with the dilution effect could create strong Zn-deficient plants (Lambert *et al.*, 1979). Also Zn availability in the plant product (cereal grain) could be reduced due to zinc-phytic acid interactions. The mycorrhizal symbiosis can possibly alleviate such forms of Zn deficiency for human nutrition (Ryan *et al.*, 2008).

11.11 Phosphate Nutrition

VAM enhances plant growth by improving mineral nutrition (Hayman,

1982; Barea and Azcon-Aguilar, 1983; Smith and Gianinazzi-Pearson, 1988). The available evidence suggests that VA mycorrhizae have significant effect on phosphorus uptake by plant. Greater soil exploration by mycorrhizal roots as a means of increasing phosphorus uptake is well known. In phosphate deficient soils immobile phosphate ions develop a phosphate depletion zone around the root. The hyphae reach beyond this zone and directly translocate nutrients from the soil to root cortex (Hayman, 1983) in arbuscules where phosphorus is transferred to plant cell in exchange for carbon. Fungal hyphae are also known to absorb phosphorus from lower concentration compared to non-mycorrhizal roots. Tracer techniques using ^{32}p labelled phosphate in the experiments showed that VAM hyphae obtain their extra phosphate from the labile pool rather than dissoluble phosphate (Raj *et al.,* 1981). Thus the increased uptake of phosphorous by plant occurs in three steps such as absorption of phosphate by hyphae from outside to internal cortical mycelia and finally the transfer of phosphate to cortical root cells, ready to be used by the plants. Although genotypic and environmental factors influence symbioses, a prolonged, stable and compatible relationship between plants and VAM is based primarily upon the transfer of phosphorus from symbiont to the host that has a net phosphorus deficit (Barea, 1991); one within the root, transfer takes place after the digestion of fungal arbuscules or hyphae. The increased plant growth inoculated with VAM fungi is also due to greater availability of other plant elements like zinc, molybdenum, copper, potassium, sulphur, iron, manganese etc. Thus the efficiency of an endophyte will depend on the extent of mycelial spread in the soil and the number of connecting hyphae.

11.12. VAM and Nitrogen Nutrition: The Efficiency

VAM effectiveness has been found to decrease as a result of fertilizer application in soil (Alexander and Fairley, 1983; Pareek *et al.,* 1990). However, VAM improves growth, nodulation and nitrogen fixation by *Rhizobium* and *Frankia* (Barea and Azcon-Aguilar, 1983, Hayman, 1986). VAM inoculation also increased nodulation and nitrogen content and grain yield of pigeon pea. The effect was equivalent to *Rhizobium* inoculation. The effect was more pronounced with combined inoculation of *Rhizobium* and *G. fasciculatum* (Singh, 1996). VAM increases nitrogen nutrition by different mechanisms as :

- Improvement in a symbiotic nitrogen fixation by supplying phosphate and molybdenum to azotrophs.
- Direct uptake of combined nitrogen by VAM fungi, and
- Facilitates N-transfer, a process by which a part of nitrogen fixed by nodulated plant benefits the non-N-fixing plant nearby (Newman 1985; Allen, 1988).

11.12.1 Plant Hormones and Water Deficits

Allen *et al.* (1982) showed that VA mycorrhiza directly affects the level of plant hormones such as cytokinins and gibberellin-like substances. The beneficial effects of VAM fungi have been reported under drought and saline conditions (Nelson and Safir 1982; Peiffer and Bloss, 1988; Singh, 1991) The important feature is that in drought conditions a mycorrhizal root has ability to tap additional water sources unavailable to non-mycorrhizal plant roots (Allen and Boosalis, 1983) Such effects are apparent under phosphate limitation as additional phosphate to non-mycorrhizal plants boosted their performance under drought and salinity (Nelson and Safir, 1982). A commercial inoculum could be a substitute for chemical fertilizer (Skujins and Allen, 1986).

11.12.2 Protection against Pathogens

Certain mycorrhizal association benefit plant health by suppression of pathogens like *Phytophthora, Pythium, Rhizoctonia* (Duchesne *et al.,* 1987) and *Fusarium oxysporum* (Singh and Singh, 1995). The inoculation of VAM also reduced the reproduction and development of root knot nematode, *Meloidogyne incognita* (Sharma *et al.,* 1995). The mechanism involved for VAM efficiency may be nutrient competition, a barrier effect caused by sheathing of root in fungal mantle or production of antibiotics effective against pathogens (Sylvia, 1983, Duchesne *et al.,* 1988), Disease suppression may also result from activation of defense mechanism (Sylvia and Sinclair, 1983) Identification of these facts from plant host and VAM fungus responsible for disease suppression may lead to genetic manipulation of VAM fungus and the plant host to maximize disease resistance potential of the host.

11.12.3 Other effects

Analytical and physiological studies have shown that mycorrhizal plants have increased rates of respiration, photosynthesis and higher levels of sugar, amino acids, RNA etc., and larger and or more number of chloroplasts, mitochondria, xylem vessels motor cells etc. (Hayman, 1983, Dyaladoss *et al.,* 1988). Changes in the root exudation and altered rhizosphere organisms beneficial to plant growth may result because of colonisalm of root by VAM fungi. Mycorrhizal colonization also augments the population of introduced inoculants like *Azotobacter, Azospirillum, Rhizobium* and PSM (Sattar and Gaur 1989, Gaur and Rana, 1990).

11.13 Interaction of VA mycorrhiza and PSM

Symbiotic micro-organisms such as VAM and *Rhizobium* are known to improve plant growth and nutrition much more efficiently as compared to free-living micro-organisms (Brown, 1974; Hayman, 1975). The mycorrhizal plants, particularly in low phosphate soils, are more efficient in phosphate uptake than non-mycorrhizal ones (Mosse, 1973) and in such soils satisfactory nodulation and nitrogen fixation are dependent on an adequate supply of phosphorus (Schreven, 1958). Hence, there must be some synergistic interactions between *Rhizobium* and VAM and such interactions have been described (Mosse *et al.*, 1976; El-Giahmi 1974).

Plant inoculated with VAM and PSB together were also able to mobilize significantly higher amounts of phosphorus than deficient soils (Barea *et al.*, 1975; Azcon *et al.*, 1976,). PSB are known to produce substances like auxins, gibberellins and cytokinins (Barea *et al.*, 1976, Sattar and Gaur 1987). Azcon *et al.*, (1978a) tried to differentiate the role of PSB cells, rhizobial cells and their cell-free supernatants on mycorrhizal infection and studied the influence of VAM and PSB on nodulation and nitrogen fixation by *Medicago sativa.* They observed positive interactions between the PSB and VAM assayed on the formation of VA symbiosis.The cell-free supernatant of PSB had a direct influence on the infection process, giving the highest percentage of VAM infection. Further, the cell-free supernatants of *Rhizobium* and PSB showed a positive influence on nodule formation by rhizobia and also exhibited a stimulatory effect on the growth of mycorrhizal plants. In turn, mycorrhizal plants were found to produce significantly more nodules than non-mycorrhizal plants. The inoculation of *Rhizobium* and PSB together gave highest yield and maximum N and P absorption by plants (Azcon *et al.*, 1978b). These observations show definite synergistic interactions among *Rhizobium,* PSB and VAM.

Interaction of PSM and VAM in Rhizosphere soil of Wheat Crop

The findings with wheat crop indicated that microbial inoculation augmented their rhizosphere count taken at 60 days of plant growth and at the harvest stage (Sattar and Gaur, 1984). The maximum count was recorded at the flowering stage. PSF were more than the PSB. The PSF were less when the soil was inoculated with *Glomus fasciculatum* and with PSB but the PSB were more with *G. fasciculatum* and less with PSF. Phosphatic fertilizers and farmyard manure increased their number considerably. At the flowering stage, rock phosphate at 90 kg P_2O_5/ha with farmyard manure recorded maximum PSF, but at harvest stage the maximum was obtained with farmyard manure only. Maximum PSB were observed with combined application of rock phosphate and superphosphate in farmyard manure series at both the stages of plant growth.

11.14 Ectomycorrhizal Fungi for Forestry

Ectomycorrhizal fungi associated with most trees are important for forestry production (*Pinales, Fagales, Myrtaceae* and *Dipterocarpaceae*) and are capable of nutrient uptake from both inorganic and organic recalcitrant sources (e.g., Read and Perez-Moreno, 2003; Hoffland *et al.,* 2004). Trappe (1977) and Marx (1980) reviewed methods and ectomycorrhizal species for inoculation of plantation tree seedlings in nurseries, although it is noted that if enough inoculum is present in the plantation field soil, nursery inoculation may not be needed. Nursery inoculation is especially needed in the case of exotics if compatible inoculum is lacking. Plantations of pine in the southern hemisphere were initially established after soil with ectomycorrhizal inoculum was introduced from the northern hemisphere (Netherlands, Great Britain). Subsequent plantations were established with soil with inoculum from existing populations. Currently, however, with increased inoculum availability, pines behave as invasive plants in the southern hemisphere.

Virtually all plantation trees rely on ectomycorrhizal fungi for their nutrition and maintaining a good source of inoculum and high ectomycorrhizal diversity is vital for plantation production (Perry *et al.,* 1987). Similarly as for agriculture, the diversity of ectomycorrhizal fungal communities in forest soils is affected by management practices such as harvesting, liming, fertilizing with nitrogen to enhance tree growth, choice of tree species and monoculture versus mixed species composition (Teste *et al.,* 2012), whereas AMF are sensitive to high phosphorus availability, ectomycorrhizal fungi are sensitive to high nitrogen availability. Both the direct pathway (negative effects of soil nitrogen on the symbiosis) and the indirect (tree-mediated) pathway regulate the fungal response to nitrogen. Like in AM the direct pathway is more important. A decline in ectomycorrhizal fungi has been noted in the last decades in forests saturated with N due to high anthropogenic N deposition over the past decades (Lilleskov *et al.,* 2011; Kuyper, 2013). Decline of ectomycorrhizal fungi and a poorer functioning of the symbioses results in poorer tree performance due to nutrient deficiencies and unbalances (especially the N:P ratios), increased sensitivity to drought and enhanced susceptibility to pathogens. Recovery of ectomycorrhiza under conditions of lowered nitrogen loads is a slow process.

11.15 Role of Ectomycorrhiza in Tree Species

Ectomycorrhizae rely on simpler carbohydrates for their nutrition since they are unable to use cellulose or lignified materials as carbon sources. Thus they depend on their hosts for sustenance. Many of them are unable to form fruit bodies when detached from their hosts (Laiho, 1970). It has also been observed that most species of Ectomycorrhizal fungi are not host specific. The ability of a single species of host to form mycorrhizae of more than one type is of wide occurrence under natural conditions. They may form either ectomycorrhiza or VAM. Pinus sp. may form ecto or ectendomycorrhiza.

Mycorrhizal fungi are perennial in nature and need a permanent nutrition from where they can translocate food to their growing tips. They use products of current photosynthesis. The non-specificity to particular hosts is of advantage to the fungus for its survival in that particular ecological site. It is quite common to find that the same plant host has a number of symbionts. Trappe (1962) reported that at least eight different species of fungi could form the mycorrhizal association with *Betula pendula* and *B. pubescens*. Giltrap (1979) extended the list to forty species of Basidiomycetes that form mycorrhizal association with *Betula*. Likewise, *Pinus sylvestris* and *P. strobus* were reported to be associated with more than dozen species of fungi belonging to several genera. However, *Alnus* is one such plant where only five out of 30 or more fungi present formed mycorrhizal association (Molina, 1979).

Molina and Trappe (1982) reported sporocarp specific host associations. This could be restricted to a particular species involving narrow host range, a single genus or a wider group (wide host range). This has been, therefore referred to as ecological specificity. *Laccaria laccata* and *Pisolithus tinctorius* are conifer-specific fungi whereas *A. diplophoeus*is specific to *Alnus*. Rhizopogon fruits only with members of pinaceace. A large number of fungi form mycorrhizal associations in young plantations, but many fungi are eliminated as the plantations age. This may be caused due to shift in the root exudates, accumulation of litter and leachates from them, depletion of nutrients, progressive accumulation of phenolics and toxic compounds etc. These types of changes have far reaching influence, especially in replanting programmes when establishment of seedlings becomes problematic unless inoculated with appropriate symbiont.

11.16 Tree Responses to Ectomycorrhiza

Ashton (1956) reported that growth of *Eucalyptus regans*increased substantially in 17th months growth period when inoculated with *Meshophellia* sp. Pryor (1956) observed 88% increase in the growth of 3 different species of *Eucalyptus* after 6 months of inoculation with *Scleroderma* sp. *Laccarialaccata* promoted growth of *Eucalyptus globulus* seedling by more than 200 % (Burges and Malajczuk, 1989). Increased nutrients uptake due to ectomycorrhiza by tree plants was reported (Lee, 1982). Total N and P contents of *Pinus densiflora* seedlings inoculated with *P. tinctorius* increased by 188 and 144% respectively over control. Positive response of Caribbean pine to inoculation with *Rhizopogon leuteolus* was reported by Ekwebelam (1979). Similar results were obtained by inoculation with *P. tinctorius* and *L. laccata* (Lee, 1982 and Valdes *et al.,* 1983).

Quintos and Valder (1987) reported that 90% of the *P.tinctorius* inoculated pine seedlings were mycorrhizal compared to 60% in the uninoculated control.

Heinrich and Patrick (1985) obtained higher phosphorus uptake in *Eucalyptus* due to inoculation with ectomycorrhizal fungi. The composite culture of ectoand endo-mycorrhizae was more beneficial than single type of infections presents in forests trees. Combined effect of *G. mosseae* and *Paxillusinvolutus* increased dry matter yields and P content of hybrid poplars more than the either fungus (Lopazaguillon and Garbaye, 1989). In *Eucalyptus* VAM fungi were more prevalent on young seedlings where ectomycorrhizae took over as the plant aged (Chilvers *et al.,* 1987). However, both types of symbionts may be present simultaneously in the cortex of the same root. The VAM fungi colonized the inner part of the cortex whereasecto-mycorrhizae were found in the outer cell layer (Boundarga *et al.,* 1990). In *Eucalyptus* sp., *Glomus fasciculatum* and *P. tinctorius* form active and efficient mycorrhizae (Boudarga and Dexheimer, 1988).

11.16.1 Inhibition of Root Pathogens

Certain ecto-mycorrhizal fungi inhibited *Pythium* sp. *Rhizoctonia solani, Fusarium oxysporum* (Sasek and Musilek, 1968). *Scleroderma aurantium*has also been shown to reduce the incidence of disease caused by *Fusarium* and *Pythium. Suillus bovinus* was effective against *R. solani,*while *Amanita citrina, Lactarius helves* and *Russulla fragilis*were effective against *P. debaryanum. Fomesannosus* was inhibited by antibiotics produced by several ectomycorrhizal fungi (Sasek and Musilek 1968). Water-soluble inhibitory compounds produced by *Boletus bovinus,* a fungi symbiont of *Picea abies* (Norway spruce) protected the seedlings from *Fomesannosus* (Hyppel 1968a'b). Marx (1969a) found that *Leucopaxillus cerealis var. piceina* inhibited *Eannosus* slightly while severely inhibited *Cylindrocladium scoparium,* many species of *Pythium, Rhizoctonia, Phytophthora* and *Sclerotium bataticola.* However, it was not effective against *Fusarium oxysporum, Armillaria mellia, Rhizoctonia crocorum* etc.

11.16.2 Production of Antibiotic and its Role

The antibiotic produced by *Leucopaxillus cerealis* var. *peiceina* was identified as diaretyne nitrile, polyacetylene (Marx, 1969b). This could inhibit the growth of zoospores of *Phytopthora cinnamoni* at a low concentration of 50-70 ppm and could fill them at 2ppm. *Laccaralacata* inhibit sixteen species of *Pythium* and *Phytophthora.* Marx and Bryan (1969) reported that *Scleroderma,* and fungal symbiont of pine and pecan inhibited five species of *Phytophthora* species and four*Pythium* species which caused damage to feeder roots. Park (1970) reported that *Lactarius* sp., a symbiont on barswood *(Tibia americana)* inhibited feeder root pathogens like *Pythium irregulare* and *R. solani. Pinus resinota* seeds presoaked in the culture filtrate of *Lactarius* sp. has 93% germination whereas the control has only 7%. Marx (1970) observed that *Thelephoraterrestris* a widespread fungal symbiont on many trees inhibited *Pythium* sp. *Rhizopogonvinicolor,* a symbiont of doughlas-fir was reported to inhibit the growth of *Pythium* spp., *Fomusannosus* and *Poriaweirii.* Kuo and Tang (1981) reported that *Boletus* sp. and *Suillusgrevillei* reduced the mortality of pine seedlings caused by *R. solani and Fusarium oxysporum.* Sylvia and Sinclair (1983) observed that a diffusible substance produced by *Laccaria laccata* inhibited mycelial growth and delayed spore germination of *Fusarium oxysporum.* Natarajan and Govindasamy (1990) observed that *Suillusbrevipes* inhibited all the common root pathogens tested such as *Armillariamellea, Cylindrocladium parvum, C. scoparium, Fusarium oxysporum, F. solani* and *Rhizoctonia solani.*

11.16.3 Mechanisms of Protection

Stack and Sinclair (1975) considered antibiotics and competition as inadequate explanation for root protection of Douglas-fir seedlings by *Laccaralaccata* because the fungus did not inhibit *Fusarium oxysporum* in pure culture on media rich in nutrients, Park (1960) suggested that high levels of nutrients mask the antibiotic effect. Smiley and Cook (1973) and Elad *et al.,* (1980) noted that low vigour of the pathogens may reduce its tolerance to antagonism.

11.17. Strains and Functioning

Globally, more than 200 species of AMF have been described and named. However, current molecular ecological studies have indicated that the actual species number is higher. An rDNA based phylogeny indicates that currently some 350 virtual taxonomic units ('species') are known, a number that is likely to increase as more regions and plants are being studied (Öpik *et al.,* 2013). Virtually all members of the phylum Glomeromycota (the group that constitutes the AMF) can associate with plants. Most species are able, under controlled conditions, to associate with a large number of plant species, and that has given rise to the statement that there is no specificity in the mycorrhizal symbiosis. However, field investigations have shown non-random associations between plants and fungal species. Also differential benefit from a mycorrhizal fungus by different plants has been shown. While that may result in a classification of fungi as relatively co-operative or non-cooperative (Kiers *et al.,* 2011), it is not clear whether the underlying behavior is a genetic trait of the fungus or is context-dependent. An important study by Angelard *et al.* (2010) addressed this issue. These authors tried to generate genetic variation in the mycorrhizal fungal species

Glomus intraradices and were able to increase plant size (of rice plants) fivefold after inoculation of the 'best' progeny of that fungus. However, the same study did not show any genetic improvement of that fungus when inoculated with *Plantago lanceolata*. It remains unclear to what extent therefore fungal 'breeding' or 'selection' programs can contribute to a substantially higher mycorrhizal efficiency. As indicated above, the mycorrhizal symbiosis has several benefits (the symbiosis is called multifunctional as it does not only contribute to enhanced nutrient uptake, but also provides protection against pathogens, contributes to soil structure improvement, etc. Attempts have been made to link these differential benefits to the fungal phylogeny or to functional traits. It has been proposed that a major functional difference exists between species with relatively high intraradical colonization (supposed to make a major contribution to nutrient provision to the plant and pathogen protection) and species with relatively extra high extraradical colonization (with a larger role in improving soil structure) and that this distinction roughly separates to *Glomeraceae* from the *Gigasporaceae* (Hart *et al.,* 2001). However, the generality of that proposal has not yet been established. Attempts to come up with a functional classification of fungal traits of the AMF are in their infancy. Chagnon *et al.* (2013) classified functional traits and suggested they result in the distinction of three fungal strategies that coincide with the three major plant strategies recognized (ruderals, stress tolerators, competitors). Another classification of fungal species along the r-K continuum (separating fungal species with an annual life cycle with prolific spore production and a large capacity to regenerate from hyphal disruption through disturbance from those with a perennial life cycle with a lower spore production) is potentially relevant in the debate about agriculture based on annual crops or forms of agriculture of perennials crops (including agricultural systems that include perennial plants like shrubs and trees) but needs more systematic testing (De Carvalho *et al.,* 2010).

The relation between species (or functional) diversity and ecosystem functioning has also been debated for AMF (Kuyper and Giller, 2011). From the perspective of mycorrhizal application (or the use of microbial inoculants in general) the relevant implication is whether we should apply monospecific inoculum (of one best performing strain, e.g., the improved strain generated by Angelard *et al.* (2010) or whether a more diverse mixture of different mycorrhizal fungi is likely to be more beneficial. The literature in that respect has not yet come to a common conclusion. It should also be kept in mind that there could be potential trade-offs between fungal traits that result in largest beneficial effects to the plant and traits that allow the fungus to be competitively superior. Finally, because the mycorrhizal symbiosis is multi functional and different benefits are almost certainly the result of evolutionary trade-offs, a higher mycorrhizal fungal diversity is likely to be more beneficial, especially in systems of intercropping and crop rotations, because of fungal selectivity for certain plants and or specificity in fungal benefits to different plants (Kuyper and Giller, 2011). Verbruggen *et al.,* (2012a) noted a further trade-off between agricultural sustainability and crop productivity, where AMF, and especially those from organically managed agricultural fields, did not improve maize yield, but reduced phosphorus leaching.

11.18 Mass Production of VAM Fungi

The role of VAM in improving the growth and nutrition of plants has been established but the mass production for field application has been limited due to their inability to grow on artificial media. The most common technique for its limited production has been to grow symbiont with the host plant in pot culture (Mosse, 1953). The availability of the inoculum of VA mycorrhizae for field application is meager. Therefore, the propagation of axenic culture of these fungi may be used. However, *in vitro* cultivation under pure culture remains the most challenging task for the research workers (Johri and Mathew, 1989).

11.18.1 Conventional Method or Soil-based Inoculum Production

Culturing VAM fungi on living plants growing in disinfected soil has been the most frequently used technique for increasing the number of propagules. The conventional methods involve the use of starter inoculum/material and grow on soil in pots using suitable trap plants. Further to bulk up on mass scale after selecting potential VAM fungi the same can be propagated in pots followed by in raised beds (on-farm) under field conditions (Douds *et al.,* 2006).

11.18.2 Soilless Substrate based System

AMF can be cultured on plants in suitable soilless media with appropriate nutrients. Hoagland's nutrient solution is currently being used routine to grow AMF with inert materials in pots. Bark, calcined clay, expanded clay, perlite, soilrite and vermiculite have been used as inert substrates. Pre-colonized roots, spores and hyphae of AMF could be produced in hydroponics or aeroponic systems. Pre-colonized plants on sterile substrate that are needed for these systems. However, plants can be infected directly in the aeroponic system. In hydroponics inoculum can be propagated using the nutrient film technique by growing pre-colonized plants in a defined nutrient solution, which flows over the host roots.

Nutrient film technique (NFT) was adapted for the AMF inoculum production by Mosse and Thompson (1984). Culture host plants are placed on an inclined tray over which flows a layer of nutrient solution. In an aeroponic system (a fine mist of defined nutrient solutions), pre-colonized roots are suspended in air and are bathed in (Hung and Sylvia, 1998). Lee and George (2005) proposed a modified nutrient film technique for large scale production of AM fungal biomass with the help of improved aeration by intermittent nutrient supply, optimum P supply, and use of glass beads as support materials. This method of AM multiplication is free from any substrate and can be sheared which resulted in to very high propagules number.

11.18.3 Root-Organ Culture or Axenic Culture of AM Fungi

In vitro culture of AMF was achieved for the first time in early 1960s (Mosse, 1962). Since then various pioneering steps were aimed at axenic culturing of AM fungi. Becard and Fortin (1988) and Fortin *et al.* (2002) presented a detail evaluation of the root organ culture (ROC) technique and reported the basic improvements necessary for AM fungus colonization of roots. The root organ culture is a most attractive mass multiplication method for providing a pure, viable, rapid and contamination free inoculum using less space and has an advantage over the pot culture multiplication/conventional system (Fortin *et al.*, 2002; Tiwari *et al.*, 2003; Cranenbrouck *et al.*, 2005). The mono-specific strains available can be used directly as starting material for large scale inoculum production, a sole Petri dish culture being enough to generate several thousand spores and meters of hyphae within four months. The hairy roots formation occurs due to transfer of root inducing Ri-plasmid from the bacterium, *Agrobacterium rhizogenes*. This system offers a rapid, clean and more efficient method and opens up the avenues for AM fungi commercialization and field application.

Mycorrhizal inoculum can be applied in various forms. There are various formulations available and the active ingredients are typically spores, mycelium/ hyphae and colonized root pieces. The inert materials (inactive carriers), can be used as pot mix to grow and bulking up the AM inoculum. The important thing for growers is that AM formulations should be easy in application. Some of the inocula are available in active form (on-going symbiosis) where one has to take caution and strictly follow the instructions given on the packet. The inoculum should be applied in the root zone through layering technique, not on the soil surface. The inoculum applied or left on the surface will quickly die in absence of growing roots/host.

11.18.4 Axenic Culture of VAM

Surface sterilized clean spores (Hepper, 1985) are aseptically transferred to water agar and other media for suitable spore germination, Monoaxenic culture of VAM have been produced as whole plant cultures, excised root (root organ) culture or pure two member cultures (John *et al.*, 1981; Mosse, 1962). Infections were found to be the same as in case of whole plant systems. External mycelia grew vigorously, but it was not possible to subculture from external mycelium. New methods are being developed for mass production of VAM fungi which may be found in the right direction (Smith,1994; Walker and Vestberg, 1994 and Zeze *et al.,* 1994).

11.18.5 Solution Culture for Inoculum Production

Macdonald (1981) used autoclavable hydroponic culture systems for the production of axenic VAM. Surface sterilized spores germinated on agar were utilized for infecting seedlings and incubated in green house at 25^0C till lateral roots appeared. The hydroponic culture was superior to root organ culture (Nelson and Safir, 1982; Newman, 1985). The infected roots thus can be used for inoculation purposes.

11.18.6 Maintenance of VAM in Pot Culture

Rhodes grass *(Chlorisgayana)* is considered the best host, perlite: soil (1:1 vol.) mix to be the best substrate and use of pesticides such as captan, furadan at half of the recommended rates checked the growth of contaminants without any adverse effect on VA mycorrhizae. Thus this technique to produce clean mycorrhizal inoculum can be used for developing "starter cultures" for supply to farmers. The farmers can further multiply it is in their own fields for inoculating crop plants.

Inoculation methods

Various methods can be tested for introducing VA mycorrhizal inoculum in the field crops.

Transplanted crops

Seedlings are grown in sterilized and unsterilized soil inoculated with efficient selected VAM fungi in small nursery beds or other suitable containers which can be planted when mycorrhizal colonization has been established. This method has been used in important crops like chilli, finger millet, tomato, tobacco and wetland rice (Govinda Rao *et al.,* 1983, Secilia and Bagyaraj, 1994) horticultural crops such as citrus, mango, asters and marigold (Bagyaraj *et al.,* 1988) and forest tree species e.g. *Leucaena, Tamarindus indica, Acacia nilotica, Calliandra callothyrus* and *Casuarina equisetifolia* (Bagyaraj *et al.,* 1989, Reena and Bagyaraj, 1990 a, b and Vasanthakrishna *et al.,* 1995).

Directly sown plant

a) **Coating of seeds with VAM inoculum:** Seeds are coated with VAM fungi with an adhesive such as methyl cellulose. It is expected that the inoculum would stick to the seeds and subsequently infect the emerging radicals. It has proved satisfactory for large seeded plants such as citrus in field nurseries (Hattingh and Gerdemann, 1975).

b) **Mycorrhizal pellets:**The seeds are incorporated into the inoculum to form multi-seeded. These pellets (1cm in diameter) consist of soil inoculum from pot cultures stabilized with clay (Powel 1979, Hayman *et al.,* 1981) or peat inoculum.

c) **Fluid drilling:** Mixing of seeds and inoculum into uniform suspension containing 4% (w/v) methyl cellulose and then the slurry is applied in furrows. This has been reported to be successful under field conditions (Hayman *et al.,* 1981).

d) **Inoculation in furrows:**The inoculum is placed under or besides seeds sown in a furrow (Hayman *et al.*, 1981) which may be an effective method for inoculating field crops. In field experiments, mycorrhizal inoculum was placed below the seeds. Very high rates of inoculum varying from 0.8 to 167t/ ha with commonly used rates of 20 to 30 t would be difficult to put in practice due to economic and practical problems.

e) **Precropping:** Population of VAM propagules can be raised *in situ*by growing strongly mycorrhizal host plants and leaving the infected roots and the associated spores in the soil to infect the following crop, judicious crop rotations might be practiced as a means of manipulating rising of *in situ* VAM propagules.

11.19 Constrains for Commercialization

The yield of the crops would be augmented if the VAM inocula become available to farmers. Several agencies including private companies have tried to develop commercial VAM inoculants. Efficient strains have been identified (Duke *et al.,* 1986; Mahoney *et al.,* 1985). The culture methods have been developed and patented (Dixon and Buchesna 1988, Jones and Huktchison, 1988). Several inoculant formulations have been worked out and techniques for applying inoculants to nursery and field soil have been formulated and tested (Wood, 1991). Despite these developments, VAM has not been widely commercialized and used. The market for VAM inoculants exists for different agricultural and horticultural crops. The largest segment comprises of agronomic crops such as maize, soybean, wheat, cotton and others. Horticultural crops grown in fumigated soils and artificial soil mix provide another area for inoculation. At present their use is in agroforestry and reclamation of eroded soils and waste lands.

Mycorrhizal inoculants should be effective over a wide range of crops grown in diverse soils and agro climatic conditions for their widespread and routine use. The cost of the inoculation should justify the profits and benefits accrued to the inoculated crop. Some of the constrains such as ability of the inocula to complete with native mycorrhiza, variations in plant genotypes, phosphorus status of soils etc. may limit their use.

11.20 Mechanism of AM Symbiosis and Action

It is recognized that mycorrhizal association is a natural strategy that most plants have developed in their development process (Paul and Clark, 1996). Arbuscular mycorrhizal fungi are obligate biotrophs; their spores are the only plant-independent phase and they have an ability to germinate and arrest growth many times if plant derived signals are missing (Requena *et al.,* 2007). The process of the development of AM symbiosis varies with the plant and fungal species involved (Smith and Smith, 1990). AM spores germinate under appropriate water and temperature conditions, hyphae grow for 2-3 weeks and they respond to root present in their vicinity (Requena *et al.,* 2007). Also, root exudates from host root elicit hyphal growth due to presence of plant phenolic substances, in particular flavonoids (Nair *et al.,* 1991), some lypophylic molecules (Buee *et al.,* 2000) or strigolactones (Akiyama *et al.,* 2005). On contact with host root cell the fungal hyphal cells form appressoria (Paszkowski, 2006) that develop in to haustoria like structures (arbuscules) to penetrate the root cortical cell walls and form interface with host cytoplasm (Smith and Read, 1997). Here it undergoes extensive dichotomous branching in to a tree like fungal structure that may entirely fill the cortical region and provide an increased surface area for metabolic exchanges between the plant and the fungus. These arbuscules collapses after some days, leaving an intact cortical cell which is then able to host another arbuscules. The signals that induce intracellular fungal branching or arbuscules decay have not been identified (Paszkowaski, 2006).

AM fungi also produce vesicles which are structures that function as storage (lipids) and reproductive organs. They also interface directly with the soil by producing extra radical hyphae that may extend several centimeters out into the soil, which increases the potential of nutrient uptake, and possibly also water uptake (Auge, 2001). Except for species from the genera *Scutellospora* and *Gigaspora*, all AM fungi form intra-or intercellular storage organs, lipid-rich vesicles, to varying degrees in late phases of the symbiosis (Smith and Read, 1997). The arbuscules are the key features of AM and are responsible for nutrient exchange. They represent a dead end in the growth of AM fungi (Bonfante and Perotto, 1995), because they finally senesce and collapse after 4-10 days of symbiosis (Sanders *et al.,* 1977). The fungal structures are then degraded completely by the plant cell and the plant cell recovers its original morphology (Jacquelinet-Jeanmougin *et al.*, 1987). This way, cortical cells are able to allow a second fungal penetration and arbuscules formation. During colonization, the fungal arbuscules occupies a major portion of the plant cortex cell, but is separated from the cell protoplast by a part of the host plasma membrane, the peri-arbuscular membrane. This membrane completely surrounds the arbuscules, leading to up to a fourfold increase of the surface of the plasma membrane. Although it originates from the plant plasma membrane, the peri-arbuscular membrane exhibits different properties. In particular, phosphate transporters were shown to be located specifically in the peri-arbuscular membrane (Harrison *et al.*, 2002). Moreover, a high amount of $H^+_$AT Pase activity (Gianinazzi-Pearson *et al.*, 1991) accompanied by the highly acidic nature of the space separating plant and fungal plasma membranes has been found (Guttenberger, 2000). The space separating plant and fungal plasma membranes corresponds to a new apoplastic compartment and represents the symbiotic interface. It is continuous with the peripheral plant cell wall, but its structure differs from it (Peterson and Bonfante, 1994).

The mixture of primary plant cell wall components helps arbusculated cells to maintain their ability to synthesize and secrete cell wall material. This material, however, does not assemble further to build up a secondary wall (Peretto *et al.*, 1995). It has been accepted that the establishment of the symbiosis must be the result of a continuous molecular 'dialogue' between plant and fungus, as exerted through the exchange of both recognition and acceptance signals (Azcon-Aguilar and Barea, 1997). The result of this dialogue finally depends on the genome expression of both partners (Gianinazzi *et al.*, 1994). In exchange of the nutrients the fungus is supplied with photosynthates (about 20% of the plant assimilation capacity) and these assimilates are either respired or released in to the soil via transfer of sugar from the root to their endosymbiotic fungi, which later translocated this carbon in form of lipids and sugar into the external mycelium spreading in the soil (Bago *et al.*, 2003). As a result of higher carbon drain, microbial rhizosphere composition is altered.

11.21 Photosynthesis and Root Phosphatase Activity

11.21.1 Photosynthesis

The total photosynthates availability in the plant depends upon photosynthetic surface and the rate per unit surface area, minus respiration and photo respiration (Sinha *et al.*, 1988). ShantaKumari and Sinha (1972) observed that photosynthetic rate in mungbean, cowpea and chickpea attained a maximum value at flowering, it decline in post-flowering period when pod development commenced. Utilization of photosynthates and their translocation to different organs have been reviewed by many workers. Minchin and Patel (1973) showed that only 26% of the carbon-assimilated in the process of photosynthesis remained in the over ground part of the plant, 42% was transported to the nodule bearing roots and 32% entered the nodules directly. An adequate supply of carbohydrate to nodules is a major constraint in biological N_2 in legumes (Allison, 1965). In legume nodules, maximum rate of N-fixation are under conditions favorable to photosynthesis (Rainbird *et al.*,1983; Vijaylakshami, 1988; Panwar *et al.,* 1990).

Short term $^{14}CO_2$ labeling experiment indicate that carbon requirements of VAM and *Rhizobium* are supplied by the photosynthesis of host plant (Ho and Trappe, 1973; Cooper, 1984). The fungi convert host metabolites into specific fungal compounds, lipids and particularly abundant in mature arbuscules, vesicles and hyphae (Kinden and Brown, 1975; Bonfante-Fasolo, 1984). Glycogen granules are usually associated with these lipid containing structures and also found in young vacuolated hyphae and the finest arbuscules branches. The demand of host photosynthates depends upon the amount of fungus associated with the root and its metabolic rate. Mycorrhizal roots have higher respiratory activity than uninfected roots and infected root cells have higher cytoplasmic volume, number of mitochondria, soluble protein content and several enzyme activities (Oliver *et al.*, 1983; Cox and Sanders, 1974).

Mycorrhizal plants with low available P export up to 10% more photosynthates to their root than P sufficient non-mycorrhizal plants (Kucey and Paul, 1982; Koch and Johnson, 1984). This additional drain could result in reduction in yield, unless compensation mechanism operates. When P limits the rate of photosynthesis, the investment of carbon in fungal symbiosis leading to increased rate of P acquisition could in turn lead to considerably increased rate of photosynthesis. This would offset the additional carbon utilization in the tissue below ground. According to Pang and Paul (1980) the non mycorrhizal plants transferred 39% of the fixed ^{14}c beneath ground. The $^{14}CO_2$ produced by root-microbial system of the mycorrhizal *Vicia faba* was twice as great as that of the non-mycorrhizal plants. Kucey and Paul (1982) found that N-fixing nodules of non-mycorrhizal *Vicia faba* utilized 6% of the total photosynthates and those of mycorrhizal plants 12%. Nodulated plants with or without mycorrhizal infection fixed about 47% higher $^{14}CO_2$ than plants without microbial symbionts (Harris *et al.*, 1985). At early stages of mycorrhizal and *Rhizobium* infection the carbon drain, not yet compensated for, may induce a transitory negative growth response.

Bethlenfalvay *et al.* (1982) observed that during the first ontogenic stages of the association, VAM fungus was parasitic as a result of carbohydrate demand on the host by the endophyte, whereas at later stages mycorrhizal fungus and soybean expressed associated mutualistically. Negative correlation of root hexose sugar concentration with VAM colonization, nodule activity and root P-content were observed, when soybean plants inoculated with one of the endophyte after the other has colonized the host root. Pacovsky *et al.* (1986) showed that rapid colonization by a VAM fungus would result in an enhanced P-status, but would lower the level of the carbohydrate in roots. *Bradyrhizobium* strains unable to store P but capable of storing as poly-β-hydroxyl butyrate could have a competitive advantage under these conditions. The elevated CO_2 has been reported to alter the partitioning of carbon between the above and below ground components of the forest ecosystem. Rygiewicz *et al.* (1994) reported that the elevated CO_2 (525 and 700 moles mol^{-1}) concentration increased the rate of colonization, number of mycorrhizae and the presence of hyphae in the soil. The higher degree of mycorrhizal infection is likely to be associated with higher carbohydrate level with fine roots of the higher CO_2 treatments.

11.21.2 Root Phosphatase Activity

Helal (1990) studied the varietal difference in root phosphatase activity as related to the utilization of the organic phosphate. Bean varieties showed significant difference in their uptake of phosphorus from inocitolhexaphosphate with a clear dependency on the phosphatase activity of their roots at pH-5. The result suggests that root phosphatase activity is a significant factor of nutritional efficiency under limited mineral phosphorus supply. Liu *et al.* (2004) studied the Rhizosphere effect and root growth of two maize genotypes with contrasting P efficiency at low P availability. The result showed that P treatment increased root biomass, root to shoot ratio, lateral root length and acid phosphatase activity in roots and on the root surface but reduced root exudation of organic acids and pH in Rhizosphere for both genotypes. It was concluded that efficient use of P in calcareous soil is related to large root system, greater ability to acidify the Rhizosphere and positive response of acid phosphatase production and excretion to low P conditions.

Mc Lachlan and De (1982) studied acid phosphatase activity of intact roots and phosphorous nutrition in plants and its relation to phosphorous garnering by wheat and a comparison with leaf activity as a measure of phosphorous status. In actively growing plants leaf phosphatase activity was better related to total phosphorous uptake than was root phosphatase activity. At the second stage, where yields continued to increase but there was little change in the phosphorous content of the plant other than it's redistribution between component parts, root phosphatase was better related to total phosphorous uptake. Grierson and Comerford (2000) studied the non-destructive measurement of acid phosphatase activity in the Rhizosphere using nitro cellulose membranes and image analysis. Over 95% of the acid phosphatase activity of the root system of *D. sessilis* was associated with cluster roots and between 20 to 32 % of the active root surface. About 26% of the acid phosphatase activity of the root system of *P. taeda* was associated with mycorrhizal roots and unsupervised white root tips and less than 10% of the root surface was active, irrespective of root type.

Mc Lachlan (1980) studied the acid phosphatase activity of intact roots and phosphorous nutrition in plants. The phosphatase activity of wild and cultivated progenitors of modern wheat's ranged in optical density from 15.5 to 45.4 per g fresh weight. All the cultivated species had low phosphatase activities. Ohtsuka and Saka (1988) studied the undestructive measurement of acid phosphatase activity in the root of rice seedlings. The enzyme activity of the whole roots including seminal and crown ones at the 4th leaf stage plant was higher than at younger or older seedlings stage. Tarafdar and Marschner (1994) studied the phosphatase activity in the Rhizosphere and hyposphere of VAM with wheat supplied with inorganic and organic phosphorous. In the root compartment acid phosphatase activity was much higher than alkaline phosphatase activity and both were slightly enhanced by mycorrhizal infection. Throughout the hyphae compartment phosphatase activities were distinctly higher in the presence of mycorrhizal plants particular with a supply of organic P which also increased the percentage of infected root length. Phosphatase activity was strongly correlated with hyphae length which was highest with inoculation site from the root surface. Saker *et al.* (1999) studied the relationship between phosphorous fractions phosphatase activity and fertility in three tropical rain forest soils. Root phosphatase activity was greater for plants grown in soil low in available phosphorous.

Fernandez and Ascencio (1994) studied the acid phosphatase activity in bean and cow pea plants grown under phosphorous stress. Acid phosphatase activity was above the values reported for these species under P-stress; however, APA in these legumes appears not to be inducible by the low P concentration level used in the study. A higher APA was found in young leaves, and the expression of APA also showed intra specific variation.Root phosphatase activity was enhanced with the application of AM fungi at 60DAS in urdbean (Preeti, 2011). Johnson *et al.* (2005) studied the liming and nitrogen fertilization effects, phosphatase activities, microbial biomass and mycorrhizal colonization in upland grassland. A positive linear relationship was observed between root phosphatase activity and P concentration of plant shoots. Rao *et al.* (1999) studied the influence of organic amendments on the development of ectomycorrhizae and their efficiency in P uptake and seedlings growth of *Pinus kesiya*. Root phosphatase activity was more with mycobionts inoculated activity of root was observed in seedlings inoculated with *S. leteus* and grown in grass litter amended soil.

Kumar and Chandra (2008) observed the influence of PGPR and PSB on *Rhizobium leguminoserum* bv.*viciae* strain competition and symbiotic performance of lentil. Combined inoculation treatment of *Rhizobium* sp. +PSB+LK-884 was significantly superior to *Rhizobium sp.* alone in plant dry weight at different intervals and P uptake by grain and straw. Singh and Shrivastava (1985) studied the nodular physiology of urd bean as affected by urd bean mosaic virus and effect on some enzymatic activity. The enzymatic activities were found higher in the nodules of soil grown plants than in the nodules of plants grown in sand.Tarafdar and Marschner (1994) studied phosphatase activity of VAM fungi in the Rhizosphere (root compartment) and hyphosphere (hyphal compartment) in mycorrhizal and non-mycorrhizal wheat (*Triticum aestivum*) supplied with phosphorus (200 mg P Kg^{-1} soil), either in organic (Na-phytate) or inorganic [$Ca(H_2PO_4)_2$] form. They found that in root compartment, acid phosphatase activity was much higher than alkaline phosphatase activity, and both were slightly enhanced by mycorrhizal infection.

The high affectivity for P acquisition of the external hyphae is related to (i) formation of polyphosphatase in the hyphae, thus maintaining low internal P concentrations, (ii) the small hyphae diameter leading to a relatively larger soil volume delivering P per unit surface area, compared to the root surface area (Jung and Classen, 1989) and a correspondingly 2-6 times higher P-influx rate per unit length of hyphae (Jakobsen *et al.*, 1992),and (iii) production of extra-cellular acid phosphatase which catalyzes the release of P from organic complexes in the soil. Rates of translocation of phosphorus by *Glomus mosseae* in association with various host plants are in the range of 0.1- 2.0×10^{-9} mol cm^2s^{-1} (Cooper and Tinker, 1981) and are a temperature sensitive process. This is stopped by cytochalasin-B (inhibit cytoplasmic streaming), and reduced when transpiration of the associated plants is prevented. Translocation of P occurs down a concentration gradient between P source in the external hyphae and a sink in the roots. Loading and unloading of P from hyphae vacuoles, coupled with cytoplasmic streaming, are important in maintaining the high rates of translocation.

Acid phosphatase enzyme secreted mainly by the plant roots, causes hydrolysis of organic phosphates present in the soil as phosphate esters resulting in the release of inorganic P for plant uptake. Exudation of phosphatases increases when plants are grown under phosphorus stress (Radersma and Grierson, 2004). It was reported that when wheat grown in an acidic P-deficient soil amended with Fe-P, the P-efficient genotypes had a greater acid phosphatase activity in the rhizosphere than the inefficient genotype, with phosphatase activity correlating positively with growth and phosphorus uptake (Marschner *et al.*, 2006). In an another study involving wheat and rye under phosphorus stress, showed that wheat plants exuded more acid phosphatase into the rhizosphere followed by rye and triticale. A significant positive correlation was also found between acid phosphatase activity (both extra and intra cellular) and root, shoot and total plant dry weight (Pandey, 2006).

11.22 Stimulation of Assimilative Activities Essential For Plant Growth

Nitrate reductase (NR) is the first enzyme in the NO_3-assimilation pathway and probably represents the rate-limiting step in this process (Campbell, 1988). However, NR levels are drastically decreased by adverse environmental conditions such as drought stress (Sanchez and Aguirreolea, 1993). AM fungi have the gene set for assimilatory NO_3 reduction (Kaldorf *et al.,* 1994), and NR activity has been located in arbuscules-containing cells (Kaldorf *et al.,* 1998). In addition, levels of NR activity are higher in AM plants subjected to drought stress than in the non-mycorrhizal ones.

11.23 Mycorrhizal Effect on Plant Roots

Unlike *Arabidopsis*, more than 80% of higher plants associate with mycorrhizal fungi, which elicit profound changes in the root morphology of host plants (Hetrick, 1991). In particular, ectomycorrhizae suppress root elongation and induce dichotomous branching of short lateral roots, culminating in the formation of coralloid structures resulting from higherorder dichotomous branching. All of these anatomical structures are variable depending on the plant and fungal species.Once the fungus is established, root branching is suppressed, which makes the plant more dependent on the nutrients provided by the fungus (Hetrick, 1991; Price *et al*., 1989). Whether this modification of root system architecture (RSA) is a direct consequence of symbiosis or an indirect effect of improved nutrient status of the plant is not clear. However, it appears that symbionts can triggers RSA changes by promoting lateral root initiation very early in the interaction (Harrison, 2005). Moreover, the maize mutant *lrt-1* normally lacks lateral roots, but displays extensive lateral root development following inoculation with the mycorrhizae *Glomus mosseae* (Paszkowski *et al*., 2002). Notably, many micro-organisms that interact with plants can produce plant hormone analogs. Thus, symbiotic association might employ hormone signaling pathways to regulate RSA.

11.24 VAM and Plant Growth

As VAM are ubiquitous in nature, they are commonly associated with most plants important in agriculture, horticulture and forestry. In agriculture, potential hosts occur in the economically important families *viz. Gramineae* and *Leguminoseae*. It is also found in many tropical plantation crops, trees, walnuts and all temperate fruit trees. Although VAM being omnipresent, the establishment of successful entry points on host trees, establishment, development and spreading of internal and external hyphae and its response to various kinds of hosts is dependent on prevailing fungi, plant and environmental factors interaction.

Mycorrhizal development is greatly affected by the light received by the shoot of the host plant. Low irradiation and short day lengths can reduce the development of VAM (Daft and El-Giahmi, 1978; Hayman, 1974), but which also markedly reduce plant growth in uninfected plants. Johnson, (1976) noted that shading reduces development of infection in shade tolerant species but not in light demanding species. Defoliation greatly changes the soluble carbohydrate content of grass roots and result in relation of mycorrhizal development (Daft and El-Giahmi, 1978). The effect of temperature (Smith and Brown, 1979; Hetric *et al.,* 1984), light (Koide, 1985;), pH (Abbott and Robson, 1985) and moisture (Ponder, 1983) have been well documented. In addition to above factors, metals, salinity, organic matter distribution in depth and aeration also play key role in establishment and development of VA-mycorrhizal fungi.

In a symbiotic relationship between host and VAM fungus, the former is autotrophic for its carbon requirement and the latter is heterotrophic. Being an obligate endosymbiont they have very little capacity of independent growth. Harley and Smith (1983) pointed out that growth responses to mycorrhizal inoculation will be determined by movement (translocation) of soil derived nutrients to the plant through the fungus and the consumption of carbon compounds derived from host by the fungus.The main agronomic significance of VAM lies in their ability to supply the plants with extra nutrients particularly phosphorus. Crush (1974) reported that endophyte host relationship varies between parasitism and mutualism depending on the available phosphorus levels. Maximal mycorrhizal effectiveness occurs when nutritional conditions for the host are sub-optimal but these become severely deficient and the fungus becomes ineffective.

VAM system involves three components as (i) the plant (ii) the soil and (iii) the fungus. Interactions between the three determine the mycorrhizal effect on plant growth. The establishment, development and functioning is greatly affected in highly fertile or heavily fertilized soil. Under such conditions the infection more usually dies out because of high internal nutrient levels of both the nutrients (N and P) can make the plant immune to infection (Sanders, 1975). A differential tolerance to phosphate supply is observed not only in relation to development of infection (Jasper *et al.,* 1979) but also in the ability of fungal species or isolate to improve plant growth at different levels of fertilization (Gianinazzi - Pearson, 1982). Mycorrhizal endophyte responded significantly to the host plants grown in soils, containing 3.5 ppm available phosphorus (Jalali, 1988).

Phosphorus plays an important role in the control of VA-mycorrhizal infection of root systems (Mosse, 1972; Sanders, 1975; Mengeet *al.,* 1977). Sanders and Sheikh, (1983) established a correlation between the final percentage of infection and the quantity of phosphorus added to the soil. Studies with onion plants which received foliar application of P, showed that the P concentration within the plants plays a more important role than soil P levels (Sanders, 1975). Plenchette (1982) supported this finding by using a split root technique in which the two halves of individual sundan grass root systems were grown in soils with light and low P in the presence and absence of the mycorrhizal fungus, *Glomus fasciculatum.* Hetrick *et al.,* 1984) reported that mycorrhizal corn plants were significantly larger at lower concentration of 'P' and higher levels of 'P' were detrimental to the mycorrhizae. Negative correlation of differences in colonization and mycorrhizal response in differently fertilized growth media with the logarithm of the equilibrium solution P-concentrations was reported by Bierman and Linderman (1983). At extreme levels of available phosphate in the rooting medium the growth of non-mycorrhizal plants can exceed that of VA-mycorrhizal ones (Bethlenfalvay *et al.,* 1983) and for the same concentration of P in the tissues. VA-mycorrhizal plants can have lower yields than non-mycorrhizal plants (Pairunan *et al.,* 1980). Increased levels of added soluble phosphorus reduced the mycorrhizal infection and spore production. However, root, shoot and total plant dry weight were significantly greater in mycorrhizal plants, at all levels of added soluble P, than non-mycorrhizal ones (Krishna and Bagyaraj, 1982). The increase in the addition of soluble P above certain level, resulted in decreased length of infected root. This also resulted in decreased length of external hyphae per cm of infected root and the formation of vesicles (Abbott and Robson, 1984). Same *et al.,* (1983) reported a decrease in percentage of mycorrhizal infected root length with increasing P-supply. It was also accompanied by a reduction in soluble carbohydrates. Some mycorrhizal fungi tolerate higher levels of soluble phosphate than others (Pond *et al.,* 1984). In general, large P additions to the soil decrease mycorrhizal development and functioning. Adding soluble phosphate improves growth of mycorrhizal plants less than that of non-mycorrhizal ones (Khan, 1972).

11.25 VA-Mycorrhizae and Plant nutrition

11.25.1 Nitrogen

Nitrogen is an inevitable pre-requisite for increased growth following relief of phosphate stress. Mycorrhizal fungi have a potential role in acquisition of N by plants (Raven *et al.,* 1978). Mycorrhizal celery plants derived significantly more ^{15}N than control plants. Positive corrections have been established between ^{15}N in mycorrhizal plants with percent mycorrhizal fungal colonization, number of hyphal crossings through the mesh into the area of ^{15}N placcmcnt and total length of hyphae g^{-1} of soil. The mean flux of N through the hyphae of *G. mosseae* was 7.42 x 10^{-8} mol. N $cm^{-2}s^{-1}$ for the inorganic N-treatment over 30 days of period and 1.74 x 10^{-8} mol. N $cm^{-2}s^{-1}$ for the organic N treatment over on 88 days period (Ames *et al.,* 1983).Rhodes and Gerdemann, (1978 b) pointed out that N content of mycorrhizal legumes has been attributed mainly to *Rhizobium* activity or in non-legumes, to the effects of increased P-content of the mycorrhizal plants.

VAM prefers NH_4^+- N over NO_3-N (Brown *et al.,* 1981). Increases (Wallace *et al.,* 1982), decreases (Pang and Paul, 1980; Nemec and Meredith, 1981; Hey*set al.,* 1982) or no differences (Rose and Youngberg, 1981; Schenck and Hielnson, 1973) in the N-content of mycorrhizal plants relative to control have been observed. Ames *et al.* (1984) reported that mycorrhizal and non-mycorrhizal sorghum did not differ significantly in total plant N or percent utilization of the applied N at any fertilizer dose. The percent N derived from native soil sources and the 'A' values were significantly higher for mycorrhizal plants compared to non-mycorrhizal plants at intermediate fertilizer rates indicating that the mycorrhizal sorghum plants derived N from a source that was less available to non mycorrhizal plants.

11.25.2 Phosphorus

In phosphate deficient soils plants infected with VAM usually take up more phosphorus and grow better than non-mycorrhizal plants (Sanders and Tinker, 1973). The amount of phosphorus required for maximum plant yield differs greatly for different soils and for different plant species grown in the same soil (Barrow, 1975). Rorison (1968) observed that *Deschampia flexuosa* grew continuously in concentrations as low as 10^{-7}M whereas *Scabiosa columbaria* responded favourably at a concentration of 10^{-3} M where there was significant increase of phosphorus in the roots. Maximum translocation rate of phosphorus for onion was found to be 6.4x10^{-16} moles s^{-1} and the P flux in the hyphae about 2x10^{-10} moles cm^{-2} S^{-1} (Cooper and Tinker, 1978). Moreover, the phosphate in the root environment also influences the mycorrhizal establishment and effect of mycorrhizae on plant growth.

Allen *et al,*(1981) reported that in *Boutelouagracilis,* the maximum infection was in the presence of phytate (75%) whereas lower root infection levels were noted in the presence of sodium phosphate (19%). They also found higher alkaline phosphatase activity and significantly increased chlorophyll concentrations with mycorrhizal infection with no change in a/b ratio. Different species of *Endogonaceae* differ in their ability to form VA-mycorrhizal associations. There are abundant evidences of some species being more efficient than others (Hall 1976; Abbott and Robson, 1978; Kuo and Huang, 1982)

Glomus tenuis was more efficient than the indigenous fungi at a particular location (Mosse, 1977).

Sanders *et al.* (1977) suggested that different species of VA-endophyte may have different growth promoting abilities.Mycorrhizal and non-mycorrhizal roots appear to absorb phosphorus from the same labile pool of P. The specific activity of P in plants growing in ^{32}P labeled soils was the same for mycorrhizal and non-mycorrhizal plants (Hayman and Mosse, 1972; Powell, 1975). Pairuman *et al.* (1980) reported that the effect of mycorrhizae in stimulating P uptake was the same for poorly soluble sources of P. Different plant species and varieties differ in their ability to extract phosphorus from the same soil (Ballard and Dean, 1941; Hall, 1978; Powell, 1982). Cultivars of wheat and pearlmillet are known to differ in dry matter production upon mycorrhizal inoculation (Azcon and Ocampo, 1981; Krishna and Bagyaraj,1984).

Several reports are indicating that mycorrhizal inoculation increased not only the utilization of soluble phosphates like mono calcium phosphate and superphosphate (Mosse, 1973a; Powell and Daniel, 1978) but also less soluble forms like rock phosphate (Mosse, 1977) bone meal (Daft and Nicolson, 1966) and even hydroxyapetite and aluminium and iron phosphates (Nyabyenda, 1977) observed growth of mycorrhizal plants in the presence of insoluble tri-calcium phosphate. Bieleski (1973) reported that P-uptake may be influenced by the steepness of the phosphate gradient from root to leaves. In mycorrhizal plants the increased conversion of phosphate from inorganic to organic forms in the leaves could increase the inorganic phosphate gradient from root to leaves. Possibly increasing the phosphate sinks and thus enhances P-uptake (Allen *et al.*, 1981).

11.25.3 Other nutrients

Mycorrhizal infection can also increase the uptake of any soil nutrients in addition to phosphorus and nitrogen that move to plant roots, primarily by diffusion (Gildon and Tinker, 1983; Timmer and Leyden, 1980, Zn Gilmore, 1971; Swaminathan and Verma, 1979) and other heavy metals (Pacovsky *et al.,* 1986). Increased absorption of potassium, sulphate, zinc and strontium 90 (calcium tracer) by mycorrhizal plants has been reported (Gray and Gerdemann, 1973; Powell, 1975). Mycorrhizal roots can absorb zinc (Pacovsky, 1986: Thakur and Panwar, 1995) and copper (Pacovsky 1986; Thakur, 1994) from solutions at faster rate than can non-mycorrhizal P fertilized roots. Buwalda *et al.,* (1983) reported that mycorrhizal infection in wheat and barley increased the concentration of anions *viz.* chloride, sulphate and nitrate while addition of phosphorus fertilizer did not have similar effect.Tinker and Gildon, (1983) concluded that copper and zinc supplied to the plants can be markedly improved by mycorrhizal formation.

Changes in the levels of phytohormones in VA-mycorrhizal plants have been reported (Allen *et al.,* 1982). Although VAM have been shown to synthesize phytohormones and their precise role in plant growth and metabolism has not been studied so far in detail. Verma *et al.* (2004) stated that generally growth in terms of plant height, number of leaves/plants; number of primary branches/ plants were recorded higher in both boron and zinc treated seeds over control, foliar spray of 0.5% $ZnSO_4$ solution alone, followed by seed soaking plus foliar spray of $ZnSO_4$ proved effective and recorded significantly higher plant height, number of branches and number of leaves than control during all the years of experimentation. Similar results were obtained by Hemantranjan, *et al.* (2003).

11.26 Plant Growth and Phosphorus Nutrition

In soil low in available P mycorrhizal plants have higher rates of growth than non-mycorrhizal plants, root shoot ratio is often lower and shoot fresh weight, dry weight ratio is higher following infections and their beneficial effect has been reported by many workers in India and abroad (Mosse, 1973; Tilak and Diwedi, 1990; Panwar, 1991). VAM has more extension of the root absorbing surface beyond the zone explored by the root hairs allowing more water and nutrients uptake under irrigated and dry land conditions (Allen *et al.,* 1981; Panwar, 1991). VAM symbiosis can affect the stomatal behaviour of the host plant (Auge and Duan, 1991).The effects are not usually apparent for several weeks after germination. The rate of plant growth is determined by interactions between mycorrhizal infection and number of nutritional and non-nutritional aspects of the symbionts physiology. The importance of VAM in the 'P' nutrition is evidenced that the inflow of 'P' from soil in to the roots of mycorrhizal plants is faster than in to non-mycorrhizal plants (Sanders, 1975; Sander *et al.*, 1977 and Tester and Smith, 1985). On soils low in available 'P' the amount of 'P' in the orthophosphate (Pi) fraction in both shoot and root is increased in mycorrhizal plants (Gianinazzi-Pearson, 1984;Gianinazzi- Pearson and Gianinazzi, 1986). One mechanism underlying the increased rate of uptake of 'P' is the efficiency with mycorrhizal roots exploits the soil profile with hyphae extending beyond the depletion zone surrounding the absorbing root and its root hairs (Nye and Tinker, 1977). This fits with what is known of the fraction that determines rates of nutrient supply to and uptake by roots. For non-mobile nutrients such as 'P' (and to a lesser extent 'K' and $NH^+{}_4$) root growth, root radius, development of root hairs and initial concentrations in the soil solution are more important determinants of the rate of uptake than are the kinetic properties of the uptake system of the roots (Nye and Tinker, 1977; Clarkson, 1985) mycorrhizal modifications of the nutrients uptake properties of roots depends upon (i) development of extra material hyphae in soil (ii) hyphae absorption of phosphate (iii) translocation of 'P' through hyphae over considerable distance and (iv) transfer of 'P' from the fungus to the root cells.

There is clear evidence that all these processes take place. So there is immense need to compare different fungus-host combinations grown under different environmental condition.Not only 'P' but other nutrients like P, Zn, S, Ca and N can be translocated from the place exceeding the radius of any depletion zone likely to develop around and actively absorbing roots (Rhodes and Gerdemann, 1975). Although growth responses to infection frequently do not occur when soil 'P' is high, it is clear that inflow via hyphae can continue and give rise to high 'P' concentration in shoot as well as in roots, in the absence of a growth response being due to operation of a limiting factor other than 'p' availability (Smith *et al.,* 1986; Smith and Son, 1987). Luxury 'P' accumulation could provide a storage pool used in later stage of plant growth.

Improved 'P' nutrition as well as direct fungal effects may be implicated in the enhanced uptake by plants of other macro nutrients such as K and S (Powell, 1975) and of micronutrients such as Cu and Zn (Cooper and Tinker, 1978) nutritional interactions can be complex. For example, K requirement is strongly influenced by the source of N (NO_3 or NH^+_4) and by the concentration of Na. The anion and cation ratio can be altered by mycorrhizal infection. Mycorrhizal plants accumulate lower amount of organic anions in their vacuole and may therefore regulate cytoplasmic pH by disposing of a higher proportion of OH^- generated during nitrate reduction in the soil (Raven and Smith, 1976). They might, therefore, be expected to have higher rhizosphere pH than non-mycorrhizal plant assimilating this N source.Singh and Gupta (2006) studied the effect of bio-fertilizers at different levels of phosphorus on nodulation, yield and protein content in black gram and found that dual inoculation of *Rhizobium* and PSB was superior to individual inoculation of these bio-fertilizers. The highest values for number of root nodules per plant, grain and straw yields, and seed protein content were obtained when dual inoculation was combined with optimum P dose of 20 ppm.

Deka and Azad (2006) studied the cultural and biochemical characteristics and found considerable variation of culture in respect of configuration, margin, elevation and color. Dubey (2006) studied the interaction of integrating bio-agents, oil cake and fungicide in various modes of application for the better management of web blight of urdbean. Among 54 integrations evaluated soil application of *P. glabra* cake, seed treatment with integration of *T. viride*, Carboxin and *Rhizobium* were found superior as they increased seed germination by 31.6%, yield by 36% and reduced disease intensity by 94.8% and also significantly enhanced plant height, root length and number of root nodules in urdbean. Manivannan *et al.* (2003) studied the growth and growth analysis of rice -fallow-black gram as influenced by foliar application of nutrients with and without *Rhizobium* seed inoculation. They found that *Rhizobium* seed treatment and foliar application of Micro sol B recorded mark- ably higher leaf area index, dry matter production, crop yield and relative growth rate.

Prasad *et al.* (2002) studied synergism among phosphate solubilizing bacteria, *Rhizobacteria* and *Rhizobium* in urd bean. *Rhizobium* +PSB-1 +PGPR were superior in terms of plant dry weight, grain yield and N and P uptake. Mishra and Tiwari (2002) studied the response of urd bean to applied phosphorus, sulphur and *Rhizobium* inoculation on udicustochrepts. The highest leghaemoglobin content was obtained with P. The greatest P uptake was obtained with P followed by Sulphur (S) and *Rhizobium.* S uptake was greatest with S followed by P and *Rhizobium.* Tanwar *et al.* (2003) studied the effect of phosphorus and bio-fertilizers on yield and nutrient content and uptake by black gram. The crop yield, N and P contents and N and P uptake increased with increasing P rate. Inoculation with the combination of the bio-fertilizers resulted in higher yield, N and P content and N and P uptake of the grain. Tanwar *et al.* (2003) studied economics of black gram cultivations as influenced by phosphorus levels and bio-fertilizers. The seed yield increased with increase in P application.

Sinha and Sharma (2001) studied the response of black gram to phosphorus fertilization and *Rhizobium* inoculation in the hill soils of Assam. *Rhizobium* inoculation significantly increased the number and mass of the root nodules per plant compared to the control.Tanwar *et al.* (2002) studied the effects of phosphorus and bio-fertilizers on growth and productivity of black gram. The interaction between P rate and bio-fertilizers was significant with regard to the number of nodules and seed yield. Kulshreshtha and Khan (2002) studied the effect of ozone on root colonization by VAM fungi and root nodulation on black gram. The suppression of growth and other parameters were less in symbiotic plants than in non-symbiotic plants. Higher values for growth and other parameters were recorded for dual inoculated plants than for single inoculated plants.

11.27 Enhanced Phosphorus Nutrition

Resistance to water movement in mycorrhizal soybean roots was diminished 40% when compared with P-deficient non-mycorrhizal controls (Safir *et al.,* 1972). Nelson and Safir (1982a, b) reported that the hyphae do not make significant direct contributions to water uptake. Several studies have indicated that VAM-hyphae may not be able to transport the quantities of water necessary to affect shoot water economy (Cooper and Tinker, 1981; Sanders and Tinker, 1973). Mycorrhizal effect on hydraulic conductivity and transpiration of various citrus root stocks could be replaced by P-fertilization. Above studies supported the nutritional hypothesis offered by Safir *et al.* (1971) there were no significant differences between mycorrhizal plants and non-mycorrhizal plants with adequate P-contents. Koide, (1985) who arrived at the same conclusion with mycorrhizal sunflower, when plants with equivalent root lengths were compared, no effect of VAM-colonization on the intrinsic hydraulic properties of the soil/plant system over a wide range of transpiration rate was shown. Elevated transpiration rates exhibited by mycorrhizal sunflower were attributed to improved P-status.

Radin and Eidenbach (1984) suggested that low levels of foliar P limit leaf expansion in cotton by decreasing the hydraulic conductivity of the root system. They also observed that P-deficient plants displayed lower leaf water potential. Low level of soil moisture reduces the P-diffusion rate considerably so that mycorrhizal infection could improve the P-nutrition of plants even where the soil P-content is quite high. Radin (1984) reported that P-nutrition had only a very small effect or no effect on cotton leaf water relations when soil moisture was adequate. Transpirational flux, closely linked to stomatal diffusive conductance, has been reported to both decrease and increase in response to low P levels. Claims for direct hyphal participation in water uptake were made (Allen *et al.,* 1981). Mycorrhizal colonization frequently alters root-to-shoot ratio could conceivably affect foliar water relations. It is likely that potted plants with different underground biomass, will respond differently under conditions of drought, as proportionately different amounts of water would be available to transpiring plant tissues. Additionally, type, size and branching patterns of root system can affect exploitation of a given soil mass by plant through changes in hydraulic conductivity and water flow-rates (Nye and Tinker, 1977).

VAM can tolerate a wide spectrum of soil moisture, as evidenced by their presence in arid deserts and aquatic habitats (Cooper, 1984). Many workers reported that P-uptake in non-mycorrhizal plants may be diminished in soils of low water content (Nelson and Safir, 1982b). The diffusion coefficient of phosphate in soil is linearly related to soil moisture content and therefore, mycorrhizal P-supplies are likely to be much more beneficial to hosts growing in dry than in moist soils (Fitter, 1985). Colonization by *G. fasciculatum* could decrease resistance to water transport in *Bouteloua* up to 90% under drought conditions (Allen *et al.,* 1981). Leaf water potentials of mycorrhizal plants dropped more quickly than those of non-mycorrhizal plants as moisture stress increased and stomatal resistance remained lower in colonized plants. Cooper (1984) suggested that this influence of *G. fasciculatum* might have resulted from improved water uptake, increased photosynthesis or elevated cytokinin levels.

Mycorrhizal root colonization of plants subjected to drought is reported to be significantly greater than well watered plants at the lowest P level. The high P levels which reduce benefit of mycorrhizal colonization under well watered conditions may still be inadequate under severe drought conditions. Different species of VAM fungi response differently to water stress (Yocom *et al.,* 1987). Krishna and Bagyaraj (1984) suggested that recovery of mycorrhizal plants from water stress is rapid. They also noted higher relative water content; lower amount of proline and lower stomatal resistance in mycorrhizal plants, 24 hours after rewatering. Bethlenfalvay *et al.* (1987) reported that under stress conditions nodule development and activity, transpiration, leaf conductance, leaf and root dry weights were greater in soybean plants infected with VAM fungi when compared to P-fed non-VAM plants.

11.28 VAM-legume-*Rhizobium* Interaction and Nitrogen Fixation

It is well known fact that successful production of legumes depends on effective nodulation and nitrogen fixation. For maximizing crop yields, we have to select efficient strain(s) of specific *Rhizobia* and investigate the compatibility of host cultivars with the *Rhizobial* strain. Bacteriazation is particularly important in case of soybean which was recently introduced in India. Inoculation of seed with *Bradyrhizobium japonicum* at the time of sowing is imperative, since *Rhizobia* specific to soybean are lacking in soils beneficial responses of legumes to inoculation with *Rhizobial* cultures have been reviewed by several workers (Brockwell, 1977; Khalifa, 1987; Panwar and Thakur, 1995; Thakur, 1997). Interaction between mycorrhiza, nodulation and nitrogen fixation have been studied in *Phaseolus sp.* (Daft and El-Giahmi, 1974); *Medicago sativa* (Smith and Daft, 1977); Clovers (Powell, 1977); peanut (Bagyaraj *et al.,* 1979); White clover, chickpea (Subba Rao *et al.,* 1985) and mung bean (Thakur and Panwar, 1995).The coexistence of a bacterium and fungus as root endophytes of legumes establishing a tripartite symbiotic association was first described by Janse (1986).

As a consequence of a simultaneous infection with *Rhizobium sp.* and mycorrhizal fungi, legumes are benefitted because of improved N and P supplied and also from the N-P interactions (Munns and Mosse, 1980). The double symbiotic association of *Rhizobium* and VAM in legumes not only reduces the inputs of synthetic fertilizers, thereby saving energy, but also appear to reduce the cost of the system itself in terms of photosynthates drain, *Rhizobium* and VAM are known to derive carbon (C) compounds from their host plants. Additional allocation of C to mycorrhizal roots of *Vicia faba* is reported to be equivalent to about 10% of total photosynthates (Pang and Paul, 1980). Kucey and Paul (1982) found that N_2–fixing nodules of non-mycorrhizal *V.faba* utilized 6% of total photosynthates and those of mycorrhizal plants 12%. The mycorrhizae respired or incorporated about 4% of total photosynthates in both nodulated or singly infected plants. Nodulated plants with or without mycorrhizal infection fixed about 47% higher $^{14}CO_2$ than plants without microbial symbionts (Harris *et al.,* 1985).

At early stages of mycorrhizal and *Rhizobial* infection the carbon drain, not yet compensated for, may induce a transitory negative growth response. Bethlenfalvay *et al.,* (1982b) observed that during the first ontogenic stages of the association, VA-mycorrhizal fungus was parasitic as a result of carbohydrate demand on the host by the endophyte whereas at later stages mycorrhizal fungus and soybean associated mutualistically. Negative correlations of roots hexose sugar concentration with VAM colonization, nodule activity and root P content were observed, when soybean plants inoculated with one of the endophyte after the other has colonized the host root (Bethlenfalvay *et al.,* 1985). They also reported that non-symbiotic plants or dipartite association produced significantly higher dry weights than those of tripartite plants inoculated at planting.

The rates of CO_2 exchange, leaf area, P content were associated with P nutrition or the presence of VAM fungus; leaf fresh weight and root dry weight with N nutrition or the presence of *Rhizobium* and leaf N and starch content by both factors. The development of micro-symbiont structure and nodule activity was significantly low in tripartite association than in plants colonized by one endophyte only. Mycorrhizal and nodulated plants usually have lower root/shoot ratio than plants inoculated with either symbionts alone.The role of phosphorus in nitrogen fixation by soybean has been described by Bethlenfalvay and Yoder (1981). At low levels of P, no nodulation was recorded whereas at higher P regime shoot dry weight was six times higher than those grown in lowest P regime. Increase in acetylene reduction activity with increasing amount of P has been observed (Bethlenfalvay and Yoder, 1981). Bethlenfalvay *et al.,* (1983) reported that at lower levels of P mycorrhizal plants produced significantly higher rates of N_2-fixation, plant and nodule mass and P content, with increased plant growth and greater ratio of extra-to-Intraradical mycelium (index for VAM fungal effectiveness). Mycorrhizal infection in soybean roots was found reduced from 25.4 to 5.2 vesicles/4cm root length at high levels of P, but there was not effect on the number of nodules plant^{-1} and yield.Under N-deficient condition, even crops well infected with VAM may not benefit from the association increased growth of nodulating/non-nodulating isolines was observed due to mycorrhiza in the presence of N source; however, nitrate inhibited the effectiveness of mycorrhiza (Karunarathne *et al.,* 1987). Addition of combined nitrogen decreased VAM infection and nodulation in *Pisum sativum* (Lanowska, 1966) and *Trifolium subterraneum.* Bayne and Bethlenfalvay (1987) observed that different endophyte isolates have differential effect on the development and function of their co-endophytes as well as host plant. Colonization of soybean roots by VA-mycorrhizal fungi varies significantly with cultivars. Daft and El-Giahmi (1974) weight of nodules, amount of nodular tissue, rate of acetylene reduction, N and P contents and concentration of leghaemoglobin. By using labelled nitrogen sources (^{15}N), it was observed that nodulated root systems of mycorrhizal beans fixed more nitrogen than nodulated root system with non-mycorrhizal plants (Kucey and Paul, 1982). In alluvial sandy-loam soil of Delhi, it was observed that inoculation of chickpea *(Cicer arietinum)* with *Rhizobium sp.* or *Glomus fasciculatum* individually or in combination resulted in significantly higher amounts of nitrogen fixation. This increase of nitrogen fixation in plants was pronounced at 50 Kg P_2O_5/ha application (Subba Rao *et al.,* 1985). Ho and Trappe (1975) suggested that VAM fungus*(Glomus mosseae)* and *G. macrocarpum* are capable of reducing nitrate to nitrite and presumably possess the nitrate reductase system. With the capability for reducing nitrate these fungi assume increased symbiotic effectiveness in terms of nitrogen assimilation and translocation to the host. Plants did not nodulated unless their P-concentrations were at least 0.15% and also observed that mycorrhizal infection helped the plant to reach this required level.

Smith and Brown (1979) noted mycorrhizal clover roots have higher P concentration than did the non-mycorrhizal control plants. They also suggested that, the enhancement of nodule efficiency could be explained by the increase in root P, which accumulated mainly as polyphosphate. The higher rates of photosynthesis were found in symbiotic plants than non-symbiotic plants. Increased growth and yield in nodulated soybeans inoculated with the VA-mycorrhizal fungus were accompanied by improved P-uptake lower root to shoot ratios, better nodulation and nitrogenase activity and modification in the nodulation pattern during growth. Kumar and Elamathi (2007) studied the effect of nitrogen levels and *Rhizobium* application methods on yield attributes, yield and economics of black gram. The maximum pod number, plant height, number of leaves, dry weight of plant, nodule number and pod number were obtained under the application of *Rhizobium.* Kalita *et al.* (2006) studied the evaluation of native *Rhizobium* from acid soils of Assam. *Rhizobium* isolate AR1 was the most resistant against 9 antibiotics. *Rhizobium* isolate AR1 was found to be more effective in terms of number of nodules (75 and 100), nodule dry weight (0.41g and 0.38g) and percentage of nitrogen content both in black gram and green gram. Performance of green gram with all the isolates was better with respect to the parameters than black gram. *Rhizobium* isolate AR1 also showed the highest nitrogenase activity in the root nodules of black gram and green gram (4.02 and 4.25 mole/g/h). Deka and Azad (2006) did the screening for effective strain of *Bradyrhizobium.* Among the 15 strains, Gn3 was the most effective strain giving the highest acetylene reduction assay (479.01 m mole/g/h) and leghaemoglobin content (0.125 mm). On the other hand, strain C10 showed the highest antibiotic efficiency in antibiotic resistance against nine different antibiotics and wide host range infectively nodulating cow pea, urdbean, mungbean and pigeon pea. A significant variation in *Rhizobial* sensitivity was observed in different isolates from different mutant plants in contrast to native strain. Significant variations were also observed in the physiological characteristics of *Rhizobium* after gamma irradiation. The variations might have appeared due to *Rhizobial* mutation inside the host plant and a direct interaction between the *Rhizobia* and altered host physiology/phenotype occurred (Raisinghani, 2004). The nitrogen content of the fresh seeds was 5.78% in the *Rhizobium* inoculated plants, while that in uninoculated plants was only 2.72%. The plants inoculated with the AH isolate showed better nodulation and nitrogen content compared to the plants inoculated with VM isolate. However, the SG isolate completely failed to produce nodules on the black gram (Reddy and Malliah, 2002).

Chandrasekharan and Vaiyapuri (2003) studied the influence of granular insecticide carbofuran in association with *Rhizobium* on nodulation, growth and yield of black gram. Irrespective of the levels of carbofuran, *Rhizobium*-inoculated black gram showed better growth and higher pod yield (50.3g) and straw yield per pot (81.1g) than the uninoculated crop. Balyan *et al.* (2002) studied the enhancing nodulation in *Vigna mungo* by applying higher quantity of *Rhizobium* in planting furrows and PSB. The inoculation of PSB along with *Rhizobium* gave higher nodulation rate, plant dry matter, yields and N-uptake than PSB inoculation alone. The result suggested that *Rhizobium* inoculated in furrows at higher rate gave better results and the *Rhizobium* enhanced the beneficial effects of PSB. Nagrajan and Balchandar (2001) studied the influence of *Rhizobium* and application of organic amendments on nodulation and grain yield of black gram and green gram in acid soil. Seed inoculation of *Rhizobium* and application of organic amendments enhanced biomass, root nodulation and grain yield. Bio digested slurry at 5 t/ha+ *Rhizobium* gave the greatest plant height (44.2cm and 53.7cm, for black gram and green gram, respectively), nodule no. (23.3 and 24.0), nodule weight (45.3 and 42.3mg) and grain yield (758.3 and 732.0kg/ha). Shanmugam *et al.* (2001) studied the interaction of *Pseudomonas fluoresecens* with *Rhizobium* in the management of black gram root rot incited by *Macrophomina phaseolina*. Glass house and field studies showed that seed treatment and soil application of Pf-1 was the most effective treatment to reduce root incidence to increase yield.

11.29 VAM-Water Stress-Host Relationship

Water is pre-requisite for the growth of any plant. Plants growing in a natural environment are rarely free from water stress.Lack of water exerts a controlling influence on crop distribution and productivity, as well (Fischer and Turner, 1978). Bartels and Caesar (1987) found that two weeks of drought during seed filling reduced the yield of soybean by 25%. Examination of

VA-mycorrhizal symbiosis has indicated that VAM may improve host resistance to drought stress (Johnson and Hummel, 1985).The first systematic examination of mycorrhizal influence on plant-water relations was conducted by Safir *et al.* (1972) on soybean plants. The external hyphae of the mycobiont greatly enlarge the potential surface area of water and nutrient scavenging systems of plants. Under low moisture levels, water absorption capacity of roots might be enhanced by mycorrhizae. Nye and Tinker (1977) observed that hyphae ramifying into the soil are likely to increase the absorbing area for water uptake even further and may also be able to by-pass the dry zones that often surround slow growing roots during periods of drought. Other less obvious influences of endogonaceous fungi on plant-water relations are fungal derived hormonal effects on stomatal physiology, increased vasculature for water conduction and lowered radial resistance to water flow in colonized root cortical tissues (Daft and Okusanya, 1973; Allen, 1982; Safir and Nelson, 1985).

11.30 AM Fungi and Environmental Stress Management

AMF have been shown to decrease plant yield losses in saline and alkaline soil conditions (Kashyap *et al.,* 2004; Murkute *et al.,* 2006). Shoot and root dry matter yields and leaf area were higher in mycorrhizal than in non-mycorrhizal plants. Total accumulation of P, Zn, Cu, and Fe was maximum along with chlorophyll content and relative water content in mycorrhizal than in non mycorrhizal plants under both control and medium salt stress conditions (Colla *et al.,* 2007). AM fungi could also influence plant hormones (Barea and Azcon-Aguilar, 1981). Improved salt tolerance of mycorrhizal plants can be mainly related the changes in physiological processes such as increased carbon dioxide exchangeable rate, transpiration, stomatal conductance and water efficiency. This may be due to increased uptake of nutrients with low mobility, such as P, Zn and Cu (Ruiz-Lozano *et al.,* 1996) and to improved water relations (Ruiz-Lozano and Azcon, 1995). Enhanced salt tolerance in AM symbiosis was mainly related with the elevated superoxide-dismutase (SOD), catalase (CAT), ascorbate peroxidase (APX), peroxidase (POD) activity by AMF which degrade more reactive oxygen species and so alleviated the cell membrane damages under salt stress.

The contribution of the AM symbiosis to plant drought tolerance results from a combination of physical, nutritional, physiological and cellular effects. Following are the mechanisms by which AM symbiosis can alleviate drought stress in host plants (Auge, 2001). Hyphae with a diameter of 2-5mm can penetrate soil pores inaccessible to root hairs (10-20mm diameter) and so absorb water that is not available to non-mycorrhizal plants through root extension via mycelium/ hyphae association. An increase in proline accumulation has been observed in mycorrhizal plants subjected to drought (Azcon *et al.,* 1996; Goioechea *et al.,* 1998). It has also been shown that mycorrhizal colonization and drought interact in modifying free amino acid and sugar pools in roots (Auge *et al.,* 1992b). Finally, a greater osmotic adjustment has been reported in leaves of mycorrhizal basil plants than in non mycorrhizal ones during a period of lethal drought (Kubikova *et al.,* 2001). During soil drying, mycorrhizal plants often maintain higher gas exchange rates than non-mycorrhizal plants of similar size and nutrient status (Goicoechea *et al.,* 1998). Duan *et al.* (1996) have suggested that AM fungi probably increase the ability of the root system to scavenge water in dried soil, resulting in less strain on foliage, and hence higher stomatal conductance and shoot water content at particular soil water potential. On the other hand, evidence has come from other studies that arbuscular mycorrhiza may influence leaf transpiration, even after leaves have been detached (Green *et al.,* 1998). Some studies have demonstrated that AM symbiosis development induces the expression of genes encoding aquaporins that facilitate the passive movement of water molecules down a water potential gradient (Krajinski *et al.,* 2000).

11.31 AM Fungi and Pest Management

Number of studies have demonstrated the beneficial role of AM fungi in plant growth and health (Bagyaraj, 1992) and it appears that they are essential for the survival of plant species in many ecosystems (Allen, 1991). Root pathogens are a major limiting factor for plant production. Fungi belonging to the genera *Phytophthora, Fusarium, Phythium, Rhizoctonia*, some bacteria, nematodes are the most widespread among the various fruits and vegetable and pulse crops.They kill the roots or reduce their ability to absorb water and nutrients by penetrating root tissues and producing toxins. Many reports indicate an interaction between AM fungi and plant pathogenic organisms (Mukerji, 1999; Abdel-Fattah and Shabana, 2002). A host plant previously inoculated with AM fungi exhibits increased resistance to several root diseases (Linderman, 2000). Reduction of disease symptoms and severity has been reported in different fruits and vegetables with mycorrhiza-pathogen interaction for various fungal pathogens like *Phytophthora, Fusarium, Pythium, Rhizoctonia, Verticillium, Thielaviopsis, Aphanomyces, Pyrenochaeta,* bacterial pathogens like *Pseudomonas, Erwinia,* viruses and nematodes like *Radopholus, Heterodera, Rotylenchus, Pratylenchus, Tylenchulus, Meliodogyne* (Bharadwaj and Sharma, 2006 a;b).

11.32 Mycorrhiza and Soil Heavy Metal Bioremediation

AM fungi provide an attractive system to advance plant-based environmental clean-up. During symbiotic interaction the hyphal network functionally extends the root system of their hosts. Thus, plants in symbiosis with AM fungi have the potential to take up heavy metals (Pb, Cd, As) from an enlarged soil volume. At high concentrations heavy metals interfere with essential enzymatic activities by modifying protein structure or by replacing a vital element resulting in deficiency symptoms. Similarly, heavy metals are taken up via the fungal hyphae and can be transported to the plant. Thus, in some cases mycorrhizal plants can show enhanced heavy metal uptake and root-to-shoot transport (phytoextraction) while in other cases AM fungi contribute to heavy metal immobilization within the soil (phytostablilization). The result of mycorrhizal colonization on clean-up of contaminated soils depends on the plant-fungus-heavy metal combination and is influenced by soil conditions. Different mechanisms involved in heavy metal detoxification have been proposed, as chelating agents are secreted that bind metals in the soil, e.g. histidine and organic acids from the plant, glomalin from the fungus. Binding of heavy metal to cell wall components in plants and fungi, the plasma membrane was found as a living, selective barrier in plants and fungi. Specific and nonspecific metal transporters and pores in the plasma membrane of plants and fungi (active and passive import). Chelates in the cytosol, e.g. metallothioneins (plants and fungi), organic acids, amino acids, and metal-specific chaperons (shown for plants, assumed for AM fungi). Export via specific or nonspecific active or passive transport from plants or fungal cells, sequestration of heavy metal in the vacuole of plant and fungal cells and transport of heavy metal in the hyphae of the fungus. In arbuscules, metal export from the fungus and import into plant cells via active or passive transport and mycorrhizae remediate the heavy metal contaminated soils by mainly two processes namely phytostablilization and phytoextraction.

11.33 Potential Applications of AM Fungi

The AM fungi had a very wide application and can be used for improving the nutrient status and moisture availability from the deeper zone of the soil beyond the root extension and very well be used under moisture stress conditions, salt stress and drought, heavy metal bioremediation and also in increasing the capacity against pest and disease management, waste land reclamation and its potential can be applied in tissue culture too.The major role of AM fungi is to absorb nutrients (macro and micronutrients) and water from the soil and to transfer them into their hosts (Mishra, 2008b). This mechanism relies on the existence of an acidic compartment in the periphery of arbuscules presumably identical with the peri-arbuscular space separating the plant and fungal plasma membranes (Guttenberger, 2000). This acidification of the peri-arbuscular space corresponds to proton gradients across the fungal arbuscular and the plant peri-arbuscular membrane powering transport processes across these membranes. Enzymes responsible for the generation of such an acidification –H+-AT Pases- have been studied for a long time (Murphy *et al.,* 1997).Besides phosphate, a number of mineral nutrients and most notably water are transported by AM fungi to their host plants. In the case of nitrate, AM fungi play an even more active role when compared to phosphate, because they are able to liberate nitrate from complex organic material within the soil (Hodge, *2000*). Few proteins involved in nitrate transport in mycorrhizal roots have been found so far; amongst them a tomato nitrate transporter with increased transcript levels in AM colonized roots (Hildebrandt *et al.,* 2002). In addition, a fungal *(G. intraradices)* nitrate reductase has been cloned, which is expressed in AM roots and therefore might be involved in the transfer of nitrogen as well.

11.34 AM Fungi and Wasteland Reclamation through Nutrient Mobilization and Improving Soil Physical Properties

Abbott and Robson (1991) stated that the relationship between levels of mycorrhizal colonization and soil physical and chemical properties are markedly variable. Mycorrhizal fungi, specifically AM fungi have been adapted to wide range of environmental conditions (Sharma *et al.,* 1996). AM fungi are aerobic and soil aeration has considerable impact on their distribution. The AM symbiosis produces extra radial hyphae that may extend several centimeter out in to the soil and secretes organic material that are substrate for other microbes. (Singh and Tilak, 1995). These hyphae associated microbes frequently produce sticky material that causes soil material to stick together and improve soil structure. Further by nutrient uptake, they improve plant cover and root proliferation. Mycorrhiza brings about reclamation of wasteland by improving nutrient status of soils and by improving soil physical properties (Kashyap *et al.,* 2005).The cycling of macro and micronutrients in an ecosystem is influenced by multiple interactions involving soil microbial populations. Arbuscular mycorrhizal (AM) associations play important roles in this nutrient cycling through their microbial activity and their involvement in plant nutrient acquisition. The extra radical mycelium connects plant roots to the surrounding soil microhabitat increasing the soil volume exploited by host plants. Mycorrhizal hyphae transport mineral nutrients over greater distances from depleted zones than do roots. Thus, under low nutrient conditions AM-colonized roots may have an enhanced uptake of relatively immobile macro and micronutrients (Subramanyan and Charest, 1998). In addition, mycorrhizal roots often have not only increased length but also modified root architecture (Berta *et al.,* 1995). Rootcolonization often results in an enhanced uptake of relatively immobile metal-micronutrients such as Cu, Zn and Fe (Mishra, 2008b).

Soil structure is influenced by soil water content and its variation with time and plant growth can strongly influence the magnitude and frequency of wetting and drying cycles. Decreased water content typically increases contact points between primary particles and organic matter, resulting in increased soil cohesion and strength (Horn and Dexter, 1989; Horn *et al.,* 1994). Localized drying of soil, in close proximity to roots, promotes binding between root exudates and clay particles (Reid and Goss, 1982), directly facilitating micro aggregate formation. More efficient exploration of water by mycorrhizal fungi may lead to more extreme wet and dry cycles, which could have very strong consequences for soil aggregation. Additionally, because the symbiosis can allow leaves to fix more carbon during water stress (Duan *et al.,* 1996), carbon inputs into the soil would be expected to be increased, which might be especially important in more arid environments. The release of compounds from living roots can be strongly influenced by mycorrhizal fungi (Jones *et al.,*2004), because these fungi can effect plant carbon metabolism (Douds *et al.,*2000).

During development of AM, the fungal symbionts grow out from the mycorrhizal root to develop a complex, ramifying network into the surrounding soil which can reach upto 30m of fungal hyphae per gram of soil (Cavagnaro*et al.,* 2005; Wilson *et al.,* 2009). The secretion by AM fungi of hydrophobic, 'sticky' proteinaceous substances, referred to as glomalin. The combination of an extensive hyphal network and the secretion of glomalin is considered to be an important element in helping to stabilize soil aggregates, thereby leading to increased soil structural stability and quality (Bedini *et al.,* 2009).

11.35 Mechanism of Interaction between AM Fungi and Plant Pathogens

Several hypothesis have been put forth to explain the mechanism involved in reduction of disease severity from pathogenic infections by mycorrhizal fungi (Linderman, 2000). Depending on the disease and the environmental conditions, any or all mechanism may be involved. AM inoculated roots may control the infestation of disease and pest as a result of changes in the root morphology along with physiological and biochemical changes, enhanced nutrient status and root exudates inhibiting pathogens activity in the zone of root elongation enzyme production and biological interactions of mycorrhizosphere competition of carbon source and developing a plant defense response in the host plant roots.

Plants colonized by AM fungi develop extensive root system with higher number of lateral roots as compared to non-mycorrhizal plants (Berta *et al.,* 1993). Mycorrhizal infection enhances lignifications in root tissues, particularly in stellar region which may be responsible for restricting soil borne pathogens that invade the host roots (Dehne and Schonbeck, 1978). This strong vascular system which increases nutrient flow, provides greater mechanical strength and reduces the detrimental effect of pathogens (Hussey and Roncadori, 1982). There is increase uptake of water and nutrients by mycorrhizal fungi which affects pathogenesis. Through increased P nutrition, AM fungi enhance root growth, increase the capacity for absorption and affect cellular process in roots (Smith and Gianinazzi-Pearson, 1988). Such changes may explain increased tolerance of mycorrhizal plants towards pathogens. Mycorrhizal roots show increased respiration than non AM roots (Dehne, 1982). Increased respiration rate of AM roots indicates higher metabolic activity which might enable mycorrhizal plants to be more resistant against root pathogens. Increased ethylene and DNA methylation have also been reported in AM roots (Dugassa *et al.,* 1996). Similarly the arginine level in mycorrhizal roots was reported to be six fold more due to an increased specific activity of enzymes in ornithine cycle and blocking of cycle by arginase enzyme system in such roots (Dehne *et al.,* 1978). Mycorrhizal development enhances the production of isoflavonoids which play major role in natural defense of plants (Morandi, 1996).

Root exudation pattern also changes in quality as well as quantity with mycorrhizal colonization which might play an important role in pathogenesis (Graham *et al.,* 1981). Root exudates directly affect the pathogen by inducing their germination or by inhibiting saprophytic and pathogenic activity. Root exudation is maximum in zone of elongation, the portion of root growth that is most susceptible to AM fungal colonization (Buwalda *et al.,* 1984). Greater tolerance of mycorrhizal plants towards the pathogens is attributed to increased phosphate status and more vigorous growth of plants (Azcon-Aguilar and Barea, 1996). In addition to phosphates, AM fungi increase uptake of micronutrients (Fe, Zn, Mg, Mn etc).

Mycorrhizal fungi, soil fungal pathogens and plant parasitic nematodes, all occupy the same root tissue, resulting in direct competition for space which might explain the disease resistance in mycorrhizal plants. Both mycorrhizal fungus and pathogen are dependent on host photosynthates for their growth. The major substrate for microbial activity in the rhizosphere or on *Rhizoplane* is organic C released by host roots(Azaizeh *et al.,* 1995). When mycorrhizal fungus has primary access to photosynthates, the higher C demand may inhibit pathogen growth. The concept of mycorrhiza sphere is based on the fact that mycorrhizae exert the strong influence on the micro flora in the rhizosphere (Bansal *et al.,* 2000). This results in microbial shift in the mycorrhiza sphere which could influence the health of plants (Whipps, 2001, Barea *et al.,* 2002). The number and proportion of bacterial antagonists against several root pathogens increase when AM association is established. More actinomycetes antagonistic to *Fusarium solani* and *Pseudomonas solanacearum* were isolated from the rhizosphere pot cultures of *G. fasciculatum* colonized plants than from non-mycorrhizal control (Secilia and Bagyaraj, 1987).

The symbiotic host-fungus relationship is characterized by the formation of haustoria like intracellular arbuscules which are successively degraded. This process represents digestion of the fungus by the host. For this degradation to occur, the fungal cell wall of the endophyte has to be attacked by the host. Therefore, the roots colonized by mycorrhizal fungus show high chitinolytic activity. These enzymes can be effective against other fungal pathogens as well. Increased activity of peroxides associated with epidermal and hypodermal cells has also been reported in mycorrhizal roots.The response of plants to AM fungi involves temporal and spatial activation of different defense mechanisms (Garcia-Garrido and Ocampo, 2002), however, how these defenses affect the functioning and development of AM remains unclear. A number of regulatory mechanisms of plant defense response have been described during establishment of AM symbiosis. These include (a). Elicitor degradation (b) Modulation of second messenger concentration (c) Nutritional and hormonal plant defense regulation and (d) Activation of regulatory symbiotic gene expression.

Colonization of roots by AM fungi brings about biochemical changes within host tissues. These biochemical changes includes as :

- Stimulation of the phenylpropanoid pathway (Harison and Dixon, 1993).
- Changes in levels of aliphatic polyamines (El Ghachtouli *et al.*,1995).
- Activation of defense related genes (Harison and Dixon, 1993, Franken and Gnadinger, 1994; Dumas-Gaudot *et al.*, 2000).
- Enhancement of certain hydrolytic enzyme activities (Dumas-Gaudot *et al.*, 1996).
- Synthesis of protein of unknown function (Dumas-Gaudot *et al.*, 2000).
- Elicitation of plant chitinase and ß-1, 3 glucanase (Blee and Anderson, 1996) which are antifungal against pathogenic soil and root born fungi.
- Induction of chitinase isoforms by mycorrhiza. These chitinase release oligosaccharide elicitors from the chitinase AM fungal cell walls which in turn stimulates the general defense responses of plants (Cordier *et al.*, 1996).
- Increase ethylene production and DNA methylation by AM roots can be related to gene expression for higher resistance of plants against pathogens (Dugassa *et al.*, 1996).
- The level of phytoalexin medicarpin, coumestrol, daidzein, medicarpin-malonylglucoside, formononetin, showed transient increase in roots of *Medicago truncatula* days after inoculation with *Glomus* species,

thereby showing that phytoalexin and their precursors are activated by AM inoculation (Harison and Dixon, 1993), and

- Phenolics may also play a role in protecting AM roots against pathogenic fungi (Grandmaison *et al.*, 1993). A continuous increase of total soluble phenols has been found in mycorrhizal roots of peanuts (Krishna and Bagyaraj, 1984).

11.36 AM Fungi and Rooting Behavior in Tissue Culture

The symbiotic association of AM fungi with plants often produces plant growth regulating hormones, mainly auxins, affecting the plant root physiology and subsequently plant growth. The phytohormones released by this mycorrhizal association, if applied *in-vitro* in form of root extracts influences the growth of culture and rooting in shoot cultures. The root extract of AM *Morus alba* was found to be equivalent to IBA in rhizogenesis while the extracts of *Trigonella-foenum-graecum, Ricinus communis* and *Solanum melongena* were not very effective in rooting (Sharma *et al.*, 2005b).

Plant Growth and Yield

It is well known fact that successful cultivation of the legumes depends, in part, on an effective symbiosis with *Rhizobium.* Significant seed-yield increases following *Rhizobium* inoculation in blackgram (Sahu and Behera, 1972; Lehri *et al.*, 1974), cowpea (Bagyaraj and Hedge, 1978) and mungbean (Oblisami *et al.*, 1976). Inoculants prepared with more than on *Rhizobium* strains were found to be superior over single strain inoculants in some (Oblisami *et al.*, 1976), but not all studies (Ramachandran *et al.*, 1980). The high yields report in northern Nigeria suggest that groundnut nodules estimates of as much as 240kg N/ha or 80% of the plant's N-uptake (Dart and Krantz, 1977). Seth (1979) studied the field performance of five selected *Rhizobium* isolates of groundnut at three different locations in Gujarat. A yield increase was obtained with one isolate at one location and was equivalent to the yield obtained with applications of 25-50 kg N/ha to uninoculated plots. Gupta *et al.* (1982) showed a significant increase in grain yield (16.8%) and nitrogenase activity (44.4%) in an interaction study between four cultivars of chickpea and four strains of *Rhizobium* over uninoculated control.

Mycorrhizal development is greatly affected by light received by the shoot of the host plant. Low irradiation and short day lengths can reduce the development of VAM (Daft and El-Giahmi, 1978), which also reduce plant growth in uninfected plants. As VAM is an obligate endosymbiont, they have very little capacity of independent growth. Harley and Smith (1983) pointed out that growth responses to mycorrhizal inoculation was determined by translocation of soil derived nutrients to the plant through the fungus and the consumption of carbon compounds derived from host by the fungus. Maximal mycorrhizal effectiveness occurs when nutritional conditions for the host are suboptimal, but when these become severely deficient the fungus becomes ineffective.

Ratnayake *et al.* (1978) found that under low level of P, root exudation measured in terms of the net leakage of soluble amino-acids and reducing sugars was significantly higher than under the high P-levels. Increased levels of added soluble P reduced the mycorrhizal infection and spore production, however, root, shoot and total dry weight were significantly greater in mycorrhizal plants, at all levels of added soluble P, than non-mycorrhizal ones (Krishna and Bagyaraj, 1982). The effect of temperature on root growth and colonization were studied by different workers (Smith and Brown, 1979, Hetric *et al.,* 1984). Vogelzang *et al.* (1993) found that shoot dry weight and root fresh weight were highest at 30°C temperature in mungbean both in mycorrhizal and non-mycorrhizal plants. While at 30°C root length were highest in mycorrhizal but in non-mycorrhizal plants root length were more at higher temperature (38°C).

Inoculation with *G. fasciculatum* significantly increased the root dry weight in both cowpea and pigeon pea. Application of P significantly increased the dry weights of shoot and root. Plants inoculated with both *Rhizobium* + *Glomus* and supplemented with P recorded highest dry weight of root and shoot (Manjunath and Bagyaraj, 1984). Mathew and Johari (1989) found that in mungbean when a tripartite inoculation of indigenous VAM (*G. caledonicum)* + *R. phaseoli* was used, root and shoot ratio significantly increased. Application of indigenous VAM alone did not result in any significant increase in total biomass, whereas a 25% increase occurred from a tripartite inoculum. The number of pods and leaves per plant was also greater for the mixed treatment. A significant increase in leaf length and width was also observed for all the treatments. Rawat *et al.* (1991) reported that the shoot dry weight significantly increased in mycorrhizal plants than non-mycorrhizal. *Rhizobium* also affects shoot dry weight over control, but mycorrhizal and *Rhizobial* strains independently and their combinations failed to show significant differences with respect to grain yield.

Cowpea inoculated with the VAM fungus and *Rhizobium* strain pairing produced significantly more pods and higher N and Pcontent than either single inoculations or uninoculated control (Thiagarajan *et al.,* 1992). Mahdi and Atabani (1992) also reported that inoculation of soybean and lablab bean with *Glomus mosseae* increased nodulation, dry matter yield and tissue N and P-contents more than triple super phosphate fertilizer, but greater responses were obtained from *G. mosseae* combined with fertilizer. Interaction between VAM fungi and *Bradyrhizobium* affect shoot dry weight and P-content in *Colopogonium caeruleum* (Ikram *et al.,* 1993). Jain and Gupta (2002) observed the effect of

Rhizosphere fungi on nodule number, shoot and root length of *Vigna mungo*. All fungi did not significantly affect root and shoot length. Sheikh *et al.* (2003) observed the effect of *Rhizobium* and N, P, K on growth, physio-morphological traits and yield of waste water grown black gram. Waste water application increased the leaf nitrate reductase activity, carbonic anhydrate activity and total chlorophyll contents of *Rhizobium* inoculated plants. The optimum fertilizer treatment was N:P:K at 10:30:20 kg/ha and NPK treated plant showed better performance under inoculated conditions than uninoculated. Tomar*et al.*, (2003) studied the residual effect of black gram inoculated *Rhizobium*, VAM and PSB on succeeding wheat crop. *Rhizobium* combined with VAM or PSB recorded a 10% increase in plant dry weight than *Rhizobium* alone. The highest residual effect on all the parameters including the wheat grain yield was recorded with combined inoculation of *Rhizobium*+PSB +VAM. Kumar and Ganesh (2003) studied the effect of VAM fungi and *Rhizobium* on growth and nutrition of black gram and cow- pea. It was concluded that inoculation with mycorrhizal fungi and *Rhizobium spp* is more effective in increasing legume growth and biomass production.

Combined inoculation by *Glomus mossae* and *Azotobacter* improved plant growth and yield of pomegranate in field conditions (Aseri *et al.*, 2008). *Glomus clarum* and *G. etunicatum* with *Rhizobium* improved seedling growth of *Schizolobiumamazonicum* in nursery conditions (Siviero *et al.*, 2008). Improvement in fodder production and quality in silvopastoral system by AM fungi consortia and native *Rhizobium* (Mishra, 2008b). Dual inoculation by AM fungi (consortia) and *Azotabacter* (exotic) improved the acclimatization of tissue cultured saplings of *Morus alba* (Kashyap and Sharma, 2006). Combined inoculation by AM fungi consortia and native *Azotobacter* was better in terms of biomass yield of multipurpose tree sp. and soil microbial biomass (Sharma *et al.*, 2005a). *Glomus mossae* with bacterial sp. (*Azotobacter* + *Bacillus megateranium* (P solubilizers) + *Bacillus mucilaginous* (K solubilizers) improved growth and biomass of *Zea mays* (Wu *et al.*, 2005). In *Casurinaequisetifolia,* biomass yield and nutrient uptake was maximum in combined inoculation of *Glomus fasciculatum* + *Azospirillum* + Phosphobacterium + *Frankia* (Rajendran and Devraj, 2004). *G. intraradiaces* and *Rhizobium tropici* together improved plant growth and nutrient uptake in two native forage legumes (Shockley *et al.*, 2004). *Glomus fasciculatum* + *Rhizobium* + *Pseudomonas striata* improved plant yield and nutrient uptake but with *Penicillium* variable negative effect observed on all parameters (Zaidi *et al.*, 2003). AM fungi with *Azospirillum* had negative overall effect while with *Rhizobium* improved nodule no. rhizosphere microflora and nutrient uptake in alfalfa (Biro *et al.*, 2000). Decrease in spore germination, hyphal length and root infection of AM fungi observed with *Rhizobium.*

Kumar and Sharma (2006) studied the effect of *Rhizobium* and co-inoculants VAM and PSB on yield and nutrient uptake in black gram. Results showed that inoculation of the bio fertilizers significantly increased grain and straw yield compared with uninoculated control. The highest grains and straw yields (223.5 and 229.8 g/micro plot, respectively) were obtained with the mixed inoculation of all 3 bio-fertilizers under 120 g/hectare. Gunasekaran*et al.*, (2004) studied the synergism between *Rhizobium*, plant growth promoting rhizobacteria (PGPR) and phosphate solubilizing bacteria (PSB) in black gram. Inoculation of *Rhizobium* alone had given a grain yield of 655 kg/hectare whereas inoculation control had recorded 472 kg/ha only. Combination of all three organisms (*Rhizobium* +PGPR+PSB) had recorded the maximum nodules; plant biomass and grain yield (760 kg/ha) which was 61% higher than individual inoculation of *Rhizobium* alone.

Production of Glomalin by AM in Soil

AMF hyphae abundantly produce glomalin (Rillig, 2004), which is a brown to reddish-brown glycoprotein (Rilling *et al.*,2005). Glomalin is a major component of organic matter in soil and contributes to better soil aggregate formation, which is important for soil structure and stability against erosion (Wright *et al.*,1996; Wright and Upadhyaya,1998). Glomalin, due to its recalcitrance, hydrophobicity, and adhesiveness play an important role as cementing material of the soil particles, at the same time acting as a highly stable form of organic C storage that could represent an important fraction of the total organic matter of the soil (Morales *et al.,* 2005; Rillig and Mummey, 2006; Cornejo *et al.,* 2008a). As the hyphae senesce, glomalin is deposited within the soil, where it accumulates until it represents as much as of soil C (Rillig *et al.*,2001, 2003) and N (Lovelock *et al.,* 2004). Even though this compound could constitute an important global reservoir of C and N, environmental controls on glomalin are not well understood. In addition, the ecophysiological function of glomalin if any remains unknown, although (Gadkar and Rillig, 2006) have found evidence that glomalin may be related somewhat to a heat shock protein. Standing stocks of glomalin in soil are determined by its production and decomposition, and environmental conditions could affect the two fluxes independently (Rillig, 2004). Namely, glomalin production should be controlled directly by the abundance and community composition of AM fungi. In addition, standing root length, host plant availability, and plant nutrient balance might indirectly affect glomalin production by altering the allocation of photosynthesis to AM fungi. These plant characteristics are, in turn, partially influenced by the availability of inorganic resources such as CO_2, NH_4^+, NO_3^-, PO_4^{3-} and water. Glomalin decomposition, on the other hand, might be altered by soil characteristics such as nutrient availability (Nichols and Wright,

2005). The increase of glomalin levels is usually related to greater AMF activity in systems with organic substances (Oehl *et al.*,2004), or adding organic nutrient sources, such as cattle manure, compost, and crop stubble, which significantly stimulate mycorrhizal development (Douds *et al.*,1997; Joner, 2000). Results of Valarini *et al.* (2009) have also showed that, in general, compost application increased soil pH, mycorrhizal roots, mycelium length, glomalin levels, and water stable aggregates. (Lee and Yun, 2011) have also reported that the average concentrations of glomalin in the organic farming system was significantly higher than that in the conventional farming system (Lee and Eom, 2009) in order to study effect of organic farming on spore diversity of Arbuscular. Increasing in soil content of Arbuscular mycorrhizal fungi will be improving the efficiency of nutrient uptake, such as immobile phosphate ions. Which it seems that this improvement in nutrient availability could be important, especially in low nutrient status soils. Whereas glomalin is produced by Arbuscular mycorrhizal fungi, so it seems that with increasing in AM fungi content and activity, the concentration of glomalin in soil increases as well. As a consequence, high levels of glomalin result in better soil aggregate formation, which is important for soil structure and stability. In general, it seems that agricultural practices (organic farming in comparison with conventional farming) significantly can affect the AMF communities and follow that the other agents which are in associated with AMF is affected as well.

11.37 Quantitative Estimates of Nutrients to be taken up, Quantification in terms of Ecosystem Services

Quantitative estimates of AMF impact on plant nutrition are difficult to make because the non-mycorrhizal control in either laboratory or field studies is an unnatural status of the plant (Smith and Smith, 2011). Field experiments with different levels of mycorrhizal inoculum at the start of the growing season (through disturbance such as plowing, the use of selective fungicides or by pre-growing fields with non-mycorrhizal crops such as cabbage) have shown that also in the field the net benefits of AMF symbiosis on plant fitness outweigh (short-term) negative effects (Singh and Vyas, 2009). The ecosystem services delivered by the mycorrhizal symbiosis has been discussed by Gianinazzi *et al.*, (2010) and Fester and Sawers (2011). Attempts to quantify services are very rare. An early attempt was made by Miller *et al.* (1994), who described a methodology but did not provide quantitative data. Kuyper and Giller (2011) estimated phosphorus saving through increased nutrient uptake efficiency of possibly around 10%, and this amounted to US $2 billion. Further quantifiable ecosystem services could be the contribution to nitrogen fixation in legumes (as these plants need higher amounts of phosphorus, they also depend on mycorrhiza). Assuming 10% yield increases of legumes due to mycorrhiza would result in an additional service with an annual value of 10% of the ecosystem service provided by nitrogen fixers. This would amount to another US $9 billion annually. The ecosystem services related to enhanced drought tolerance, improved resistance against pests and pathogens and improved soil quality have not been quantified.

11.38 Suggestions for use in Agricultural Production

Inoculation of crops with AMF strains that are targeted at increasing crop yield, and that repress less mutualistic AMF strains that reduce yield, has shown some success according to meta-analyses by Lekberg and Koide (2005) and Hoeksema *et al.* (2010), but due to the obligate biotrophic nature of AM fungi, the development of cost-efficient large-scale production methods to obtain high-quality AM fungal inoculum is complicated, and their commercial exploitation is still in its infancy (Ijdo *et al.,* 2011). Nevertheless, Angelard *et al.* (2010) cultivated improved AMF strains that increased the growth of rice (but not of a second plant!). Still, it is unlikely that selected AMF strains will be able to display the wide variety of benefits of a diverse AMF community (Fester and Sawers, 2011).

Furthermore, Ryan and Graham (2002) and Lekberg and Koide (2005) concluded that high soil phosphate levels as a result of application of highly soluble inorganic P fertilizers (e.g., superphosphate) limit the effectiveness of AMF inoculation, so that targeted inoculation of crops with AMF may only be effective in low-fertilizer input systems (Fester and Sawers, 2011), such as organic agriculture where P availability is low (Ryan and Graham, 2002), and only rock phosphate is used, which is less soluble and therefore more slowly released than conventional P fertilizers. However, in a holistic approach taking into account benefits of AMF not related to increased nutrient uptake (pathogen protection, increased soil stability, protection from toxic levels of metals), management shifts that contribute to establishing and maintaining a (functionally) diverse mycorrhizal fungal community while maintaining high productivity levels is almost certainly much more cost-effective (Jeffries *et al.,* 2003; Gosling *et al.,* 2006; Shennan, 2008). This holistic approach aims to improve soil microbial and rhizosphere communities in general, motivated by the possibility that diverse soil microbial communities provide ecosystem functions that cannot be generated by specific organisms, species or strains (Fester and Sawers, 2011). For instance Jansa *et al.* (2005) observed synergistic effects by combining fungal strains, and mycorrhizal fungi are known to enhance nitrogen fixation by symbiotic N fixers (Smith *et al.,* 1979; Barea *et al.,* 2002).

Mycorrhizal management could also have an impact on food quality. For instance Ryan *et al.* (2008) noted that mycorrhizal plants possessed lower amounts of phytate; consequently, a higher proportion of the zinc in the plant seed was available for humans during consumption. Similarly, effects of mycorrhizal fungi on increased production of secondary compounds have been reported (Zeng *et al.*, 2013); for instance, the production of artemisin, a high-potential anti-malarial drug, was much higher in mycorrhizal than non-mycorrhizal *Artemisia annua* (Kapoor *et al.*, 2007). In intensive agriculture it may be possible to reduce negative impacts on mycorrhizae by foliar application (because the direct soil effects of P on the symbiosis are stronger than the indirect, plant-mediated effects), by micro dosing or other forms of selective placement around the growing plant. However, these topics have been rather under researched. There are also various reports about the use of AM inoculum in micro propagation systems and soilless production systems. Under those conditions mycorrhizal plants show better post-transplanting performance, probably through enhanced nutrient uptake, changed hormonal balances and pathogen protection (Kapoor *et al.*, 2008). Possible fine-tuning of nutrient management with application of mycorrhizal inoculum needs further research.

11.39 Future Intervention

AM undoubtedly plays a very important role in enhancing the nutrient and water uptake and has a wider applicability under various agro-climatic condition subjected to climatic change enhancing the tolerance in plants against various pathogens and insects, improving the soil physical conditions and wastelands managements. However, there is a need for exploitation of AM fungi with enhanced potential use such as -

- Screening of existing indigenous *species* and exotic species for better plant- soil- fungus interaction.
- Understanding the molecular characterization of plant-fungus relationship to better AM symbiosis.
- Studies related to genetic diversity, molecular mechanism and beneficial characteristic to mass culturing under laboratory conditions.
- Role of AM fungi under changing C-N ratio in soil health with CO_2 sequestration and climatic change and defense mechanism of resistance against pathogens.
- Field testing of different AM strains for crop plants, horticultural and vegetable crops, afforestation and timber plantation even at nursery level also.

- Improvement for mass culturing and inocula production techniques of efficient type of strains needs attention.
- Educating the farmers and gardeners for the potential use of different spp/strains of AM fungi.
- Detailed studies on physiological and biochemical changes brought about by the host-AM-interaction for better symbiosis needs immediate attention.

Chapter - 12

Frankia (Non-Leguminous Symbiosis)

Most trees legumes thrive in tropical regions but N_2 fixing non leguminous trees are prevalent in temperature conditions. Callatiam and coworkers from U.K. cultivated *Frankia* for the first time in 1978 in pure culture. A number of non-leguminous shrubs and trees as Alnus, Myrics, Hippophae, Casuarina and Elaeagnus are nodulated by N_2 fixing Actinomycetes, *Frankia* with Frankiaceae (Actinomycetes), the only acyinomycete reported to fix nitrogen to date. The so called actinorhizal plants include 7 orders, 8 families, 17 genera and 173 species of dicotyledons plants reported by Gaur, (2010), whereas 200 plants and 23 genera were reported by Diagne *et al.* (2013). Some of the important non-leguminous families and genera are given in Table 12.1.

Table 12.1 : *Frankia* symbiosis: Non–leguminous Nitrogen fixing angiosperms

Family	Genus	Species	Product/Some use	Country
Betulaceae	Alnus (Alder)	*Alnus nepalensis* *Alnus rubra*	Fuel wood Paper pulp	USA
Casuarinaceae	Casuarina (Australian Pine)	*C. equisetifolia* *C. junghuhniana* *C. oligodon*	Fuel wood Timber Pole Coffee shade	China, India, Thailand, Papua New Guinea
Coriariaceae	Coriaria			
Elaegnanceae	Eleagnus Hippophae Shepherdia	*E.* spp. *H. rhamnodes*	*H. rhamnodes* Fruit juice	USA, China
Rhanmaceae	Ceanothus (*California lilac*), Discaria, Colletia and Trevoa	-	-	-
Myricaceae	Myrica (*Bog myrele*)	*Myrica rubra* *Myrica jasunica*	Fruit, Bark Crater vegetation	China Java, Indonesia
Rosaceae	Carocarpu, Dryas, Purshla	-	-	-

Source : Gaur, 2010

12.1 Genus *Frankia*

Frankia is a gram (+) nitrogen-fixing actinobacterium that forms a symbiotic association with actinorhizal plants. It is a filamentous free-living bacterium (Normad *et al.,* 1996) found in root nodules or in soil (Chaia *et al.,* 2010)similar to the *Rhizobia* bacteria that are found in the root nodules of legumes in the Fabaceae family. The genus *Frankia* was originally named by J. Brunchorst (In 1886) to honor the German biologist, AB. Frank.The genus *Frankia* has been classified in the order of Actinomycetales on the basis of morphology, cell chemistry, and 16S rRNA sequences (Lechevalier, 1994).The first strain of *Frankia*, strain CpI1 (also known as HFPCpI1) was reported from *Comptonia peregrina* nodules by Callaham (1978). *Frankia* supplies most or all of the host plant nitrogen needs without added nitrogen and thus can establish N_2fixing symbiosis with host plants where nitrogen is the limiting factor in the growth of the host (Fig 12.1). Therefore, actinorhizal plants colonize and often prosper in soils that are low in combined nitrogen. Symbiosis of this kind adds a large proportion of new nitrogen to several ecosystems such as temperate forests, dry chaparral, sand dunes, mine wastes etc.

Fig. 12.1 : *Frankia*

12.2 Morphology

Frankia produce three types of cell such as(i)Vegetative hyphae (ii) Vesicles and (iii) Spores. These cells can be seen in symbiosis depending on the plant species.

- **Vegetative Hyphae:** Vegetative hyphae measures 0.3 to 0.5 Or 0.5 to 0.8μm, a filamentous stricture typical of actinomycetes. The cell walls appear two layered with cross-walls initiating from the base layer (Baker, 1980), Lipid layers are sometimes seen outside the wall that probably correspond the lipid layers so prominent on vesicles (Newcomb, 1987). The cytoplasm of hyphae contain rosette-shaped granules presumed to be glycogen as well as lipid bodies. Cytoplasmic tubules are also found in the cytoplasm.

- **Vesicles:** Vesicles are lipid-encapsulated, roughly spherical cellular structures measuring between 2 to 6mm in diameter, attached to hyphae by a short vesicle stalk.They are commonly made in response to nitrogen deprivation, and vesicles contain nitrogenase and ancillary enzymes. Vesicle clusters in plant cells form the starting material for isolation attempts. Vesicles are the site of nitrogenous activity and are generally formed when nitrogen is very limited in the medium (Zhang and Benson, 1992). Due to the presence of resistant structure in culture, *Frankia* inoculum is easier to conserve than *Rhizobium* inoculum (Brunck *et al.,* 1990).
- **Sporangia:** Sporangia are developed on separate hyphae and give raise to non-motile spores (Obertello *et al.,* 2003). The spores are sporangiospores borne in multilocular sporangia that are formed either terminally, or in an intercalary position on the hyphae.

12.3 Growth Media

The Actinomycete could reduce acetylene was a further advance, when grown in complex media containing fixed N compound. *Frankia* is difficult to isolate directly from soil, so most strains originate from root nodules. Two factors limit success viz. (i) slow growth of *Frankia* strains, and (ii) fast-growing contaminants. To minimize the fast-growing contaminants, nodules are disinfected with dilute sodium hypochlorite (NaOCl) and then peeled. Vesicle clusters can be separated from plant tissue by differential screening or density centrifugation. Clusters are best pour-plated on a variety of media and followed microscopically until they begin to grow over a period of 1-3 weeks. Monitoring the outgrowth of hyphae microscopically improves the chances of obtaining a monoculture.

Frankia strains are heterotrophic aerobes having generation times of 15 or more hrs. Some strains, like HFCC13, prefer lower O_2 levels than others and grow poorly in vigorously aerated media, particularly at low cell density. As a consequence of their filamentous morphology, the growth kinetics of *Frankia* strains generally consist of a stationary phase after transfer, followed by a short exponential phase, and then by a slower increase in biomass over time. Problems typical of growing other filamentous organisms apply to *Frankia* strains. Care must be taken to avoid nutrient and waste gradients across mycelia and a flocs or pellet formation should be avoided by frequent homogenization (Benson and Schultz, 1990). The medium used in isolating new *Frankia* strains is important but universal, or selective, media have not been reported. Effective media range from defined propionate media, to the complex QMod medium of Lalonde and

Calvert (Lalonde, 1979). Antifungal agentslike cycloheximide or nystatin, can minimize fungal contamination. Other additions include root-lipid extracts to stimulate growth during isolation (Quispel, 1983). Virtually all *Frankia* strains isolated require no growth factors, and thus grow well in defined minimal medium. Some are inhibited by undefined media additives such as yeast extract (Lechevalier, 1990).

12.4 Genome Structure

This microaerophilic bacterium is characterized by a high GC% content and a slow growth rate (Benson and Silvester, 1993). The %G+C contents of *Frankia* strains is around 66-75 Mol% which is typical of the filamentous Actinobacteria (Fernandez, 1989). *Frankia* strains have a Type III cell wall [meso-diaminopimelic acid, alanine, glutamic acid, muramic acid and glucosamine (Lechevalier, 1990)] commonly found amongst aerobic actinomycetes. Whole cell sugars include fructose, ribose, xylose, madurose, mannose, galactose, glucose, rhamnose, and 2-O-methyl-D-mannose (Lechevalier, 1990; Mort, 1983; St-Laurent, 1987). Phospholipid analysis places *Frankia* strains into group PI containing phosphatidyl inositol, phosphatidyl inositol mannosides, and diphosphatidyl glycerol; nitrogen-containing phospholipids are absent (Lechevalier, 1989). This grouping distinguishes them from Geodermatophilus (group PII) which also produce multilocular sporangia, and were once considered close relatives of *Frankia* (Hahn, 1989). In the few strains that have been analyzed, a menaquinone pattern consisting mainly of MK9 (H4) (menaquinone with nine isoprene units, four of which are hydrogenated) with lesser amounts of MK9 (H6 and H8) was found (Lechevalier, 1987).

As minimum criteria for including strains within the genus, morphology in culture, together with isolation from an actinorhizal root nodule, and usually an analysis of the 16S rRNA gene are sufficient. However, designating species has been problematical. Becking (1970) named species according to plants infected. This approach suffers from the fact that some strains can nodulate plants from different plant orders and some strains cannot reinfect their source plant. Three *Frankia* genomes have been sequenced, and wide fluctuations have been noted in their sizes, despite a divergence in the 16S rRNA genes that, by traditional standards, might place all three strains in the same species (>97% sequence identity).

12.5 Isolation of *Frankia*

Attempts to isolate *Frankia* earlier were misleading because isolated endophyte could not reproduce root nodulation in experimental plants.However, very recently, the endophytes from *Alnus crispa* var. *Mollis* and *Comptonia peregnina* have been isolated in pure culture employing micro dissection and enzyme degradation techniques. Isolate is slow growing and could include nodulation in Comptonia seedlings in sand cultures. *Frankia* have been grouped in two types based on serological characteristics and homology for nodulation between host species. The following four procedures may be adopted.

a) Nodule homogenates centrifuged in sucrose density gradient. This separates vesicles clusters from contaminating micro-organisms Plate out separated fraction on nutrient agar.

b) Surface sterilized nodules> homogenization> micro-dissection of the homogenate to separate vesicle clusters> pass through changes of sterile water> transfer to liquid media.

c) Grown plant axenically > Crush nodules > pour in to nutrient agar.

d) Sterilize outermost side of nodules with 3% osmium tetroxide > fragment nodules in liquid medium > sub culture pragonents. Great caution has to be used in this method because of highly toxic osmic tertroxide.

12.6 Nodules and Nitrogen Fixation

Genomic studies have been undertaken to characterize *Frankia* (Normand *et al.,* 2007; Persson *et al.,* 2011) and to better understand the functioning of actinorhizal symbiosis. Among species of Frankiaceae, the capacity to fix nitrogen is restricted to *Frankia* (Wall, 2000; Franche*et al.,* 2009). However, *Frankia* morphology in the nodule varies according to host plant (Fig 12.2). The symbiosis between actinorhizal plants and *Frankia* induces the formation of a perennial root organ called nodule, wherein bacteria is hosted and nitrogen is fixed (Perrine *et al.,* 2011; Tromas *et al.,* 2012). In the field, actinorhizal nodule can have variable forms and colours (Bargali, 2011). Comparison of actinorhizal and leguminous nodules shows that morphology, anatomy, origin, and functioning of nodules are different for these two N_2fixing plants (Hocher *et al.,* 2012). Two types of nodule formation occur in actinorhizal symbiosis viz. (a) intercellular and (b) extracellular infection (Pawlowski and Demchenko, 2012).

Nodules have a range of colors from light yellow to orange, to various hues of brown and even red. In size, they range from less than a millimeter for very young nodules up to more than ten centimeters for older nodules that have developed in loose sandy soil. The larger nodules (more than 1cm) generally only have active nitrogen-fixing tissue on the tip of young nodule lobes. The older portion of the nodule in the interior of the cluster is frequently quite senescent and inactive. Nodules are perennial and can continue to develop from season to season over several years. Actinorhizal nodules appear in a variety of forms in the field. They all resemble clusters of lateral roots that increase in size by repeated branching of the nodule lobe tips, leading to the final structures shown above. Some nodules have negatively geotropic nodule roots growing from the tips of nodule lobes, such as in the *Morella pensylvanica* shown above, whereas others may be supplied with lenticels on the surface of the nodule, presumably to enhance gas exchange. The position of the nodules on roots varies with the plant and environment in which the plant lives i.e. alder trees growing near water tend to bear nodules quite near the surface such that merely brushing away the leaf litter beneath an alder plant reveals dozens of nodules. By contrast, some of the *Rosaceous* and *Rhamnaceous* species that dwell in dry rocky soil or chaparral tend to have nodules that are quite deep and may be difficult to find in rocky soil.

(a) (b) (c) (d)

Fig. 12.2 : Root nodules in non-leguminous plants (a) *Casuarina glauca* (b) *A. incana* subsp. *rugosa* (c) *Casuarina* (d) *Morella pensylvanica*.

Frankia has symbiotic relationships with numerous dicot plants and is said to be responsible for 15% of the biologically fixed nitrogen in the world. One type of symbiotic relationship including plant, mycorrhizae, and *Frankia* is called a tripartite relationship and is a complex, multi-layered community of organisms that protect and support each other. Other symbiotic relationships include *Frankia*, one or multiple plants, and other bacteria. Specific role *Frankia alni* in these relationships is to infect the roots of plants: it deforms root hairs of its plant host by going into the cortical cells and causing the formation of nodules. Vegetative hyphae colonize the nodules and then differentiate into diazovesicles. Diazovesicles are the thick-walled (containing hopanoids), spherical cells in which reductive nitrogen fixation occurs.

Frankia N_2 fixing activity has been shown in a free living stage *in vitro* but its actual contribution to soil nitrogen economy comes mainly through its capacity to establishing N_2 fixing symbiosis with over 200 plants belonging some 23 genera. The nitrogen fixed in the root nodules of this symbiosis is in part translocated eventually to the rest of the plant. Then the nitrogen goes back to the soils in the annual decomposition of the plant material and through root exudates, both important for the growth of associated plant communities. *Frankia* symbiosis also provides other nutrients to the plants, which contribute to soil fertility. Deciduous trees conserve nitrogen by mobilizing leaf nitrogen in autumn with subsequent storage of the nutrient in twigs and branches, from where it is made available for growth the following spring. Such a phenomenon is not observed in alder plants probably owing to the fact that they can fix atmospheric nitrogen and obtain lesser benefits from such conservation mechanisms.

N_2 fixation occurs in terminal swellings of the actinomycete hyphae, called vesicles. Each infected cell contains one vesicle, cluster containing several hundred spherical vesicles. Nodule size can range from few millimeters to several centimeters in diameter. They are not pink in colour, but a legHb like molecule has recently been detected in nodules from most but not all actinorhizae investigated. It is recently proved possible to grow *Frankia* explant in pure culture. Although it is often described as a microaerophile, N_2 fixation in pure culture appears to be relatively insensitive to oxygen. All the isolations have been made from nodules. To date no-one has succeeded in isolating *Frankia* from soil stage in its life cycle. All the isolates produce stores in large sporangia (upto 60mm in diameter) when grown in pure culture, but only some of them produce spores inside the nodules. Several species of non-leguminous higher plants were listed to fix atmospheric nitrogen (Werner, 1992). The most predominate belong to Alnus (47 species), Casuarina (90 species) Elaegnus (38 species), Hippophae (2 species), Shepherdia (2 species), Myrica (38 species) and Ceanothus (31 species). A low number of families Rosaceae also reported to fix atmospheric nitrogen.

12.7 Field Estimates of Nitrogen Fixation

N_2 fixing potential of actinorhizal plants varies widely depend on several factors. It is as low as 18kg N/ha/yr in *Myrica faya* to modest 40-60kg/ha/yr in *C. equisetifolia* and an impressive 320 kg/ha/yr in *Alnus rubra* (Table 12.2). Casuarina are the most important plant species for semiarid condition but under irrigated condition the quantity of nitrogen fixation can go upto around 80KgN/ha/yr (Dommergues *et al.,* 1990). Non-leguminous nitrogen systems are of modest agricultural importance at present as their potential as nitrogen source is not widely taped. However, their nitrogen input globally may be enormous. Alnus alone can add 100kg N/ha/yr to soil from leaf fall alone so that the overall contribution could be valuable in agroforestry. Sometimes, soil gets enriched with nitrogen due to the growth of these trees. It was reported that soil under *Alnus rugosa* registered 0.46% nitrogen while the control soil contained 0.24% (Voigt and Steucek, 1969). *Alnus glutinosa* benetified non-nodulating trees (Delver and Post, 1968) by supplying fixed nitrogen.

Table 12.2 : Nitrogen Gains by some non-leguminous trees

Plant type	Nitrogen fixed kg/ha/yr	References
Alnus spp.	20-179	Newton *et al.*, (1968)
Alnus rubra	320	Alexander, 1985
Hippophae	27-179	Dommergues (1987)
Casuariana equisetifolia	60-10012-8040-60	Gauthier *et al.*, (1985)
Myrica spp.	10-18	Alexander (1985)
		Itousek and Walker (1989)

Source : Gaur, 2010

12.8 Actinorhizal Plants

Actinorhizal plants are grouped in the clade of Rosid I (Soltis and Soltis, 2000). Actinorhizal plants are mostly trees or woody shrubswith the exception of two species belonging to Datiscaceae family,.They are found on all continents with the exception of Antarctica (Baker Casuarina and Alnus are the most important and widely spread actinorhizal plants due to their uses in soil reclamation, agroforestry systems, dune stabilization, and windbreaks (Dommergues, 1997). They are generally pioneer species that colonize disturbed environments with low soil fertility and facilitate the establishment and development of subsequent plant communities (Gtari and Dawson, 2001).

Among the actinorhizal plants, Casuarinaceae trees are widely distributed in tropical areas (He and Critchley, 2008). These plants are originated from Australia, South-East Asia, Malaysia, Melanesian, and Polynesian regions of the Pacific,NewGuinea (Zhong and Zhang, 2003). Casuarina species are able to grow well under a range of stresses such as drought, flooding, salinity, and sites polluted by heavy metal (Bargali, 2011). Casuarina plantations improve physical and microbiological quality of degraded soils (Izquierdo, 2005). These plants have an important ecological and economical role in tropical countries.They contribute to the improvement of soil fertility by fixing nitrogen and producing thick leaf litter fromneedles that can be used as compost by farmers (Sougoufara *et al.*, 1989; Forrester *et al.*, 2006).

Casuarina trees are widely used for the rehabilitation of degraded lands in South Africa, Kenya.They are used as windbreaks to protect adjacent crops and fix sand dunes in Senegal, Tunisia, Egypt, China, and India (Mailly *et al.*, 1994; Sayed, 2011). In agroforestry system, these trees are used to improve soil fertility and increase crop yield e.g. intercropping with legumes in China and India (Zhong *et al.*, 2010). Casuarina trees are also used in the production of smokeless fuelwood with a high calorific value, and hardwood in the construction of houses in Benin and in the production of paper pulp wood in India (Jain and

Mohan, 2011). In Asia, shelter belts formed by Casuarina trees have played a major protective role during typhoons and tsunami (Zhong *et al.,* 2010). Association with *Frankia* increased Casuarina growth and biomass [(Shah *et al.,* 2007). Furthermore, in this symbiotic relationship, bacteria confer to plants a high resistance to abiotic and biotic stresses. In India, about 50×10^5ha are planted with *C. equisetifolia* and produce 10MT of pulpwood.Recognizing the importance of symbiotic association between Casuarina-*Frankia*, farmers have begun to use these biofertilizers through inoculation with crushed nodules (Karthikeyan*et al.,* 2009). The positive role of Casuarina and its ability to be propagated by seed, cutting, and tissue culture makes it an ideal species for the reclamation of degraded lands. However, *C. equisetifolia* clones have been reported to show marked variation in their ability to tolerate salt stress (Tripathi *et al.,* 2001; Mathish *et al.,* 2010). Thus, identifying stress tolerant clones that can grow in degraded lands are crucial for the rehabilitation of these lands. Like Casuarina, Betulaceae contains members that play an important role in improving soil fertility. They are used in the production of firewood, pulp, and timber. These species are also used in land reclamation, agroforestry, and as windbreak to avoid erosion and promote the establishment of the more nutrient demanding plants (Roy *et al.,* 2008). Inoculation with *Frankia* increases the growth and biomass of *Alnus* (Quoreshi *et al.,* 2007). Furthermore, alder growth is higher when plants are inoculated with more *Frankia* strains. Inoculation with *Frankia* increases the leaf nitrogen content of *Alnus glutinosa* (Oliveira *et al.,* 2005). Alder plants inoculated with *Frankia* have a higher productivity, a higher shoot length, root length, dry matter production, and chlorophyll content. Under nursery conditions, *Frankia* provides good alder plants to be used in reforestation programs. Inoculation with *Frankia* increases the survival rate of transplanted seedlings in the field. A plantation of *Alnus acumita* increases soil N content by about 279 kg/hareported by Carlson and Dawson (1985). Some other species of actinorhizal plants are also used as pioneer plants such as Coriaria and Datisca. When compared to legume-rhizobia symbiosis, actinorhizal symbiosis fixes at similar high rate of N_2 estimated at around 240–350 kg/h/yr (Wall, 2000).

The group of Alder tree which fixes adequate amount of nitrogen shows its rapid growth and development in nitrogen deficient soils. The trees grown under field conditions typically possess nodules that may attain larger size when compared with the size of tennis ball. The soils under Alder tree are enriched with nitrogen when compared with the surrounding soils. The *Alnus* type of root nodules corralled root nodules, by dichotomous branching is formed. The nodules lobes are in the axil of lateral roots whose apical meristem grows slowly or it's completely inhibited. In Myrica/Casuarina type of root nodules, the apex of each nodule lobe produces a normal but negative geostrophic root. Thus, the root nodules mass gets covered with upward growing rootlets.

12.9 Actinorhizal Symbiosis and Plant Growth

Inoculation with N_2-fixing bacteria *Frankia* improves the nutrient status and enhances actinorhizal plant development (Sayed, 2011).However, the successful establishment and growth of actinorhizal plants in nutrient deficient and/or marginal soils depends on the formation of the symbiotic relationship between plant and N_2fixing bacteria. Thus, to optimize association between the actinorhizal/*Frankia*, an efficient combination in nitrogen fixation well adapted to environmental conditions is recommended. In fields, inoculation with *Frankia* is commonly carried out with crushed nodule, Frankia suspension, *Frankia* enrobed in alginate bead, soil containing Frankia, or leaf litter from around nodulated plants (Mansour *et al.,* 1990; Sougoufara *et al.,* 1989). The response to *Frankia* inoculation is strongly linked to factors such as provenance source, *Frankia* strain, and nutrients status of the site such as nitrogen. Frankia inoculation in nursery and field conditions is beneficial to Casuarina given that it reduces transplantation shock (Landis and Dumrose, 2006).

12.10 Actinorhizal Symbiosis and Environment Stresses

Actinorhizal plants are generally tolerant of abiotic stresses. This tolerance can be improved when plants are associated with *Frankia* (Belanger *et al.,* 2011). The actinorhizal plant-*Frankia* system is widely used for reclaiming lands affected by abiotic stresses. Most of the *Frankia* strains were resistant to an elevated level of several heavy metals and also to salinity (Srivastava*et al.,* 2012). In association with actinorhizal plants, *Frankia* strains resistant to abiotic stresses confer and/or increase capacity of the plant partner to tolerate abiotic stresses. Generally, *Frankia* strains are absent or rare in stressed soils, which indicate the requirement of inoculation of the host plant before transplantation for rehabilitation of saline soil (Hafeez *et al.,* 1999). It has been demonstrated that nodulation occurred under saline conditions until 300mm, approximately 28dS/m, and the symbiotic association between *C. equisetifolia* and *Frankia* can be widely used for the recovery of saline soils (Tani and Sasakawa, 2003).

Frankia and host plant is a key for obtaining the right actinorhizal plant-*Frankia* for rehabilitation of salt affected soils (Ng, 1987). Actinorhizal symbiosis is a biological tool generally used for the remediation and revegetation of soils affected by salt, heavy metal, oil.Besides, the important role of Casuarina on lands reclamation, the symbiotic relationship between *Frankia* actinorhizal plant can also be used as a biocontrol toolagainst diseases such as cataplexy (*Rhizoctonia* sp.), bacterial wilt (*Ralstonia*), powdery mildew (*Oidium*sp.), Hexenbesen (mycoplasma-like organism; bacteria-like organism), and canker (Huang *et al.,* 2011). Actinobacterium is a group of bacteria that generally produces antagonistic compounds against pathogens; the establishment of actinorhizal symbiosis could mitigate the development of plant disease (Zheng, 1996). In nursery experiments, Casuarina plants inoculated with *Frankia* are more resistant to bacterial wilt disease than noninoculated plantsreportd by Kang (1999). *Frankia* strains could counter balance root rot of *C. equisetifolia* caused by *Rhizoctonia* sp. reported by Gopinathan (1995) (Table 12.3 and 12.4). There is a positive correlation between the dose of *Frankia,* the efficiency of disease control, and also nodule formation. Taken together, these results highlight the positive role of *Frankia* as an eco-friendly tool for the control of plant disease and the improvement of agricultural productivity (Mahdi, 2010).

Ecosystem degradation leads to soil infertility and crop losses therefore leading to decreased food security. Plantation of pioneer trees such as actinorhizal plants that are well adapted to such harsh conditions is a promising tool for restoring these degraded lands. However, for the successful establishment of the plantation, care should be taken to select appropriate tree species for a specific environment. To reduce the transplantation shock during reforestation programs and to increase the productivity of these plantations, *Frankia* inoculation must be carried out during the nursery stage. Association between actinorhizal plant and *Frankia* must be optimized by selecting a more efficient symbiosis. Symbiotic microorganism *Frankia* significantly improves performance of actinorhizal plantations. It becomes very important to enhance *Frankia* production particularly in arid and semiarid areas for largescale adoption by farmers.

Table 12.3 : Effect of inoculation with *Frankia* on plant height in field trials of *Casuarinaceae*

Species	Treatments	Plant height (cm)	References
Casuarina equisetifolia	Control	64.8094.60	Karthikeyan
(2 yrs old field plantation)	*Frankia*		*et al.* (2009)
C. cunninghamiana			
(6-8-yrs old field plantation)	Control *Frankia*	580 680	
C. glauca			
(6-8 yrs old field plantation)	Control *Frankia*	780 890	Bulloch (1994)
C. obesa(6-8 yrs old field plantation)	Control		
	Frankia	500-550	
Allocasuarina verticillata			
(15 yrs old field plantation)	Control *Frankia*	580 700	

Source : Diagne *et al.,* (2013)

Table 12.4 : Effect of inoculation with *Frankia* on plants biomass and N contentin field trialsof *Casuarinaceae*

Species	Treatments	Dry weight (g)	Total N (g/plant)	References
Casuarina equisetifolia (3 yrs old field plantation)	Control *Frankia*	1000 1356	15.57 23.35	Sougoufara *et al.* (1989)
Casuarina equisetifolia (2 yrsold field plantation)	Control *Frankia*	300 680	1.02 3.8	Karthikeyam *et al.* (2009)
Alnus crispa (1.5 yrs old field plantation)	Control *Frankia*	111.8 154.1	3.5 4.8	Le francios *et al.* (2007)

Source : Diagne *et al.*, (2013)

12.11 Bio-product of *Frankia*

A biofertilizer product called '*N-fixer*' (the actinomycete *Frankia* isolated from the root nodules of *C. equisetifolia* and cultured in an artificial nutrient medium) was developed by the Division of Forest Protection, IFGTB, Coimbatore. Under experimental conditions the biofertilizer N fixer results 100% of root nodule formation in the seedlings and cuttings of *C. equisetifolia* and *C. junghuhniana* by application of @ 5ml/seedling. It contains $2x10^7$/ ml. This nitrogen fixer was suggested to inoculate particularly for Casuarina to enhance growth and nutrient improvement particularly nitrogen. Shelf life of this product is 3 months in refrigerated condition and one month under room temperature.

12.12 Hapanoids

Frankia are Gram-positive, aerobic, nitrogen-fixing bacteria. The membranes of *Frankia*, as well as the membranes of some other bacteria like *Bradyrhizobium*, *Rhizobium*, and *Streptomyces*, contain lipid components called hapanoids. Hapanoids, which are amphiphilic, pentacylic triterpenoid lipids, condense membrane lipids therefore stabilizing the membranes in a similar way to which sterols do in higher organisms. There are many structural variants of hapanoids like polyol and glycol derivatives that can be found within various bacteria; the specific functions of each hopanoid derivative have not been found in *Frankia* specifically, membranes containing hopanoids surround vesicles called diazovesicles that contain nitrogenase, an oxygen-sensitive enzyme. While hopaniods thicken and stabilize the walls which help to keep oxygen away from the nitrogenase, some suggest that the hopanoids themselves play a more specific and molecular role in the oxygen protection mechanism of nitrogenase.

CHAPTER - 13

Vermiculture and Vermicomposting

Earthworms have been on the earth for over 20 million years. In this time they have faithfully done their part to keep the cycle of life continuously moving. Their purpose is simple but very important. They are nature's way of recycling organic nutrients from dead tissues back to living organisms. Many have recognized the value of these worms. Ancient civilizations, including Greece and Egypt valued the role earthworms played in soil. The Egyptian Pharaoh, Cleopatra said, "Earthworms are sacred."

13.1 Description of Earthworms

Earthworms belong to Phylum Annelida of animal Kingdom. They are long and cylindrical shape, length of earthworms vary from a few mm to nearly 3mtr as noted in case of *Megascolex* species found in Australia. The body of earthworms has a large number of grooves. Due to these grooves the body of the earthworm is divided into number of compartments called segments. In mature earthworms, few anterior segments appear swollen due to glandular thickening. This part is called as clitellum. The segments are not clearly seen in this part. There are 2500 to 3,000 species of earthworms in the world which are adapted to range of environment. More than 350 species have been identified in India. The species of earthworms are differentiated mostly on the basis of external features such as length of earthworms, number of segments, position of clitellum and external aperture present on the body.

13.2 Life cycle of Earthworms

Earthworms are hermaphrodite. However, single earthworm cannot usually self-fertilize itself and two mature earthworms are required to propagate. At the time of egg lying, the clitelium is transformed into hard, girdle like capsule called cocoon. Shedding of cocoon takes place about 10 days after fertilization.Cocoons are small, spindle shaped and translucent structures. The size of cocoon varies with the species of earthworms. The fertilization takes places inside the cocoon. Although, number of fertilized ova in each cocoon ranges from 1 to 20, only a few of them survive and hatch. The juvenile earthworms crawl out of cocoons. The cocoons have an incubation period of 18 days after which they hatch into earthworm. The entire life cycle from hatching of cocoons, maturing of juveniles and again formation of cocoons take a period of 50 to 60 days. Normally the average life span of an earthworm is about one year during which they reproduce about 10 times.

13.3 Type of Earthworms

Ecologically, earthworms are classified into following three types (Table 13.1) such as (a) Epigeics (b) Anecics, and (c)Endogeics (Card *et al.,* 2004).

13.3.1 Epigeics (Greek for "upon the earth")

These "decomposers" are the type of worm used in vermicomposting. These are surface feeders and feed on leaf litter and other plant wastes. They do not have permanent burrows. The do not mix the soil and generally have no effect on soil structure etc. They have affinity for nitrogen rich organic matter, and live in more unstable environment. They resemble the stages of ecological development, characterized by small body size, high metabolic rate, high, fecundity and short life cycle. These earthworms are normally from forest floor community in tropical countries. With loss of natural forests, they have moved into plantations. When such earthworms are chosen to work on waste organic matters their major contribution is in reducing the particle size of organic matter.

13.3.2 Anecics (Greek for "out of the earth")

These are burrowing worms that come to the surface at night to drag food down into their permanent burrows deep within the mineral layers of the soil. They feed on leaf litter and soil of upper layers. They move along vertically into the soil forming vertical furrows. Vermicasts are liberally produced on the soil surface.

13.3.3 Endogeic (Greek for "within the earth")

These are also burrowing worms but their burrows are typically more shallow and they feed on the organic matter already in the soil, so they come to the surface only rarely. They live within the soil and ingest large quantities of organically rich soil. The generally make horizontal furrows and deposit their casts within the furrow.

Epigeic and anecic type of earthworms are important in management of organic wastes e.g. compost formation. Epigeic such as *Eisenia foetida* and *Eudrilus euginiae* are exotic and *Perionyx excavatus* is native which are being used for vermicomposting. There are three varieties of earthworms available in India which is called manure worms. *Eisenia foetida* and *Eudrilus euginiae* are exotic species and used in composting. *Perionyx excavatus* and *Lampito mauritti,* epigeic and anecic forms respectively have also been successfully used in vermiculture and vermicomposting. Anecic such as *Lampito mauritti* is a native one and can play a role *in situ* biodegradation of organic wastes of soil. Endogeic commonly include *Metaphire posthuma* and *Octochaetonia thurstoni* which are indigenous and used for vermicomposting. Research work have been done by Kaplan *et al.* (1980) to show that it grows in a wide range of agricultural wastes. *Eisenia fetida* worms have also been shown to have a wide tolerance range for temperature from 0° to 40°C (Phillip. 1981) and suitability for vermicomposting in terms of temperature (Reinecke *et al.*, 1992) and moisture (Reinecke and Venter, 1987).

Table 13.1 : Comparative Chart of the Major Characteristics of three Earthworm Genera

Sl. No.	Characteristics	A	B	C
		Perionyx ***excavatus*** (Hallatt *et al.*, (1990)	***Eudrilus eugeniae*** (Viljoen and Reinecke, 1989)	***Eisenia foetida*** (Venter and Reinecke, 1988)
1.	Growth rate (mg/worm/day)	3.5	12.0	7.00
2.	Maximum body mass (mg/worm)	600.0	4294.0	1500.00
3.	Maturation attained at age (days)	21.0	40.0	150.00
4.	Start of cocoon production (days)	24.0	46.0	55.00
5.	Cocoon production (worm/day)	1.1	1.3	0.35
6.	Cocoon incubation period (days)	18.7	16.6	23.00
7.	Mean number of hatchings	1.1	2.7	2.70
8.	Number of hatchings ($cocoon^{-1}$)	1.3	1.5	1.90
9.	Duration of life-cycle (days)	46.0	60.0	70.00

Source : Gaur (2010)

13.4 Role of Earthworms in Nature

13.4.1 Earthworms and Soil Formation

Darwin (1981) first recognized that earthworms are common in the A-l horizon of soil, and was called asvegetable mould. These result from the burrowing, casting and subsequent mixing of soil mineral and organic materials by earthworms. In most soils, earthworms burrow and cast throughout the A and B horizons and are sometimes below the B horizon. They constantly mix the minerals and organic materials that make at least A and B horizons of soils. Earthworms are usually in high numbers near the soil surface while the numbers are relatively low in deeper soil layers.

13.4.2 Earthworms on Soil Hydraulic Properties

By burrowing through soil, earthworms alter the soil fabric which, in turn, changes the physical characteristics of soil. *Allolobophorac aliginosa* and *Allolobophora rosea* giveburrow formation on soil hydraulic properties and shows saturated hydraulic conductivity as well as percolation rates was higher when compared to the control.

13.4.3 Earthworms and Soil Amelioration

Earthworms play an important role in the process of soil amelioration and maintenance of soil fertility. They incorporate organic matter and turn over large amount of soil by burrowing, feeding and casting. It leads to improved soils structure (Hoogerkamp *et al.,*1983; Strewart and Scullion. 1988).Application of vermicompost significantly improves water holding capacity and porosity which facilitates the root respiration and growth of crops (Lee, 1992). Earthworm activities are advantageous for the stabilization of reformed soil aggregates. The organic carbon content was increased by 4.1 to 21.0% for burrow wall material and by 21.2 to 43.0% for casts. The total content of polysaccharides was increased by 35 to 87% for casts and by 33 to 46% for the burrow-wall material of all earthworm species (Zhang and Schrader, 1993).

13.4.4 Earthworms and Recycling of Organic Matter

The role of worms in organic matter recycling process and their effect on acceleration of decomposition process (Heungens, 1969) is gaining importance.

13.4.5 Earthworms and Organic Carbon Turnover

Allobophora caliginosa alters the biomass and the metabolic activity of the edaphic microflora over a wide range of soil, and these alterations are highly soil and resource specific and the microbial carbon turnover in soils by *A. caliginosa* is significantly different from the microbial carbon turnover in aged casts and burrows (Wolters and Joergensen, 1992). Vermicomposting resulted in significant reduction in C/N ratio and increase in mineral nitrogen after 90 days of composting over treatments uninoculated with earthworms (Bansal and Kapoor, 2000). Different earthworm species are impacted differently by C/N ratio and feed mixture type (Ndegwa and Thompson, 2000).

13.4.6 Earthworms in Land Improvement/Plant Nutrient Availability

The introduction of earthworms into soils is usually beneficial in terms of plant growth and crop yields. Stockdilland Cossens (1966) introduced earthworms into pastures in New Zealand, and found that with the establishment of the earthworm population the organic matter at the base of the grass disappeared and compacted soils with poor structure were transformed into deep friable top soil. Earthworms increased significantly in number and the rate of growth and yield of plants growing on the inoculated soils. Edwards and Lofty (1980) demonstrated that increased root growth after earthworm inoculation was primarily due to the formation of burrows that were responsible for more available nutrients than the surrounding soil. They have clearly shown that burrows provided ideal channels which promoted the nutrient availability for buffer as well as maximum root growth. Better root growth is also an index for efficient utilization of available nutrients which in turn reflect on the biomass productivity of crops.

13.4.7 Earthworms on Availability of N and P to Plants

Earthworms have an important role in transfer of nitrogen from decaying plant tissues by releasing it in chemical form that can be easily taken up and recycled for plant growth. The significance of earthworms in the nitrogen economy of soils, especially the role of excreted nitrogen contained in mucus and that derived from the decomposition of dead earthworm tissues has received considerable attention. Phosphorus is limiting element for plant growth and especially for fruit and seed production. Much of the phosphorus in soils is bound to organic matter in forms that are unavailable to plants. It is now well established that passage through the gut of some lumbricid earthworms results in some of this phosphorus being converted to forms that are available to plants. This process involves uptake of phosphorus by bacteria and fungi from a pool of phosphorus that is fixed by soil minerals and subsequent ingestion of these organisms by soil fauna like earthworms. Some release of P in forms available to plants is then mediated by the enzyme phosphatase(s) that are produced within the earthworm gut, while further release of P may be induced by micro-organisms in casts after their excretion (Lee, 1992). Vermicomposting may prove to be an efficient technology for providing better P nutrition from different organic wastes (Gosh *et al.,* 1999).

13.4.8 Earthworms on Exchangeable K

Casts of earthworm contained two to three times more available K than the surrounding soil (Bezbordov and Khalbayeva, 1990; Hullegalle and Ezumah, 1991). Basker *et al.* (1993) reported that the changes in the K dynamics of soil resulting from ingestion by worms must be due to atleast in part to the changes in the distribution of K between exchangeable and non-exchangeable forms.

13.4.9 Earthworm Guts Fungi on P Solubilization

Soil micro-organisms particularly bacteria and fungi associated in the root regions of plant play an important role in supply of phosphorus (Gaur and Mathur, 1990). Earthworm casts are reported to improve the soil fertility (Krishnamoorthy and Vajranabhaiah, 1986).The soil fertility factor in these casts might be due to microbial enzymes which help in the availability of nutrients. Phosphate solubilization by fungi isolated from earthworm gut was studied by employing dicalcium and tricalcium phosphates. *Choane phoracucur bitarum* and *Verticillium chlamydosporum, Neurospora crassa* were responsible for liberation of maximum phosphorus form dicalcium and tricalcium phosphates. Fungi differ in their capacity to solubilize phosphorus form dicalcium and tricalcium phosphates. *Verticillium chlamydosporum* was most efficient and dissolved maximum amount of phosphate in soil and made available to plants (Dorcasloy *et. al.* 1991).

13.4.10 Earthworms and Microbial Enzymatic Activities

The interaction between earthworms and micro-organisms are of major importance in the degradation of organic matter and release of mineral nutrients into soil (Lee, 1985). Certain micro-organisms are killed during passage through gut (Yeates, 1981) whereas other proliferates. Efficient dispersal of beneficial organisms like *Rhizobium, Pseudomonas* (Madsen and Alexander, 1982), *Frankia* spp. (Reddell and Spain, 1991) and hyphal filament spores of VAM fungi (Rebatin and Stinner, 1988) and possibility of biocontrol agent application through earthworms food were reported by Duble *et al.* (1994). Earthworms have also been reported to increase the availability of plant nutrient in soil (Krishnamoorthy and Vajranabhiah, 1986) increase the production of plant growth regulators (Tomati *et al.,* 1988) and also increase foliar concentration of Ca, Na, Mn, Cu, Fe and Al in wheat (Stephens *et al.* 1994). An increase in the density of microbes and nitrogen fixer due to vermicompost application was reported by Kale *et al.* (1992).The worm could not digest micro fungal species like *Aspergillus, Penicillium* and *Fusarium* (Nivadita *et al.,* 1989). Binet *et al.* (1988) observed the stimulatory effect of earthworms on soil microbial activity under micro-organisms controlled condition.

13.5 Vermiculture and Vermicomposting

Vermiculture is a mixed culture containing soil bacteria mixed and an effective strain of earth worms (NIIR Board, 2008). Earthworm has efficiency to consume all types of organic rich waste material including vegetable waste, industrial and other organic waste. Vermicomposting refers to the production of plant nutrient rich excreta of worms. Vermicomposting is a simple biotechnological process of composting, in which certain species of earthworms are used to enhance the process of waste conversion and produce a better end product. Vermicomposting differs from composting in several ways (Gandhi *et al.*, 1997). It is a mesophilic process, utilizing microorganisms and earthworms that are active at 10–32°C (not ambient temperature but temperature within the pile of moist organic material). The process is faster than composting; because the material passes through the earthworm gut, a significant but not yet fully understood transformation takes place, whereby the resulting earthworm castings (worm manure) are rich in microbial activity and plant growth regulators, and fortified with pest repellence attributes as well! In short, earthworms, through a type of biological alchemy, are capable of transforming garbage into 'gold' (Vermi, 2001; Crescent, 2003). Vermicomposting is the process of turning organic debris into worm castings (composting of organic matter). The worm castings are very important to the fertility of the soil. The castings contain high amounts of N, P, K, Ca and Mg. Castings containsfive times the available nitrogen, seven times the available potash, and one and half times more calcium than found in good topsoil. Worm castings can hold close to nine times their weight in water. "Vermiconversion," or using earthworms to convert waste into soil additives, has been done on a relatively small scale for some time. A recommended rate of vermicompost application is 15-20%. Several researchers have demonstrated that earthworm castings have excellent aeration, porosity, structure, drainage, and moisture-holding capacity. The content of the earthworm castings, along with the natural tillage by the worms burrowing action, enhances the permeability of water in the soil. Vermicomposting is done on small and large scales. In the 1996 summer olympics, the Australians used worms to take care of their tons and tons of waste.They then found that waste produced by the worms could be very beneficial to their plants and soil. People in the U.S. have commercial vermicomposting facilities, where they raise worms and sell the castings that the worms produce. Then there are just people who own farms or even small gardens, and they may put earthworms into their compost heap, and then use that for fertilizer.

13.5.1 Utilization of Vermicompost

Vermicompost is nothing but the excreta of earthworms, which is rich in humus and nutrients. We can rear earthworms artificially in a brick tank or near the stem/trunk of trees (specially horticultural trees). By feeding these earthworms with biomass and watching properly the food (bio-mass) of earthworms, we can produce the required quantities of vermicompost.

13.5.2 Materials for Preparation of Vermicompost

Several types of biodegradable wastes as crop residues, weed biomass, vegetable waste, leaf litter, hotel refuse, waste from agro-industries and biodegradable portion of urban and rural wastes are used for the preparation of vermicompost.

13.5.3 Phase of Vermicomposting

Phase 1 : **Processing:** involving collection of wastes, shredding, mechanical separation of the metal, glass and ceramics and storage of organic wastes.

Phase 2 : **Pre-digestion of organic waste**: Pre digestion of organic waste for twenty days by heaping the material along with cattle dung slurry. This process partially digests the material and fit for earthworm consumption. Cattle dung and biogas slurry may be used after drying. Wet dung should not be used for vermicompost production.

Phase 3 : **Preparation of earthworm bed**: A concrete base is required to put the waste for vermicompost preparation. Loose soil will allow the worms to go into soil and also while watering; all the dissolvable nutrients go into the soil along with water.

Phase 4 : **Collection of earthworm after vermicompost collection:** Sieving the composted material to separate fully composted material. The partially composted material will be again put into vermicompost bed.

Phase 5 : **Storing the vermicompost**: Storing the vermicompost in proper place to maintain moisture and allow the beneficial microorganisms to grow.

13.5.6 Essentials Compost Worms need Five Basic Things

1. An hospitable living environment, usually called "bedding".
2. A food source.
3. Adequate moisture (greater than 50% water content by weight).
4. Adequate aeration.
5. Protection from temperature extremes.

These five essentials are discussed in more detail below :

13.5.6.1 Bedding

Bedding is any material that provides the worms with a relatively stable habitat. This habitat must have the following characteristics (Table 13.2):

- **High absorbency:** Worms breathe through their skins and therefore must have a moist environment in which to live. If a worm's skin dries out, it dies. The bedding must be able to absorb and retain water fairly well if the worms are to thrive.
- **Good bulking potential:** If the material is too dense to begin with, or packs too tightly, then the flow of air is reduced or eliminated. Worms require oxygen to live, just as we do. Different materials affect the overall porosity of the bedding through a variety of factors, including the range of particle size and shape, the texture, and the strength and rigidity of its structure. The overall effect is referred to in this document as the material's bulking potential.
- **Low protein and/or nitrogen content (high C: N):** Although, the worms do consume their bedding as it breaks down, it is very important that this be a slow process. High protein/nitrogen levels can result in rapid degradation and its associated heating, creating inhospitable, often fatal, conditions. Heating can occur safely in the food layers of the vermiculture or vermicomposting system, but not in the bedding.

Table 13.2 : Common Bedding Materials

Bedding Material	Absorbency	Bulking Pot.	C:N Ratio
Corn stalks	Poor	Good	60 - 73
Corn cobs	Poor-Medium	Good	56 - 123
Corn Silage	Medium-Good	Medium	38 - 43
Hay – general	Poor	Medium	15 - 32
Straw – general	Poor	Medium-Good	48 - 150
Straw – oat	Poor	Medium	48 - 98
Straw – wheat	Poor	Medium-Good	100 - 150
Corrugated cardboard	Good	Medium	563
Lumber mill waste — chipped	Poor	Good	170
Paper fibre sludge	Medium-Good	Medium	250
Paper mill sludge	Good	Medium	54
Newspaper	Good	Medium	170
Paper from municipal waste stream	Medium-Good	Medium	127 - 178

Contd...

Bark – hardwoods	Poor	Good	116 - 436
Bark — softwoods	Poor	Good	131 - 1285
Sawdust	Poor-Medium	Poor-Medium	142 - 750
Shrub trimmings	Poor	Good	53
Hardwood chips, shavings	Poor	Good	451 - 819
Softwood chips, shavings	Poor	Good	212 - 1313
Leaves (dry, loose)	Poor-Medium	Poor-Medium	40 - 80
Horse Manure	Medium-Good	Good	22 - 56
Peat Moss	Good	Medium	58

Worms can be enormously productive (and reproductive) if conditions are good; however, their efficiency drops off rapidly when their basic needs are not met (see discussion on moisture below). Good bedding mixtures are an essential element in meeting those needs. They provide protection from extremes in temperature, the necessary levels and consistency of moisture, and an adequate supply of oxygen. Fortunately, given their critical importance to the process, good bedding mixtures are generally not hard to come by on farms. The most difficult criterion to meet adequately is usually absorption, as most straws and even hay are not good at holding moisture. This can be easily addressed by mixing some aged or composted cattle or sheep manure with the straw. The result is somewhat similar in its bedding characteristics to aged horse manure. Mixing beddings need not be an onerous process; it can be done by hand with a pitchfork (small operations), with a tractor bucket (larger operations), or, if one is available, with an agricultural feed mixer.

13.5.6.2 Worm Food

Compost worms are big eaters. Under ideal conditions, they are able to consume in excess of their body weight each day, although the general rule-of-thumb is ½ of their body weight per day. They will eat almost anything organic (plant or animal origin), but they definitely prefer some foods to others (Table 13.3). Manures are the most commonly used worm feedstock, with dairy and beef manures generally considered the best natural food for Eisenia, with the possible exception of rabbit manure. The former, being more often available in large quantities, is the feed most often used.

Table 13.3 : Common Worm Feed Stocks

Sr. No.	Food	Advantages	Disadvantages
1.	Legume hays	Higher N content makes these good feed as well as reasonable bedding	Moisture levels not as high as other feeds, requires more input and monitoring
2.	Fresh food scraps (e.g., peels, other food prep waste, leftovers, commercial food processing wastes)	Excellent nutrition having good moisture content, and possibility of revenues from waste tipping fees	Pre-composting is required
3.	Pre-composted food wastes	Good nutrition	Nutrition less than with fresh food wastes
4.	Seaweed	Good nutrition due to high in micronutrients and beneficial microbes	Salt must be rinsed off
5.	Biosolids (human waste)	Excellent nutrition and possibility of waste management revenues	Heavy metal and/or chemical contamination
6.	Cattle manure	Good nutrition	Weed seeds make pre-composting necessary
7.	Poultry manure	High N content results in good nutrition and a high-value product	High protein levels can be dangerous to worms
8.	Sheep/Goat manure	Good nutrition	Require pre-composting (weed seeds) and small particle size can lead to packing
9.	Hog manure	Good nutrition and produces excellent vermicompost	Usually in liquid form, therefore must be dewatered or used with large quantities of highly absorbent bedding
10.	Rabbit manure	N content second only to poultry manure therefore contains very good mix of vitamins & minerals	Must be leached prior to use because of high urine content
11.	Corrugated cardboard (including waxed)	Excellent nutrition due to high-protein glue used to hold layers together	Must be shredded (waxed variety) and/or soaked (non-waxed) prior to feeding
12.	Fish, poultry offal; blood wastes; animal mortalities	High N content provides good nutrition; opportunity to turn problematic wastes into high-quality product	Must be pre-composted until past thermophillic stage

13.5.6.3 Moisture

The bedding used must be able to hold sufficient moisture if the worms are to have a livable environment. They breathe through their skins and moisture content in the bedding of less than 50% is dangerous. With the exception of extreme heat or cold, nothing will kill worms faster than a lack of adequate moisture. The ideal moisture-content range for materials in conventional composting systems is 45-60%. In contrast, the ideal moisture-content range for vermicomposting or vermiculture processes is 70-90%.

13.5.6.4 Aeration

Worms are oxygen breathers and cannot survive anaerobic conditions (absence of oxygen). When factors such as high levels of grease in the feedstock or excessive moisture combined with poor aeration conspire to cut off oxygen supplies, areas of the worm bed, or even the entire system, can become anaerobic. This will kill the worms very quickly. Not only are the worms deprived of oxygen, they are also killed by toxic substances (ammonia) created by different sets of microbes that bloom under these conditions. This is one of the main reasons for not including meat or other greasy wastes in worm feedstock unless they have been pre-composted to break down the oils and fats.

13.5.6.5 Temperature Control

Controlling temperature to within the worms' tolerance is vital to both vermicomposting and vermiculture processes. This does not mean, however, that heated buildings or cooling systems are required. Worms can be grown and materials can be vermicoposted using low-tech systems, outdoors and year-round.

- **Low temperatures:** Eisenia can survive in temperatures as low as 0°C, but they don't reproduce at single-digit temperatures and they don't consume as much food. It is generally considered necessary to keep the temperatures above 10°C (minimum) and preferably 15°C for vermicomposting efficiency and above 15°C (minimum) and preferably 20°C for productive vermiculture operations.
- **High temperatures:** Compost worms can survive temperatures in the mid-30s but prefer a range in the 20°C. Above 35°C will cause the worms to leave the area. If they cannot leave, they will quickly die. In general, warmer temperatures (above 20°C) stimulate reproduction.

- **Worms's response to temperature differentials**: Compost worms will redistribute themselves within piles, beds or windrows according to temperature gradients. In outdoor composting windrows in wintertime, where internal heat from decomposition is in contrast to frigid external temperatures, the worms will be found in a relatively narrow band at a depth where the temperature is close to optimum. They will also be found in much greater numbers on the south-facing side of windrows in the winter and on the opposite side in the summer.
- **Effects of freezing:** Eisenia can survive having their bodies partially encased in frozen bedding and will only die when they are no longer able to consume food.

13.5.7 Other Important Parameters

There are a number of other parameters of importance to vermicomposting and vermiculture:

- **pH:** Worms can survive in a pH range of 5 to 9. Most experts feel that the worms prefer a pH of 7 or slightly higher. In general, the pH of worm beds tends to drop over time. If the food sources are alkaline, the effect is a moderating one, tending to neutral or slightly alkaline. If the food source or bedding is acidic (coffee grounds, peat moss) than the pH of the beds can drop well below 7. This can be a problem in terms of the development of pests such as mites. The pH can be adjusted upwards by adding calcium carbonate. In the rare case where they need to be adjusted downwards, acidic bedding such as peat moss can be introduced into the mix.
- **Salt content:** Worms are very sensitive to salts, preferring salt contents less than 0.5%. If saltwater seaweed is used as a feed (and worms do like all forms of seaweed), then it should be rinsed first to wash off the salt left on the surface. Similarly, many types of manure have high soluble salt contents (up to 8%). This is not usually a problem when the manure is used as a feed, because the material is usually applied on top, where the worms can avoid it until the salts are leached out over time by watering or precipitation. If manures are to be used as bedding, they can be leached first to reduce the salt content. This is done by simply running water through the material for a period of time. If the manures are pre-composted outdoors, salts will not be a problem.
- **Urine content:** If the manure is from animals raised or fed off in concrete lots, it will contain excessive urine because the urine cannot

drain off into the ground. This manure should be leached before use to remove the urine. Excessive urine will build up dangerous gases in the bedding. The same fact is true of rabbit manure where the manure is dropped on concrete or in pans below the cages.

- **Another toxic components:** Different feeds can contain a wide variety of potentially toxic components. Some of the more important are as (a) De-worming medicine in manures, particularly horse manure (b) Detergent cleansers industrial chemicals, pesticides: These can often be found in feeds such as sewage or septic sludge, paper-mill sludge, or some food processing wastes, and (c) Tannins.

13.5.8 Vermicompost Production Methodology

Production of Vermicompost is completed in following stages as-

13.5.8.1 Selection of Suitable Earthworm

For vermicompost production, the surface dwelling earthworm alone should be used. The earthworm, which lives below the soil, is not suitable for vermicompost production. The African earthworm (*Eudrillus engenial),* Red worms (*Eisenia foetida)* and composting worm (*Peronyx excavatus)* are promising worms used for vermicompost production (Fig. 13.1A, B & C). All the three worms can be mixed together for vermicompost production. The African worm is preferred over other two types, because it produces higher production of vermicompost in short period of time and younger ones in the composting period.

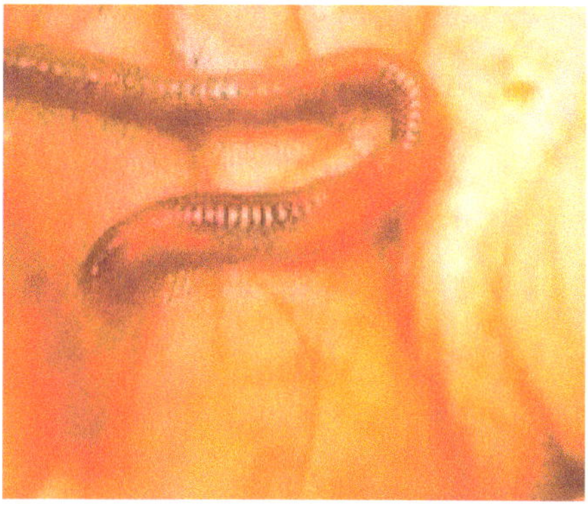

Fig 13.1A : African Earthworm (*Eudrillus euginiae*)

Fig 13.1B : Tiger Worm or Red Wrinkle (*Eisenia foetida)*

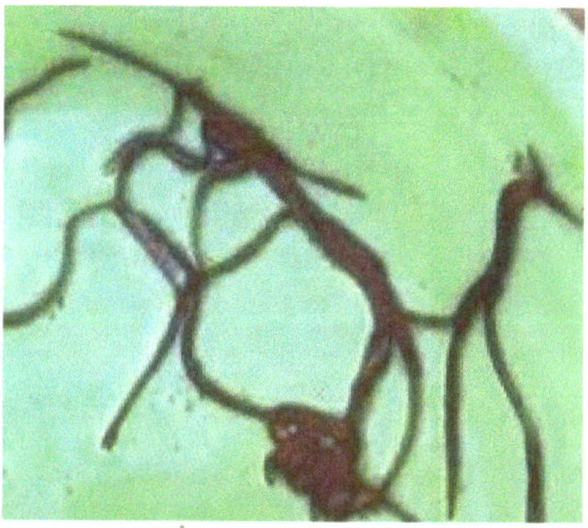

Fig 13.1C : Asian Worms (*Perionyx ecavatus*)

13.5.8.2 Selection of Site for Vermicompost Production

Vermicompost can be produced in any place with shade, high humidity and cool. Abandoned cattle shed or poultry shed or unused buildings can be used. If it is to be produced in open area, shady place is selected. A thatched roof may be provided to protect the process from direct sunlight and rain. The waste heaped for vermicompost production should be covered with moist gunny bags.

13.5.8.3 Containers used for Vermicompost Production

A cement tub may be constructed to 2½ feet height and a of 3 feet breadth. The length may be fixed to any level depending upon the size of the room. The bottom of the tub is made to slope like structure to drain the excess water from vermicompost unit. A small sump is necessary to collect the drain water.In another option over the hand floor, hollow blocks/bricks may be arranged in compartment to a height of one feet, breadth of 3 feet and length to a desired level to have quick harvest. In this method, moisture assessment will be very easy. No excess water will be drained. Vermicompost can also be prepared in wooden boxes, plastic buckets or in any containers with a drain hole at the bottom.

13.5.8.4 Preparation of Vermiculture Bedor Worm Bed

Vermiculture bed (3cm) can be prepared by placing after saw dust or husk or coir waste or sugarcane trash in the bottom of tub/container. A 3cm layer of fine sand should be spread over the culture bed followed by a 3cm layer of garden soil. All layers must be moistened with water.

13.5.8.5 Selection of the Raw Material for Vermicompost Production

Cattle dung (except goat, poultry and pig,), crop residues, vegetable market waste, flower market waste, , farm wastes, fruit market waste ,agro industrial waste and all other bio degradable waste are suitable for vermicompost production. The cattle dung should be dried in open sunlight before used for vermicompost production. All other waste should be predigested with cow dung for twenty days before put into vermibed for composting.

13.5.8.6 Putting the waste in the Container

The predigested waste material should be mud with 30% cattle dung either by weight or volume. The mixed waste is placed into the tub/container upto brim. The moisture level should be maintained at 60%. Over this material, the selected earthworm is placed uniformly. For one-meter length, one-meter breadth and 0.5-meter height, 1 kg of worm (1000 Nos.) is required. There is no necessity that earthworm should be put inside the waste. Earthworm will move inside on its own.

13.5.8.7 Watering on the Vermibed

Daily watering is not required for vermibed, but 60% moisture should be maintained throughout the period. If necessity arises, water should be sprinkled over the bed rather than pouring the water. Watering should be stopped before the harvest of vermicompost.

13.5.8.8 Harvesting Vermicompost

13.5.8.8.1 Tub Method : In the tub method of composting, the castings formed on the top layer are collected periodically. The collection may be carried out once in a week. With hand the casting will be scooped out and put in a shady place as heap like structure. The harvesting of casting should be limited up to earthworm presence on top layer. This periodical harvesting is necessary for free flow and retains the compost quality. Otherwise, finished compost get compacted when watering is done.

13.5.8.8.2 Small Bed Type method : In small bed type of vermicomposting method, periodical harvesting is not required. Since the height of the waste material heaped is around one foot, the produced vermicompost will be harvested after the process is over.

13.5.8.9 Harvesting of the Earthworm

After the vermicompost production, the earthworm present in the tub/ small bed may be harvested by trapping method. In the vermibed, before harvesting the compost, small, fresh cow dung ball is made and inserted inside the bed in five or six places. After 24 hrs, the cow dung ball is removed. All the worms will be adhered into the ball. Putting the cow dung ball in a bucket of water will separate this adhered worm. The collected worms will be used for next batch of composting.Worm harvesting is usually carried out in order to sell the worms, rather than to start new worm beds. Expanding the operation (new beds) can be accomplished by splitting the beds that is, removing a portion of the bed to start a new one and replacing the material with new bedding and feed. When worms are sold, however, they are usually separated, weighed, and then transported in a relatively sterile medium, such as peat moss. To complete this, the worms must first be separated from the bedding and vermicompost. There are three basic categories of methods used by growers to harvest worms such as manual, migration, and mechanical.

13.5.8.9.1 Manual Methods : Manual methods are the ones used by hobbyists and smaller-scale growers, particularly those who sell worms to the home-vermicomposting or bait market. In essence, manual harvesting involves hand-sorting, or picking the worms directly from the compost by hand. This process can be facilitated by taking advantage of the fact that worms avoid light. If material containing worms is dumped in a pile on a flat surface with a light above, the worms will quickly dive below the surface. The harvester can then remove a layer of compost, stopping when worms become visible again. This process is repeated several times until there is nothing left on the table except a huddled mass of worms under a thin covering of compost. These worms can then be quickly scooped into a container, weighed, and prepared for delivery.There are several minor variations and/or enhancements on this method, such as using a container instead of a flat surface, or making several piles at once, so that the person harvesting can move from one to another, returning to the first one in time to remove the next layer of compost. They are all labourintensive, however, and only make sense if the operation is small and the value of the worms is high.

13.5.8.9.2 Migration (Self-Harvesting) Methods : These methods, like some of the methods used in vermicomposting, are based on the worms tendency to migrate to new regions, either to find new food or to avoid undesirable conditions, such as dryness or light. Unlike the manual methods described above, however, they often make use of simple mechanisms, such as screens or onion bags.The screen method is very common and easy to use. A box is constructed with a screen bottom. The mesh is usually ¼", although 1/8" can be used as well. There are two different approaches. The downward-migration system is similar to the manual system, in that the worms are forced downward by strong light. The difference with the screen system is that the worms go down through the screen into a prepared, pre-weighed container of moist peat moss. Once the worms have all gone through, the compost in the box is removed and a new batch of worm-rich compost is put in. The process is repeated until the box with the peat moss has reached the desired weight. Like the manual method, this system can be set up in a number of locations at once, so that the worm harvester can move from one box to the next, with no time wasted waiting for the worms to migrate.

The upward-migration system is similar, except that the box with the mesh bottom is placed directly on the worm bed. It has been filled with a few centimeters of damp peat moss and then sprinkled with a food attractive to worms, such as chicken mash, coffee grounds, or fresh cattle manure. The box is removed and weighed after visual inspection indicates that sufficient worms have moved up into the material. This system is used extensively in Cuba, with the difference that large onion bags are used instead of boxes. The advantage of this system is that the worm beds are not disturbed. The main disadvantage is that the harvested worms are in material that contains a fair amount of unprocessed food, making the material messier and opening up the possibility of heating inside the package if the worms are shipped. The latter problem can be avoided by removing any obvious food and allowing a bit of time for the worms to consume what is left before packaging.

13.5.8.10 Nutritive Composition of Vermicompost

The nutrients content in vermicompost vary depending on the waste material that is being used for compost preparation. If the waste materials are heterogeneous one, there will be wide range of nutrients available in the compost. If the waste materials are homogenous one, there will be only certain nutrients are available. The common available nutrients in vermicompost are given in Table 13.3.

Table 13.3 : Common Available Nutrients in Vermicompost

Nutrients	Amounts
Organic carbon	9.5 – 17.98%
Nitrogen	0.5 – 1.50%
Phosphorous	0.1 – 0.30%
Potassium	0.15 – 0.56%
Sodium	0.06 – 0.30%
Calcium and Magnesium	22.67 to 47.60 meq/100g
Copper	2 – 9.50 mg/kg
Iron	2 – 9.30 mg/kg
Zinc	5.70 – 11.50 mg/kg
Sulphur	128 – 548 mg/kg

13.5.8.11 Storing and Packaging of Vermicompost

The harvested vermicompost should be stored in dark, cool place. It should have minimum 40% moisture. Sunlight should not fall over the composted material. It will lead to loss of moisture and nutrient content. It is advocated that the harvested composted material is openly stored rather than packed in over sac. Packing can be done at the time of selling. If it is stored in open place, periodical sprinkling of water may be done to maintain moisture level and also to maintain beneficial microbial population. If the necessity comes to store the material, laminated over sac is used for packing. This will minimize the moisture evaporation loss. Vermicompost can be stored for one year without loss of its quality, if the moisture is maintained at 40% level.

13.5.9 Advantages of Vermicompost

- Vermicompost is rich in all essential plant nutrients, and provides excellent effect on overall plant growth encourages the growth of newshoots/leaves and improves the quality and shelf life of the produce.
- Vermicompost is free flowing, easy to apply, handle and store and does not have badodour.It improves soil structure, texture, aeration, and waterholding capacity and prevents soil erosion.
- Vermicompost is rich in beneficial micro flora such as a fixers, P-solubilizers, cellulose decomposing micro-flora etc. in addition to improve soil environment.
- Vermicompost contains earthworm cocoons and increases the population andactivity of earthworm in the soil.

- Vermicompost is free from pathogens, toxic elements, weed seeds etc.
- Vermicompost minimizes the incidence of pest and diseases.
- Vermicompost enhances the decomposition of organic matter in soil, and contains valuable vitamins, enzymes and hormones like auxins, gibberellins etc.
- Vermicompost prevents nutrient losses and increases the use efficiency of chemical fertilizers, and neutralizes the soil protection.

13.5.10 Pests and Diseases of Vermicompost

Compost worms are not subject to diseases caused by micro-organisms, but they are subject to predation by certain animals and insects (red mites are the worst) and to a disease known as "sour crop" caused by environmental conditions.

- **Moles:** Earthworms are moles' natural food, so if a mole gets access to your worm bed, you can lose a lot of worms very quickly. This is usually only a problem when using windrows or other open-air systems in fields. It can be prevented by putting some form of barrier, such as wire mesh, paving, or a good layer of clay, under the windrow.
- **Birds:** They are not usually a major problem, but if they discover your beds they will come around regularly and help themselves to some of your workforce. Putting a windrow cover of some type over the material will eliminate this problem. These covers are also useful for retaining moisture and preventing too much leaching during rainfall events. Old carpet can be used for this purpose and is very effective.
- **Centipedes:** These insects eat compost worms and their cocoons. Fortunately, they do not seem to multiply to a great extent within worm beds or windrows, so damage is usually light. If they do become a problem, one method suggested for reducing their numbers is to heavily wet (but not quite flood) the worm beds. The water forces centipedes and other insect pests (but not the worms) to the surface, where they can be destroyed by means of a hand-held propane torch or something similar.
- **Ants:** These insects are more of a problem because they consume the feed meant for the worms. Ants are particularly attracted to sugar, so avoiding sweet feeds in the worm beds reduces this problem to a minor one.
- **Mites:** There are a number of different types of mites that appear in

vermiculture and vermicomposting operations, but only one type is a serious problem: red mites. White and brown mites compete with worms for food and can thus have some economic impact, but red mites are parasitic on earthworms. They suck blood or body fluid from worms and they can also suck fluid from cocoons. The best prevention for red mites is tomake sure that the pH stays at neutral or above. This can be done by keeping the moisture levels below 85% and through the addition of calcium carbonate, as required.

- **Sour crop or protein poisoning:** This "disease" is actually the result of too much protein in the bedding. This happens when the worms are overfed. Protein builds up in the bedding and produces acids and gases as it decays. According to some workers: "when you see a worm with a swollen clitellumor see one crawling aimlessly around on top of the bedding, you can just bet on sour crop and act accordingly, but fast". The solution is a "massive dose of one of the mycins, such as farmers give to chicken or cattle". Farmers wishing to avoid these or similar antibiotics should work to prevent sour crop by not overfeeding and by monitoring and adjusting pH on a regular basis. Keeping the pH at neutral or above will preclude the need for these measures.

13.5.11 Precautions

a. Do not cover vermicompost system with plastic sheets as it may trap heat and gases.

b. Do not overload the vermibed with organic matter because rise to thermophilic temperature will adversely affect their numbers.

c. Avoid addition of higher quantities of acid rich substances like tomatoes or citrus fruit wastes.

d. Proper maintenance of moisture is extremely important. Dry conditions kill the worms and waterlogging condition drives them away.

e. Precaution against attacks by red ants who may also feed on cocoons is required by spraying a suitable decoction around the enclosure. A decoction containing 20L water, 100g each of chili powder, turmeric powder and salt and a little soap powder may be sprayed around the units on the soil to keep the red ants under control, Neem oil (0.5%) may also be used (Ismail, 2000).

13.6 Vermiwash

Earthworms play a vital role in plant growth. It is a quite possible to effect quick change over for sustainable agriculture by harnessing brand new vermicompost technology to the soil. Now days, the commercial vermin culturists have started promoting a product called vermiwash Vermiwash (clear and transparent, pale yellow coloured fluid). This vermiwash would have enzymes, secretions of earthworms which would stimulate the growth and yield of crops and even develop resistance in crops receiving this spray. Such a preparation would certainly have the soluble plant nutrients apart from some organic acids and mucus of earthworms and microbes (Shivsubramanian and Ganeshkumar, 2004). But so far there are no experimental evidences to quantify the effect of such spray. Microbes in the environment significantly influence the biogeochemical cycle of phosphorus. The organic phosphorous compounds are decomposed and mineralized by enzymatic complexes like phosphatases produced by microbes. In the ecosystem, a mixed population of microbes is essential to promote enzymatic degradation of naturally occurring phosphorous compounds (Trivedi and Bhatt, 2006). Vermiwash are transported to the leaf, shoots and other parts of the plants in the natural ecosystem as foliar spray.Shweta and Singh (2007) reported that presence of plant growth promoting substance in vermicompost and in an article published in The Hindu Newspaper by Subasashri (2004), Vermiwash was reported for an effective biopesticides. It contains nitrogen fixing bacteria like *Azotobacter* sp., *Agrobactericum* sp. and *Rhizobium* sp. and some phosphate solubilizing bacteria (Zambare *et al.,* 2008) and other nutrients as given in Table 13.4a, b & c)

13.6.1 Principle of Vermiwash Production

The basic principle of Vermiwash preparation is simple. Worm worked soils have burrows formed by the earthworms. Bacteria richly inhabit these burrows, also called as the drilospheres. Water passing through these passages washes the nutrients from these burrows to the roots to be absorbed by the plants. This principle is applied in the preparation of vermiwash. Vermiwash can be produced by allowing water to percolate through the tunnels made by the earthworms on the coconut leaf - cow dung substrate kept in a plastic barrel. Water is allowed to fall drop by drop from a pot hung above the barrel into the vermicomposting system. But barrels are not a must for Vermiwash preparation. Vermiwash units can be set up either in barrels or in buckets or even in small earthen pots (Fig. 13.2).

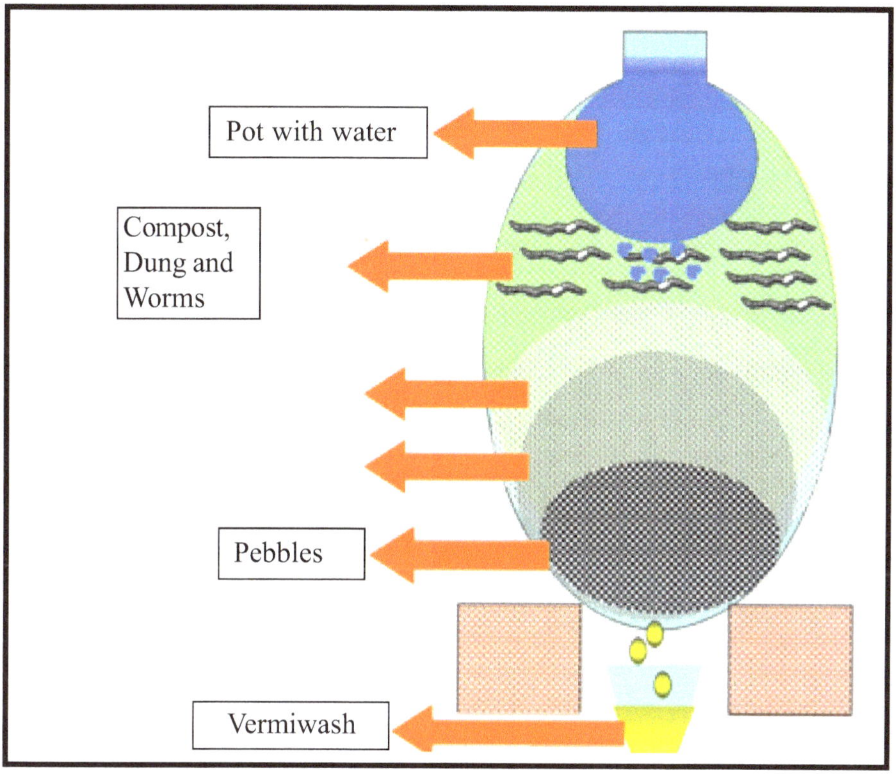

Fig. 13.2 : Vermiwash

Table 13.4a : Composition of Vermiwash

1.	pH	7.48 ± 0.03
2.	Organic carbon %	0.008 ± 0.001
3.	Nitrogen %	0.01 ± 0.005
4.	Available phosphorous %	1.69 ± 0.05
5.	Potassium (ppm)	25 ± 2

Table 13.4b : Micro-nutrients in Vermiwash

Sl. No.	Nutrient	Quantity (ppm)
1.	Sodium	8 ± 1
2.	Calcium	3 ± 1
3.	Copper	0.01 ± 0.001
4.	Ferrous	0.06 ± 0.001
5.	Magnesium	158.44 ± 23.42
6.	Manganese	0.58 ± 0.040
7.	Zinc	0.02 ± 0.001

Table 13.4c : Microbial composition of Vermiwash

Sl. No.	Microbes	Microbial count (CFU/ml)
1.	Total heterotrophs	$1.79 x 10^3$
2.	*Nitrosomonas* sp.	$1.01 x 10^3$
3.	*Nitrobacter* sp.	$1.12 x 10^3$
4.	Total fungi	$1.46 x 10^3$

13.6.2 Method of Preparation of Vermiwash

Vermiwash preparation is completed into following steps such as:

- Select one sufficiently large container made of concrete or plastic bucket.
- Drill a hole on the base of the container to fix a tap to it. A base layer of gravel or broken small pieces of bricks are placed upto height of 10-15cm.
- On the coarse sand layer place 40-45cm pre-decomposed organic wastes and moistens the different layers by using water.
- Introduce about 2000 earthworms into the container. To get vermiwash continuously suspend a mud pot or a small bucket with some holes. Cotton wicks/or bamboo sticks are place in the holes so that water can trickle down.
- Fill the container with 4-5 liters water every day.
- After 10 days vermiwash starts forming in the container.
- Every day about 3-4 liters of vermiwash can be collected.

13.6.3 Benefits of Vermiwash

- A mixture of Vermiwash (1litre) with cow urine (1litre) in 10 liters of water acts as bio-pesticides and liquid manure.
- Vermiwash increases the rate of photosynthesis in crop and rate of decomposition of compost, therefore enhance the crop yield.
- Vermiwash increases the number of micro-organisms in the soil, and enhance the resistance to pest and diseases.

13.6.4 Dosage for Use

a) **Root dip/Stem dip:** The seedlings before transplanting are dipped in Vermiwash solution which is diluted 5 times with water for 15-20 minutes and then transplanted. Similarly the cuttings can also be dipped in the solution.

b) **Foliar spray:** Vermiwash is diluted in water 5 times and sprayed on the foliage of crops. It provides the plant with vital nutrients but also helps tocontrol plant disease.

c) **Soil drench:** Vermiwash is diluted 10 times with water and the soil is drenched with the solution to prevent some of the soil borne pathogens.

13.6.5 Precautions

- The tap should be always kept open to collect the washings.
- The unit starts yielding good quality Vermiwash after ten days.
- Do not allow to compact the contents.
- Do not mix un-decomposed materials while, watering, and add any green material.
- Water should be poured slowly.
- Vermiwash should be stored in cool dry place.

CHAPTER - 14

Liquid Biofertilizers

14.1 Introduction

The green revolution brought amazing consequences in food grain production but with insufficient concern for agricultural sustainability. The availability and affordability of fossil fuel based chemical fertilizers at farm level in India have been ensured only through imports and subsidies which are largely dependent on GDP of the country. Dependence on chemicals for future agricultural needs would result in further loss in soil health, possibilities of water contamination and calculated burden on the fiscal system. Indiscriminate synthetic fertilizer usage has polluted the soil, water basins, destroyed micro-organisms and eco-friendly insects, made the crop more susceptible to diseases and depleted soil fertility at the primary levels as of today (Pindi, 2012). India is an agricultural based country. In order to feed the ever growing populations, India has to increase the per unit area productivity. According to United Nations Food and Agriculture Organization (FAO) estimations, the average demand for the agricultural commodities will be 60% higher in 2030 than present time and more than 85% of this additional demand will be from developing countries (Mia andShamsuddin, 2010). For over half a century, the world has relied on the concept of increasing crop yields to supply an ever increasing demand of food. Therefore, vertical expansion of food production is necessary. In order to increase the unit area productivity of agricultural land, the role of different crop nutrients in contributing increased crop yield is vital. Among the crop nutrient, nitrogen as well as phosphorus play an important role in increasing the crop productivity. Further, the nitrogenous chemical fertilizers are manufactured industrially using nonrenewable petroleum products under high temperature and high pressure. Increase in petroleum cost day by day effects the cost of the chemical fertilizers. In addition, more than 50% of the applied N-fertilizers are somehow lost through different agricultural processes which not only lead to economical loss to the farmers and polluted environment consequently.

Feedstock/fossil fuels depletion and increasing fertilizer cost are making marginal farmer unaffordable. Growing concern about environmental hazards, increasing threat to sustainable agriculture are some of the other reliable reasons for the biofertilizer promotion. However, plant nutrients like N, P and K are highly essential for plant growth and metabolism. It is also evident that plants utilizes nutrients in greater amounts from soil in modern intensive cultivation and needs replenishment (Pindi and Satyanarayana, 2012). Under such conditions Micro-organisms offer good alternative technology to replenish crop nutrients. In agricultural eco-system, microorganisms have vital role in fixing/ solubilizing/ mobilizing/nutrient recycling. These microorganisms occur in soils naturally, but their populations are often scanty. In order to increase the crop yield, the desired microbes from rhizosphere are isolated and artificially cultured in adequate count and mixed with suitable carriers or as they are in suitable combinations.

14.2 Estimated Demand and Supply of Biofertilizers

India is one of the important countries in biofertilizer production and consumption. In order to encourage the organic agriculture by biofertilizers, five biofertilizers namely *Rhizobium*, *Azotobacter*, *Azospirillum*, Phosphate solubilizing bacteria and *Mycorrhiza* have been incorporated in the FCO, 1985. The average consumption in the country is about 45,000 tons per annum while the production being less than the half. At present in India there is a gap of about 10MT of plant nutrients between removal by crops and replenishment through fertilizers. It is evident that there is a tremendous gap between the annual demand and production of the biofertilizers globally especially in India. Hence, the judicious combination of chemical fertilizers and biofertilizers is also encouraged considering economic and ecological concerns. It is estimated that the present level of biofertilizer use is quite low and there is a substantial potential to increase it to 50 $x10^3$to 60 $x10^3$ tons by 2020 (Pindi and Satyanarayana, 2012).

14.3 Liquid Biofertilizer Formulation (LBF) Vs. Carrier Based Biofertilizers (CBBF)

LBF is the promising and updated technology of the conventional carrier based production technology which inspite of many advantages over the agrochemicals, left a considerable dispute among the farmer community in terms of several reasons major being the viability of the organism. Shelf life is the first and foremost problem of the carrier based biofertilizers which is up to 3 months and it does not retain throughout the crop cycle, LBF on the other hand facilitates the long survival of the organism by providing the suitable medium which is sufficient for the entire crop cycle. CBBF are not so tolerant to the temperature which is mostly unpredictable and uncertain in the crop fields while temperature tolerance is the other advantage of the liquid biofertilizers. The range of possible contamination is very high as bulk sterilization does not provide the desirable results in the case of CBBF, whereas the contamination can be controlled constructively by means of proper sterilization techniques and maintenance of intensive hygiene conditions by appropriate quality control measures in the case of LBF. Moisture retaining capacity of the CBBF is very low which does not allow the organism viable for longer period and the LBF facilitates the enhanced viability of the organism (Mahdi *et al.,* 2010). However, the administration of LBF in the fields is comparatively easier than CBBF. The other disadvantages of CBBF like poor cell protection, labor intensity, and dosage controversy, limited scope of export, expensive package and transport, very slow adaptation by the farmer community are some of the strongest problems which are being solved by the Liquid biofertilizers very effectively. Therefore, LBF are believed to be the best alternative for the conventional carrier based biofertilizers in the modern agriculture research community witnessing the enhanced crop yields, regaining soil health and sustainable global food production.

CBBF has already proved to be the best over the agro chemicals and have been showing the tremendous effect on the global agriculture productivity since the past two decades. Rectifying the disadvantages of the carrier based biofertilizers, liquid biofertilizers have been developed which would be the only alternative for the cost effective sustainable agriculture. The article focuses on Liquid Biofertilizer Technology providing reliable reasons for their necessity, specificity and emphasizes that "Use of agriculturally important microorganisms in different combinations as Liquid microbial consortium (LMC) is the only solution for restoration of soil health". Even though biofertilizers are being produced and distributed constantly by private agencies, NGO's, State and Central Government production units for the last three decades, their corresponding usage is not in the satisfactory proportions. To cope with the rising demands for food commodities, serious efforts are being made by the State and Central Governments (under the National Projects) for the sufficient agricultural production by popularizing biofertilizers and making them available to the farmer community. In spite of these efforts, the rate of consumption of biofertilizers is not to the optimum level in comparison with the agrochemicals.

14.4 Liquid Biofertilizers (LBF)

The earlier products of bio-fertilizers were carrier (solid) based where lignite is usually added as a carrier material. Lignite is hazardous to the production workers. Also, the shelf life of carrier based bio-fertilizers is only 6 months and is difficult to transport. LBFs on the other hand have ashelf life of minimum one year, with no health hazards to production workers and are easy to transport. Additionally, LBF can be used in drip irrigation and as a component of organic farming. The liquid Bio fertilizers are suspensions having agriculturally useful micro-organisms, which fix atmospheric nitrogen and solubilize insoluble phosphates and make it available for the plants. The use of this biofertilizer is environment friendly and gives uniform results for most of the agricultural crops and directly reduces the use of chemical fertilizer by 15 to 40%. The shelf life of the liquid bio fertilizer is higher (in the range of one to two years) compared to that of solid matrix base biofertiliszer. Liquid biofertilizers are liquid formulation containing the dormant form of desired micro-organisms and their nutrients along with the substances that encourage formation of resting spores or cysts for longer shelf life and tolerance to adverse conditions. The dormant form on reaching the soil, germinate to produce fresh batch of active cells. These cells grow and multiply by utilizing the carbon source in the soil or from root exudates.

14.4.1 Concept of Liquid Biofertilizers

"A preparation comprising requirements to preserve organisms and deliver them to the target regions to improve their biological activity orA consortium of micro-organisms provided with suitable medium to keep up their viability for certain period which aids in enhancing the biological activity of the target site".

Liquid biofertilizers are the microbial preparations containing specific beneficial micro-organisms which are capable of fixing or solubilizing or mobilizing plant nutrients by their biological activity. Liquid formulation is a budding technology in India and has very specific characteristics and uniqueness in its production methods. They are broadly classified into three groups such as-

1) Nitrogen Fixing Microbes (NFM)
2) Phosphorus Solubilizing Microbes (PSM) and Phosphate Mobilizing Microbes (VAM) and
3) Potash Mobilizing Microbes (*Frateuria aurentia*)

Among the three groups 'Phosphorus solubilizing microbes' usage is to the larger extent in the current agro market. The commercial production of Potash mobilizing bacteria has been developed recently. Solubilization or mobilization of micronutrients by exploiting the biological activity of microbes like as :

1) Zinc and Sulphur Solubilizing Bacteria (*Thiobacillus* spp.)
2) Manganese solubilizer (*Pencillium citrinum*)

The biggest challenge of the biofertilizer production units is the viability of organisms up to field application. Some of the beneficial organisms are very effective *in vitro*, but may fail at some stage of cropping, even at the stage of harvesting. The common and potential causes of this demise are poor stability of the biofertilizer (by means of the viability) which may not be constant from prior storage to field application, very low potential ingredients reaching the active site of the plant practically and rapid degradation of the fertilizer by means of acclimatization problems. Liquid formulations of the beneficial organisms (LBFs) or bioinoculants are the promising inputs which can effectively combat the present agro adverse Scenario, resolving the low viability of the organisms, maximizing the efficacy of the inputs and satisfying the concept of cost-effectiveness. Anand Agricultural University (AAU), Anand, Gujarat is one of the universities that has developed a liquid formulation of bio-fertilizers that are safe and eco-friendly alternative to chemical fertilizers. LBFs are sold to farmers under the brand name "Anubhav liquid Bio-fertilizers" by the University (Fig. 14.1). Anubhav LBF is based on native cultures of bacteria viz. *Azotobacter chroococcum*, *Azospirillum lipoferum* and *Bacillus coagulans*. To extend its reach to the farmers, the AAU has licensed the technology of LBF for commercialization to three companies in Gujarat through its Business Planning and Development Unit (BPDU) under Public Private Partnership (PPP) mode. BPDU is a special project at AAU, Anand under the World Bank funded scheme of National Agricultural Innovation Project (NAIP), Indian Council of Agricultural Research (ICAR), New Delhi. AAU has supplied LBFs to the tune of 50,000 litres to the Government of Gujarat for distribution to farmers as a part of Krishi kit during *Krishi Mahotsav,* a mass agricultural technology dissemination programme of the Department of Agriculture, Government of Gujarat. The response of Gujarat farmers on use of LBF in different crops such as Cotton, Banana, Potato, Rose, Turmeric, Papaya etc. reported better yield and quality M/s Kumar Krishi Mitra Bio Products Ltd., Pune, has already commenced production. There is a growing demand for organic foods in the global market. The use of these liquid bio-fertilizers would help the Indian farmers to produce organic crops so as to compete in the global market.

Fig. 14.1 : Anubhav Liquid Biofertilizer (*Source*:http://www.icar.org.in/en/node/5667. Technology for Liquid Bio-Fertilizers Commercialized)

14.4.2 Issue of Formulations

Formulation comprises aids to preserving organisms and to deliver them to their target fields and – once there-to improving their activities. A technical concentrate of an organism that has been achieved by a particular process is called as a formulation, or a product, which may be stored and put on sale commercially. A product often does not fully serve all the requirements of organisms up to the field. There are varieties of formulation both in liquid and solid. The main types currently used for biofertilizers have been classified into dry products (dusts, granules and briquettes) and suspensions (oil or water-based and emulsions). A wider range of formulations with additives are available in market. Liquid formulations are sprayable propagules of the microbial agents suspended in suitable liquid medium (or) A consortium of microorganisms provided with suitable medium to keep up their viability for certain period which aids in enhancing the biological activity of the target site. Basic characteristics of formulation is given as :

- Protect the micro-organism from harmful environmental factors at the target site (field), thereby increasing persistence.
- Stabilize the organism during production, distribution and storage.
- Enhance activity of the organism at the target site by increasing its activity, reproduction, contact and interaction with the target crops.
- Easily delivered to the field in the most appropriate manner.

Liquid biofertilizers are special liquid formulation containing not only the desired beneficial microorganisms and their biological secretions, but also special cell protestants or substances that encourage the formation of dormant spores or cysts for longer shelf life and tolerance to adverse conditions. Although, India boasts to be the largest biofertilizers producer in the globe the creditability of biofertilizers is slow among farmers. Apart from the carrier based inoculums, considering above said limitations in carrier based Regional Centre of Organic Farming (RCOF), Bangalore formulated the liquid based biofertilizers in year1998 for all types of biofertilizers and biopesticides. This technology is an alternative solution to carrier based biofertilizers, biopesticides and bio-control agents and this is not the usual broth culture from the fermentor or a water suspension of the carrier based biofertilizers.

14.4.2.1 Dormant Aqueous Suspensions

Several current commercial products available in market follow the dormant technology. Generally, they areusing growth suppressants, contaminant suppressant like Sodium azide (NaN_3), Sodium benzoate ($C_7H_5NaO_2$), Butanol (C_4H_9OH), Acetone $(CH_3)_2CO$, Fungicides, Insecticides etc. for the long term viability,while these formulation in particular is not suitable for organisms whose fast actions are required immediately for short duration crops. It was noted, the *Rhizobium* dormant liquid formulations when used after 8 months reduce the size of nodules and effect N_2 fixation process due to the need of long duration of activation time. In case of *Azospirillum, Azotobacter*, PSM, it was observed that these bacteria crossed extreme dormant stage.

14.4.2.2 Dormant Oil Suspension

Micro-organisms can be suspended in oil at high concentration in various degrees of dehydration and remain viable. This formulation delivers organisms in a physiologically dormant state and does not encourage the growth of contaminants during storage. The bacteria/ fungus has been successfully dried by continuous aeration as a suspension in oil to provide inoculants with shelf lives of several years. Oil suspension also has the limitation in reactivating the bacteria/fungus. It generally takes 10 to 20 days to regenerate and the initial requirement of the plant is not fulfilled by them.The liquid suspension contains particular micro-organisms about 40%,suspender ingredient 1-3%,dispersant 1-5%, surfactant 3-8% and carrier liquid (water) 35-65% by weight. Viscosity roughly equals the setting rate of the particles. This is achieved by the use of colloidal clays, polysaccharide gums, starch, cellulose or synthetic polymers.

14.4.3 Biosurfactant for Improving Liquid Formulation

For improving technology in liquid formulation, efforts are being made to add biosurfactant. These are surface active biomolecules produced by a variety of micro-organisms. These amphiphilic are generally produced on microbial cell surfaces but also through extracellular secretion (Muthuswamy *et al.*, 2008). These surfactants are used as emulsifier and are safer than chemical emulsifiers. The common biosurfactants are Glycolipids, Lipopeptide Rhamnolipids and Sophorolipids (Table14.1). There is enough scope to use such biosurfactants for the development of liquid biofertilizers.

Table 14.1 : Some Important Biosurfactants for Improving the Quality of Liquid Formulation

SlNo.	Type of biosurfactants	Produced by	Substrate used
1.	Glycolipids	*Candida apicola*	Oil refinery waste
2.	Lipopeptide	*Serratiamarcescens*	Sunflower oil
3.	Rhamnolipids	*Pseudomonas aeruginosa*	Curd whey and distillery waste
4.	Sophorolipids	*Candida lipolytica*	Babassu oil waste

14.4.4 Factors affecting the Liquid Formulation

14.4.4.1 Temperature

Temperature is important for the shelf life of microbial products and it can affect their activity before or after application. Colonization proceeds at field temperatures in the cropping season, but slows at lower temperature. Strains used in liquid formulation normally grow at 37⁰C, (Chandra *et al.,* 1999) and able to tolerate temperature up to 45⁰C for two year or more. Whereas, solid base shelf life is hardly up to 3 months as raise in temperature beyond 35⁰C and start rapid decline of organisms.

14.4.4.2 Acclimatization

Effect of environment on bio fertilizers has been studied by many workers. It is reported that it has negative effect on different strains, but all the research was done on the carrier base biofertilizers. It is observed that the efficiency of liquid is almost same in all environments, but efficiency may reduce 20-25% in different climatic conditions in case of solid base. Normally in RCOF liquid formulation of an organism remain active for a long time after application, ideally throughout the period of the crop (max. 2 years), or in soil throughout the crop cycle. Temperature, humidity, leaf surface exudates and competitors, etc are the critical factors which inactivates the microbes. They may be lost physically from the target location by the action of wind, rain or leaching depending on the fact why and where the product is used.

14.4.4.3 Effect of Humectants

The effect of moisture content on storage stability, some organisms may need moisture for its activity this need is fulfilled by liquid inoculum but in case of carrier base inoculum bacteria get stressed, when carrier become dry during transport and storage. Bacteria used for plant growth need the plant surface to be wet in order to establish them. These needs can be overcome by only liquid formulation as product contains humectant. In general, there is little direct effect of relative humidity on the spore forming bacteria in liquid form.

14.4.4.4 Sunlight Intensity

Generally microbes are sensitive to temperature. The most harmful rays UVB (280-320nm) and UVA (320-400nm) reaching the earth's surface have direct effect on microbes. Of course some are less damaging where as some have the maximum effects. However, microbes are sensitive to the wave-lengths outside this range. To counter the harmful effects of high temperature sunscreens are added to a formulation. Sunscreens act by reflecting and scattering physically or by absorbing radiation selectively, converting short wavelengths to harmless longer ones. However, no such types of sunscreens are available in solid base to resist the effect of sunlight.

14.4.4.5 pH Value

The pH of a product plays a vital role in liquid inoculum preparations. It must be stabilized within certain ranges. Very low and high pH conditions will normally inactivate agents. A buffer therefore is maintained by adding some additives which render the better shelf life in liquid. Maintenance of optimal pH improve shelf-life of some of microorganisms like *Azotobacter, Azospirillum*, Phosphorus Solubilizing bacteria (PSM), Potash mobilizing bacteria (KMB) *Frateuria aurentia* (Chandra *et al.,* 1999).

14.4.4.6 Plant Exudates

Plant exudates may influencethe efficiency of liquid biofertilizerin various ways. These include chemicals present in the soil, as well as plant secretions and chemicals on the phylloplane. For example, on cotton and many other xerophytic plants, high levels of magnesium (Mg), calcium (Ca) and manganese (Mn) as carbonates and bicarbonates are exuded onto the leaf surface; this can raise the pH of the leaf surface to values as high as 10 or 11. The presence of these ions orthe associated pH value, when wetted by dew or water spray, can inhibitor inactivate some organisms. Toovercome with these problems only liquid proper formulations take care as it has a buffering property which neutralizethe effect of temporary pH effect. However, Potash Mobilizing Bacteria (KMB) as *Frateuria aurentia* has no effect as it can tolerate even at pH 11.

14.4.4.7 Stabilization of Liquid Inoculum

The longest period of time in the life of a product elapses during storage, i.e. the time between manufacture and eventual use in the field. This period can range from several weeks to years. Organisms are required to remain viable during storage, with minimum loss of potency/activity and without loss or breakdown of the desired formulation properties. The authors treat the liquid formulations for improving stability by appropriate growth conditions during production by appropriate storage prior to formulation by appropriate processing after production. In addition, or alternatively, additives are included to improve stability. Depending on the organisms involved, different problems must be addressed to improve stability. It is suggested during preparation of base of liquid some additives that inhibit germination of the organisms after application must be avoided.

14.4.4.8 Application of Liquid Inoculum

Addition of surfactant or oil and emulsifier to water, or use of pure oil, forms drops of more even size than those of water alone, with consequently better controlled spray. Oil is preferred for ultra-low-volume (ULV) sprays, water is normally used as the diluentat higher volumes. Liquid formulations contain high count ranges from 10^9 to 10^{10} cells/ ml which optimize spray droplet size to maximum coverage of the target and ensures maximum number of effective bacteria for action.

14.4.4.9 Handling and Application of Liquid Inoculums

Liquid inoculum ensures that the product is easy to handle and apply. For example, in suspensions thickeners or suspenders added by the authors which help in maintaining even distribution of the organism. The liquid prevent clumping of the micro-organism and ensure its ready re-suspension after prolonged storage. Dusts and wettable powders not able to maintain uniformity. It is shown that in case of *Azospirillum* the population come down up to 10^5 at six months durationat room temperature where as in liquid survives up to 2 years and population maintain up to 10^8/ml. Similarly in *Azotobacter*, KMB, *Rhizobium* solid maintain the shelf life up to 6 months except PSM which services up to 8 months but in the case of liquid formulations in *Azotobacter*, PSM and KMB survives up to 2 years followed by *Rhizobium* only 14 months (Table. 14.2). This shows the superiority of liquid formulation over carrier base formulations.

Table 14.2 : Liquid Inoculum Vs. Carrier based Inoculum-Shelf life; L-Liquid (Liquid formulation without preservative was used); S-Solid (lignite carrier mesh size 150µ pre-sterilized in autoclave used).

Inoculum	**Months**											
	0	2	4	6	8	10	12	14	16	18	20	24
Azo sp. (L)	10.0	10.0	10.0	10.0	9.0	9.0	9.0	9.0	8.0	8.0	8.0	8.0
Azo sp. (S)	9.0	8.0	7.0	5.0	NIL	NIL	NIL	NIL	NIL	NIL	NIL	NIL
Azoto (L)	9.0	9.0	9.0	9.0	9.0	8.0	8.0	8.0	8.0	8.0	7.0	7.0
Azotor (S)	8.0	8.0	7.0	6.0	NIL	NIL	NIL	NIL	NIL	NIL	NIL	NIL
PSM (L)	9.0	9.0	9.0	9.0	9.0	9.0	9.0	9.0	9.0	9.0	9.0	9.0
PSM (S)	9.0	8.0	8.0	8.0	7.0	NIL	NIL	NIL	NIL	NIL	NIL	NIL
KMB (L)	9.0	9.0	9.0	9.0	9.0	9.0	9.0	8.0	8.0	8.0	8.0	8.0
KMB (S)	8.0	8.0	8.0	7.0	5.0	NIL	NIL	NIL	NIL	NIL	NIL	NIL
Rhizo (L)	9.0	9.0	9.0	9.0	9.0	9.0	8.0	7.0	5.0	NIL	NIL	NIL
Rhizo (S)	9.0	8.0	8.0	7.0	NIL	NIL	NIL	NIL	NIL	NIL	NIL	NIL

Source : Chandra *et al.*,(2005)

14.4.5 Advantages of Liquid Biofertilizers

The advantages of liquid biofertilizer over conventional carrier based biofertilizers are given below:

- Longer shelf life as 12- 24 months.
- No loss of properties due to storage upto 45^0C since no effect of high temperature.
- High populations can be maintained more than10^9 cells/ml up to 12 months to 24 months.
- No need of running biofertilizer production units throughout the year.
- Identified by typical fermented smell.
- Cost saving on carrier material, pulverization, neutralization, sterilization, packing and transport.
- Quality control protocols are easy and quick.
- Better survival on seeds and soil and easy to use by the farmers.
- Dosages are10 times less than carrier biofertilizer.
- High commercial revenues and export potential.
- Very high enzymatic activity since contamination is nil.

Further, it is very necessary to understand that, "liquid biofertilizers are not merely usual broth cultures from fermenters which are specially prepared with desired agriculturally important micro-organisms along with proper nutritional base to facilitate protection of organisms from adverse conditions and resting spore formation or cysts etc". Many researchers still have doubts about the liquid formulations. Although, in commercial all bacterial inoculum being well adopted by the users

14.5 *Rhizobium* as Liquid Biofertilizers

It belongs to bacterial group and the classical example of symbiotic nitrogen fixation. The bacteria intact the legume root and form root nodules within which they reduce molecular nitrogen to ammonia which is readily utilized by the plant to produce valuable proteins, vitamins and other nitrogen containing compounds. The site of symbiosis is within the root nodules. It has been estimated that 40-250kg N/ha/year could be fixed by different legume crops by the microbial activities of *Rhizobium* (Table 14.3). It is dull white in colour and has no bad smell and foam formation and requires pH 6.8 to 7.5.

Table 14.3 : Quantity of Biological N fixed by Liquid Rhizobium in different Crops

Sl. No.	Host Group	*Rhizobium* Species	Crops	N fix kg/ha
1.	Pea Group	*Rhizobium leguminosarum*	Gren pea, Lentil	62-132
2	Soybean group	*R. japonicum*	Soybean	57-105
3	Lupini group	*R. lupini orinthopus*	Lupinus	70-90
4	Alfalfa group	*Rhizobium meliloti, Medica gorigonella*	Melilotus	100-150
5	Beans group	*Rhizobium phaseoli*	Phaseoli	80-110
6	Clover group	*Rhizobium trifolii*	Trifolium	1 3 0
7	Cowpea group	*Rhizobium* sp.	Moong, gram, Cowpea, Groundnut	57-105
8	Cicer group	*Rhizobium* sp	Bengal gram	75-117

Source: Chandra *et al.*,(2005)

14.5.1 Use of Fungicides/Pesticides/Herbicides with Liquid Rhizobium inoculum

The effect of chemicals on nodulation and nitrogen fixing *Rhizobium* is negligible. Seed dressing fungicides such as Thiram and mercury have some adverse effect on the viability of *Rhizobium*, which can be overcome by doubling the doze of *Rhizobium* culture. Studies suggest that *Rhizobium* strain MRL3 may be exploited as a bioinoculant to augment the efficiency of lentil exposed to insecticide-stressed soils (Ahmed and Khan, 2011). On the other hand fungicides such as Dithane M-45 and Bevistin do not have any adverse effect on *Rhizobium* and the minimum tolerance doses with different fungicides, corbandazium, Dithane M-45.

14.5.2 Use of Chemical Fertilizer along with Liquid Rhizobium

Most pulses and legume oil seed crops require30-40kg nitrogen and 40kg Phosphorus/ha for better growth and yield. If crop is provided chemical nitrogen only 5-10kg/ha as a basal dose; is essential to fulfill the initial nitrogen requirement of crop up to 20-25 day i.e. till the *Rhizobium* nodules start their functioning.Phosphorus is essential for better *Rhizobium* inoculation so 30-40kg of phosphorus/ha is required as basal dose, if PSM is used the dose of chemical Phosphorus can be 15-20kg/ha. Potassium is also essential for better *Rhizobium* inoculation so 30-40kg of Potassium/ha is required as basal dose, if KMB is used the dose of Potassium chemical can be 10-15kg/ha.

14.5.3 Use of Sticky Materials for Seed Treatment

Generally sticky materials are used for the easy adhesion of the inoculums onto the target plant parts and subsequently increase bacterial population on the seeds for maximum nitrogen fixation. The germination and nodulation tests were performed on the pelleted seeds and results showed that although the sugar and molasses initially bound less inoculant and lime to the seed, the number of surviving *rhizobia* was similar to that obtained by the Gum Arabic treatment after storage at 27.50°C for five days for fast growing *Rhizobia*. The inoculated pelleted seeds of slow growing *Rhizobia* showed promising results in germination and nodulation in acidic soil.

14.6 *Azospirillum* as Liquid Biofertilizers

Azospirillum belongs to bacteria and is known to fix considerable quantity of nitrogen in range of 20-40kg N/ha in the rhizosphere in non-leguminous plants such as cereals, millets, oilseeds, cotton etc. *Azospirillum* is considered as the efficient biofertilizers because of its ability of inducing abundant roots in several plants like rice, millets and oilseeds even in upland conditions. An estimated amount of 25-30% chemical nitrogen fertilizer can be saved by the appropriate use of *Azospirillum* inoculants. The genus *Azospirillum* has three species viz. *A. lipoferum, A. brasilense* and *A. amazonense*. These species have been commercially exploited for the use of nitrogen supplying biofertilizer. One of the characteristics of *Azospirillum* is its ability to reduce nitrate and denitrify. Table 14.4 presents occurrence of *Azospirillum* and their nitrogen fixing capacity in different plants. Both the species *A. lipoferum* and *A. brasilense* may comprised of strains which may denitrify actively or weakly or reduce nitrate to nitrite. Therefore, for inoculation preparation, it would be necessary to select strains which do not possess these characteristics. *Azospirillum lipoferum* present in the roots of some tropical forage grasses such as Digitaria, Panicum, Brachiaria, Maize, Sorghum, Wheat and Rye.

14.6.1 Physical Features of Liquid Azospirillum

- Colour of the liquid may be blue or dull white.
- Bad odors confirm improper liquid may be broth.
- Production of yellow gummy colour materials confirms the quality product.
- Acidic pH always confirms no *Azospirillum* bacteria present in liquid.
- Yellowish green colour of leaves.

Indicate no bacteria in the roots of several plants and the amount of nitrogen fixed by them.

Table 14.4 : Occurrence of *Azospirillum* and N_2 fixing capacity bacteria in the roots of several plants and the amount of nitrogen fixation

Plant	**g N_2 fixed/g of substrate**
Paddy (*Oryza sativa*)	28
Sorghum (*Sorghum bicolor*)	20
Maize (*Zea mays*)	20
Panicum Sp.	24
Cynodondactylon	36
Setaria Sp.	12
Amaranthus spinosa	16

Source : Chandra *et al.*,(2005)

14.6.2 Production of Growth Hormones

Azospirillum cultures synthesize considerable amount of biologically active substances like vitamins, nicotinic acid, indole acetic acid, gibberellins. All these hormones/ chemicals help plants in better germination, early emergence, and better root development.

14.6.3 Field Applications of Liquid Azospirillum

- Stimulates growth and produces green color characteristics of a healthy plant.
- Aids utilization of potash, phosphorus and other nutrients
- Encourages plumpness and succulence of fruits and grains and increases protein percentage.

Inoculants of *Azospirillum* have been proved beneficial in a variety of crops including cereals, forage and other crops. The crop response however, varies greatly with types of crop and variety, location, season, soil fertility level; native micro-organism interaction etc., the ability to fix atmospheric nitrogen and to produce phytohormones determines the crop response and yield. A series of experiments have been conducted on a range of crop varieties to evaluate the crop response to *Azospirillum* inoculants and a wide variation in production up to 40% have been observed. Non-functioning of *Azospirillum* in the fields can be identified by yellowish green color of leaves which indicates very poor or no nitrogen fixation which in turn results in arresting the growth promotion.

14.7 *Azotobacter* as Liquid Biofertilizers

Azotobacter belongs to the genus diazotrophic bacteria which is a free-living organism whose resting stage is a cyst. It is abundantly found in neutral to alkaline soils, in aquatic environments, and on some plants. *Azotobacter* is capable of performing several metabolic activities, including atmospheric nitrogen fixation by conversion to ammonia. *Azotobacter* spp. has the highest metabolic rate of any organisms. It serves as potential biofertilizer for all non-leguminous plants especially rice, cotton, vegetables etc. *Azotobacter* population is high in rhizosphere region and the rhizoplane does not contain *Azotobacter* cells. The lack of organic matter in the soil is a limiting factor in the proliferation of *Azotobacter* in the soil.

14.7.1 Physical Features of Liquid Azotobacter

The pigmentation is produced by *Azotobacter* in older culture is melanin due to oxidation of tyrosine by tyrosinase an enzyme which has copper. The colour can be noted only in liquid forms as (a) brown-black pigmentation in *A. chroococcum* (b) yellow-light brown in *A. beijerinchii* (c) green fluorescent in *A. vinelandii, A. agilis* and *A. paspali* (d) Produces, pink pigmentation in *A. macrocytogenes* (e) grayish-blue in *A. insignis* in liquid inoculum.

14.7.2 Liquid Azotobacter in Tissue Culture

- The performance of *Azotobacter* liquid inoculants was comparatively better than all the treatments in 10% MS medium followed by *Azospirillum*.
- The performance of *Azotobacter* liquid inoculants was comparatively better than all the treatments followed by *Azospirillum* for the growth of polybag sugarcane seedlings.

14.7.3 Liquid **Azotobacter** *as Bio-control Agent*

Azotobacter have been found to produce some effective antifungal substance which inhibits the growth of some soil fungi like *Aspergillus, Fusarium, Curvularia, Alternaria, Helminthosporium,* and *Fusarium* etc.

14.8 *Acetobacter* as Liquid Biofertilizers

Acetobacter (a sacharophilic bacteria) associated with sugarcane, sweet potato and sweet sorghum, coffee plants is acid-producing bacteria which fix atmospheric nitrogen. In fact, the *A. diazotrophicus*-sugarcane relationship is beneficial symbiotic relationship between grasses and bacteria which fixes about 30kg N/ha/year of nitrogen. This bacterium is being exploiting and commercialized extensively for sugarcane crop reducing the primary dependency in the chemical nitrogen fertilizer. It is known to increase cane yield 10-20ton/ acre and sugar content by 10-15%.

14.8.1 Liquid Acetobacter diazotrophicus in Sugarcane

Application of *Azospirillum* and Phosphobacteria for the cash crop sugarcane has been a regular practice for the past few years which is saving nearly 20% chemical nitrogen and phosphate applications in South India. It is evident from various studies that *Acetobacter diazotrophicus* present in sugarcane stem, leaves, soils have a capacity to fix up to 300kg of nitrogen. This bacterium first reported in Brazil where sugarcane is cultivated in very poor sub-soil fields which are fertilized with phosphate, potassium and micro elements alone could produce a yield for three consecutive recorded harvests of 182 to 244 tons per hectare, without the use of any chemical nitrogen fertilizer. This manifests the assumption that active nitrogen fixing bacteria must be associated within the plant regularly. An extensive research work has been conducted by the scientists of Brazil and Australia on bacteria like *Acetobacter diazotrophicus, Herbaspirillum seropedicae* and *Herbaspirillum rubisubalbicans* in leaves, stems and roots and revealed the fact that they are largely responsible for atmospheric nitrogen N-fixation.

14.8.2 Acetobacter pasteurianus for Sulphur

The physical structure of sulphur consists of several amino acids, including methionine and cystine, which are essential components of plant and animal proteins. Sulphur is important in synthesis of nitrogen in plants; sulphur acts more like nitrogen than any other essential plant nutrients. Indeed, sulphur deficiency symptoms in plants closely resemble nitrogen deficiency symptoms. Moreover, leaching losses of sulphur from the soil are very much like those of nitrogen caused by percolation waters through light-textured sandy and salty soils. Thus, there is need for annual supplies of sulphur to meet the requirements of growing plants. Although, we are accustomed to use sulphur in the form of fertilizers but number of such farmers are very less, due to its cost factors etc. It was estimated that about 70-90% of total surface soil sulfur is found in organic matters. The remainder of the sulphur is present as sulfides and as soluble and insoluble sulfates. Usually, only 5-10% of the sulfates are found in the surface furrow slice of humid region, cultivated soils. In humid region soils, most of the sulfates are of 12-14 inches deep and are associated with oxides of iron and aluminum, especially in acid soils having kalononitic type clay. These bacteria help in converting this non-usable form to usable form. It was experienced that use of bacteria 200 ml/acre influenced in level of sulfur in crops like vegetable, cabbage, turnip, onion, cotton, fruits, oilseeds, spices etc.

14.9 Phosphorus Solubilizing Micro-organisms (PSMs)

Microbial mineralization, solubilization and mobilization are the most important aspects of the phosphorus cycle, besides chemical fixation of phosphorus in soil. The enzymatic activity of microorganisms is responsible for the mineralization of organic phosphorus which is left over in the soil after harvesting, or added as plant or animal residues to soil (Bot and Bonetis, 2005; Eckert, 2012).Phosphorus solubilizing bacteria and fungi play a vital role in persuading the insoluble phosphatic compound such as rock phosphate, bone meal and basic slag and particularly the chemically fixed soil phosphorus into available form. These special types of microorganisms are known as Phosphate Solubilizing Microorganisms (PSM) which includes different groups of microorganisms such as bacteria and fungi that convert insoluble phosphatic compounds and fixed chemical fertilizers into soluble form. The species of *Pseudomonas, Micrococcus, Bacillus, Flavobacterium, Penicillium, Fusarium, Sclerotium, Aspergillus* and some other are considered as active in biophosphorus conversion.

These bacteria and fungi can be grown in the media containing $Ca_3(PO_4)_2$, $FePO_4$, $AlPO_4$, apatite, bone meal, rock phosphate or similar insoluble phosphate compounds are the sole source of phosphate. Such organisms not only assimilate phosphorus but also trigger the release of an excess amount of soluble phosphate than their actual requirements. Several rock phosphate dissolving bacteria, fungi, yeasts and *actinomycetes* were isolated from soil samples collected from rock phosphate deposits and rhizosphere soils of different legume crops. Some of the isolates solubilized and made available phosphorus to crop from rock phosphate and many isolates solubilized a very high quantity of tricalcium phosphate.

The most efficient bacterial isolates were identified as *Pseudomonas striata, Pseudomonas rathonis, Bacillus polymyxa, B. megatherium.* Apart from these, fungal isolates such as *Aspergillus awamori, Penicillium digitatum, Aspergillus niger, Schwanniomyces occidentalis* are identified as the potential phosphorous solubilizers. These efficient microorganisms have shown consistency in their capability to solubilize chemically fixed soil phosphorus and rock phosphate from different sources. PSM were found to mineralize organic phosphorus into soluble form by their enzymatic activity and save P_2O_5 up to 30-50 kg/ha.

14.9.1 Physical Features of Liquid PSM

High turbidity is observed in the appearance of the solution. The color may be creamy mostly and it varies along with the strain. The odor may be pleasant but sometimes mild. The pH is 4.5 or buffer, but buffer with 6.5-7.0 ensures more than 2 years viability.

14.9.2 Effect of Liquid PSM under Field Conditions

- Encourages early root development.
- PSM produced organic acids like malic, succinic, fumaric, citric, tartaric and alpha ketoglutaric acid which hastens the maturity and thereby increases the ratio of grain to straw as well as the total yield.
- Increases the compatibilityof other beneficial microbes with plants.
- Stimulates formation of fats, convertible starches and healthy seeds.
- Helps rapid cell development in plants and consequently increases resistance towards diseases.
- Increase micro-nutrients in soil like Mn, Mg, Fe, Mo, B, Zn and Cu etc.depending upon the nutrient presents in non-available form.

Inoculants of phosphate solubilizing bacteria as fertilizer increases P uptake by the plant and enhance crop yield. There are many strains of different genera which promote the phosphate solublization among which, strains from the genera *Pseudomonas, Bacillus* and *Rhizobium* are the most powerful phosphate solubilizers. Phosphate solubilization by the organisms mainly involves the production of organic acids which is the principal mechanism for solubilization, and the organic phosphorous in the soil is mineralized by acid phosphates which is a vital step. Cloning of several phosphatase encoding genes which are isolated is done and characterized for the better phosphate solubilization. Genetic manipulation of phosphate-solubilizing bacteria followed by their expression in selected rhizobacterial strains is the interesting approach among the scientific community for the optimum phosphate solubilization (Rodrýìguez and Fraga, 1999). The PSB has the greatest potential of utilization of India's abundant deposits of rock phosphates, much of which is not enriched (Ghosh).

14.10 Phosphate Mobilizing Micro-organisms (PMMs)

Compared with the other major nutrients, phosphorous is the least mobile and available to plants in most of the soils. In spite of its abundance in the most of the soils both in organic and inorganic forms, phosphorous is prime limiting factor for plant growth. Moreover plant species, nutritional status of the soil and ambient soil conditions determine the bioavailability of soil inorganic phosphorus. To circumvent the phosphorus deficiency to the soil phosphate solubilizing microorganisms (PSM) could play an important role in supplying phosphate to plants in a more suitable and eco-friendly manner (Khan *et al.*, 2009).

The later one endomycorrhizae are known as AM, which possess special structures known as vesicles and arbuscules, the latter helping in the transfer of nutrients from soil into root system. These fungi are classified on the basis of their spore morphology in seven genera namely Glomus, Gigaspora, Acaulospora, Sclerocystis, Scutellospora, Entrophospara, Archeospora as para Glomus. In general, two typical types of mycorrhizal fungi are found in associations with plant roots namely as *ectomycorrhiza*, *endomycorrhiza* and *ectendomycorrhiza*. Mycorrhizae increase the surface area of the plant root system for better absorption of nutrients from soil especially in phosphorus deficient soils. The *endomycorrhiza*, also known as *Arbuscular mycorrhiza*, aids in transfer of nutrients from soil into root system through specialized structures known as vesicles and arbuscules. The potential role of *mycorrhiza* in mobilizing & phosphate uptake and plant growth is being widely recognized in the recent times and AM are the most common type which is being highly exploited. Advancement in biotechnology has promoted experimentation in artificial inoculation of plants with AM. Trials have mainly involved broadcasting and raking-in inocula over plant-growing substrates rather than application of inocula to seed before sowing. Since large-scale production of AM in axenic culture is not yet attained, inocula have been produced in pot cultures or in small field plots on plants grown under carefully controlled conditions, to avoid contamination by plant pathogens. Such inocula have comprised of infected roots or spores and hyphae trapped in soil, peat and clay carriers. Introduction of AM into field on a large scale has tended to use inoculum at excessively high and impractical rates in order to guarantee rapid infection by the introduced fungi. Application of inoculum directly to seed may reduce the amounts of inoculums required. Experimental inoculation has involved coating spores, mycelium and infected root fragments, wet-sieved from soil, onto seeds using methyl cellulose as sticker.

According to the studies of root nodulation as well as N and P, uptake is improved by green gram plant on the application of favorably interacting rhizospheric microorganisms as the inoculants and hence, yield is increased in the phosphorous deficient soils (Zaida *et al.,* 2004) *Bradyrhizobium* sp. (*Vigna*) and phosphate solubilizing microorganisms (PSM), *Pseudomonas striata* or *Penicillium* variable have shown a significant effect in enhancing the nutrient uptake and plant yield. Combined inoculation treatment of AM fungus, *Glomus fasciculatum* with *Bradyrhizobium* sp. (*Vigna*) + *P. striata* further augmented the nutrient uptake and plant yield (Khan *et al.,* 2009).

Benefits and Constraints

As the fungal hyphae facilitate increased surface area of absorption of nutrients like P, Zn, Cu etc., up to an extent of 8cm the biochemical reactions in plant cycle are not interrupted promoting the plant growth. Moreover, it is rather easy to inoculate on crops which are normally grown on nursery beds or root trainers or poly bags because of the difficulty in mass producing AM fungi. Nearly 25 to 50% of phosphatic fertilizers can be saved on a year with the inoculation of efficient AM fungi.

In spite of many advantages, AMF have some constraints which causes their limited usage. As it is an obligate symbiotic organism it need a host to survive, and it cannot be cultured in synthetic media in laboratory for large scale production. Variations and specifications in plant genotypes and soil nature make it difficult to use appropriate *mycorrhizal* strains for the phosphorous mobilization. Slow growth and development of *mycorrhizal* fungi association with the plant roots and poor resistance to the fungicides are some of the other potential constrains which are to be overcome by the rigorous field research with different approaches and optimizations.

14.11 Potash Mobilizing Bacteria (KMB) as Liquid Biofertilizers

In addition to the regular biofertilizers like *Rhizobium, Azospirillum, Azotobacter, Acetobacter* and Phosphate solubilizing/mobilizing and microbes to meet N&P nutrition in plants, microbes responsible for Potassium mobilization has been isolated and developed from Banana rhizosphere which is having the ability to mobilize the elementary or mixture of potassium which can be easily absorbed by plants. It is estimated that 50-60% of potash chemical fertilizers usage can be reduced by using *Frateuria aurentia*, a new bacterial species (species conformation by IMTECH (Chandigarh) as a bioinoculant. This new bacteria belonging to the family *Pseudomonaceae* have the extra ability to mobilize K in almost all types of soils especially, low K content soils, soils of pH 5-11 and it survives in the temperature up to 42°C. This potash mobilizing biofertilizers can be applied in combination with *Rhizobium, Azospirilum, Azotobacter, Acetobacter*, PSM etc. Potash mobilizing bacterial based product containing *Frateuria aurentia* producing plant growth promoting substances which offers plant a multifaceted benefits in terms of growth, by mobilizing potash and making it available to crops. It also enhances the efficiency of chemical fertilizer (Patel, 2011). Since large-scale production of AM in axenic culture is not yet attained, inocula have been produced in pot cultures or in small field plots on plants grown under carefully controlled conditions, to avoid contamination by plant pathogens. Such inoculais comprised of infected roots or spores and hyphae trapped in soil,peat and clay carriers. Introduction of AM into field on a large scale has tended to use inoculum at excessively high and impractical rates in order to guarantee rapid infection by the introduced fungi. Application of inocula directly to seed may reduce the amounts of inocula required. Experimental inoculation has involved coating spores, mycelium and infected root fragments, wet-sieved from soil, onto seeds using methyl cellulose as sticker.

14.11.1 Physical Features of Liquid KMB

- Solution always brown in colour.
- Smell of H_2S or chocolate.
- pH-4.5 but buffer 6.5-7.0 ensures more than2 years expiry.

14.11.2 Constraints in AM Fungi Efficacy

- Obligate symbionts – needs host for multiplication.
- Cannot be cultured on synthetic media in laboratory.
- Problems in mass cultures.
- Difficult to apply.
- Variation between plant genotypes in mycorrhizal.
- Affected by pesticides especially fungicide.
- Variation in soil nutrient status.
- Competition with indigenous population of mycorrhizal fungi.
- Phosphorus inhibition of mycorrhizal development.
- Slow growth of mycorrhizal fungi and slow development of mycorrhizal association.
- Incompatibilities with established cultural practices.

14.11.3 Mechanism of Improved Plant Growth due to AMF Application

- Nutrient uptake.
- Production of growth promoting substances.
- Beneficial interactions between soil microorganisms.
- Drought tolerance.
- Disease resistance.

14.11.4 Advantages of AMF Application in Micropropagated Plants

- Corrects the physiological defects of tissue culture plantlets.
- Enhances the absorption and transport of water and nutrients.
- Hormonal effects on shoot and root development.
- Resistance to pathogenic infection.

- Better survival and establishment.
- Reduces the hardening period.
- Healthy and vigorous plantlets.

14.11.5 Effect of Liquid Potassium Mobilizer with AM

Optimization studies on the liquid potassium mobilize revealed following facts. *Arbuscular mycorrhizae* (soil base) and potassium mobilizing bacteria, individually, as well as in combination with organic compost and phosphocompost shown better results in capsicum, *Hungarian yellow* variety when compared to the controls (plain soil). Potassium mobilizing bacteria (*Frateuria aurentia*) enriched phosphocompost shows the highest results in growth parameters next to all the biofertilizers applied together. This may be because of significant increase in mobilization and uptake of K leading to release of growth hormones by the K-mobilizer. Improves plant growth is attributed to better uptake of nutrients like P, Zn, Cu etc. This is mainly because of increased surface area of absorption of nutrients by the hyphae. Mycorrhizal plant roots can extend up to 8cm. Because of the difficulty in mass producing AM fungi, the best way to utilize AM fungi for crop production would be to concentrate on crops which are normally grown on nursery beds or root trainers or poly bags, where they could easily be inoculated with desired AM fungi and transplanted to the field. Nearly 25-50% of phosphatic fertilizers can be saved through inoculation with efficient AM fungi.

14.12 Zinc and Sulphur Solubilizing Bacteria

Thiobacilus ferrooxidans also known as *Acidithiobacillus ferrooxidans* and *Thiobacillus thiooxidans* and *Thiobaccilus acidophilus* is an acidophilic bacteria present in water or soil. The bacteria produce number of organic acids. They are able to produce and tolerate highly acidic conditions below upto pH 1.5 with the action of acid they solubilize large amount of Zn, Cu, Fe, Co etc. Whatever, micro- nutrient present in the soil in elementary or complex form. The liquid inoculums of above bacteria are very effective in all types of soils. During multiplication, bacteria produce large amount of acid, hence, the pH become low up to 2.5-4.5. Use of 100ml/acres in field along with 500kg of FYM/cow dung or 100ml inoculum with 100ml its water spray on field or through drip irrigation improves the quality/yield of crops.

14.13 Manganese Solubilizer (*Pencillium citrinum*)

Pencillium citrinuma fungal culture present in soilis knownto be solubilize manganese from the low grade manganese ores and from the soil if it is present. The fungus produces reductive compounds such as organic acids (oxalic acid, citric acid and which help in solubilizing of manganese. Temperature required for multiplicationis 32$\pm$2^{0}C. During multiplication, pH became low upto 5.5.

14.14 Role of Liquid Biofertilizers in Various Crops

- Increased crop yield.
- Reduces chemical N, P & K fertilizers.
- Increases seed quality of the crops.
- Benefits the next crop too, due to its residual effect.
- Enhances soil health and soil fertility.
- Output returns economically profitable.
- Disease and pest occurrence reduced.

14.15 Limitations of Carrier-Based Biofertilizers

- Shorter Shelf life (3 months) and sensitivity to temperature.
- More chance of contamination and unavailability of good carrier in local area.
- Lack of identifiable character and instant visual effects on application.
- Restriction on use as a measure of conservation (wood charcoal).
- Quality inconsistency of carriers.
- Bulk sterilization problem in terms of economic and facilities.
- Poor moisture retention capacity and cell protection.
- Labour intensive and takes more time for quality control.
- Even less contamination in packets the enzymatic activity of beneficial bacteria disturbed.
- Problem of proper packing and high transport cost.
- Dosage contradictory to farmers.
- Very slow adoption by farmers.
- Less scope for export and no commercial value.

14.16 Methods Used for the Testing of Quality of Liquid Biofertilizers

14.16.1 General Tests

Prescribed marking on the Biofertilizer bottle like name of the product, crop for which intended, name and address of the manufacturer, batch number, date of manufacture, date of expiry, methods of use and handling instructions etc.

14.16.2 Qualitative Tests

a) **Serial Dilution Technique by Plate Count** : Total number of viable bacteria present in the Biofertilizer bottle could be known by this method, and it is completed in to following steps:

Usually 10ml of Liquid based Biofertilizer is removed from the bottle to be mixed in 90ml of sterile water and shaken thoroughly. After homogenization, 1ml is taken and added with 9ml of sterile water. Likewise, serially it is diluted under sterile conditions till 10^{-10}. Finally, usually 10^{-6} to 10^{-10} dilutions are chosen and out of each dilution 1.0ml of aliquot is inoculated in sterile solidified enrichment media specific to the test micro-organism and spread uniformly. Petri plates are inverted and incubated in BOD incubator at $+28^{\circ}C$. The plates are observed periodically for the appearance of bacterial colonies.

b) Colonies appeared in the plates are observed for colony morphology, shapes like either round, round with scalloped margin, round with raised margin, wrinkled, concentric, irregular and spreading, filamentous, round with radiating margin, fill form, complex etc. The margin of appeared colony like either smooth, wavy, lobate, irregular, ciliate, branching, wooly, is observed. The elevation of the colony either like flat, raised, hilly, convex, drop like is observed.

c) One of the isolated colonies is removed and spread in a loop on the slide and gram stained.

d) The colonies observed are stained and its morphological characters like cellular shape, size, flagella, attachment of flagella etc. are studied preferably under a phase-contrast microscope and strain is identified.

e) Specific confirmation tests for each micro-organism are done to check its authenticity.

14.16.3 Quantitative Tests

a) Through plate counts, the total number of colonies appeared arecounted in an electronic colony counter and depending on the dilution factor, the colonies formed are enumerated.

b) From the broth, the total microbial populations could be counted directly under a microscope using a Petroff-Hauser Counter otherwise through a Haemocytometer and populations are enumerated. This test does not confirm the purity.

c) From the broth, samples are taken and its optical density is measured in a Spectrophotometer against its grown media as check and indirectly, the population is extrapolated through its standard.

d) The nitrogen fixing potentiality/phosphate solubilizing ability of the strains could be tested using standard procedures.

However, in the absence of adequate laboratory facilities to carry out sophisticated micro-biological tests a simple "Grow out" test could be conducted to test its ability for the respective crop marked in the packet through different parameters of crop response and indirectly the quality of the biofertilizer packet is partially assessed.

Precautions

a) Keep biofertilizers bottles away from direct heat and sunlight, and store it in cool and dry place.

b) Sell only biofertilizers bottles which contain batch number, the name of the crop,the date of manufacture and expiry period. If the expiry period of the bottles is over, discard it. It will not be effective.

c) Keep bottles away from fertilizer or pesticide containers should not mixed directly.

d) Don't mix with fungicides, insecticides, herbicides and chemical fertilizers.

14.17 Liquid Biofertilizer Application Methodology

Following methods are to be used for using liquid biofertilizers such as seed treatment, root dipping and soil application method.

14.17.1 Seed Treatment Method

Seed treatment is almost common, effective and economic method as adopted for all types of inoculants. It is used for small quantity of seeds up to 5kg quantity only. The coating can be done in a plastic bag. For this purpose,a plastic bag having size (21"x10") or big can be used. The bag has to be filled with 2kg or more seeds.The bag has to be shut in such a way to trap the air as much as possible. The bag has to be twisted for 2 minutes or more until all the seed were uniformly wetted. The bag have to be opened, inflated the bag again and shaked gently. Stop shaking after each seed got a uniform layer of culture coating. The bag has to be opened and spread the seed under a shade for 20-30 minutes for dry. For large amount of seeds counting can be done in a bucket and inoculant can be mixed directly with the hand.

Seed treatment with *Rhizobium, Azotobacter, Azospirillum*, KMB along with PSM. Seed treatment can be done with any of two or more bacteria. There is no side (antagonistic) effect. The important things has to be kept in mind that the seeds must be coated first with *Rhizobium* or *Azotobacter* or *Azospirillum* when each seeds get a layer of above bacteria then PSM and KMB inoculant has to be treated on outer layer of the seeds. This method will provide maximum number of population of each bacterium required for better results. Mixing any of two bacteria and the treatment of seed will not provide maximum number of bacteria of individuals.

14.17.2 Root Dipping Method

This method is needed for the application of *Azospirillum*/KMB/PSM with the paddy transplanting/vegetable crops. The required quantity of *Azospirillum*/KMB/PSM has to be mixed with 5-10 litres of water at one corner of the field and all the plant roots have to be dipped for minimum 1hr before sowing.

14.17.3 Soil Application Method

PSM has to be used as as oil application use 200ml of PSM per acre. Mix PSM and KMB with 400-600kg of Cow dung, FYM along with half bag of rock phosphate if available. The mixture of PSM and KMB, Cow dung and rock phosphate have to be kept under any tree or ceiling for overnight and maintain 50% moisture. Use the mixture as a soil application in rows or during leveling of soil.

Table 14.5 : Recommended Dosage of Liquid Biofertilizer in Different Crops

Field Crops	Recommended Biofertilizer	Application Method	Quantity to be Applied
Pulses			
Chickpea, pea, Ground nut, Soybean, beans, Lentil, lucern, Berseem, green gram, Black gram, cowpea, Pigeon pea.	*Rhizobium*	Seed treatment	200ml/acre
Cereals			
Wheat, oat, barley	*Azotobacter/ Azospirillum*	Seed treatment	200ml/acre
Rice	*Azospirillum*	Seedling treatment	200ml/acre

Oil Seeds			
Mustard, Seasum Linseeds, Sunflower, Castor	*Azotobacter*	Seed treatment	200ml/acre
Millets			
Pearl millets, Finger	*Azotobacter*	Seed treatment	200ml/acre
Maize & Sorghum	*Azospirillum*	Seed treatment	200ml/acre
Forage Crops and Grasses			
Barmunda grass, Sudan grass, Napier	*Azotobacter*	Seed treatment	
Grass, Paragrass and Star grass etc.		200ml/acre	
Other Miscellaneous Plantation Crops			
Tobacco	*Azotobacter*	Seedling treatment	500ml/acre
Tea, Coffee	*Azotobacter*	Soil treatment	400ml/acre
Rubber, Coconuts	*Azotobacter*	Soil treatment	2-3ml/acre
Agro-Forestry/Fruit Plants			
All fruit/agro-forestry (herb, shrubs, annuals & perennial) plants for fuel wood, fodder, fruits, gum, spice, leaves, flowers, nuts & seeds purpose	*Azotobacter*	Soil treatment	Soil 2-3 ml/ plant at nursery
Leguminous plants / Trees	*Rhizobium*	Soil treatment	Soil 1-2 ml/ plant

Source: Chandra *et al.,* (2005)

Note: Dosage recommended when count of Inoculum is 1x10^9cells/ml. If the population is less than 10^8 cells/ml doses will be ten times more besides above said Nitrogen fixers, Phosphate solubilizers and Potash mobilizers at the rate of 200ml/acre could be applied for all crops.

14.18 Crisis of Efficient Strains

Unavailability of potential regional strains is one of the major reasons. The specificity and competitive ability of the strain is the key point on which the efficacies of the organism relay with respect to the hosting soil and plant variety. The ability to fix nitrogen and survival capacity are the limiting factors of the liquid as well as the carrier based biofertilizers.

14.19 Possible Genotypic Changes

During the production of the fertilizers the organism may get interacted with other organisms which may leads to change in basic character of the organisms. Apart from this during fermentation the strains may undergo mutations which may alter the efficacy and viability leading to the economic loss.

14.20 Lack of Awareness

In spite of many ongoing projects on the development of biofertilizers, proper attention towards the technology is still needed in order to manifest the results at field level. Communication gap and miscommunication of the farmers by some commercial producers also is a considerable constraint. Lack of storage and marketing facilities, consistent awareness in the farmers by conducting community programs is still welcome (Revellin *et al.*, 2001). As the production and distribution of biofertilizers is seasonal, the commercial production units may suffer from lack of demand. Moreover, the establishment of microorganisms resisting the antagonistic activity of the native microbes is always challenging. Soil nature, temperature, humidity, pH, local insect population, etc., are some other acclimatization problems (Ngampimol and Kunathigan, 2008).

14.21 Conclusion

Human has been relaying on agriculture throughout the evolution and would be depending on it forever. The impact of globalization, transition in the technology is creating a negative shade on agriculture in the developing countries like India and declined the percentage of the farming community drastically resulting in raising demand for food commodities. Growing global population demanding the safe and sufficient food for the survival. Soil health has become the greatest assertion for the scientific community in this ever growing polluted globe. Uncertainty in the agro climatic conditions (edapho-climatic factors), monsoon failures by the priceless human activities, lack of proper awareness among the farming community are the direct causes for the agriculture failure in the developing countries like India. The rising demand for the fields like food processing, packing industries, ready to eat foods etc., witness the demand for raw materials in agriculture sector. Crises of agricultural land day by day, vertical increase in the cost of agriculture input technologies are leading to transitions in farming community. In such an agro critical scenario, a multi-faced solution for different constraints in agro industry is necessary. It is evident that Biofertilizer technology has inaugurated a new era in biological input technology and recorded a tremendous raise in the annual agriculture production particularly in the past two decades. To combat the threat of global food crises the alternative technologies in the agriculture like liquid biofertilizers are obligatory. Liquid biofertilizer of course have the capacity to replace the traditional chemical fertilizers & carrier based biofertilizers and plays a major role in restoring the soil health, but a lot of measures in terms of technology, government support, subsidies, and constructive awareness by well trained technicians among the agrarians are emphasized.

14.22 Future Strategies

- Identification and characterization of potential organisms (unexplored) and their effective exploitation in the field of Agriculture is urgently needed.
- Selection and application of suitable bio-inoculants with respect to soil nature, agro-climatic condition, crop variety under proper agriculture practices is needed.
- Genotypic study of the strains and molecular characterization of the plant parts is necessary to understand the plant mechanisms.
- Study of soil texture and compatible studies with respect to microbial interventions.
- Exploring the novel soil bacteria and maintaining the genomic libraries for future exploitation.
- Suitable combinations of microbial formulations (liquid microbial consortium) with optimized field results are preferable for the sustainable production.
- Soil analysis, crop rotation, organic manure usage, maintenance of proper moisture content, regular sterilization practices are emphasized which are necessary to maximize the biofertilizer efficacy.
- Global standards in the research & development should be maintained during the production and storage of the formulations.
- Development of new strains with enhanced capabilities by genetic engineering techniques and rDNA technology is needed to maintain an eco-friendly and sustainable agriculture.
- Constructive awareness and technical support by microbiologists and agricultural professionals must be provided to the Agrarians.

CHAPTER - 15

Production, Quality and Marketing of Biofertilizers

Biofertilizer is still an unclear term. It can be easily found that biofertilizers are identified as plant extract, composted urban wastes, and various microbial mixtures with unidentified constituents, and chemical fertilizer formulations supplemented with organic compounds. Likewise, the scientific literature has a very open interpretation of the term biofertilizer representing everything from manures to plant extracts. Quality control of biofertilizers involves several major aspects such as isolation and identification of various strains, screening of the pure isolated strains, fermentation and quality control and finished product.

15.1 Isolation and Identification of Bacterial Strains

Different bacteria vary in their growth requirements. Depending upon the requirements, synthetic media are prepared and sterilized in the laboratory. The bacterial strains which can be used as BF are isolated from various sources using these cultures are isolated from the rhizosphere of the plant root system, while rhizobium cultures are isolated from well developed, healthy nodules developed on the root system of leguminous plants. Soil samples and root samples are collected from different regions of Gujarat state for isolation of all the four cultures. The cultures are also brought from various research institutions like IARI and ICRISAT. On the basis of morphological colonial and biochemical characteristics the strains are identified and differentiated. The pure cultures of the same are preserved for the screening of the efficient strains.

15.2 Screening of the Pure Isolated Strains

The screening of the pure isolated strains carried out at two different but interrelated levels. viz. *In vitro* and *In vivo* level.

15.2.1 In vitro

Azotobacter and *Azospirillum* cultures are treated in a special device called Gibson's Tubes. In Gibson's tube, the seeds of various cereal crops are grown on nitrogen free media and the culture treatment is also given. Depending upon the nitrogen fixing ability of the culture, the growth of the plant will vary.

Rhizobium cultures are tested in a device called Leonard jars for nodulation studies. Nitrogen fixation studies of all these cultures are carried out by Acetylene- Ethylene Reduction Assay. The phosphate cultures are studied on Tricalcium Phosphate Agar. When this culture is grown on Tricalcium phosphate agar, it liberates certain organic acids which are diffused in to the medium. The acids attack on the bound phosphates i.e. Tricalcium phosphate and the phosphate is released from the complex into the medium. The phosphate solubilization is observed in the form of zone of clearance around the colony of phosphobacteria developed on Tricalcium Phosphate agar. The strains showing satisfactory results are further tested at field level.

15.2.2 In vivo

Field level testing of all the cultures is carried out every season in pots, microplots and on our own Research and Development (R&D) Farm. Trials are also taken on farmer's field, government seed farms, agricultural universities and various research institutions. In micro plot trials the cultures tested in the soils brought from different regions of Gujarat. Trials are also taken on farmer's fields. Govt. Seed farms, Agricultural Universities and various Research institutions. The trails conducted at various sugar factories in Gujarat indicate increase in number of tillers, box pol reading and purity percentage with the use of *Azotobacter* and PSB cultures, the phosphate content is also increased from 221mg/lit to 250mg/lit. With the increase in phosphate crop production is increased and at the same time sugar recovery is also increased and in another experiment on groundnut conducted at Chikhli Distt. Bulsar an appreciable increase (from 24.8 to 62.8kg/acre) in available phosphorus was recorded with the treatment of PSB culture. The parameters studied in all the experiments are germination percentage, grain weight and crop yield etc. The strains which are proved efficient after series of tests at lab level and field level are selected for production.

15.3 Fermentation

Azotobacter, Azospirillum, Rhizobium and Phosphate cultures have their own growth requirements. The reasons for adopting shake flask technology are as follows:

- Low capital and operational investment.
- Easy operation.
- Reduced contamination chances.

- Contamination in the flask can be visualized through glass.
- In case of contamination no need to discard the whole batch, only contaminated flask are required to be discarded.
- In different flasks different cultures can be fermented at a time.
- Small batches can be taken in case of less demand.
- Fermented batches can be kept aside in controlled temperature room if production schedule is changed.

Strict quality control measures are observed right from the agar slants to bulk fermentation on shakers in terms of sterility and purity at each step. Periodically calibration of the sterilizers is also carried out with the use of bacteriological indicators. Fermented broth is strictly checked quantitatively for count i.e. numbers of bacterial cells/ml of broth and qualitatively for purity by staining the bacterial cells before further processing.

15.4 Finished Product

After the fermentation for required hours, when the optimum count of pure desired bacteria is achieved; the broth is mixed with sterile carrier i.e. lignite and calcium carbonate (calcium carbonate is added for pH neutralization of lignite). It is mixed in a specific proportion to maintain the proper moisture content i.e. minimum 10/g of the finished product. This mixing is carried out in a specialized mixing machine. All the necessary precautions are taken to reduce the chances of contamination. This reason for using lignite as a carrier in biofertilizer production are that is cheaper as compare to other carrier material and easily available. It has got higher water holding capacity and good adhesion to the seeds, since it inert by nature and hence contamination interference risk is not there. To avoid any further chance or contamination, filling and sealing of the culture mixed carrier material is carrier out on specially designed, advanced automatic form fill seal machine. Weighting of the material bag forming, filling and sealing operations are performed automatically on the same machine at with utmost precision. The maximum capacity of this machine is 15-17 bags/ minute, are pre-printed with all the necessary information like name of the culture, name of the crop for which it is to be used, the method of application, the dose of application and precautions required to be taken. Manufactures date, expire date, batch number and price are also printed on each bag, these bags are further packed in corrugated boxes and each box contains 50 packets. All the boxes are stored in temperature controlled room. Finished product is checked periodically for viability and purity of the bacterial cells as well as confirmation of shelf-life.

15.5 Production Constraints

15.5.1 Raw Material

Peat is an ideal carrier for biofertilizers, which is not abundantly available in India and so lignite is the option left with manufactures. Lignite again is not locally available and so the transportation cost goes up.

15.5.2 Bacterial Strain

Performance of bacterial strain in use depends on various factors like humidity and temperature of the atmosphere, soil pH, soil temperature, chemical and moisture content of the soil and native micro flora of the soil with which the introduced culture has to struggle a lot for the survival. Nature of soil plays a significant role in nitrogen fixation and phosphate solubilization, so bacterial strains should be such that they should be effective in varied agro climatic conditions.

15.5.3 Economic Viability

To have a quality product, the instruments and machines with advanced technology are required which increases the production cost.

15.5.4 Production Process

Even, after utmost care exercised for contamination control, limitation comes at the time of mixing of fermented broth with sterile carrier and at the time of filing, sealing and packaging of the base manually.

15.5.5 Shelf Life

Because of the limited shelf life of the product only limited production can be taken well in advance.

15.6 Proposed Production Technology

There should be institutional agency to properly streamline the production technology which could yield biofertilizer of the desired quality in maximum quantities within a shortest time span which is normally experienced by biofertilizer manufactures. The production technology encompasses various stages as-

15.6.1 Establishment of Efficient Culture Bank

Culture bank which consists of suitable strains of *Rhizobia, Azotobacter, Azospirillum, Pseudomonas, Aspergillus, VA-mycorrhizae* and several other strains of diazotrophs, P-Solubilizers and also of late, composting cultures should be established. These cultures are essentially obtained from various research organizations through active, liaison with scientists and experts working in the laboratory for efficacy and suitability for a particular micro environment in green house and pot culture experiment.

Strains are particularly for those strains which have been the desired traits on their plasmids. Computerized information in the form of catalogue is maintained regarding the sources, target crops, temperature, pH and maximum efforts are done to produce the biofertilizer which are tailor made according to suitability of use in particular crop and area of application. There should be active collaborative support with different agricultural universities in the country, like Indian Agricultural Research Institute (IARI) New Delhi, International Crops Research Institute for the Semi-arid Crops (ICRISAT) Hyderabad, Bhabha Nuclear Research Centre (BNFRC) Bangkok and Nitrogen Fixation by Tropical Agricultural Legumes (NifTAL) Hawaii from were strains are regularly procured authenticated, mass produced and also back tested. Non-government organization should have active liaison with National Biofertilizer Development Centre, Ghaziabad and Regional Centres throughout the country.

15.6.2 Research and Development (R&D)

Research and Development efforts in the unity should be basically focused on optimization of the fermentation process coupled with formulation trials and of late for increasing longevity on the shelves of retailers and also in the store rooms of end users. With the commonly known fact that biofertilizer could not attain popularity in the long ago due to lack of prominent response to naked eyes as is evident by use of chemical fertilizers. Institution should compound the effect of micronutrients and hormones to accelerate and amplify the effect of either diazotroph or P-solubilizing activities of microbial cultures. A part from these regular efforts for optimizing the mass production through fermentation in the shortest possible time, formulation of media which could reduce the production cost, development of carriers and stickers suitable for increased shelf life and ease of application on fields are of comparable quality as that of other primary institutions/biofertilizers manufactures. Efforts should have to be made to increase the shelf life of biofertilizer by immobilizing the useful strains of microbes in caalginate beads. Initial results obtained are quite encouraging and can increase expiry time of our products to a minimum of 18 months compared to 6 months recommended currently.

Research should also be undertaken to standardize/optimize non-conventional methods for biofertilizers applications like that of liquid inoculum and drenching of biofertilizers in the standing crops during inter culture operations is also underway. This has particular helped those farmers who missed the opportunity of using biofertilizers at the time of sowing or transplanting. This could on one hand insure application of biofertilizer on all crops at every given stage and on the other hand could ensure regular demand/ production for the biofertilizer manufactures. Apart from this it has also helped in finding possibility for use of biofertilizers where the soil temperatures at the time of sowing are higher and also where prevalent farming practices do not use necessary implements for isolation of seeds, fertilizers and pesticides.

15.6.3 Mass Production

Mass production of different biofertilizers is undertaken in 200liters capacity fermenters. A scale up process staring from 200ml of broth and 500ml capacity flasks is scaled up in the shortest time under the logarithmic growth phase under the constant vigil for purity. Fermentation technology should be employed in the country. The facility created by it has been widely appreciated by Indian/UNDP experts coming from NifTAL and Australia. Novel methods for mass producing some of the fungal cultures in solid state and also in tissue cultures are actively underway. Scientists at national level in the right earnest undertook the development of different inoculants by applying latest scientific techniques from eighties onwards with the ultimate objectives of producing biofertilizers on commercial scale for the benefit of the farming community. After through screening from in vitro to in vivo bacterial strains for mass multiplication were finalized in the year of 1984. At present India is producing and marketing (a) *Azotobacter* (b) *Azospirillum* (c) *Rhizobium* (d) Phosphate culture and (e) Super culture biofertilizers. The first three provide the nitrogen while the fourth solubilizes the unavailable soil phosphorus and the 5th one has the additive effect over and above the supply of nitrogen to crops.

15.7 Quality Management

Quality management is very essential, and must be performed continually to control the microbial products in favor of the customers. The guidelines used for evaluating quality are limited to the density of the available microorganisms and viability and preservation of the guaranteed microorganisms. It is important to set control plots that do not contain available microorganisms, but whose other compositions are the same as the final microbial products. Also it is highly desirable that the biofertilizer manifests the major effects for quality management of the final biofertilizer products. The major effects are used as indicators for the biofertilizer. Also, the effects are included as guaranteed activities of the biofertilizer. It is an indispensable requirement to distinguish between the available microorganisms and the supplementary compositions on the effects of the biofertilizer guaranteed by the suppliers. If the final results of the two experimental plots are the same or cannot be confirmed statistically, then the product is only an organic matter. This means that the effects of microbial products have to originate from the guaranteed microorganisms, and the target of the matters should be presented in details as a prescription. It is essential to evaluate precisely the functions under the given usage manifested by the applicant (Fig. 15.1).

Biofertilizers, known as microbial products, act as nutrient suppliers and soil conditioners that lower agricultural burden and conserve the environment. Good soil condition is imperative to increased crop production, as well as human and/or animal health welfare. Thus, the materials used to sustain good soil condition, are treated as environmental matters. However, as mentioned earlier, there are still some problems to be met on the use of microbial products. More precise quality control must be made in favour of the customers. With this in mind, we will do our best to develop better production techniques and to improve the management system for microbial products. Although the effects of biofertilizers are different among nations due to variances in climate and soil conditions, the importance of biofertilizer on environmental conservation in the 21st century must not be ignored. In the same manner, various biotechnologies should be accepted for increasing the biofertilizer effects with concern for the environment. Biofertilizers lessen the environmental burden emanating from the chemical compounds. Our viewpoints on biofertilizers are the same for bio-control and bioremediation, because we are members of an ecosystem related to the world wide web of foods.

15.8 Procedures for Quality Control of Biofertilizer

15.8.1. Rhizobium

Quality checks on *Rhizobium* biofertilizer can be divided into three parts:

1. Mother culture test
2. Broth test
3. Peat test

15.8.1.1. Mother Culture Test

Before producing *Rhizobium* biofertilizer, the mother culture should be checked on the growth, purity and Gram staining.

- **Growth:** By streaking a mother culture on yeast mannitol + Congo red agar (YMA) plates, checking the growth of *rhizobia*. Fast–growing *rhizobia* colonies will appear in 3-5 days, and slow growing *rhizobia* will appear in 5-7 days.

- **Purity:** Check purity by streaking culture on glucose peptone agar plate, and incubate for 24hrs at 30°C. No growth or poor growth should be obtained on GPA. Good growth and color changes can be expected from contaminants.
- **Gram staining:** A loop of mother culture is checked by Gram staining. Rhizobial cell is Gram–negative, retains safranin color. Cells should appear red and not violet when observed under the microscope.

15.8.1.2. Broth Test

The following qualities of the broth samples must be checked to make sure that the broths are in good conditions as pH, staining, optical density, total count and viable number.

- **pH:** Slow–growing rhizobia such as *rhizobia* for soybeans, mungbean and peanut produce a little basic compounds. After incubation, the pH will increase (i.e. pH before growing 6.0, after growing pH 6.1-6.2). If broth pH decreases, it means some contaminants occur; lower pH indicates presence of contaminants.
- **Staining** (Gram stain or Fuchsin stain): Rhizobial cells are stained for observation of shape and size of the cells. Cells of *rhizobia* are rod–shaped, with one or two cells sticking together. They do not appear in long–chain. Long–chained cells are indicative of contaminants. Gram-stained cells should appear red, not violet. Fuchsin staining is an easier and faster method. Rhizobial cells can be routinely checked using Fuchsin stain (Table 15.1).

Table 15.1 : Gram Stain

Solutions	Reaction and Appearance of Bacteria	
	Gram–positive	**Gram-negative**
I. Crystal violet (CV)	Cells stain violet	Cells stain violet
II. Iodine solution (I)	CV-I formed within cells	CV-I formed within cells
	Cells remain violet	Cells remain violet
III. Alcohol	Cell walls dehydrated	Lipid extracted from cell walls
	Shrinkage of pores occurs	porosity increases,
	Permeability decreases,	CV-I is removed from cell
	CV-I complex cannot pass	
	Out of cells, cells remain Violet	
IV. Safranin	Cells not affected	Cells take up this stain,
	Remain violet	Become red.

- **Optical Density:** Broth culture with active rhizobial growth will become turbid in 3-4 days. Broth turbidity or optical density using spectrophotometer (at 540 nm) will show readings of 0 –to 1.0 optical density (O.D.) The value of O.D. correlates to number of cells. If O.D. values are high then cells number are also high. We can measure cells from 10^7 to 10^9 per milliliter; if the numbers of cells are low this method is not accurate. This method has its limitations: (i) It gives a direct count (viable + dead cells), (ii) Polysaccharide production in different media gives different results, and (iii) Limitation from the instrument itself.
- **Total Count:** Total count includes viable cells and dead cells by using Petroft-Hausser counter at least 10 small squares all around the total area are counted, and not only in one large square.

  Precautions

  a) Cells have to be homogeneous.

  b) Clumping of cells (use non–ionic detergent).

  c) It gives total count only.

  d) Petroft, cover slip must be properly positioned to get uniform depth.
- **Viable Count:** The number of living cells is counted by spread plate or drop plate methods. Doing spread plate by making serial dilutions from 10^{-1} to 10^{-6} or 10^{-7} (depend on concentration) then three replicates of 0.1mm of broth from 10^{-6} and 10^{-5} is spread over the YMA+CR plates. Plates are incubated in incubator (28–30°C) or at room temperature for 7 days. Colonies of rhizobial cells are round, opaque and have smooth margin. They are white and do not absorb red color as well as the other bacteria. Calculation of the number of rhizobia per ml from below given formula-

$$\text{Number of Cells/ml} = \frac{\text{Number of Colonies} \times \text{Dilution Factor}}{\text{Volume of Inoculum}}$$

$$\text{For example, Number of Cells/ml} = \frac{32 \times 10^6}{0.1} = 32 \times 10^7$$

15.8.1.3. Peat Test

For the peat inoculant, we check these qualities as pH, moisture content, viable number and plant infection method (MPN).

- **pH:** Maintain neutral pH for the inoculant. Since peat is acidic the pH has to be increased with $CaCO_3$. Weigh 10g of inoculant, pour 20ml of distilled water, mix well with glass rod, incubate at least 30 minutes, and then measure with pH meter.
- **Moisture Content:** The optimum moisture content of peat-inoculant is between 40 to 50%. At low moisture rhizobia will die rapidly. If moisture is high, inoculant may stick to the plastic bag and, thus, not good for rhizobial growth.
- **Viable Number:** The number of viable rhizobia is counted by spread-plate method as in the broth test. It is more difficult when analyzing non-sterile peat. Colonies may sometimes be contaminated by other bacteria. The well trained staff is needed to conduct this microbiological analysis.
- **A Plant Infection Analysis using Most Probable Number (MPN) Method:** This is an indirect method of assessing plant infection on nodulation. It is widely used when peat is not sterile. It takes more time than spread plate method (because we have to grow plants). We usually do MPN to compare the results with a spread plate method. Some laboratories conduct the MPN analysis and not by the spread plate method. This method is based on the assumptions that:
    - If viable *rhizobia* are inoculated on its specific host, nodules will develop on those roots.
    - Nodulation on that inoculated plant is a proof of the presence of infective *rhizobia*.
    - Absence of nodule is a proof of the absence of infective *rhizobia*.
    - Uninoculated plants are used as control, with absence of nodule.
- **Estimation of MPN:** To calculate the MPN of organism sample, select as P1 the number of positive tubes in the least concentrated dilution in which all tubes are positive or in which the greatest number of tubes is +ve tubes in the next two higher dilutions. Then find the row of numbers in the MPN table (Appendix I) in which P1 and P2 correspond to the values observed experimentally. Follow that row of the numbers across the table to the column headed by the observed value of P. The figure at the point of intersection is the MPN of organism in the quality of original sample represented in the inoculum added in the second dilution. Multiply this figure by the appropriate dilution factor to obtain MPN Value.

$$\text{Count/g of Carrier} = \text{MPN Table Value} \times \frac{\text{Dilution Level}}{\text{Dry Mass of Product}}$$

15.8.2 Non-Symbiotic N_2 Fixer

In the laboratory, microbial growth may be represented by the increment in cell mass, cell number or any cell constituent. Utilization of nutrients or accumulation of metabolic products can also be related to growth of the organism. Growth, therefore, can be determined by numerous techniques based on one of the following types of measurement: (a) cell count, directly by microscopy or by an electronic particle counter, or indirectly by colony count, (b) cell mass, directly by weighing or measurement of cell nitrogen, or indirectly by turbidity; and (c) cell activity, indirectly by relating the degree of biochemical activity to the size of the population. The multiplication of *Azospirillum* is expected to have reached its maximal at 3-5 days after inoculation. Inoculants in autoclaved carriers are not expected to contain many inoculants. The recommended counting technique for BIO-N inoculant utilizing known volume of serial dilutions is the drop-plate method (Miles and Misra, 1938). Plate dilutions are ranging from 10^{-4} to 10^{-7}. If proper aseptic procedures are not fully observed, contaminants may be accidentally introduced during the injection of the broth culture and during serial dilution and plating. Such contaminants will usually be detectable on these indicator media and their number should be reported together with their number of viable cells as additional measure of the quality.

Procedure

A. Dilution

a. Weigh 10g of BIO-N inoculant and inoculate it on 95ml of distilled water.

b. Shake vigorously and set aside.

c. Make serial dilution of the 95ml inoculated with diluted BIO-N. To achieve this, set out 7 tubes each containing 9ml of sterile diluents.

d. Use a fresh pipette tips for each dilution.

B. Plating

a. Use sterile Enriched Nutrient Agar plates which are at least 3 days old or have dried at 37°C for 2hrs.

b. Plate dilutions 10^{-4}, 10^{-5}, 10^{-6} and 10^{-7}.

c. Allow the drops to dry by absorption into the agar; then invert and incubate at room temperature. Wrap the plates with sterile paper.

d. After 3-5 days of incubation with daily observations count the colonies of the respected organisms of the BIO-N inoculant.

e. Preferred counting range should be 10-30 colonies.

C. Computation

Example

If the average number of colonies per drop is 30 at 10^{-5} dilution, the number of viable cells is:

$1/0.03 \times 30 \times 10^{-5} = 1{,}000 \times 10^{-5} = 1 \times 10^{-8}$ ml

15.8.3 Arbuscular Mycorrhizal Fungi (AMF)

Quality control in the production of AMF inoculum is essential for product consistency, reliability and reproducibility. This is applied to the laboratory, preparation room, growth room, storage room and the greenhouses, taking care into the design, to achieve the most efficient control in inoculum production.

15.8.3.1. Laboratory Quality Control

a) Spores are extracted from selected batches of monospecific spore cultures in the preparation room.

b) Spores are transported in petri dishes to the laboratory and placed in a refrigerator before examination.

c) Petri dishes are examined under stereoscopic microscopes.

d) Descriptions of the spores from each petri dish are recorded.

e) Petri dishes are then cleaned and dried.

15.8.3.2 Preparation Room Quality Control

a) This room has to be isolated from the greenhouse and growth room, and should not receive unsterilized soil or potting media samples.

b) Stored materials (cultures; sterilized growth media) are clearly labeled and placed in specific containers.

c) Floor should always be clean, avoiding sweeping, which encourages distribution of dust.

d) Benches and other surfaces are cleaned with wet towels.

e) Containers are surface-sterilized with 10% sodium hypochlorite.

15.8.3.3 Growth Room Quality Control

a) The growth room should be temperature controlled (22°C), and air is exhausted to the outside (no recycling of stale air).

b) Bench tops should be painted with anti-microbial paint.

c) All surfaces should be sterilized periodically e.g. monthly.

d) All samples are checked for contaminants and pathogens.

e) Watering is done manually, with great care to avoid cross-contamination.

15.8.3.4 Storage Room Quality Control

a) All sampled stored are placed in plastic bags, with proper labeling, and surface of bags should be wiped clean before storage.

b) Floors and bench tops are wiped regularly, and dusting or sweeping should be avoided to prevent generation of dust.

15.9 Occupational Health and Safety for Biofertilizers

The ethical manufacture of good-quality biofertilizer products requires both safe production and safe use in agriculture. This demands that the manufacturing procedures and selected micro-organisms must have passed a rigorous risk assessment, considering the following factors:

- All biofertiliser micro-organisms used should be known to be non-pathogenic to humans, animals or plants. Obviously, this requirement demands that they be identified by genus and species as described elsewhere in this manual. In normal practice, this will mean that if there is significant doubt about the identification of any microbial strain, this strain should not be used, irrespective of how successfully it promotes plant growth and crop yield.
- All micro-organisms to be included in biofertiliser products should be shown to be compatible with each other, even if they are grown separately in production.
- Information on whether there are incompatibilities between biofertiliser micro-organisms and other beneficial strains of microbes that farmers may need to use (such as *rhizobia* for inoculating legumes) should

also be prominently displayed, preferably on the label.

- Inert carrier materials should also be shown to be of good quality, free from organic contaminants, faecal coliforms, heavy metals or other undesirable properties (e.g. acidity adjust with limestone). Such materials should be handled to minimize dust generation with effective face masks worn by workers.
- Safeguarding human health should be a strict operating principle in biofertiliser factories, including protection from heavy lifting and other hazards.

15.10 Laboratories Notified (Approved) under the FCO for Biofertilizer Analysis

Biofertilizer samples can be analyzed for their quality parameters in any well-equipped microbiological laboratory by well qualified microbiologists. Such laboratories exist in virtually all agricultural/horticultural universities, a number of ICAR institutes, research institutes of other organizations and laboratories operated by reputed and quality conscious biofertiliser producers. However, the FCO has recognized/approved the following seven centrally notified biofertiliser testing laboratories for quality control purposes. These are the same laboratories which were part of the erstwhile National Biofertilizer Development Center. These are not only reference laboratories; their results are also acceptable in case of legal disputes. List of the centres of organic farming is below given

1) National Centre of Organic Farming 204 B wing, CGO complex-II, Kamla Nehru Nagar Ghaziabad-2012002, Uttar Pradesh. Covers the area of Uttar Pradesh, Uttarakhand, Delhi and Rajasthan.
2) Regional Centre of Organic Farming, State Agriculture Farm, Mantripukhri, Imphal, 795002. Manipur. Covers the area of Assam, Arunachal Pradesh, Meghalaya, Mizoram, Manipur, Nagaland, Tripura and Sikkim.
3) Regional Centre of Organic Farming 34, II Main Road, Near Baptist Hospital, Hebbal, Bangalore 560024. Covers the area of Karnataka, Kerala, Tamilnadu, Pondicherry and Lakshadweep.
4) Regional Centre of Organic Farming, New Secretariat Building (East Wing), Civil Lines, Nagpur 440001, Maharashtra. Covers the area of Maharashtra, Gujarat, Andhra Pradesh, Goa, Daman & Diu, Dadra and Nagar Havel.
5) Regional Centre of Organic Farming, GA-114, Niladri Vihar (Near KV-4), PO Sailashree Nagar, Bhubaneswar-751007, Orissa. Covers

the area of Bihar, Orissa, West Bengal and Andaman & Nicobar.

6) Regional Centre of Organic Farming, 21 Hira Bhawan, New Chungi Naka, Adhartal, Jabalpur 482004, Madhya Pradesh. Covers the area of Madhya Pradesh, Chhattisgarh and Jharkhand.

7) Regional Centre of Organic Farming, 798, Patel Nagar, Opp. CID Colony, Hissar 125001, Haryana. Covers the area of Haryana, Himachal Pradesh, Punjab and Jammu & Kashmir.

(*Source* : http://krishi.bih.nic.in/contacts_farming_centres.pdf)

15.11 Other Aspects of Biofertilizer Quality Control

15.11.1 Reporting of Analytical Results

According to the FCO, findings of quality analysis of biofertilizer are to be reported in a specific format, an example of which is provided in Appendix II.

15.11.2 Monitoring Biofertilizer Quality

One of the reasons for the poor acceptance and low consumption of biofertilizer is its poor/uncertain quality which can erode the confidence of farmers in the usefulness of these products. The above seven notified laboratories (Section 15.10) have been analyzing samples of biofertilizers and organic fertilizers for quality purposes for several years. A summary of their findings indicates that the number of sub-standard samples is fairly large in most cases and years (Table 15.2). In most of the years, almost one out of four samples or more analyzed were found to be substandard, that is deviated from the quality standard laid down after taking the tolerance limit into account. During 2009-2010, one out of every three samples tested was found to be sub standard both in case of biofertilizers and organic fertilizers. This is alarming. Earlier, some NPK biofertilizers under the name of consortium biofertilizers or poly biofertilizers are being produced and marketed. However, their quality standards are lacking. This requires appropriate attention to safeguard the interest of the farmers.

Table 15.2 : Trends in the Quality of Sampled Biofertilizers

Year	Samples Analyzed	% Samples found Sub-standard
2005-06	2007	23.1
2006-07	1705	25.2
2007-08	1907	0.9
2008-09	1234	27.0
2009-10	983	35.6

Source : Bhattacharyya and Tandon (2012)

A survey of 800 biofertiliser samples conducted by Kabi and Poi (2004) has brought out the following:

- Thickness of the packing material was below the ISI specification in 60% samples.
- Most of the wood charcoal samples (carriers) were found to be substandard.
- Only 9% *Rhizobium* samples had required cell count 15 days before the stated date of expiry and 52% samples of *Azotobacter* were found to be substandard.
- Packing material of about 50% test samples was found to be LDPE.
- Out of 800 samples, only 195 (about 25%) were found to be free of contaminants at the desired level of dilution.
- Only 176 out of 800 *Rhizobium* samples tested were found to be showed sero-positive.

15.11.3 Improved Diagnosis Techniques for Quality Control

There are several biotechnology tools which can be applied for quality testing of biofertilisers. The more common ones are (a) serological methods like agglutination, (b) Fluorescent Antibody Technique (FA) (c) Enzyme Linked Immune-sorbent Assay (ELISA) (d) Use of Immunoblotting Technique (e) Use of Genetic Marker (f) PCR- Finger Printing Technique (g) Application of Oligonucleotide Probe and (h) Use of Fluorescent Dye.

15.11.4 Development of Quality Testing Kit

The Department of Agricultural Microbiology at CSS Haryana Agricultural University, Hissar, Haryana has developed a biofertiliser quality testing kit. This is based on molecular technology and is under testing and evaluation. Now a day, mycorrhiza had been included under biofertiliser in the FCO. Depending on newer initiatives and developments, periodic modifications are made in the FCO.

15.12 Standards for Biofertilizers

The potential range of biofertilizers is too large for one standard to be set. Rather, a standard for each product or product type may be required. In the case of biofertilizers, unlike legume inoculants, there is often no observable effect. Therefore, carefully designed trials are required to measure responses. Often, the mode of action of biofertilizers is unknown, and investigation of a nil response is difficult. Because many of the organisms are part of the normal population of soil organisms without distinct characteristics, it is often difficult to count their numbers in the plant rhizosphere. Again, this is a disadvantage when investigating the failure of a crop to respond, or when trying to set inoculation doses to ensure good rhizosphere colonization. Therefore, to set standards for biofertilizers we need to:

- Accumulate response data from field experiments.
- Carefully characterize the trial sites.
- Use identified and characterized strains.
- Check that cultures are pure.
- Count the cfu/g of active bacteria.
- Calculate the inoculum potential.

Then, after accumulating sufficient data, perhaps obtained by different laboratories or field stations, it should be possible to judge the requirements of each biofertiliser matched to crop and environment in terms of:

- Under what conditions was there a positive response?
- What inoculum potentials are associated with inoculation successes and failures?
- What other crop inputs are associated with successes and failures?

15.12.1 Standards for BioGro

BioGro is a multistrain plant growth-promoting biofertiliser that has been the subject of several collaborative research projects in Australia and Vietnam since the late 1990s. In Vietnam, standards already exist for biofertilisers, but these are based on functional tests (e.g. phosphate solubilisation or nitrogen fixation) and are not specific to selected strains. For BioGro, standards for each particular species have been determined based on both agronomic performance and numbers per g achievable at commercial-scale production. These are expressed as colony forming units (cfu) per gram of product. The current standards for each species, suggestions for selective media and confirmatory tests and media recipe for plate-counting BioGro strains are given in Table 15.3, 15.4 and 15.5 respectively.

Table 15.3 : Current Standards for Each Species in BioGro

BioGro Strain	cfu/g of carrier	kg/ha	Nominal cfu/ha
1N	10^7	100	10^{12}
HY	5×10^6	100	5×10^{11}
B9	10^7	100	10^{12}
E19	10^7	100	10^{12}

Table 15.4 : Suggestions for Selective Media and Confirmatory Tests for BioGro Strains

Strain	Selective/ Differential Media	Confirmatory Tests
1N	Tryptic soy or nutrient agar with antibiotics (vancomycin and cycloheximide)	Gram stain Antibody reaction using colony immunoblotting or ELISA
HY	Pikoskaya's medium for P-solubilizing colonies (look for zone of clearing around colony) or selection using nutrient agar with tetracycline	Gram stain Antibody reaction using colony immunoblotting or ELISA Sensitivity to cycloheximide
B9	Nutrient or modified nutrient agar and heat selection at 80°C, or colony morphology	Gram and spore stain Ability to suppress growth of *Fusarium* sp. but not *Bradyrhizobium japonicum* CB1809
E19	Nutrient or modified nutrient agar and colony morphology	Gram and spore stain Ability to suppress growth of *B. japonicum* CB1809

Table 15.5 : Media Recipes for Plate-Counting BioGro Strains

Strain	Recipe for 1L Media
HY	Pikoskaya's recipe, Nutrient agar and Antibiotics (tetracycline, 10mg/L)
1N	Nutrient agar Antibiotics (vancomycin, 10mg/L and cycloheximide, 100mg/L)
B9 and E19	Nutrient agar supplemented with 5g yeast extract, 0.5g glucose, 0.5g sucrose

15.13 Specifications of Biofertilizers and Raw Materials required for Biofertilizers Production

Specifications of Biofertilizers as *Rhizobium, Azotobacter, Azospirillum*, PSB & Raw Materials required for biofertilizers production are given in Table 15.6, 15.7, 15.8, 15.9 and 15.10 respectively.

Table 15.6 : Specifications of *Rhizobium*

Sl. No.	Characteristics	Description
a.	Base	Carrier based* in form of moist/dry powder or granules, or liquid based
b.	Viable cell count	CFU minimum $5x10^7$ cell/g of powder, granules or carrier material or $1x10^8$ cell/ml of liquid.
c.	Contamination level	No contamination at 10^5 dilution
d.	pH	6.5 – 7.5
e.	Particle size in case of carrier based material	All material shall pass through 0.15-0.212 mm IS sieve
f.	Moisture percent by weight, maximum in case of carrier based	30-40%
g.	Efficiency Character	Should show effective nodulation on all the specieslisted on the packet.

* *Type of Carrier* : Carrier material such as peat, lignite, peat soil, humus, wood charcoal or similar material favoring growth of the organism.

Table 15.7 : Specifications *Azotobacter*

Sl. No.	Characteristics	Description
a.	Base	Carrier based* in form of moist/dry powder or granules, or liquid based
b.	Viable cell count	CFU minimum $5x10^7$ cell/g of carrier material or $1x 10^8$ cell/ml of liquid.
c.	Contamination level	No contamination at 10^5 dilution
d.	pH	6.5 – 7.5
e.	Particle size in case of carrier based material	All material shall pass through 0.15-0.212 mm IS sieve
f.	Moisture percent by weight, maximum in case of carrier based	30-40%
g.	Efficiency character	The strain should be capable of fixing at least 10 mg of nitrogen per g of sucrose consumed

* *Type of Carrier* : Carrier material such as peat, lignite, peat soil, humus, wood charcoal or similar material favoring growth of the organism.

Table 15.8 : Specifications *Azospirillum*

Sl. No.	Characteristics	Description
a.	Base	Carrier based* in form of moist/dry powder or granules, or liquid based
b.	Viable cell count	CFU minimum $5x10^7$ cell/g of carrier material or $1x 10^8$ cell/ml of liquid.
c.	Contamination level	No contamination at 10^5 dilution
d.	pH	6.5 – 7.5
e.	Particle size in case of carrier based material	All material shall pass through 0.15-0.212 mm IS sieve
f.	Moisture percent by weight, maximum in case of carrier based	30-40%
g.	Efficiency character	Formation of white pellicle in semisolid nitrogen free bromothymol blue media.

**Type of Carrier* : Carrier material such as peat, lignite, peat soil, humus, wood charcoal or similar material favoring growth of the organism.

Table 15.9 : Specifications Phosphate Solubilizing Bacteria

Sl. No.	Characteristics	Description
a.	Base	Carrier based* in form of moist/dry powder or granules, or liquid based
b.	Viable cell count	CFU minimum $5x10^7$ cell/g of carrier material or $1x 10^8$ cell/ml of liquid.
c.	Contamination level	No contamination at 10^5 dilution
d.	pH	6.5 – 7.5 for moist/dry powder granulated carrier based and 5.0-7.5 for liquid based.
g.	Particle size in case of carrier based material	All material shall pass through 0.15-0.212 mm IS sieve
g.	Moisture percent by weight, maximum in case of carrier based	30-40%
h.	Efficiency character	The strain should have phosphate solubilizing capacity in the range of minimum 30%, when tested Spectrophotometrically. In terms of zone formation,Minimum 5mm solubilization zone in prescribed media having at least 3mm thickness.

* *Type of Carrier* : Carrier material such as peat, lignite, peat soil, humus, wood charcoal or similar material favoring growth of the organism

Table 15.10 : Specifications of Raw Materials required for Biofertilizers Production

Sl. No.	Parameters	Purpose	Specifications
a)	Wood Charcoal	Carrier	50-22 Mesh, Good Quality, pH 6.5-7.5, 120-140% Water Holding Capacity, 40% Moisture Percentage Ash Content 5-10%
b)	Raw Lignite Powder	Carrier	150-200 Mesh, 35% Moisture Percentage, finely powdered, Duly packed in HDPE bag of 40-50kg
c)	Gum Arabic	Adhesive	150-200 Mesh Dry Powder
d)	HDPE Polythene Bag	Packing	1st Grade Milky white high density polythene having thickness of 300guage and sealed from one end, Standard packet size 15x25cm for 200g biofertilizer
e)	Instafix Plastic Tape Gummed	Packing	5cm wide tape for fixing and sealing the corrugated card board boxes.
f)	LDPE Plastic Strips	Packing	For packing the card board made of first grade LDPE, 5cm wide thick in 200m rolls
g)	Iron Clips	Packing	For packing strip of 5cm

15.14 Procedure for Sampling of Biofertilizers

15.14.1 General Requirements of Sampling

a) In drawing, preparing and handling the samples, the following precautions and directions shall be observed.

b) Sampling shall be carried out by a trained and experienced person as it is essential that the sample should be representative of the lot to be examined.

c) Samples in their original unopened packets should be drawn and sent to the laboratory to prevent possible contamination of sample during handling and to help in revealing the true condition of the material.

d) Intact packets shall be drawn from a protected place not exposed to dampness, air, light, dust or soot.

15.14.2 Scale of Sampling

a) **Lot:** All units (containers in a single consignment of type of material belonging to the same batch of manufacture) shall constitute a lot. If a consignment consists of different batches of the manufacture the containers of the same batch shall be separated and shall constitute a separate lot.

b) **Batch:** All inoculant prepared from a batch fermenter or a group of flasks (containers) constitute a batch.

c) For ascertaining conformity of the material to the requirements of the specification, samples shall be tested from each lot separately.

d) The number of packets to be selected from a lot shall depend on the size of the lot and these packets shall be selected at random and in order to ensure the randomness of selection procedure given in IS 4905 may be followed (Please visit on https://law.resource.org/pub/in/bis/S07/is.4905.1968.pdf for details).

15.14.3 Drawl of Samples

a) The Inspector shall take three packets as sample from the same batch. Each sample constitutes a test sample.

b) These samples should be sealed in cloth bags and be sealed with the Inspector's seal after putting inside Form P. Identifiable details such as sample number, code number or any other details which enable its identification shall be marked on the cloth bags.

c) Out of the three samples collected, one sample so sealed shall be sent to incharge of the laboratory notified by the State Government under clause 29 or to National Centre for Organic Farming or to any of its Regional Centres. Another sample shall be given to the manufacturer or importer or dealer as the case may be. The third sample shall be sent by the inspector to his next higher authority for keeping in safe custody. Any of the latter two samples shall be sent for referee analysis under sub clause (2) of clause 29B.

d) The number of samples to be drawn from the lot as given below-

Lot/Batch	Number of Samples
Upto 5,000	Packets 03
5,001-10,000	Packets 04
More than 10,000	Packets 05

15.15 Tolerance Limit of Organic Fertilizer

A sum total of nitrogen, phosphorus and potassium nutrients shall not be less than 1.5% in city compost and shall be not less than 2.5% in case of vermicompost. Tolerance limit of the city compost and vermicompost us given in Table 15.11 and 15.12 respectively.

Table 15.11 : Tolerance Limit of City Compost

Sl. No.	Characteristics	Description
a)	Moisture, % by weight	15.0-25.0
b)	Colour	Dark brown to black
c)	Odour	Absence of foul odour
d)	Particle size	Minimum 90% material should pass through 4.0mm IS sieve
e)	Bulk density (g/cm^3)	<1.0
f)	Total organic carbon, % by weight, minimum	12.0
g)	Total Nitrogen (N), % by weight, minimum	0.8
h)	Total Phosphates (P_2O_5), % by weight, minimum	0.4
i)	Total Potash (K_2O), % by weight, minimum	0.4
j)	C:N ratio	<20
k)	pH	6.5 - 7.5
l)	Conductivity (dsm^{-1}), not more than	4.0
m)	Pathogens	Nil
n)	Heavy metal content, (as mg/Kg), maximum	
	Arsenic (As_2O_3)	10.11
	Cadmium (Cd)	5.11
	Chromium (Cr)	50.11
	Copper (Cu)	300.00
	Mercury (Hg)	0.15
	Nickel (Ni)	50.00
	Lead (Pb)	100.00
	Zinc (Zn)	1000.00

Table 15.12 : Tolerance Limit of Vermicompost

Sl. No.	Characteristics	Description
a)	Moisture, % by weight	15.0-25.0
b)	Colour	Dark brown to black
c)	Odour	Absence of foul odour
d)	Particle size	Minimum 90% material should pass through 4.0 mm IS sieve
e)	Bulk density (g/cm^3)	0.7 -0.9

f)	Total organic carbon, % by weight, minimum	18.0
g)	Total Nitrogen (N), % by weight, minimum	1.0
h)	Total Phosphates (P_2O_5), % by weight, minimum	0.8
i)	Total Potash (K_2O), % by weight, minimum	0.8
j)	Heavy metal content (mg/Kg), maximum	
	Cadmium (Cd)	5.0
	Chromium (Cr)	50.00
	Nickel (Ni)	50.00
	Lead (Pb)	100.00

(b) In Part B, under the heading 'Tolerance Limit of Organic Fertilizers', for the figures and words "0.1 unit for combined nitrogen, phosphorus and potassium nutrients", the figures and words "A sum total of nitrogen, phosphorus and potassium nutrients shall not be less than 1.5% in City Compost and shall be not less than 2.5% in case of vermicompost", shall be substituted.

15.16 Methods of Analysis of Organic Fertilizers

15.16.1 Estimation of pH

a) Make 25g of compost into a suspension in 50ml of distilled water and shake on a rotary shaker for 2hrs.

b) Filter through Whatman No. 1 or equivalent filter paper under vacuum using a Buchner funnel.

c) Determine pH of the filtrate by pH meter.

15.16.2 Estimation of Moisture

Method

Weigh to the nearest mg about 5g of the prepared sample in a weighed clean, dry petri dish. Heat in an oven for about 5hrs at $65^0 \pm 1^0C$ constant weight, Cool in a desicator and weigh. Report the percentage loss in weight as moisture content.

Calculation

$$\text{Moisture Percent by Weight} = \frac{100\,(B-C)}{B-A}$$

A = Weight of the Petri dish

B = Weight of the Petri dish plus material before drying

C = Weight of the Petri dish plus material after drying

15.16.3 Estimation of Bulk Density

Requirements

100ml measuring cylinder, Weighing balance, Rubber pad (1sq foot; 1 inch thickness), Hot air oven.

Method

a) Weigh a dry 100 ml cylinder (W_1 g).

b) Cylinder is filled with the sample upto the 100ml mark. Note the volume (V_1 ml).

c) Weigh the cylinder along with the sample (W_2 g).

d) Tap the cylinder for two minutes.

e) Measure the compact volume (V_2 ml).

Calculation

$$\text{Bulk Density} = \frac{\text{Weight of the Sample Taken } (W_2 - W_1)}{\text{Volume } (V_1 - V_2)}$$

15.16.4 Estimation of Electrical Conductivity

Requirements

250ml flask, Funnel (OD-75mm), 100ml beaker, Analytical balance, Potassium chloride, Filter paper, Conductivity meter with temperature compensation system.

Method

a) Pass fresh sample of organic fertilizer through a 2-4mm sieve.

b) Take 20g of the sample and add 100ml of distilled water to it to give a ratio of 1:5.

c) Stir for about an hour at regular intervals.

d) Calibrate the conductivity meter by using 0.01M potassium chloride solution.

e) Measure the conductivity of the unfiltered organic fertilizer suspension.

Calculation

Express the results as millimho's or ds/cm at 25^0C specifying the dilution of the organic fertilizer suspension viz. 1:5 organic fertilizer suspension.

15.16.5 Estimation of Organic Carbon

Requirements

Silica/Platinum crucible 25g (cap) and Muffle Furnace

Method

a) Accurately weigh 10g of sample dried in oven at 105^0C for 6hrs, in a pre-weighed crucible and ignite the material in a Muffle furnace at 650-700^0C for 6-8hrs.

b) Cool to room temperature and keep in Desiccator for 12hrs.

c) Weigh the contents with crucible

Calculation

Calculate the total organic carbon by the following formulae:-

$$\text{Total Organic Matter \%} = \frac{\text{Initial Weight} - \text{Final Weight}}{\text{Weight of Sample taken}} \times 100$$

$$\text{Total C\%} = \frac{\text{Total Organic Matter}}{1.724}$$

15.17 Marketing

The essential part of marketing of biofertilizers lies in its role in extension and farmer education activities. As is true of marketing of any agricultural inputs, and particularly for a low shelf life product as biofertilizer, extension activities for farmer education is critically essential. The history of production, marketing and use of biofertilizer in the country has shown that this very essential requirement was never looked properly by the manufactures resulting in large scale failure of this product and reverse publicity. It then required a "PUSH* Strategy" by Government to popularize biofertilizer in the minikit/subsidy oriented programmes. Government turn wide resources tried to maximize the efforts on

* **PUSH Strategy:** A push strategy involves taking the product directly to the customer via whatever means, ensuring the customer is aware of brand/product at the point of purchase— "Taking the product to the customer."

the lines of extension activities for educating the end users for appropriate time and method of application. Precautions good result what they could except from a good quality product. Through classical methods of extension as that of interest, awareness in trial and final adoption, thee officers are engaged in training of farmers through conducting fields trials on their farms, arranging field days, discussing the methodology and analyzing their reactions, conducting exhibition for mass awareness: they aim to satisfy the farmers towards the utility of biofertilizer for partial supplement of their chemical fertilizers needs and thereby net saving on the cost of inputs on the one hand and to the net profit generated in terms of increased yield, adding to potency of their soil in an ecofriendly way. This has resulted in the active "PULL" type of demand from the farmers wherever National Agricultural Cooperative Marketing Federation of India (NAFED), New Delhi has entrusted their products, In addition extension trials are also undertaken with the help from State and Central Government Organization like National Resource Conservation Services (NRCS) Indore, Directorate of Oilseeds, Hyderabad through its centers located in different parts of the country.

The marketing and distribution of biofertilizers has multitier arrangements ranging from central or state Government distribution programmes in mini kits to retailing of biofertilizer through own sale counters. Necessary support for proper storage and distribution of biofertilizers is extended to each and every channel so that could function properly and efficiently. State and central governments have taken exclusive efforts in providing support of publicity and advertisement through Department of Publicity materials providing audio visual media for increasing popularity of biofertilizers. This include providing of 16mm projector slides, video films, audio cassettes and publicity media like wall paintings, lucrative trade discounts and replacement policies have been offered to various distribution functionaries so as to give proper service margins to those agencies which are active engaged in distribution. Biofertilizers unit is a unique example in the country which survive on commercial lines with only biofertilizers as commodity basically, being service oriented that of being profiteering motto and still commercially viable. Active support and useful direction from management has resulted in the progress of several units. A dynamic marketing philosophy should be formed for the market of biofertilizers with the support of development and promotional programmes and by offering before and after sales services to farmers. It resulted into the increase in the sales of various biofertilizers.

15.17.1 Farm Information Centers-cum-Depots

Farm Information Centers cum Depots should located in rural areas and manned by experienced agricultural graduates to provide the technical knowhow, conduct the demonstration and give before and after sales services to farmers.

15.17.2 Distributions

In order to provide biofertilizers upto the village level, Gujarat State Fertilizer Corporation has appointed its own Distribution network for marketing various types of biofertilizers.

15.17.3 Co-operative and Institutional Agencies

Priority is also given to the Co-operative societies and other Institutions Agencies for distribution of the biofertilizers.

15.17.4 Subsidy

In order to promote the biofertilizers, Govt. of Gujarat and various states has been offering subsidy to the extent of 50% to the farmers through FIC-cum Depots and Co-operative Societies. It has been helpful in pushing the biofertilizers in rural areas.

15.17.5 Farmer's Survey

In order to know the farmers response to Sardar Biofertilizers use, the survey was carried out in 5 talukas of Vadodara Distt. In all 80 farmers were contacted. About 62% farmers used Sardar Biofertilizers as they were convinced about their benefits, around 78% farmers found the prices as lower and reasonable, nearly 60% opined that they had gone for them due to better quality, 80% informed that they would continue to use them, 87% indicated about their timely availability and lastly about 90% were fully satisfied about their use.

15.17.6 Quality Control

In case of biofertilizers, the bacteria are not visible with naked eyes and hence there is every possibility on the part of producers to give only carrier material without inoculation and to avoid this there is a great need to have strict quality control measures. The Government of India should maintain a Culture Bank of all the Biofertilizers which should be periodically at Lab level for measuring their N_2 Fixing capacity. These cultures should also be tested in pots, microplots and at field level in every season to check their beneficial effect and on the basis of this; the best efficient strains are selected by commercial production. As necessary precautions are also taken during the whole production procedures to avoid contamination, besides, selected bacterial strain is periodically checked at inoculum level as well as bulk fermentation level microscopically, to maintain the desired bacterial count and purity. Finished Products are also checked frequently for a viability of the bacterial strains used. The shelf life fixed is also confirmed periodically.

15.18 Constraints

Biofertilizers, being new to the procedures, distributors and also farmer's consumers, have the following constraints affecting the consumption adversely:

15.18.1 Pricing Policy and Packaging

No definite pricing policy exists at present for biofertilizers. Packing standards area also not fixed. This results in the supply of very poor quality biofertilizers having variation in price and quantity as well.

15.18.2 Lack of Awareness

Inspite of the emphasis given on training, education knowledge, adoption and perception, most of the farmers do not have inclination for the use of biofertilizers. They yet need extensive education and practical training on applied aspects of biofertilizers through aggressiveness promotional programmes as well as before and after sales services.

15.18.3 Inadequate Shelf-life

The biofertilizers presently available have shelf life of 3 to 6 months. Farming in most of the areas fully depends on nature and such variation in agro-climatic conditions results in the change in purchasing power of farmers affecting the procurement of inputs which leads to the unsold stock of biofertilizers at the distribution centers and sales points with shorter shelf life thereby affecting its overall efficiency.

15.18.4 ISI Mark

The Indian Statistical Institute (ISI) Mark certainly improves the credibility in respect of quality of biofertilizers but the whole procedure is so complicated that the small manufactures cannot afford it in respect of time, manpower and finance and hence they are not coming forward for the same. It is therefore there is a great need to simplify the whole procedures along with reduction of charges so that it may be in the reach of small procedures.

15.19 Bio-fertilizers Market trend of All India

There are over 100 biofertilizer units operational in the country. These units produced about 20040MT bio-fertilizers against the installed capacity of over 86000MT during 2009-10. The year wise capacity, production and sales trend of biofertiliser is given in Table-15.13.

Table 15.13 : All India Capacity, Production & Sales Trend of Biofertilizers (Qty. in MT)

Year	Capacity	Production Utilization	% Capacity	Sales Production	% Sales of Growth	% Sales
2000-01	16446.0	6242.6	37.95	6138.6	98.3	12.5
2001-02	15439.0	7390.0	47.86	6876.1	93.0	12.0
2002-03	18679.5	8643.0	46.27	7000.0	81.0	1.8
2003-04	18632.0	10500.0	56.35	8500.0	80.9	21.4
2004-05	N.A.	10479.0	N.A.	10428.0	99.5	22.6
2005-06	N.A.	11752.0	N.A.	11358.0	96.6	8.9
2006-07	43495.0	15871.0	36.48	15745.0	99.2	38.6
2007-08	67162.0	20111.0	29.94	20100.0	99.9	27.6
2008-09	68804.0	25065.0	36.42	25000.0	99.7	24.3
2009-10	86078.0	20040.3	23.28	20000.0	99.8	(-)20.0

Source : Specialty Fertilizer Statistics, FAI (2011), MOA and NCOF, Ghaziabad

The trend, as appears from the above table is that there has been significant addition in the capacity of production of bio-fertilizers in recent years and one of the reason may be Government concern towards deteriorating soil health, backend subsidy support on installation of production unit linked with National Bank for Agriculture and Rural Development (NABARD) and inclusion of private sector also for subsidy support. However, the production remained within a specific limit with a growth rate ranging 12% to 35%, which is basically demand driven. It is also evident that the sales used to be almost 100% of the production, which also indicates that production used to be demand driven, however, a significant growth has been witnessed in the sales growth during recent past, which is a positive sign (http://www.kribhco.net/images/pdf/PAPER_2013.pdf)

15.19.1 Marketing Challenges

In spite of being cost effective input, the Bio-fertilizers have not been accepted by the farmers completely till now. Some of the reasons/constraints for low acceptance of biofertilizer are narrated below. However, the product modification as "Liquid form" has overcome few limitations and has provided opportunities for Marketers.

1) Biofertilizers are live microorganisms which dies increase of high temperature.
2) Biofertilizers are used before sowing and delay in dispatches leads to inventory carry over and expiry of product.
3) Some of the bio-fertilizers are crop specific as well as location specific

and therefore its efficacy does not remain same at different locations due to difference in agro-climatic conditions and soil edaphic factors.

4) Soil characteristics like high nitrate, low organic matter, less available phosphate, high soil acidity or alkalinity, high temperature as well as presence of high agro-chemicals or low micro-nutrients contribute to failure of inoculants or adversely affect its efficacy.

5) Supply of sub-standard or spurious material by some of the manufacturers also adversely affects the credibility of the biofertilizers, being a new product.

6) Some firms are selling organic manures as biofertilizers. Some organizations mention shelf life as two years/one year despite norm of maximum 3- 6 months.

7) Naturally occurring soil micro flora and fauna also often inhibits the growth of introduced inoculums due to competition.

8) Shelf life of bio-fertilizer is limited to 6-12 months in powder form.

9) Change in cropping pattern by farmers also adversely affects the sales.

10) Lack of awareness of the farmers regarding benefits of bio-fertilizer.

11) There is no magic effect of bio-fertilizer & its impact is not visible in standing crop and therefore farmer is not convinced with the benefits of bio-fertilizer use.

12) Some of the issues have been taken care of after inclusion of bio-fertilizers in FCO (Fertilizer Control Order) by GOI. Moreso, the product innovation and new product development like Liquid bio-fertilizer has also overcome some of the limitations.

15.19.2 Option- Switching over to Liquid Biofertilizers

There are so many options which switch over to the liquid biofertilizers and show as they are superior to powder based

1) Longer Shelf life to the extent of 12-24 months.

2) Easy identification by typical fermented smell.

3) High population can be maintained more than 10^9 Cells /ml up to 12-24 months.

4) Cost saving on carrier material, pulverization, neutralization, sterilization, packing and transport.

5) No Contamination, no effect of high temperature as tolerant up to 45°C without any property losses and greater potential to fight with native population.
6) Better survival on seeds and soil and very easy to use by the farmer.
7) Quality control protocols are easy and quick.
8) No need of running biofertilizer production units throughout the year.
9) High commercial revenues and high export potential.
10) Very high enzymatic activity since contamination is nil.
11) Dosage is 10 times less than carrier based powder based biofertilizers.

15.20 Strategic Marketing

Strategic marketing is a process that can allow an organization to concentrate its limited resources on the greatest opportunities to increase sales and achieve a sustainable competitive advantage. Competitive advantage is the strategic advantage one business entity has over its rival entities within its competitive industry. Achieving competitive advantage strengthens and positions a business better within the business environment. The rational business decisions would entirely depend upon marketing research, understanding customers and their requirements, product modifications or innovations and defining an appropriate communication strategy with developing a connect with the customers. The rural marketing perhaps does not allow all of them or is being expensive and prohibits many small players to remain out of the game or restrict to a limited geographical area. Some of the marketing stratagems as suggested below may work strongly in the marketing of biofertilizers:

15.20.1 Field Demonstration

The farmers do what they see because 'Seeing believes' and therefore result as well as method demonstration are very effective tools in promoting biofertiliser usage. The producers may synergize their efforts on this front as bio-fertilizers are new and it is very crucial to show the impact of bio-fertilizer usage to farmers and educate them the economics/returns. So, demonstration farm may be developed jointly, at different locations, defining a catchment area, which could be shown to farmers at different crop stages.

15.20.2 Market Segmentation and Product Positioning

The segmentation is primarily dividing market into various groups of buyers. The bio-fertilizer market can be segmented by specific crop grower (Vegetable/ Fruits/Pulses/Cereals/Oilseed/ Sugarcane), institutional buyers (Cotton/Oilseeds/

Pulses/Cane/Tea/Coffee/Federations & Research-farms, SFCI, Agro-industries etc.) and the customer size (major/minor), geographical location (low/high consuming area and accessibility), and product application (supplementary/ exclusive). Once the market is segmented, it is important to target the market and concentrate on the most profitable one. Positioning starts with a product, but positioning is not what one does to a product; rather it is what one does to the mind of a prospective customer. Thus, the product is being positioned in the mind of the customer, i.e. how he perceives the product. In an over communicated society, the marketer must create distinctiveness. The appropriate Unique Selling Proposition (USP) needs to be identified and propagated widely by (i) save cost through reduced dosage of chemical fertilizers (ii) improves resistance power against disease (iii) enhance sugar recovery percent in sugarcane etc.

15.20.3 Pricing

As we know, rural markets are quite price sensitive and particularly biofertilizers, being technical & new to farmers with lot of constraints, does not fall under the category of 'Zero elasticity of demand' and needs more 'PUSH' in view of lack of 'PULL'* strategy. Mostly companies generally determine price of a product on the basis of marketing objectives. Here, it is important to understand how bio-fertilizer is perceived in terms of value offered for money spent by customers. The bio-fertilizers have derived demand and So far, it has not really been perceived by farmers as giving those economic returns by reduction in quantity of chemical fertilizers used. Unless, farmers are convinced about substantial savings in cost of production through reduced usage of chemical fertilizers and getting similar productivity, probably biofertiliser manufacturers will not be able to apply pricing strategies.

15.20.4 Sales and Usage Promotion

There is a great need to promote the product, both from the point of view of sales as well as usage. The channel members (dealer or distributors) need to be motivated by offering them tangible benefits. Similarly, consumer also needs to be attracted by offering them coupons, premiums, contests, buying allowances etc. based on customer characteristics/ buying behavior. The progressive farmers/ village leaders besides dealers may also be identified for the purpose of conducting demonstrations and should be appropriately compensated.

15.20.5 Publicity and Training

Point of Sales (POS) material must be made available to all dealer/

* **PULL Strategy :** A pull strategy involves motivating cusutomers to seek out brand/produce in an active process—"Getting the customer to you."

distributors and also needs to be ensured that product is displayed visibly. Wider publicity through broadcast and educational films screening also needs to be taken up vigorously. Orientation and training programmes for field sales force and dealers/distributors also needs to be chalked out and free distribution of biofertilizer must be avoided during meeting of farmers. There is a need of exclusive team of extension administrators for promoting biofertilizers with constant visits and developing a close connect with farmers and undertaking marches with its replication in nearby villages.

15.20.6 Product Modification and Introduction of Innovative Products

The basic need of the modern marketing is to regularly keep a track of the consumers behaviour and adapt immediately to the requirements or the benefits sought by the consumers. As far as biofertilizers are concerned, it has been consistently argued for over a decade that there are tremendous product as well as market related constraints, however the marketing organization have not been able to adapt to the business environment needs. Biofertilizers in powder form had several constraints, which could be overcome to a great extent by product modification from powder form to liquid form, which have marvelous superior benefits. The product innovation is another step forward towards tackling farmers' issues and some of them are zinc and sulphur solubilisers like *Thiobacillus* species, potash mobilisers like *Frateuria aurentia* and manganese solubilizer fungal culture like *Pencillium citrinum*, which have been identified for marketable operations and are highly useful and cost-effective for enhancing agricultural productivity.

15.20.7 Marketing Linkages

Marketing linkages with technology providers (i.e. Drip Irrigation producers) may be initiated as biofertilizers have got tremendous potential as its application through this technology. Some states like Uttarakhand and Sikkim have been declared as "Organic States" and therefore agreement may be signed with their Department of Agriculture, State Agriculture University for promoting biofertilizers usage. Sugar factories may also be a bulk buyer for biofertilizers as Acetobacter and Phosphorus solubilize mobilizers, Potassium solubilize mobilizer and Zinc & Sulphur solubilisers.

15.21 Outlook

Keeping in view the trust areas indentified and the strategy worked out for increasing the consumption and the marketing of the biofertilizers, the following suggestions are given by workers.

- Periodical review of ISI specifications of all types of biofertilizers is must. The whole ISI procedures should be simplified and charges to be reduced considerably to make it more popular amongst the small producers.
- Strict quality control standards in production, storage, distribution and use must be maintained.
- Uniform packaging and pricing policy be set up and subsidy may be provided on capital as well as revenue expenditure.
- Along with manufactures own marketing channels, Government channels may be set up.
- State and central Government should have facility for testing of biofertilizers as assurance to farmers.
- On the line of Chemical fertilizers the "Biofertilizer Control Order" should be worked out to ensure the quality.

Experiments, trials and demonstrations convincingly conclude that the increase in crop production to the extent of 15 to 20% with the use of biofertilizers and the increase was higher on farmers' fields. The additional increase in the crop production by the use of biofertilizers over and above the use of chemical fertilizers may indicate the improvement in fertilizer use efficiency. Inspite of the fact that biofertilizer is a low input in increasing crop production, certain limitations and constraints particularly related to shelf-life, high temperature, transportation, timely distribution in remote areas etc. do restrict rapid growth of biofertilizers market.

Besides, uniform pricing policy, standards packaging and quality test by Govt. of India/State Government will help to develop healthy market in the time to come. The last but not the least, there is a great need to work out the aggressive promotional and marketing strategy for increasing the use of biofertilizers as a low cost input supplementary and complimentary to chemical fertilizers for getting higher crop production and profit. Future strategies aim basically in changing the overall appearance of Biofertilizers industry from the conventional carrier based, short expiry products to products of international quality/potency with an increased longevity of at least 12 to 18 months which can ensure steady state demand from agri-input dealers/traders. As the problems of dumping unsold materials could be overcome on the one hand and financial loss to the manufactures on account of expired/returned materials are minimized saving the overall national loss, on the other hand. The immediate priority is to educate the farmers effectively to realize the potential of this promising biotechnology in order to reduce the use of chemical fertilizers and save valuable foreign exchange for country as a whole, yet meeting the challenges for increased food production simultaneously. Since agriculture is the base of Indian economy and the result of preceding revolution such as **"Green Revolution"** and **"White Revolution"** had been encouraging the time is ripe enough to make **"Biofertilizer Revolution"** as the next important break-through for the Indian agriculture, this is the future endeavor.

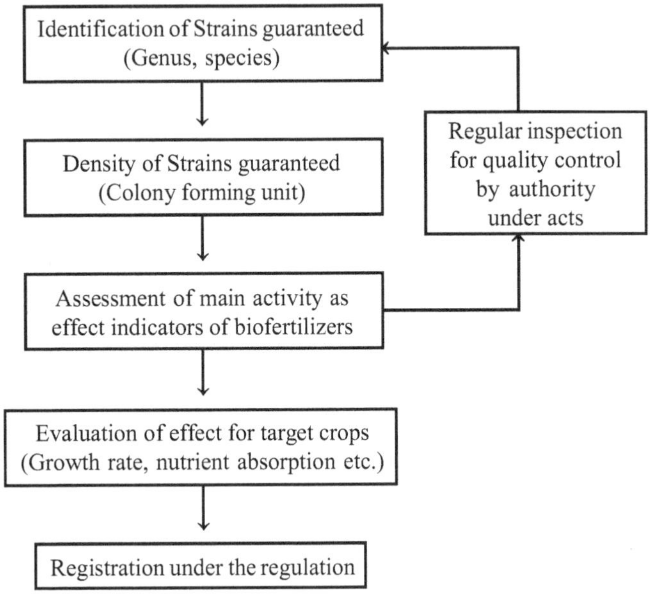

Fig. 15.1 : Procedure of Biofertilizer Quality Control

PUSH and PULL Marketing Strategies

Promotional strategies to get product or service to market can be roughly divided into two separate strategies such as :

PUSH Strategy: A push strategy involves taking the product directly to the customer via whatever means, ensuring the customer is aware of brand/product at the point of purchase—"Taking the product to the customer." A push promotional strategy involves taking the product directly to the customer via whatever means, ensuring the customer is aware of your brand at the point of purchase.

Example of Push Strategies :

a) Trade show promotions to encourage retailer demand.

b) Direct selling to customers in showrooms or face to face.

c) Negotiation with retailers to stock your product.

d) Point of sale displays.

e) Efficient supply chain allowing retailers an efficient supply

f) Packaging design to encourage purchase.

The term "push strategy" describes the work a manufacturer of a product needs to perform to get the product to the customer. This may involves setting up distribution channels and persuading middle men and retailers to stock your product. The push technique can work particularly well for lower value items such as fast moving consumer goods (FMCGs), when customers are standing at the shelf ready to drop an item into their baskets and are ready to make their decision on the spot. This term now broadly encompasses most direct promotional techniques such as encouraging retailers to stock your product, designing point of sale materials or even selling face to face. New businesses of ten adopt a push strategy for their products in order to generate exposure and a retail channel. Once your brand has been established, this can be integrated with a pull strategy.

PULL Strategy : A pull strategy involves motivating customers to seek out your brand in an active process : "Getting the customer to come to you.".

Examples of Pull Strategies :

a) Advertising and mass media promotion.

b) Sales promotions and discounts.

c) Word of mouth referrals.

d) Customer relationship management.

'Pull strategy' refers to the customer actively seeking out your product and retailers placing orders for stock due to direct consumer demand. A pull strategy requires a highly visible brand which can be devloped through mass media advertising or similar tactics. If customers want a product, the retailers will stock it - supply and demand in its purest foorm, and this is the basis of a pull strategy. Create the demand, and the supply channels will almost look after themselves.

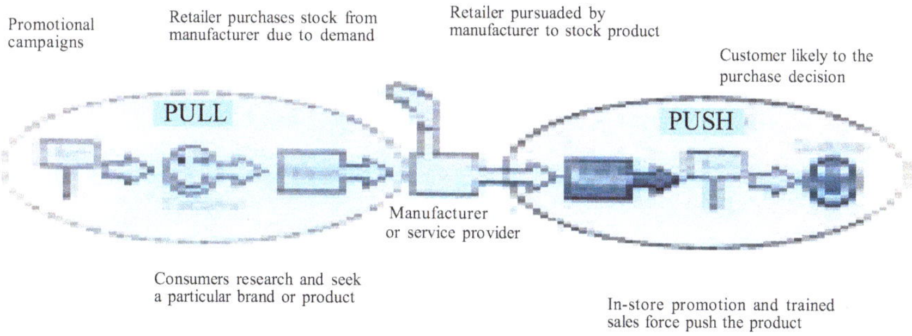

Fig. 15.2 : Differences between Push and Pull Promotional Strategy

A successful strategy will usually have elements of both the push and pull promotional methods. If we are starting a new commercial and intend to sell a product retailers, we will almost certainly need to persuade outlets to purchase and stock your product. We will also need to raise brand awareness and start building valuable word of mouth referrals. If we have designed a product around the customer and have considered all elements of the marketing mix, both of these aspects should be feasible.

Chapter - 16

Some Important Media

In nature, microbes are found as a mixed population. The types that encounter the most favourable environmental and other growth conditions emerge in abundance while others that find the environment less favourable are fewer in number. However, for any kind of study on them or for their use as a pure species, they have to be separated from other species. The nutritional media or culture media are different for different species. As the medium is specific for certain species, it checks or restricts the growth of other microbes for which it may not be suitable. However, some growth of and contamination by unwanted species may occur.

16.1 Definition

Culture medium is an aqueous solution, have all the necessary nutrients required for complete growth.

16.2 Types of Culture Medium

16.2.1 On the Basis of Physical Properties

There are three types of culture medium such as-

(a) **Solid or Synthetic Medium**

In this type of medium, 5-7% agar-agar or 10-20% gelatin is added the liquid broth for solidification. Such media are used for making agar slants or slope or agar stab.

(b) **Liquid or Broth Medium**

In this type of medium, agar-agar is not added at the time of preparation. After inoculation and later incubation the growth of cells become visible in the form of small mass on the top of broth.

(c) **Semi Solid or Floppy Medium**

These types of medium are prepared by adding half quarter of agar.

This type of medium may promote the growth of one organism and retards the growth of other organisms on the other hand; there are differential media under which some to different organism grow together.

16.2.2 On the Basis of Composition

There are three types of culture medium such as-

(a) **Natural or Empirical Medium**

The exact chemical composition of a natural medium is not known. The natural culture media include milk, urine, diluted blood, vegetable juices, meat extracts and infusions, peptone (digested protein rich in amino acids that provide nitrogen and carbon). Now days, group of living cells as tissue or callus or an organ are being used for viruses or rocketries.

(b) **Semi Synthetic or Undefined Medium**

Those media whose chemical composition is partially known are called semi-synthetic media. Any medium which contains agar becomes a semi-synthetic medium. Potato dextrose agar, Czapek-Dox agar, oat meal agar, beef peptone and nutrient agar are the examples of some of the semi synthetic media.

(c) **Synthetic or Chemically defined Medium**

Synthetic or a chemically defined culture medium is composed of special substances of known composition.

- **General Purpose Medium:** The synthetic medium may be a general purpose medium used for a wide variety of micro-organisms.
- **Selective Media:** The synthetic medium may be used for a selected microbe or Isolation of a particular species of bacteria from a mixed inoculum which is possible with the help of selective medium.
- **Assay Media:** Media of prescribed composition used for assay of vitamins, amino acids and antibiotics.
- **Differential Media:** These media distinguish colonies of specific microbes from others, when a culture medium containing certain substance helps to distinguish the different properties of different bacteria, is called differential media.

- **Selective-Differential Media:** Some media have both properties e.g. Mac Conkey agar. It is selective as it contains crystal violet and bile salts which inhibit the growth of bacteria other than coliforms. It is differential as it contains neutral red and lactose. Lactose is degraded by coliforms to acid which is detected due to change in pH. At acidic pH colourless neutral red becomes red causing the colonies to develop red colour. *Shigella* and *Salmonella* colonies remain colourless hence can be easily distinguished.

16.3 Components of Culture Media and Their Role

16.3.1 Nutrients

There are several types of nutrients which play an important role such as-

- **Carbohydrates:** Glucose, fructose, sucrose, mannitol etc. are used as a carbon source and provide energy and also indicate fermentation reactions.
- **Yeast Extract:** It is available in the form of powder. It is a rich source of vitamins B and organic nitrogen as well as carbon compounds.
- **Peptone:** It is a product resulting from digestion of protein rich materials like meat, casein, gelatin, etc. It is a source of nitrogen as well as vitamins and growth factors.
- **Beef Extract:** It is an aqueous extract of lean beef tissue concentrated to a paste. It contains water-soluble substances of animal tissue which include carbohydrates, organic nitrogenous compounds, water-soluble vitamins and salts.
- **Casein Hydrolysate:** It contains amino acids obtained by hydrolysis of milk protein casein. It can be used instead of peptone. Its constitutions are clearly defined than other peptones.
- **Malt Extract:** It consists of soluble substances extracted from barley sprouts at 55^0C. The obtained liquid is concentrated by heating at 55^0C to form a thick brown viscous material containing maltose, starch, dextrin, glucose, 5% proteins and protein breakdown products, mineral salts, growth factors and inositol.
- **Serum, Whole Blood and Aseptic Fluid:** Promote growth of less hardy organisms.

16.3.2 Agar

Agar is a natural polysaccharide produced by marine algae *Gelidium*. It is the most commonly used solidifying agent added to media. Silica gel can be used as a replacement solidifying agent. Solid media are useful for identifying bacteria by colony characteristics. It remains liquid at 100^0C (easy to pour) and solidifies a 40^0C (incubation temperatures).

16.3.3 Water

Tap water with low mineral content, distilled water of de-mineralized water.

16.3.4 Other media supplements

Substances such as dyes, vitamins, amino acids, growth factors and antibiotics are added to media for a definite purpose.

16.4 Preparation of Culture Media

There are specific media for specific types of microbes. No single medium can help grow all types of microbes. In preparing the culture medium, the contents specified for the purpose are weighed and mixed in a conical flask. Water is added and then heated to dissolve the contents, and the solution is stirred. The pH value is adjusted to the desired level. Precipitation is avoided during the preparation. In the case of a solid medium (one containing agar), the container is placed in such a way that the maximum surface area is available for the growth of microorganisms. For this purpose, slanting tubes, conical flasks and petri dishes are considered suitable. The culture medium and the glassware are sterilized in an autoclave in which the steam under pressure is maintained at 1–1.25kg/cm^2 and the heating is done for 15–30 minutes at a temperature of 121°C. For a medium containing heat-sensitive materials (e.g. sugars, amino acids and vitamins), sterilization is done by passing the medium through different types of filters (asbestos or sintered glass) that are capable of retaining bacteria (where present as a contaminant). Certain chemicals, such as salts of heavy metals ($AgNO_3$, $HgCl_2$, and $KMnO_4$), halogens (chlorine, bromine, iodine and their salts) and organic compounds are also used for sterilization of the medium.

16.5 Growth Media for Bacteria General

Thorton's Medium for Bacteria General (Thornton, 1922)

- Mannitol-1.0g
- Asparagine- 0.5g
- K_2HPO_4-1.0g

- KNO_3-0.5g
- $MgSO_4.7H_2O$-0.2g
- $CaCl_2$-0.1g
- NaCl-0.1g
- $FeCl_3$-0.002g
- Agar-15.0g
- Distilled water - 1000ml
- pH-7.4

16.6 Growth Media for *Rhizobium*

16.6.1 Congored Yeast Extract Mannitol Medium –CRYEMA (Vincent, 1970)

- Mannitol-10.0g
- K_2HPO_4-0.5g
- $MgSO_4.7H_2O$-0.2g
- NaCl-0.1g
- Yeast Extract-1.0g
- Agar-20.0g
- Distilled water-1000ml
- Congored (1% aqueous solution)-2.5ml

To prepare the medium, dissolve the salts in water, add agar, adjust the pH to 7.0, and autoclave at 121^0C for 30 minutes at 1-1.25 kg/cm^2 pressure, cool, and store for use.

16.6.2 Hofer's Alkaline Medium (Hofer, 1935)

The reaction of yeast extract Mannitol medium is raised to pH 11 with 1N NaOH of 1.6% Bromothymol blue per liter is added. In contrast to *Rhizobium* the allied genus *Agrobacterium* can grow in this medium, a characteristics feature which serves as a useful criteria to separate the two bacteria on agar plates during isolation of root nodules bacteria.

- Mannitol-10.0g
- K_2HPO_4-0.5g

- $MgSO_4.7H_2O$-0.2g
- NaCl-0.1g
- $CaCO_3$-0.05g
- Yeast Extract-1.0g
- Distilled water-1000ml

To prepare the medium, dissolve the salts in water, adjust the pH to 11-13, and autoclave at 121^{0}C for 30 minutes at 1-1.25 kg/cm^2 pressure, cool, and store for use.

16.6.3 Glucose Peptone Agar Medium

- Glucose-5.0g
- Peptone-10.0g
- Agar-15.0g
- Bromocresol Purple (1% Alcoholic solution)-10.0ml
- Distilled water-1000ml

To prepare the medium, dissolve the salts in water, add agar, adjust the pH to 7.0, and autoclave at 121^{0}C for 30 minutes at 1-1.25 kg/cm^2 pressure, cool, and store for use.

16.6.4 Norris and Date's Liquid Medium (Norris and Date, 1976)

- Mannitol-10.0g
- Yeast Extract-1.0g
- K_2HPO_4-0.5g
- $MgSO_4.7H_2O$-0.8g
- NaCl-0.2g
- $FeCl_3.6H_2O$-0.01g
- Distilled water-1000ml
- pH-7.0

To prepare the medium, dissolve the salts in water, add agar, adjust the pH to 7.0, and autoclave at 121^{0}C for 30 minutes at 1-1.25 kg/cm^2 pressure, cool, and store for use.

16.7. Media used for *Frankia*

16.7.1 Frankia Broth (Baker and Torrey, 1979)

- Yeast extract-0.5 % (W/v)
- Dextrose-1.0 % (W/v)
- Casamino Acids-0.5 % (W/v)
- Vitamin B_{12}-1.6mg^{-1}
- H_3BO_3-1.5mg^{-1}
- $ZnSO_4.7H_2O$-1.5mg^{-1}
- $MnSO_4.H_2O$-4.5mg^{-1}
- $NaMoO_4.2H_2O$-0.25mg^{-1}
- $CuSO_4.5H_2O$-0.04mg^{-1}
- Agar if required-0.8% (W/v)
- pH-6.4

16.7.2 QMOD Medium (Lalonde and Calvert, 1979)

- Yeast Extract-0.5g
- Glucose-10.0g
- Bactopeptone-5.0g
- K_2HPO_4-0.3g
- $MgSO_4.7H_2O$-0.2g
- KCl-0.2g
- Ferric Citrate-0.01g/mg
- K_2BO_3-1.5g
- $MnSO_4.7H_2O$-0.8g
- $ZnSO_4.7H_2O$-0.6g
- $(NH_4)_2$ $MoO_4.4H_2O$-0.2g
- $CuSO_4.7H_2O$-0.01g
- Agar if required-15.1g
- L-lecithin-0.01g
- Adjust to pH and add 0.1/g $CaCO_3$-6.8-7.0

16.7.3 BAP Medium (Murry et al., 1984)

- Casamino Acids-1.0g
- Sodium Propionate-1.0g
- Biotin-0.0002g
- Fe-EDTA-0.01g
- K_2HPO_4-1.0g
- $NaH_2PO_4.2H_2O$-0.67g
- $CaCl.2H_2O$-0.1g
- $MgSO_4.7H_2O$-0.2g
- Trace Elements as in QMOD
- pH-6.8

16.7.4 Bennett's Medium (Bennett, 1950)

- Yeast Extract-1.0g
- Beef Extract-1.0g
- Glucose-10.0g
- Casamino acids-2.0g
- Agar-15.0g
- pH-7.3

16.8 Selective Media for Blue Green Algae

16.8.1 Prigsheim's Medium (Prigsheim, 1951)

- KNO_3-0.02%
- $MgSO_4.7H_2O$-0.001%
- $(NH_4)HPO_4$-0.002%
- $CaCl_2.6H_2O$-0.0005%
- $FeCl_3$-0.00005%

16.8.2 Chu's Medium No. 10 (Chu, 1942)

- $Ca(NO_3)_2$-0.004%
- $MgSO_4.7H_2O$-0.0025%

- K_2HPO_4-0.0005 to 0.001%
- Na_2CO_3-0.002%
- Na_2SiO_3-0.0025%
- $FeCl_3$-0.0008%

16.9 Selective Media for *Azotobacter*

16.9.1 Ashby's Mannitol Agar Medium (Ashby, 1907)

- Mannitol-20.0g
- K_2HPO_4-0.2g
- $MgSO_4.7H_2O$ 0.2g
- NaCl-0.2g
- K_2SO_4-0.1g
- $CaCO_3$-5.0g
- Agar-15.0g
- Distilled water-1000ml

16.9.2 Jensen's N free Medium (Jensen, 1942)

- Sucrose-20.0g
- K_2HPO_4-1.0g
- $MgSO_4.7H_2O$-0.5g
- NaCl-0.5g
- $FeSO_4$-0.1g
- Na_2MoO_4-0.005g
- $CaCO_3$ -2.0g
- Agar-15.0g
- Distilled water-1000ml

16.9.3 Beijerincika's Medium (Becking, 1959)

- Sucrose-20.0g
- KH_2PO_4-0.8g
- K_2HPO_4-0.2g

- $MgSO_4.7H_2O$ -0.5g
- FeCl3-0.1g
- Na_2MoO_4-0.005g
- Agar-15.0g
- Distilled water-1000ml
- 6.5

16.9.4 Derxia (Campelo and Dobereiner, 1970 and modified by Lipman, 1903)

- Starch-20.0g
- KH_2PO_4-0.15g
- K_2HPO_4-0.05g
- $MgSO_4.7H_2O$- 0.20g
- $CaCl_2$-0.02g
- $FeCl_3$-0.1g
- Na_2MoO_4-0.002g
- Bromothymol Blue-0.5ml (0.5% in Absolute alcohol)
- $NaHCO_3$-1.0g
- Unrefined Agar-20.0g
- Distilled water-1000ml

16.10 Selective Media for *Azospirillum*

16.10.1 Semi Solid Malate Medium (Dobereiner et al., 1976)

- KH_2PO_4-0.4g
- K_2HPO_4-0.1g
- $MgSO_4$-0.2g
- NaCl-0.1g
- $CaCl_2$-0.02g
- $FeCl_3$-0.01g
- Na_2MoO_4-0.002g
- Sodium Malate-5.0g

- Bromothymol Blue-0.5ml (0.5% in Ethanol)
- Agar-1.75g
- Distilled water-1000ml
- pH-6.8

16.10.2 Okon's Medium (Okon et al.,1977)

- KH_2PO_4-4.0g
- K_2HPO_4-6.0g
- $MgSO_4$-0.2g
- NaCl-0.1g
- $CaCl_2$-0.02g
- NH_4Cl-1.0g
- DL Malic acid-5.0g
- NaOH-3.0g
- Yeast Extract (Difco)-0.1g
- $FeCl_3$-10.0g
- Na_2MoO_4-2.0mg
- $MnSO_4$-2.1mg
- H_3BO_3-2.0mg
- $Cu(NO_3)_2$-0.04mg
- $ZnSO_4$-24mg
- Bromothymol Blue-2ml (0.5% in Ethanol)
- Agar-1.5 % to 1.8%
- Distilled water-1000ml
- pH-6.8

Mixed in 100ml water and then autoclaved separately followed by mixing with clod medium later.

16.10.3 Nitrogen Free Bromothymol Blue Medium (Dobereiner et. al., 1976)

- Mannitol-5.0g
- KOH-4.0g

- K_2HPO_4-0.5g
- $FeSO_4.7H_2O$-0.05g
- $MnSO_4.7H_2O$ -0.01g
- $MgSO_4.7H_2O$-0.01g
- NaCl-0.02g
- $CaCl_2$-0.01g
- Na_2MoO_4-0.002g
- Bromothymol Blue-2.0ml (0.5% Alcoholic solution)
- Distilled water-1000ml
- pH-6.6-7.0

16.11 Selective Media for Phosphate Solubilizing Organisms

16.11.1 Katznelson and Bose Medium (Katznelson and Bose, 1959)

- Soil extract-100ml
- Glucose-1.0g
- Agar-2.0g

Sterilize in 100ml lots, cool and add 5ml of 10% k_2HPO_4 and 10 ml of 10% $CaCl_2$ to each flask. Adjust pH to 7.0 with sterile N/10 NaOH. Pour the plates immediately and allow solidifying. The bacteria are inoculated on to the agar and inoculated at room temperature for 7 days. A clearing develops around the colonies of those capable of solubilizing phosphate.

16.11.2 Pikovskay's Medium (Pikovskaya, 1948)

- Glucose-10.0g
- Tricalcium Phosphate-5.0g
- $(NH_4)_2SO_4$-0.5g
- $MgSO_4.7H_2O$-0.1g
- KCl-0.2g
- $MnSO_4$-traces
- $FeSO_4$-traces
- Yeast Extract-0.5g

- Agar-1.75g
- Distilled water-1000ml

To prepare the medium, dissolve the salts in water, add agar, adjust the pH to 7.0, and autoclave at 121^0C for 30 minutes at 1-1.25 kg/cm^2 pressure, cool, and store for use.

16.12 Selective Media for Isolation of *Pseudomonas fluorescens* (A Bio-control Agent) King's B Medium (Kings *et al.,* 1954)

- Peptone-20g
- K_2HPO_4-1.5g
- $MgSO_4$-1.5g
- Glycerol-10ml
- Agar-20g
- Ampicillin-50mg
- Cyclohexamide-12.5mg
- Chloramphenicol-12.5mg

Note: Observe under UV lamp and pick up fluorescing colonies

16.13 Selective Medium for Isolation of *Trichoderma* (An Antagonistic Fungus)

- K_2HPO_4-0.9g
- $MgSO_4$-0.2g
- KCl-0.15g
- NH_4NO_3-1.0g
- Glucose-3.0g
- Dicriplicin-10g
- Bavistin plus Kavach-200mg
- Rose Bengal-150mg
- Agar-20g

16.14 Medium for Cellulolytic Micro-organism

16.14.1 Dubos's Medium (Dubos, 1928)

- $NaNO_3$-0.5g
- KH_2PO_4-1.0g
- $MgSO_4.7H_2O$ -0.5g
- KCl-0.5g
- $FeSO_4.7H_2O$-Traces
- Distilled water-100ml

16.14.2 Martin's Bengal Agar Medium (Martin, 1950)

- Glucose -10.0g
- Peptone-5.0g
- KH_2PO_4-1.0 g
- $MgSO_4.7H_2O$-0.5g
- Rose Bengal-33mg
- Streptomycin-3ml (10% solution)
- Agar-15.0g
- Distilled water-1000ml

16.14.3 Cellulose Peptone Agar Medium (Allen, 1953)

- Cellulose-2.5g
- Peptone-5.0g
- K_2HPO_4-0.2g
- $MgSO_4$-0.2g
- K_2CO_3-0.4g
- $CaCl_2$-0.02g
- $Fe_2(SO_4)_3$-0.02g
- NaCl-0.02g
- Distilled water-1000ml

16.15 Growth Media Formulation of Liquid *Rhizobium* Biofertilizers (NifTAL, Hawaii)

16.15.1 For Formulation G-5(g/liter)

- Mannitol-1.0g
- K_2HPO_4-0.5g
- $MgSO_4.7H_2O$-0.2g
- NaCl-0.1g/liter
- Yeast Extract-1.00g
- Glucose-1.0g
- Arabinose-0.5g
- PVP+40*-20g
- Glycerol-4.0ml
- FeEDTA-400μm

16.15.2 For Formulation G-6(g/liter)

- K_2HPO_4-0.8g
- $MgSO_4.7H_2O$-0.5g
- NaCl-0.6g/liter
- Yeast Extract-2.00g
- PVP+40*-20g
- Glycerol-12ml
- $CaCl_2$-0.14g
- $FeCl_3$-8.4mg
- NH_4Cl-0.6g
- NaOH-0.16g

*PVP+40 (Polyvinyl Pyrolodone with Mol.Wt. 40,000)

16.16 Preparation of Nutrient Medium

1) To prepare 1000ml of nutrient broth, warm 500ml water in a round bottom flask.
2) Dissolve beef extract, peptone and other ingredients in it by continuous shaking.
3) Make up the volume 1000ml by adding distilled water.

4) Check the pH of the medium, and adjust the pH with NaOH or acid as required.

5) Plug with cotton plug and autoclave at 121^0C temperature and 15lbs pressure for 15 minutes.

6) To make solid medium add desired quantity of agar to the broth and boil until the entire agar dissolved.

16.17 Precautions at the Time of Preparation of Media

1) The agar should be added by sprinkling hot water, otherwise lumps will be formed.

2) Make up the final volume of medium in last.

3) Never fill up the culture tube by medium. Always leave two-third of culture tube empty for effective aeration.

4) Never use the medium immediately after autoclaving.

5) Make sure that the cotton plugs are tight because loose plugs promote contamination.

6) Certain ingredients like sugars, vitamins, hormones, etc. should be added to the medium after autoclaving as heating destroys them.

7) Before autoclaving always check the pH of the medium. At times the medium does not solidity due to very high or very low pH.

8) Any change in colour, consistency or growth in medium after incubation means some kind of contamination.

16.18 Precautions in Handling

Inoculants are mass of living bacterial cells and must be handled with carefully. Following points help in achieving good results as-

1) Culture should be purchased from authentic place and avoid the use of low quality culture.

2) Culture should be kept in cool place

3) Direct sunlight and warm air should be avoided.

4) It should not be mixed with strong insecticides or chemical fertilizers.

5) Treated seed should be grown (after drying) as quickly as possible after treatment.

6) It should be used before expiry date.

References

Chapter 1 : Organic Farming

Acharya, C.N. (1948). *Indian Journal of Agricultural Sciences* 9:741-44.

Acharya, C.N. (1993). *Indian Farming* 9:125.

Anonymous (2014a). *The World of Organic Agriculture: Statistics and Emerging Trends 2014, published by the Research Institute of Organic Agriculture (FiBL) and the International Federation of Organic Agriculture Movements (IFOAM).* July 2014.

Anonymous (2014b). *Employment News Vol. XXXIX (23): 48./http://www.apeda.gov.in*

Bhaskaran, T. (1957). *Indian Journal of Agricultural Sciences* 27:91.

Biswas, T.D.; Jain, B.L. and Mandal, S.C. (1917). *Journal of Indian Society of Soil Science* 19:31-37.

Blair, M.R. (1952*). Public Health Johannesberg* 15:70.

Chang, Y. and Hudson, H.J. (1967).*Journal of Ecological Studies* 50: 649-666.

Chang, H.No., Chen--C.L.; and Kirk, T.K. (1980). In *"Lignin Biodegradation: Microbiology, chemistry and potential Applications."* (T.K. Kirk, Higuchi, and H.M. Change, eds), Vol. 1, pp. 215-230, CRC fress, Boca Ratan, Florida.

Cooper, R.C. and Goluek, C.G. (1977). *Compost Science* 18:8.

Crawford, R.L.; Robinson, L.E.; and Cheh, A.M. (1982). *In: Lignin Biodegradation (Ed. TK Krick, T.Higuchi, and H.M. Chang)* Vol. pp.61-76. CRC Press Ration Fl.

Davey, C.B. (1953). *Proc: Soil Science Society, Amestardum* 17:512-516.

De Bertoldi M and Zucconi F (1980). *Ingegneria Ambientale* 9: 209-216.

Eberhardt, D.T. and Pipes, W.O. (1972). *In: Large Scale Composting (Ed. MJ Satrinae)* Noyes Data Corp. Park Ridge, New Jersey.

Gaind, S. and Gaur, A.C.(2000). *GEOBIOS* 27:21-24.

Gaur, A.C. (1987). *Resources and Conservation* 13:154-74.

Gaur, A.C. (1999). *Microbial Technology for Composting of Agricultural Residues by Improved Methods,* ICAR, New Delhi pp.78.

Gaur, A.C. and Sadasivam, K.V. (1980). *Recycling of Organic Wastes in Agriculture, Research* Bull. Vol. 21, IARI, New Delhi pp.31-49.

Gaur, A.C.; Neelkanthan, S. and Dargan, K.S. (1995). *Organic Manures* ICAR New Delhi pp.159.

Gaur, A.C.; Sadasivam, K.V.; Magu, S.P. and Mathur, R.S. (1982). *Agriculture Wastes* 4:453-68.

Glogowski, M.E. (1992). *Canadian Patent Application CA2030,* 282 pp.34.

Goleuk, G.C. (1954). *Applied Microbiology* 2:45-53.

Gong, C.S. and Tsao, G.T. (1979). *Annual Rep. Ferment. Processes* 3:111-40.

Gotaas, H.B. (1956). *Composting Sanitary Disposal and Reclamation of Organic Waste, WHO, Geneva.*

Gupta, M.D.; Chattopadhya, G.N. and Baral, K. (1986). *Annals Agricultural Research* 7:346.
Haimi, J. and Huhta, V. (1987). *Pedobiologica* 30:137.
Harvey, P.J.; Schoemaker, H.E.; Bown, R.M. and Palmer, J.M. (1985). *FEBS Letters* 103:13-16.
Hooda, I.S. (2005). *In:National Symposium Management of Organic Wastes for Crop production,* CCSHAU, Hissar, pp.30-31.
Jasem, A. and Alireza, S. (2014). *Intl. J. Farm. & Alli. Sci.* 3(1): 1-6.
Keeney, D.R. (1982). *In: Nitrogen in Agricultural Soils (Ed. F.J. Stevenson)* pp.605-649.
Ketcheson, J.W. (1980). *Canadian Soil Sci.* 60:403-10.
Ketcheson, J.W. and Beauchamp, E.G (1978). *Agronomy Journal* 70:792-97.
Lacey, J. (1979). *The Microflora of straw and its assessment in straw decay and effect on disposal and utilization (Ed. E. Grossbard)* pp. 57-64. Willey and Sons, New York.
Lal, R. and Kang, B.T. (1982). *In: Transactions 12th International Congress of Soil Science* pp.152-78.
Lal, S. and Mathur, B.S. (1988). *Journal of Indian Society of Soil Science* 36: 113-19.
McGill, W.; Cannon, R.; Robertson, J.A. and Cook, F.D. (1986). *Canadian Journal of Soil Science* 66:1-9.
Morgan, M.T and McDonald, F.W. (1969). *J. Environmental Health* 32:101.
Nandkishore, T.G (1980). *Studies on Urban Solid Waste Composting. M.Sc. Thesis, Univer. Afric.Sciences, Banglore.* pp. 166.
Parr, J.F.; Epstain and Wilson, GB. (1978). *FAO Soils Bulletin 36 FAO,* Rome 301.
Parr, J.F.; Papendick, S.B.; Hornick, S.B. and Meyer, R.E. (1992). *American Journal of Alternative Agriculture* 7:5-11.
Reese, E.T. (1977). *Recent Adv. Phytochem.* 11:311-67.
Reese, E.T.; Singh, R.GH.; and Levinson, S.H. (1950). *Journal of Bacteriology* 59:485-97.
Schoemaker, H.E.; Harbvey, P.J.; Bown, R.M.; and Palmer, J.M. (1985). *FEBS Letter* 103:7-12.
Sharpley, A.N. and Menzel, R.G (1987). *Advanced Agronomy* 41:297-324.
Sidhu, G (2005). *In: National Symposium Management of Organic Wastes for Crop Production* CCSHAU, Hissar, pp.39-40.
Tien, M. and Kirk, T.K. (1985). *Science* 221:661-662.
Toogood, T.A. and Lynch, D.L. (1959). *Canadian Journal of Soil Science* 39:151-56.
USDA, (1978). *Improving Soil with Organic Wastes,* USGPO.
Van Vuren, J.P.J. (1950). *Compost Bull. No. 310. Dept. Agric. Union of South Africa* pp. 39.
Velayutham, M. and Bhardwaj, K.K.R. (1994). *Indian Farming* 44:39.
Wilde, S.A. (1958). *Forest Product Journal* 8:323.
Wiley, J.S. (1957). In : *Proc. 12th Industrial Waste Conference. Purdue University Engr. Bulletin* pp. 17:596.
Wiley, J.S. and Pierce, G.A. (1955). *In : Proc: American Society of Civil Engineer. Journal of Sanitary Engg.Division* 81:846.
Yamin, M.A. and Trager, W. (1979). *Journal of General Microbiology* 113:417-20.

Chapter 2 : Basic Information on Biofertilizers

Anonymous (2012). *The Complete technology Book on Biofertilizer and Organic Farming pp.3* NIIR Project Consultancy Services, Kamla Nagar, Delhi.

Bhattacharya, P. and Tandon, H.L.S. (2012). *Biofertilizer Hand book.Fertilizer Development and Consultaion Organization* 204-204A Banot Corner, Pamposh Enclave, New Delhi, India.

Biswas, B.C.; Yadav, D.S. and Maheshwari, S. (1985). *Fertilizer News* 30(10):20-28.

Bowen, G.D. (1979). *Academic Press, New York,* pp. 209-227.

Bowen, G.D. (1980). *Academic Press, New York,* pp. 145-160.

Brahmaprakash.G.P. and Sahu, P.K. (2012). *Journal of the Indian Institute of Science* 92 (1): 37-62.

Chanway, C.P., Turkingston, R. and Holl, F.B. (1991). *Advanced Ecological Research* 21:121-69.

Cook, R.J.; Thomanshow. In. L.S.; Weller, D.M.; Fujimoto, D.; Mazzola, M.; Bangara, G. and Kim, D.S., (1995). *In : Proc. National Academy of Science, USA,* 92: 4197-201.

Curl, E.A. and Truelove, B., (1986). *The Rhizosphere, Springer-Verlag New York..*

Dommergues, Y.R. (1978). *Interactions between Nonpathogenetic Soil Micro-organisms and Plants. Elsevier Scientific Publishing New York,* pp. 1-37.

Gaur, A.C. (1984). *In : Proc. of International Conference on Organic Matter and Rice.* IRRI, Manila pp. 503-14.

Gaur, A.C. (1986). *Seriya Biologicheskaya.* 2: 200-17.

Gaur, A.C. and Subba Rao, R.V. (1976). *Indian Journal of Agricultural Sciences* 45:866-69.

Gaur, A.C.; Sadasivam, K.V.; Vimal, O.P. and Mathutr, R.S. (1971). *Plant and Soil* 34: 17-28.

Gaur, A.C.; Subba Rao, R.V.; Sadasivam, K.V. (1972). *Labdev Journal of Sciences and Technology.* 10B: 55-56.

Glick,B.R. (1995). *Canadian Journal of Microbiology* 41:109–117.

Hussain, A. and Vancura, V. (1970). *Folia Microbiologica* 15:353-78.

Jackson, R.M.; Brown, M.E. and Burlingham, S.K. (1964). *Nature,* London 191: 1323.

Jarvis, S.C. (1996). *Plant and Soil* 181: 47-56.

Johri, B.N.; Rao, Ch.; V.S. and Goel, R. (1997). *Biotechnological Approaches in Soil Micro-organisms for Sustainable Crop Production. Dadarwal KR (Ed)* Scientific Publ. Jodhpur, pp. 192-221.

Katyal, J.C.;Venkatashwarlu,B.,and Das,S.K. (1994). *Fertilizer News* 39(4):27-32.

Katznelson, H. (1965). *Nature and Importance of Rhizosphere. Ecology of Soil Borne Pathogens.* University of California Press Berkeley.

Katznelson, H., and Sirois, J.C. (1961). *Nature,* London.191:1323.

Kraffczyk,I.; Tolldeneir, G., and Beringer, H.(1984). *Soil Biol. Biochem.* 16:315-22.

Lynch, J.M. (1990). *The Rhizosphere,* John Wiley and Sons New York.

Mahdi, S.S.; Hassan, G.I.; Samoon, S.A.; Rather, H.A.; Showkat, A.; Dar and Zehra, B. (2010). *Journal of Phytology* 2(10): 42-54.

Miller, R.H. and Schmidt, E.L., (1965). *Soil Science* 100:323.

Panwar, J.D.S. and Vijayluxmi, (2005). *In: Biological Nitrogen Fixation in Cereals and Pulses, Developments in Physiology, Biochemistry, and Molecular Biology of Plants,* Eds Bandana and A. Hèmantaranjan Vol. 1 pp. 125-128. New India Publishing Agency, New Delhi.

Panwar, J.D.S.; Jain, A.K,; and Kumar, S. (2008). *In:Advances in Plant Physiology* Vol. 11 pp. 507-564 Scientific Publishesrs Jodhpur, Rajasthan India.

Penrose, D.M. and Glick, B.R. (1997). *Indian Journal of Experimental Biology* 35: 1-17.

Rajasekaran, S.; Sundaramoorhty, P.; Sankar Ganesh, K. and Baskaran, L. (2009): *Geobios,* 36: 83-85.

Rajasekaran, S.K.; Sankar Ganesh, K. Jayakumar, M. Rajesh, C. Bhaaskaran and P. Sundaramoorthy, (2012). *International Journal of Environment and Bioenergy* 4(3): 176-195.

Rovira, A.D.; Foster, R.C. and Martin, J.K. (1979). *The Soil Root Interface pp.* 1-4, (Eds Harley JL and Rusell RS) Academic Press, London.

Rupela, O.P. (2005). *In: National symposium management of organic wastes for Crop Production.* CCS Haryana Agricultural University, Hissar pp. 43-45.

Sequeera, L. and Williams, P.H. (1964). *Phytopatho.* 54: 1240.

Thomashow, L.S. and Weller, D.M., Bonsall, R.F. and Pierson, L.S. (1990). *Applied Environmental Microbiology* 56: 908-12.

Vancura, V. (1961). *Nature* London, 192:88.

Vancura, V. and Macura, J. (1960). *Folia Microbiologica* 5: 293.

Venkataraman, G.S. and Shanmugasundaran, S. (1992). *Algal biofertilizers technology for rice.DBT CentreforBGA.Bio-fertilizer,* Madurai Kamraj University, Madurai, 625021,T.N. pp. 1-24.

Venkatashwarlu, B. (2008). *Role of bio-fertilizers in organic farming: Organic farminginrainfed agriculture:* Central institute for dryland agriculture, Hyderabad. pp.85-95.

Wani, S.P. and Lee, K.K. (1995). *Microorganisms as biological inputs for sustainable agriculture in Organic Agriculture* (Thampan, P.K.ed.) Peekay Tree Crops Development Foundation, Cochin, India pp-39-76.

Winter, A.G. (1961). *Symoosia Society of Experimental Biology,* 15: 229.

Chapter 3 : Classification of Biofertilizers

Alloway, B.J. (2008). *In: Zinc in soils and crop nutrition.* Second edition, IZA and IFA publishers, Brussels, Belgium and Paris, France. pp. 21-22.

Alvarez, M.I.; Sueldo, R.J.;Barassi,C.A. (1996). *Cereal Research Communications* 24:101–107.

Bakanchikova, T.I.; Lobanok, E.V.; Pavlova-Ivanova, L.K.; Redkina, T.V.; Nagapetyan, Z.A. and Majsuryan, A.N. (1993). *Mikrobiologiya* 62: 515–523.

Baset Mia, M.A. and Shamsuddin, Z. H. (2010). *African Journal of Biotechnology* 9(37): 6001-6009.

Bashan,Y and; de-Bashan, L.E. (2002b). *Applied and Environmental Microbiology* 68:2637–2643.

Benizri,E.;Baudoin,E. and Guckert,A. (2001). *Biocontrol Science and Technology* 11:557–574.

Bhattacharya, P. and Tandon, H.L.S. (2012). *In: Biofertilizer Handbook,* FDCO, New Delhi. pp.29.

Brahmaprakash, G.P. and Sahu, P.K. (2012). *Journal of the Indian Institute of Science* 92:1.

Brown,M.E. (1974). *Annual Review of Phytopathology.* 12:181–97.

Cabello,M.; Irrazabal, G.; BucsinszkyA.M.; Saparrat M.; and Schalamuck. S. (2005). *Journal of Basic Microbiology* 45:182–189.

Chernin,L. and Chet,I. (2002). *In: Microbial enzymes in biocontrol of plant pathogens and pests'* In : R.G.Burnsand R.P. Dick (Ed.), *Enzymes in the environment: activity, ecology, and applications. Marcel Dekker, New York.* pp.171–225,

Creus,C.M.; Sueldo,R.J. and Barassi,C.A. (1997). *Plant Physiology and Biochemistry* 35:939–944.

Creus,C.M.; Sueldo, R.J. and Barassi,C.A. (1998). *Canadian J. of Bot.* 76:238–244.

Defago,G. (1993). *Plant Pathology* 42:311–312.

Dobereiner, J. (1997). *Soil Biology and Biochemistry* 29:771–774.

Frankowski, J.; Lorito, M.; Scala, F.; Schmidt, R.; Berg,G.; Bahl, H. (2001). *Archives of Microbiology,* 176:421–426.

Gholami,A.;Shahsavani,S.;Nezarat.S. (2009). *World Academy of Science, Engineering and Technology* pp. 49.

Glick,B.R. (1995). *Canadian J. Microbiol.* 41:109–117.

Goenadi, D. H. ; Siswanto and Sugiarto, Y. (2000). *Soil Science Society of America Journal* 64:927–932.

Goldstein,A.H. (1986). *American Journal of Alternate Agriculture,* 1:57–65.

Herandez,A.T.; Bustilos-cristales, M.R.; Jimnez-salgado,T.; Caballero-Mellado, J. and Fuentez-Ramirez,L.E. (2000). *Microbial Ecology* 39:49–55.

Holguin,G. and Bashan,Y. (1996). *Soil Biology and Biochemistry* 28:1651–1660.

Jain, A.K. (2009). *In: Plant Diseases Management for Sustainable Agriculture, Editor:* Shahid Ahmad, Sher-e-Kashmir University of Agricultural Sciences and Technology, (Regional Agricultural Research Station, Rajouri, Jammu, Published by Daya Publishing House, New Delhi pp. 299-327.

Jain, A.K.; Kumar, S. and Panwar, J.D.S. (2007). *Advances of Plant Sciences* 20 (2): 337-339.

Jain, A.K.; Kumar, S. and Panwar, J.D.S. (2008). *Legume Research* 30 (3): 201-204.

Jimnez-Salgado, T.; Fuentes-Ramirez, L.E.; Tapia-Herandez, A.; Mascarua,M.A.; Martinez-Romero, E. and Caballero-Mellado, J. (1997). *Applied and Environmental Microbiology* 63: 3676–3683.

Kasiamdar, R.S. ; Smith, S. E. ; Smith, F.A. and Scott, E.S.: (2001). *Plant and Soil.* 238:235–244.

Katyal, J.C. ; Venkatashwarlu, B. and Das, S.K. (1994). *Fertiliser News* 39(4):27-32.

Kloepper, J.W. (1992). *In : Soil Microbial Ecology. Applications in Agricultural and Environmental Management* (Ed.F.B.MettingJr.). 'Marcel Dekker, New York, USA, pp.255–274.

Kloepper, J.W. (1994). *In:Azospirillum/PlantAssociation'*(ed.J.Okon). BocaRaton, FL: CRC Press. pp.137–154.

Kucey, R.M.N. ; H.H. Janzen and Legget, M.E. (1989). *Advances in Agronomy* 42:199–228.

Kulkarni, S. and Nautiyal, C.S. (2000). *Current Microbiolology* 40:221-226.

Kulkarni, S.; Surange, S. and Nautiyal, C.S. (2000). *Current. Microbiolology* 41: 402-409.

Kumar, G.N. (2003). *Improvement of PGPR by Transformation. Prospectus and Challenges.* 6th International PGPR workshop held at IISR Calicut.

Kumara Swamy, C.A.; Raghunandan, B.L.; Chandrashekhar, M.; Brahmaprakash, G.P. (2010). *Indian Journal of Science and Technology* 3(7):689–692.

Loganathan. P. and Nair. S. (2003). *Biotechnology Letter.* 25:497–50.

Loganathan. P. ; Sunitha. R. ; Parida. A. K. and Nair. S. (1999). *Applied Microbiology* 87:167–172.

Lovelock, C.E.; Wright, S.F.; Clark, D.A. and Ruess, R.W. (2004). *Journal of Eco*logy 92:278–28.

Lutcht, J.M, and Bremer, E. (1994). *Microbiol.Rev.* 4:3-20.

Mahdi,S.S.; Hassan,G.I.; Samoon S.A.; Rather, H.A.; Dar, Showkat A. and Zehra, B. (2010). *Journal of Phytology* 2(10): 42–54.

McGonigle, T.P. and Miller M.H.(1999). *Applied Soil Ecology* 12:41–50.

Meenakshisundaram, M. and Karrupagnaniar, S. (2010). *International Journal of Biology and Medical Research* 1(4):298–300.

Mishra,M.; Kumar., U..; Mishra, P.K. and Veeru Praka (2010). *Advances in Biological Research* 4(2):92–96.

Mosse, B., Stribley, D.P. and Le Tacon, F. (1981). *Advances in Microbial Ecology* 5:137-210.

Oliveira, R.G.B. and Drozdowicz, A. (1987). *Zentralblattfür Mikrobiologie.* 142:387–391.

Olsson, P.A.; Thingstrup, I.; Jakobsen, I. and Baath, E. (1999). *Soil Biology and Biochem*istry 31:1879–1887.

Panwar, J.D.S. and Sirohi, G.S. (1989). *Farmers and Parliament* 24:23-24.

Panwar, J.D.S. and Swarnalakshmi (2005). *In: Biological Nitrogen Fixation in Cereals and Pulses, Developments in Physiology, Biochemistry and Molecular Biology of Plants,* Eds Bandana and A. Hemantrajan, Vol. 1 pp. 125-158, New India Publishing Agency, New Delhi.

Panwar, J.D.S. and Swarnalakshmi (2005). *In: Developments in Physiology, Biochemistry and Molecular Biology of Plants,* Eds Bandana and A. Hemantrajan, Vol. 1 pp. 159-180, New India Publishing Agency, New Delhi.

Panwar, J.D.S. and Thakur A.K. (1995). *In : Proc: Mycorrhizae biofertilizers for the future.* Tata Energy Res., Institute pp. 13-15.

Ponmurugan, P. and Gopi, C. (2006). *Journal of Agronomy* 5:600–604.

Rajasekaran,S., and Sundaramoorthy, P. (2010). *J. Ecotoxicol. Environ. Monit.* 20: 161-168.

Ramarethinam, S. and Chandra, K. (2006). *Pestology* 30(11): 35-39.

Rangrajan, S.; Loganathan, P.; Saleena, L.M. and Nair, S. (2001). *J. Appl. Microbiol.* 91:742-749.

Rangrajan, Saleena, L.M.; Vasudevan, P. and Nair, S. (2003). *Plant and Soil* 251:73-82.

Rathi, B.K.; Jain, A.K.; Kumar, S. (2009). *Journal of Indian Botanical Society* 88(1&2): 1-3.

Rillig, M.C.; Wright, S.F.; Nichols, K.A.; Schmidt, W.F.; Torn,M.S.(2001). *Plant and Soil.* 233:167–177.

Roger, P.A.and Ladha, J.K (1992). *Plant and Soil* 141:41-5.

Saravanan, V.S.; Madhaiyan, M.; Osborne, Jabez.; Thangaraju, M. and Sa, T.M. (2008): *Microboiol Ecology.* 55:130–140.

Sarma, Y.R.; Rajan, P.P.; Paul, D.; Beena, N. and Anandaraj, M., (2000). *In: Seminar on Biological Control with Plant Growth Promoting Rhizobacteria for sustainable Agriculture,* University of Hyderabad, Hyderabad.

Sawers, J.H. Caroline, G. and Paszkowski, U. (2007*). A study about cereal mycorrhiza, an ancient symbiosis in modern Agriculture. Department of Plant Molecular Biology,* University of Lausanne, Biophore Building, Lausanne.

Sevilla, M. and Kennedy, C. (2000). *Colonization of Rice and other cereals by Acetobactor diazotrophicus, an endophyte of sugarcane. In : The Quest for Nitrogen Fixation in Rice* (Eds. J.K. Ladha and P.M. Reddy). International Rice Research Institute, Manila, Philippines, 15: 1-165.

Somers, E.; Ptacek, D.; Gysegom, P.; Srinivasan, M.; Vanderleyden, J. (2005). *Applied and Environmental Microbiology* 71: 1803–1810.

Sudhakar, P.; Gangwar,S.K.; Satpathy, B.; Sahu, P.K.; Ghosh, J.K.; Saratchandra, B. (2000). *Indian Journal of Sericulture.* 39:9–11.

Sugumaran, P. and Janathanam, (2007). *World J. Agricultural Sciences* 3(3): 350-355.

Sundaravarathan, S.; Kannaiyan,S. (2002). *Biotechnology of Biofertilizers* (Ed.S.Kannaiyan). pp.251–225.

Suslov,T.V.: (1982). *In:Phytopathogenic Prokaryotes* '(Eds.M.S.London Mount and G.H. Lacy). Academic Press. pp.187–223.

Tapia-Hernández, A.; Mascarúa-Esparzá, M.A. and Caballero-Mellado, J.(1990). *Microbios.* 64:73–83.

Tripathi, A.K..; Mishra, B.M. and Tripathi, P. (1998). *J. Bio. Sci.* 23: 463-47.

Verma,L..N. (1993). *In:Organics in Soil Health and Crop Production* (Ed.P.K.Thampan). Peekay Tree Crops Development Foundation, Cochin, India. pp.152–183.

Whipps, J.M. (2001). *Journal of Experimental Botany* 52(1): 487–511.

Yu, D. and Kennedy, I.R. (1995). *Soil Biol. Biochem.* 27: 459-462.

Zemen, A.M.M.; Tchan. Y.T.; Elmerieh, C. and Kennedy, I.R. (1992). *Research in Microbiology Institute, Pasteur Elsevier* 143: 847-855.

Zhu,Y.G. and Miller, R.M. (2003). *Trends Plant Sciences* 8:407–409.

Chapter 4 : Nitrogen Fixation

Bethenod, O.; Prioul, J.L.; and Deroche, M.E. (1984). *Physiol., Veg.* 22: 565-570.

Burris, R.H. (1976). *In Plant Biochemistry, 3rdEd. J. Bonner and J. Varner, Eds.,* Academic Press, New York, pp. 887-908.

Carlson, R.W.; Forsberg, L.S.; Price, N.P.J.; Bhat, U.R.; Kelly, TM and Raetz, CRH (1995).*In progress in Clinical and Biological Research, Vol 392; Bacterial Endotoxins: Lipopolysaccharides from Gene to Therapy: Proceedings of the 3rd Conference of the International Endotoxin Society,* held in Helsinki, Finland on August 15-18, 1994, J. Levin et al., Eds, John Wiley and Sons, New York, pp.25-31.

Denison, R.F.; and Harter, B.L. (1995). *Plant Physiol.* 107:1335-1364.

Dong, Z.; Canny, M.J.; McCully, M.E.; Roboredo, MR.; Cabadilla, C.F.; Ortega, E.; and Rodes, R. (1994). *Plant Physiol.* 105:1139-1147.

Fitzler, M.E.; Kalsi, G.; Ewing, N.N.; Roberts, N.J.; Day, R.B. and Murphy, J.B. (1999). In : Proc. Natl. Acad. Sci. USA 96:5856-5861.

Heidstra, R. and Bisseling, T. (1996). *New Phytol.* 133:25-43.

Kuzma, M.M.; Hunt, S. and Layzell, D.B. (1993). *Plant Physiol.* 101:161-169.

Lazarowitz, S.G.; and Bisseling, T. (1997). *Plant Cell* 9:1884-1990.

Ludwig, RA.; and de Vries, GE. (1986). *In: Nitrogen Fixation, Vol. 4: Molecular Biology,* W.I. Broughton and S. Puhler. Eds., Clarendon, Oxford pp. 50-69.

Marschner, P.; Pawlowski, K. and Bisseling, T. (1995). *Plant Cell* 7:869-885.

Phillips, DA.; and Kapulnik, Y. (1995). *Trends Microbiol.* 3:58-64.

Reis, V.M.; Baldani, J.I.; Baldani, V.L.D. and Dobereiner, J. (2000). *Crit. Rev. Plant Sci.* 19:227-247.

Rolfe, B.G.; and Gresshoff, P.M. (1988). *Annu. Rev. Plant Physiol. Plant Mol. Biol.* 39:297-320.

Singh, G.R.; Sharma, S.R. and Kumar, A. (1986). *Arab. J. Plant Prot.* 10:53-58.

Stokkermans, T.J.W.; Ikeshita, S.; Cohn, J.; Carlson, R.W.; Stacey, G.; Ogawa, T. and Peters, N.K. (1995). *Plant Physiol.* 108:1587-1595.

Timmers, A.C.J.; Auriac, M. and Truchet, G. (1999). *Development* 126:3617-3628.

Van Rhijn, P.; Goldberg, R.B. and Hirsh, A. M. (1998). *Plant Cell* 10:1233-1249.

Vande Broek, A.; and Vanderleyden, J. (1995).*Crit. Rev. Plant Sci.* 14:445-466.

Chapter 5 : Rhizobium

Abaidoo, R.C.; Keyser, H.H.; Singleton, P.W.; Dashiell, K.E. and Sanginga, N. (2007). *Appl. Soil Ecol.* 35:57-67.

Belane, A.K. and Dakora, F.D. (2010). *Biol. Fertil. Soils* 46:191-198.

Bhattacharya, P. and Tandon, H.L.S. (2012). *Biofertilizer Handbook,* FDCO, New Delhi. pp.30.

Biswas, J.C., Ladha J.K. and Dazzo, F.B. (2000). *Soil Sci. Soc. Am.* J. 64:1644.

Bonilla, I. and Bolaños, L. (2010). *In: Organic Farming, Pest Control and Remediation of Soil Pollutants, Sustainable Agriculture Reviews,* pp. 253-274, E. Lichtfouse (ed.), Springer Netherlands.

Brahmaprakash, G.P. and Hegde, S.V. (2005). *In: Advances in Pigeonpea Research* (Eds. M. Ali and Shivkumar). IIPR, Kanpur. pp. 214–228.

Brahmaprakash, G.P. and Sahu, P.K. (2012). *J. Indian Institute of Science* 92:1.

Brockwell J.L.; Bottomley P.J. and Theis T.E. (1995). *Plant and Soil* 174:143-180.

Bucher, M., (2007). *New Phytol.* 173:11-26.

Bush, D.S. (1995). *Appl. Environ. Microbiol.* 66:5437-5447.

Corby, H.D.L., (1988). *Kirkia* 13:53-123.

Dadarwal, K.R. and Sen, A.N. (1974). *In: Proc. Indian National Sciences Academy* 40B:548-555.

De Faria, S.M.; Lewis, G.P.; Sprent J.I. and Sutherland, J.M. (1989). *New Phytol.* 111:607-619.

Figueiredo, M.; do V.B., A.C. do E.S. Mergulhão, J.K. ;Sobral,. Junior M., de A.L and de Araújo, A.S.F. (2013). *In: Plant Microbe Symbiosis: Fundamentals and Advances,* pp. 267-289, N.K. Arora (ed.), Springer India.

Fred, EB; Baldwin JL, and McCoy E (1932). *Root Nodule Bacteria and Leguminous Plants.* University of Wisconsin Press, Wisconsin, USA.

Gaur, Y.D. and Sen, A.N. (1974). *Zbl. Bakt. Abt. II* 129:369-72.

Gibson, A.H. (1963). *Austrian Journal of Biological Sciences* 16:28-42.

Giller, K.E. (2001). *Nitrogen Fixation in Tropical Cropping Systems 2nd Ed.* CABI.

Gwata, E.T.; Wofford, D.S.; Boote, K.J.; Blount A.R and Pfahler, P.L. (2005). *Crop Sci.* 45:635-638.

He, X.H., C. Critchley and Bledsoe, C. (2003). *Crit. Rev. Pl. Sci.* 22:531-567.

Herridge, D., and Rose, I. (2000). *Field Crops Res.* 65:229-248.

Herridge, D.F. (2008). *In: Nitrogen-fixing Leguminous Symbioses, Nitrogen Fixation: Origins, Applications, and Research Progress,* pp. 77-115, M.J. Dilworth, E.K. James, J.I. Sprent and W.E. Newton (eds.), Springer Netherlands.

Herridge, D.F.; Marcellos, H.; Felton, W.L.; Turner G.L. and Peoples, M.B. (1995). *Soil Biol. Biochem.* 27:545-551.

Hungria, M., and Vargas, M.A.T. (2000). *Field Crops Res.* 65:151-164.

Hungria, M., Chueire, L.M.O.; Megías, M.; Lamrabet, Y.; Probanza, A.; Guttierrez-Mañero F.J. and Campo, R.J. (2006). *Plant and* Soil 288:343-356.

Jauhri, K.S. and Koshi, P. (1984). *Zbl. Micriol.* 139:93-103.

Jensen, E.S., Peoples, M.B.; Boddey, R.M.; Gresshoff, P.M.; Hauggaard-Nielsen, H.; Alves, B.J.R and Morrison, M.J. (2012). *Agron. Sustain. Developm.* 32:329-364.

Kiers, E.T., Hutton, M.G. and Denison, E.F (2007). *Proc. R. Soc.* London B 274:3119-3126.

Koele, N,; Thomas, W.; Kuyper and Bindraban, P.S. (2014). *In: Beneficial Organisms for Nutrient Uptake. Virtual Fertilizer Research Centre (VFRC)* Report 2014/1, Washington, D.C., USA.

Kuyper, T.W. and Giller, K.E. (2011). *In: Agro biodiversity management for food security,* pp. 134-149, J.M. Lenné and D. Wood (Eds.), CABI, Wallingford.

Larimer, A.L.; Bever, J.D. and Clay, K. (2010). *Symbiosis* 51:139-148.

Lowther, W.L., and Loneragan, J.F. (1968). *Plant Physiol.* 43:1362-1366.

Matiru, V.N., and Dakora, F.D. (2004). *Afr. J. Biotechnol.* 3:1-7.

Mpepereki, S.; Javaheri, F.; Davis, P. and Giller, K.E. (2000). *Field Crops Res.* 65:137-149.

Munns, D.N. (1970). *Plant and Soil* 32:90-102.

Musiyiwa, K.; Mpepereki, S. and Giller, K.E. (2005). *Soil Biol. Biochem.* 37:1169-1176.

Norris, D.C. and Date, R.A. (1976). *Legumes Biotechnology.Commonwealth Bureau of Pasture and Field Crops.* Bull No. 57. CABI. England.

Ojiem, J.O.; De Ridder, N.; Vanlauwe, B. and Giller, K.E.(2006). *Int. J. Agricult. Sustain.* 4:79-93.

Peoples, M.B., and Herridge, D.F. (2002). *In: Nitrogen fixation: From molecules to crop productivity, current plant science and biotechnology in agriculture,* pp. 519-524, F.O. Pedrosa, M. Hungria, G Yates and W.E. Newton (Eds.), Springer Netherlands.

Quispel, A. (1974). *In: Biology of Nitrogen Fixation.* North Holland Press, Amsterdam. pp. 748.

Raychaudhuri, S.; Yadav, A.K.; Raychaudhuri, M. (2007). *Biofertilizer News Letter* 15(1), pp. 3–10.

Redondo-Nieto, M.; Wilmot, A.R.; El-Hamdaoui, A.; Bonilla I. and Bolaños, L. (2003). *Plant Cell Environ.* 26:1905-1915.

Redondo-Nieto, M.; Wilmot, A.R.; El-Hamdaoui, A.; Bonilla I. and Bolaños, L. (2001). *Funct. Plant Biol.* 28:819-823.

Rivas, R., Garcia-Fraile, P. and Velazquez, E. (2009). *Microbiol. Insights.* 2:51-69.

Rodriguez-Echevarria, S. (2010). *J. Biogeogr.* 37:1611-1622.

Rodriguez-Echevarria, S.; Fajardo, S.; Ruiz-Diez B. and Fernandez-Pascual, M. (2012). *Oecologia* 170:253-261.

Roughley, R.J. (1970). *Plant and Soil* 32:675-701.

Saint-Macary, H.; Beunard, P.: Scaglia, J.A.; Hakizimana A. and Pandou, J. (1992). *In: Biological nitrogen fixation and sustainability of tropical agriculture,* pp. 343-350, K. Mulongoy, M. Gueye and D.S.C. Spencer (Eds.), Wiley, London.

Silvester, W.B. (1975). *In: 1st International Symposium on Nitrogen Fixation. (Eds. W.E. Newton and C.J. Nyman).* Washington State Univ. Press, Washington, pp. 489–586.

Sprent, J.I. (2001). *Nodulation in Legumes. Ed. Royal Botanic Gardens,* Kew.

Sprent, J.I. (2007). *New Phytol.* 174:11-25.

Stein, M.; Brumfield E.P.S. and Dye, M. (1982). *Annals of Applied Biology.* 101:261-67.

Thompson, J.A. (1991). *In: Report of the Expert Consultation on Legume Inoculant Production and Quality Control,* (Ed. J.A. Thompson). 'Food and Agriculture Organization of the United Nations, Rome, pp. 107–111.

Thuita, M.; Pypers, P.; Herrmann, L.; Okalebo, R.J.; Othieno, C.; Muema, C. and Lesueur, D. (2012). *Biol. Fertil. Soils* 48:87-96.

Vargas, L.K.; Lisboa, B.B.; Giongo, A.; Beneduzi, A.; and Passaglia, L.M.P. (2010). *In: Microbes for legume improvement,* pp. 137-155, D.M.S. Khan, P.D.J. Musarrat and D.A. Zaidi (Eds.), Springer Vienna.

Vessey, J.K. (2003). *Plant and Soil* 255:571-586.

Vincent, J.M. (1956). *Journal of the Australian Institute of Agriculture Sciences.* 20:247-49.

Vitousek, P.M. (2002). *Biogeochemistry* 57:1-45.

Vitousek, P.M.; Menge, D.N.L.; Reed S.C. and Cleveland, C.C. (2013). Phil. Trans. R. Soc. B 368:2013.0119.

Wani, S.P.; Rupela O.P. and Lee, K.K. (1995). In: *Management of biological nitrogen fixation for the development of more productive and sustainable agricultural systems, developments in plant and soil sciences,* pp. 29-49, J.K. Ladha and M.B. Peoples (Eds.), Springer Netherlands.

Weir, B.S. (2012). The *current taxonomy of rhizobia. NZ rhizobia website, <http://www.rhizobia.co.nz/taxonomy/rhizobia>* [accessed October 31, 2013].

Weisskopf, L.; Akelio, P.; Milleret, R.; Khan, Z.R.; Schulthess, F.; Gobat, J.M. and Le Bayon, R.C. (2009). *Plant and Soil* 319:101-114.

Whitelaw, M.A. (2000). *Adv. Agron.* 69:99-151.

Wopereis, E.; Triplett, M.; Umali-Garcia, J.A;. Anarna, B.G;. Rolfe, J.K.; Ladha, J.; Hill, R. Mujoo, P.K. N.G and Dazzo, F.B. (2001). *Funct. Plant Biol.* 28:845-870.

Yanni, Y.G., Dazzo F.B. and Zidan, M.I. (2011). *In: Bacteria in Agrobiology: Crop Ecosystems,* pp. 265-294, D.K. Maheshwari (ed.), Springer Berlin Heidelberg.

Zengeni, R. and Giller, K.E. (2007). *Symbiosis* 43:129-135.

Chapter 6 : *Azolla*

Ali, M.and Leeson, S. (1994). *World's Poultry Science Journal* 50:237-51.

Ali, M. and Leeson, S. (1995). *Animal Feed Science Technology* 55:227-37.

Anonymous (2012). *Azolla Eating up as Multi animal Feed in the district.* Hindu News Paper, 16th August 2012.

Bhuvaneshwari, K. and Kumar, A. (2013). *Science Research Reporter* 3(1):78-82.

Grilli M.; Caiola, C.; Fornic and Castagnola, M. (1988). *Bacteria in the Azolla- Anabaena* association *Symbiosis* 2:185-198.

Hates, B.; Frank, O.; Angells, B.D. and Feingold, S. (1980). *Nutr. Rep. Int.*, 21: 531-536.

Jayaraman R.; Parathasarthy, R. and Chandrabose, B. (1995). *Indian Veterinary Journal* 72: 478-80.

Kannaiyan, S. (1989). *Azolla biofertilizer for rice.* Proceeding National Seminar on Biotechnology, Molecular biology, Kottayam, Kerala, pp.15.

Kannaiyan, S. (1990). *Blue Green Algae Biofertilizers. The Biotechnology of Biofertilizers for Rice Crops.* (Ed) S. Kannaiyan, Tamil Nadu Agric. University Publications, Coimbatore, T.N.India. pp.225.

Kannaiyan, S. (1993). *In: Proc. Indian National Science Academy* 59B:21-31.

Kannaiyan, S.; Thangraraju, M. and Oblisami, G. (1981). *Tamilnadu Agriculture University Newsletter,* 8:2.

Kannaiyan, S.; Thangraraju, M. and Oblisami, G. (1982). *In: Proc. National Seminar on Biological Nitrogen Fixation,* IARI, New Delhi, pp. 451-60.

Kikuche, M.; Watanabe, I. and Haws, LD. (1984). *Organic Matter and Rice.* IRRI. Manila, pp. 569-92.

Kushari, D.P. and Teheruzzaman,Q. (1991). *Biological Nitrogen Fixation Associated with Rice Production (Eds., Dutta SK and Sloger C)* Oxford and IBH Publishing Co. New Delhi. pp. 109-18.

Kushari, D.P. and Watanabe, I. (1991). *Soil Science and Plant Nutrition* 37: 271-82.

Kushari, D.P. and Watanabe, I. (1992). *Soil Science and Plant Nutrition* 37: 65-73.

Lumpkin, T.A. and Pluckmett, D.L. (1980). *Azolla- Botany, Physiology and use as a green manure, Econ. Bot.* 34: 111-153.

Manna, A.B. and Singh, P.K. (1988). *Plant and Soil* 107: 165-71.

Peters, G.A. and Calvert, H.E. (1983). *The Azolla-Anabaena, Symbiosis,* Cambridge Univ. Press.Cambridge, pp. 109-145.

Peters, G.A.; Evans, W.R.; Crist, D.R.; Mayne, B.C. and Poole, R.E. (1980). *Plant Cell Environ.* 3: 261-269.

Reynaud, P.A. and Franche, C. (1986). *Azolla pinnata var. African from Molecular Biology to use as green manure,* Dakar, Senegal pp.15.

Roger, P.A. and Renaud, P.A. (1979). *Ecology of Blue Green Algae in Paddy Fields in Nitrogen and Rice* IRRI, Los Benos. pp. 289-309.

Sah, R.N.; Goyal, S.S. and Ranis, S.W. (1989). *Crop Science.* 29:1033-37.

Singh, AL and Singh, P.K. (1989). *Nitrogen Fixation in Indian Rice fields (Azolla and BGA),* Agrobotanical Publisher, Bikaner, India pp. 236.

Singh, P.K. (1979). *Nitrogen and Rice,* IRRI, Manila pp. 407-18.

Singh, P.K. (1989). *Applied Agriculture Research* 4: 149-61.

Singh, P.K. and Singh, D.P. (1990). *Azolla and Rice Cultivation Biofertilizers* (Ed) Bhandari LL Saxena and Vyas KK pp 39-45 Scientific Publication, Jodhpur.

Singh, P.K., Singh, D.P. and Singh, R.P. (1992). *Biochemistry Physiological Pflanzen.* 88:121-27.

Singh, P.K. and Singh, D.P. (1997). *In: Biotechnological Approaches in Soil Microorganisms for Sustainable Crop Production* (Ed) K.R. Dadarwal pp. 93-107. Scientific Publishers, Jodhpur.

Singh, P.K.; Singh, D.P.; Manna, A.B.; Singh, R.O. and Bisoyi, R.N. (1991). *In: Biological Nitrogen Fixation Associated with Rice Production* (Ed) Datta SK and Slogge C pp. 95-107. Oxford and IBH Calcutta pp. 95.

Sundaravarathan, S. and Kannaiyan, S. (2002). *Influence of Azolla and Sesbania rostrata application on changes in microbial population and enzymes in rice soils. Biotechnology of Biofertilizers* (Ed. S. Kannaiyan). pp. 251–225.

Svenson, H.K. (1944). *The new world species of Azolla Tam .Fern. J.* pp. 34-69.

Tamang, Y. and Samanta, G. (1993). *Indian Journal of Animal Science* 63:226-28.

Tung, H.F. and Watanab, I. (1983). *New Phytol.* 93:423-43.

Wagner, G.M. (1997). *The Botanical Review* 63(1): 26.

Watanab, I. and Berja, N.S. (1983). *Aquatic Botany* 15:175-82.

Watanabe I., Berja, N.S. and Rosari, Del D.C. (1980). *Soil Sciences Plant Nutrition* 26:301-07.

Chapter 7 : *Azotobacter*

Beijerinck, M.W (1901). *Zentralblatt fur Bacteriology* Abt II, 7:561.

Brown, M. and Burlingham, S.K. (1968). *Journal of Genetic Microbiology* 53:135-44.

Darzneik, Yo.O. (1961). *Microbiologiya* 30:1042-44.

Das, A.C. and Saha, D. (2007). *Journal of Crop Weed* 3: 69-74.

Dashadi, M.; Khosravi, H.; Moezzi, A.; Nadian, H.; Heidari, M. and Radjabi, R. (2011). *American-Eurasian J. Agric. & Environ. Sci.,* 11 (3): 314-319.

Doberainer, J. (1966). *Pesquisa Agropecuaria Brasileira* 1: 357-365.

Dutta, S. and Singh, M.S. (2002). *Indian Journal Hill Farming* 15: 44-46.

Essam and Lattief, A.E. (2013). *International Journal of Agronomy and Agricultural Research* 3: 67-73.

Gaur, A.C. and Misra, K.C. (1978). *Zbl. Bakt. Abt.* 11, 33:357-61.

Gaur, A.C. and Rai, S.N. (1982). *National Symposium on Facets of Indian Chemistrey in 2000AD.* University of Allahabad Symposium Handbook pp. 28-29.

Gaur, A.C.; Sadasivam, K.V.; Kavimandan, S.K.; Mathur, R.S. and Vimal, O.P. (1975). *Science and Culture* 41: 136-38.

Jackson, R.M.; Brown, M.E. and Burlingham, S.K. (1964). *Nature,* London 203:851-52.

Karunakaran, A.R., Dhanasekaran.S., Hemalatha, K.; Monika, R.; Shanmugapriya, P. and Sornalatha, T. (2014). *International Journal of Latest Research in Science and Technology* 3:79-81.

Kundu, K.B.S and Gaur, A.C. (1980). *Indian Journal of Microbiology* 20:225-29.

Lipman, J.G. (1903). *Report on the New Jersey Agricultural Experiment Station* 24: 217-285.

Lipman, J.G. (1904). *Report on the New Jersey Agricultural Experiment Station* 25: 237- 289.

Page, W.J. and Shivprasad S. (1991). *Int. J. Syst. Bacteriol., 41: 369-376. Ann. Microbiol.* 51: 145-158..

Paul, S. and Verma, O.P. (1999). *Indian J. Microbiol.* 39: 249-251.

Rodales, B. (1999). *Appl. Soil Ecol.* 12:51-59.

Shende, S.T. and Apte, R.G. (1982). *Azotobacter inoculation as remunerative input for agricultural crops. Proceedings of National Symposium on Biological Nitrogen Fixation.* New Delhi. pp. 532-43.

Shende, S.T.; Apte, R.G. and Singh, T. (1977). *Indian Journal of Genetic Plant Breeding* 35:314.

Sturz, A.V.; Christie, B.R. and Nowak J., (2000). *Critical Reviews in Plant Sciences* 19: 1-30.

Thompson, J.P.; and Skerman V.B.D. (1981). *Validation list No. 6. Int. J. Syst. Bacteriol.* 31.

Tilak, K. and Sharma, K.C. (2007). *Indian Farming Digest* 9: 25-28.

Wani, S.A.; Chand, S. and Ali, T. (2013). *Current Agriculture Research Journal* 1(1):35-38.

Chapter 8 : Azospirillum

Anand, R.C.; Ger, R.; Hazija, M. and Kundu, B.S. (1999). *Indian J. Microbiol.* 39(2): 121-123.

Andrew, J.W.; Jonathan, D.; Andrew, R.; Lei, S.; Katsaridou, N.N.; Mikhail, S. and Rodionov, A.D. (2007). *Biometals* 20:501-511.

Arun, K.S. (2007). *In: Biofertilizers for sustainable Agriculture, Mechanism of P solubilization. 6th Edition,* Agribios Publishers, Jodhpur, India. pp. 196-197.

Baldani, J.I.; Baldnai, V.L.D.; Selvin, L. and Dobereiner (1986). *International Journal of Systematic Bacteriology* 36(1):80-85.

Barber, L.E.; Tjepkema, Y.; Russel, S.A. and Evans, H.J.(1976). *Applied Environmental Microbiology* 32:108-113.

Baubu, S.; Thangaraju, M. and Santhanakrishnan, P. (2002). *J. Microbiol. World* 4:51-58.

Burns, R.C. and Hardy, R.W.F. (1975). *In:Nitrogen Fixation in Bacteria and Higher Plants.* Springer-Verlag, New York.

De-Polli, H.; Bohlool, B.B. and Dobereiner, J.(1980). *Archives Microbiology* 126: 217-22.

Dobereiner, J. and Day, J.M. (1974). *Symposium on N_2 fixation interdisciplinary discussion,* Pullman, Washington, USA pp. 518-29.

Dobereiner, J.; Marriel, I.E. and Nergy, M. (1976). *Canadian Journal of Microbiology* 22:1464-73.

Fillingere, S.; Chaeroche, M.; Van Dick, P. and De Enfert, C. (2001). *Microbiol.* 147:1851-1862.

Fulchieri, M. and Frioni, L. (1994). *Soil Biol. Biochem.* 26:921-924.

Gadagi, S.; Ravi, P.; Krishnaraj, U.; Kulkarni, J.H. and Tongmin, S.A. (2004). *Scientia Horticulturae* 100(1-4): 323-332.

Gaur, A.C. (2010) *In: Biofertilizers in Suistainble Agriculture,* ICAR, New Delhi pp.143.

Gonzalez, L.J.B,; Rodelas, C.; Pozo, V.; Salmeron, M.V.; Mart, Nez. and Salmron, V. (2005). *Amino Acids* 28: 363-367.

Hartmann, A.; Fu, H.A. and Buriss, R.H. (1988). *Appl. Environ. Microbiol.* 54(1): 87-93.

Heulin, T.; Rahman, M.; Omar, A.M.N,; Rafidison, Z.; Pierrat, J.C. and Balandreau, J.(1989). *J. Microbiol. Methods* 9:163-173.

Hurek, T.; Reinhold, B.; Neimann, G. and Fendrik. (1988). *In: Azospirillum IV. Genetics, Physiology and Ecology.* W. Klingmuller (Ed.), Springer-Verlag, Berlin.

Hylemon, P.B.; Wells, J.S.; Krig, N.R. and Janna Sch, H.W. (1973). *International J. Systematic Bacteriology* 23:340-80.

Lakshmikumari, M.; Kavimandan, S.K. and Subba Rao, N.S., (1976). *Indian J. Exptl. Biol.* 14:638-639.

Magalhaes, L.M.S.; Neyra, C.A. and Doberenier, J. (1978). *Archives Microbiol.* 19:247-52.

Mishra, D.J.; Rajvir, S.; Mishra, U.K. and Kumar, S.S. (2013). *Research Journal of Recent Sciences* 2: 39-41

Neyra, C.A.; Dobereiner, J.; Lalande, R. and Knowles, R. (1977). *Canadian J. Microbiol.* 23:300-05.

Neyra, C.A.; Dobereiner, J.; Lalande, R. and Knowles, R. (1977). *Canadian J. Microbiol.* 23:306-10.

Nur, I.; Okon, Y. and Henis, Y.(1980b). *Canadian J. Microbiol.* 26:714-18.

Ogawa, Y. (1963). *Plant and Cell Physiology* 4:227-37.

Okon, Y. and Gonzalez, L. (1994). *Soil Biol. Biochem.* 26:1591-1601.

Okon, Y.; Albrecht, S.L. and Burris, R.H.(1976a). *Journal of Bacteriology* 127: 1248-54.

Okon, Y.; Albrecht, S.L. and Burris, R.H. (1977). *Applied Environmental Microbiology* 33:85-88.

Omay, S.H.; Schmidt, W.A.; Martin, P. and Bangertti (1993). *Canadian J. Microbiol.* 24:734-742.

Rao, A.V. and Venkateswarlu, B. (1985). *Acta Microbiology (Hungary)* 32: 222-24.

Saikia, S.P.; Jain, V.; Khetrapal, S. and Aravind, S. (2007). *Current Science.* 93(9):1296-1299.

Saranraj, P.; Sivasakthivelan, P. and Siva Sakthi, S. (2013). *African Journal of basic and Applied Sciences* 5(2):85-101.

Saxena, V. (1983). *Plant Growth Substances and Nitrogen Fixation by Azotobacter and Azospirillum species, Ph.D Thesis,* IARI, New Delhi.

Schank, S.C.; Weiwer, K.L. and Mac Rac I.C. (1981). *Applied Environmental Microbiology* 41:342-45.

Sivasakthi, S.; Kanchana, D.; Usharani, G. and Saranraj, P. (2013). *International Journal of Microbiology Research* 4(3): 227-233.

Smith, R.L.; Schank, S.C.; Bouton, J.H. and Quesenberry, K.H. (1978). *Ecological Bull.* 26:380-85.

Tarrand, I.J.; Kreig, N.R. and Dobereiner J. (1978). *Canadian J. Microbiol.* 24:967-80.

Tien, T.M.; Gaskin, M.H. and Hubell, D.H. (1979). *Applied Environmental Microbiology* 37: 1012-24.

Umaligarcia, M.; Hubbel, D.N. and Gassons, M.H.(1980). *Applied Environmental Microbiology* 39:219-26.

Chapter 9 : Blue Green Algae

Desikachary (1959). *Chemosphere* 70: 1919-1929.

Dhar, D.W.; Prasanna R. and Singh, B. (2007). *Journal of Sustainable Agriculture* 30: 41-50.

Gaur, A.C. (2010). *In: Biofertilizers in Sustainable Agriculture,* ICAR, New Delhi pp.150.

Kaushik, B.D. (1998). *Soil Plant Microbe Interactions in Relation to Nutrient Management (Ed. B.D. Kaushik)* Venus Publishers, New Delhi, pp. 55-63.

Mishra D.J.; Singh Rajvir, Mishra U.K. and Kumar, S.S. (2013). *Research Journal of Recent Sciences* 2:39-41.

Mishra, S. and Kaushik, B.D. (1989). *Indian National Sci. Acad.* B55: 295-300.

Nain, L.; Rana, A.; Joshi, M.; Jadhav, S.D.; Kumar, D.; Shivay, Y.;, Paul. S. and Prasanna, R.(2010). *Plant and Soil* 331: 217-230.

Schaefer, (1986). *In: Progress Report Soil Technologies Inc.* Fairfield, Iowa, USA.

Tripathi, R.D.; Dwivedi, S.; Shukla, M.K.; Mishra, S.; Srivastava, S. and Singh, R. (2008). *Role of blue green algae biofertilizer in ametiorating the nitrogen demand and fly-ash stress to the growth and yield of rice plants. Chemosphere* 70(10) : 1919-29.

Venkatraman (1992). *Algal Biofertilizer Technology for Rice,* Sankar Printer, Madurai, India.

Chapter 10 : Phosphorus Solubilizing Micro-organisms (PSMs)

Abdol Amir, Yousefi, Kazem Khavazi, Abdol Amir Moezi, Rejali, R. and Habib Allah Nadian. (2011). *World Applied Sciences Journal* 15 9: 1310-1318.

Abril, A.; Zurdo-Pineiro, J.L.; Peix, A.; Rivas, R.; Velazquez, E. (2007). *Solubilization of phosphate by a strain of Rhizobium leguminosarum bv. Trifolii isolated from Phaseolus vulgaris in El Chaco Arido soil (Argentina). In: Velazquez E, Rodriguez-Berrueco C (Eds) Developments in Plant and Soil Sciences.* Springer, The Netherlands, pp. 135–138.

Agnihotri, V.P. (1970). *Can. J. Microbiol.* 16:877-80.

Ahuja, A.; Ghosh, S.B. and D'Souza S.F. (2007). *Bioresour. Technol.* 98:3408–3411.

Alia, Aftab Afzal Shahida N. Khokhar, Bushra Jabeen, and Saeed A. Asad (2013). *Pak. J. Bot.* 45(S1): 535-544.

Altomare, C.; Norvell, W.A.; Borjkman, T, and Harman, G.E. (1999). *Appl. Environ. Microbiol.* 65:2926–2933.

Ana, M.; Baya Boethling, R.S.; and Ramos Cormenzana, A.(1981). *Soil Biology and Biochemistry* 13:527-31.

Armarger, N. (2002). *Biochimie.* 84:1061–1072.

Arora, D. and Gaur, A.C. (1978). *Indian Journal of Microbiology* 10:315.

Arora, D. and Gaur, A.C. (1979). *Indian Journal of Experimental Biology* 17:1258-61.

Asea, P.E.A..; Kucey, R.M.N. and Stewart, J.W.B. (1988). *Soil Biol. Biochem.* 20:459–464.

Atlas, R, and Bartha, R. (1997). *Microbial Ecology.* Addison Wesley Longman, New York.

Azam, F. and Memon, G.H. (1996). *Soil Organisms. In: Bashir E, Bantel R. (Eds) Soil Science.* National Book Foundation, Islamabad, pp. 200–232.

Azcon, G.; De Aguliar, C. and Barera, J.M.(1978). *Canadian J. Microbiol.* 24:520-24.

Bajpai, P.D. and Sundara Rao, W.V.B. (1971). *Soil Sci. Plant Nutr.* 17:46–53.

Banik, S. and Dey, B.K. (1983). *Zentralblatt Microbiology* 138:17–23.

Barber, S.A. (1984). *Soil Nutrients Bioavailability,* Wiley and Sons, New York.

Barber, S.A. (1995). *Soil Nutrient Bioavailability. A Mechanistic Approach,* Wiley, New York.

Bardiya, M. C. and Gaur, A.C. (1972). *Indian Journal of Microbiology* 12:269-71.

Bardiya, M. C. and Gaur, A.C. (1974). *Folia Microbiology* 19:386-89.

Barea, JM,; Navare, E. and Montoya, E. (1976). *Journal of Applied Microbiology* 40:129-34.

Bar-Yosef, B.; Rogers, R.D.; Wolfram, J.H. and Richman, E. (1999.) *Soil Sci. Soc. Am. J.* 63:1703–1708.

Bashan, Y.; Kamnev, A.A. and de Bashan, L.E. (2013a). *Biol. Fertil. Soils* 49:1–2.

Bashan, Y.; Kamnev, A.A. and de Bashan, L.E. (2013b). *Biol Fertil. Soils* 49:465–479.

Bashan, Y.; Moreno, M. and Troyo, E. (2000). *Biol. Fertil. Soils* 32:265–272.

Blake, L.; Mercik, S.; Koerschens, M.; Moskal, S.; Poulton, P.R.; Goulding, K.W.T., Weigel, A. and Powlson, D.S. (2000). *Nutr. Cycl. Agroecosyst.* 56:263–275.

Bolan, N.S.; Currie, L.D. and Baskaran, S. (1996). *Biol. Fertil. Soils* 21:284–292.

Brown, M.E. (1972). *Journal of Applied Microbiology* 35:443-451.

Bunt, J.S. and Rovira. (1955). *Journal of Soil Science* 6:119-28.

Butterly, C.R.; Bunemann, E.K.; McNeill, A.M.; Baldock, J.A. and Marschner, P. (2009). *Soil Biol. Biochem.* 41:1406–1416.

Carrillo, A.E.; Li, C.Y. and Bashan, Y. (2002). *Naturwissenschaften* 89:428–432.

Chabot, R.; Antoun, H. and Cescas, M.P. (1993). *Canadian J. Microbiol.,* 39:941-47.

Chaiharn, M. and Lumyong, S. (2009). *W. J. Microbiol. Biotechnol.* 25: 305-314.

Chandini, T.M. and Dennis, P. (2002). *For. Ecol. Manage.* 159:187–201.

Chandra. S.; Choure, K.; Chaubey, R.C. and Maheshwari, D.K. (2007). *Br. J. Microbiol.* 38:124-130.

Chen, C.R.; Condron, L.M.; Davis, M.R. and Sherlock, R.R. (2003). *Forest. Ecol. Manag.* 117:539–557.

Chen, Y.P.; Rekha, P.D.; Arun, A.B.; Shen, F.T.; Lai, W.A. and Young, C.C. (2006). *Appl. Soil Ecol.* 34:33–41.

Collavino, M.M.; Sansberro, P.A.; Mroginski, L.A.and Aguilar, O.M. (2010). *Biol. Fertil. Soils* 46:727–738.

Cordell, D.; Drangert, J.O. and White, S. (2009). *Glob Environ Chang* 19:292–305.

Criquet, S.; Ferre, E.; Farner, E.M. and Le Petit, J. (2004). *Soil Biol. Biochem.* 36:1111–1118.

Crowley, D.E. (2007). *In: Barton LL, Abadia J (Eds) Iron Nutrition in Plants and Rhizospheric Microorganisms.* Springer, Dordrecht, pp. 169–198.

De Freitas, J.R.; Banerjee, M.R. and Germida, J.J. (1997). *Biol. Fertil Soils* 24:358–364.

Dighton, J. and Boddy, L. (1989). *In: Boddy L, Marchant R, Read D (Eds) Nitrogen, phosphorus and sulfur utilization by fungi.* Cambridge University Press, Cambridge, pp. 269–298.

Di-Simine, C.D.; Sayer, J.A. and Gadd, G.M. (1998). *Biol. Fertil. Soils* 28:87–94.

Donahue, R.L.; Milller, R.W. and Shickluna, J.C.(1990). *Soils: An Introduction to Soils and Plant Growth.* Prentice Hall of India Private Limited, New Delhi, 110001. pp. 222-4.

Duponnois, R.; Kisa, M. and Plenchette, C. (2006). *J. Plant Nutr. Soil Sci.* 169:280–282.

Eivazi. F. and Tabatabai, M.A. (1977). *Soil Biol. Biochem.* 9:167–172.

Fabre, B.; Armau, E.; Etienne, G.; Legendre, F. and Tiraby, G. (1988). *J. Antibiot.* 41:212–219.

Fankem, H.; Nwaga, D.; Deube, A.; Dieng, L.; Merbach, W. and Etoa, F.X. (2006) . *Afr. J. Biotechnol.* 5:2450–2460.

Fenice, M.; Seblman, L.; Federici, F. and Vassilev, N. (2000). *Bioresour. Technol.* 73:157–162.

Fraga, R.; Rodriguez, H. and Gonzalez, T. (2001). *Acta Biotechnol.* 21:359–369.

Fretias, I.R and Germida, J.J. (1990). *Indian Journal of Microbiology* 36:265-72.

Gaind, S. and Gaur, A.C. (1991). *Plant and Soil* 133:141-49.

Galal, Y.G., J.A. El-Gandaour and El-Ake, F.A.. (2001). *In: (Ed.): W.J. Horst. Plant Nutrition-Food Security and Sustainability of Agroecosystems,* pp. 666-667.

Gaur A.C. (2010) *In: Biofertilizers in Sustainable Agriculture,* Directorate of Information and Publications of Agriculture, Indian Council of Agricultural Research, New Delhi.

Gaur, A.C. (1982). *A Practical Manual of Rural Composting FAO, Rome* pp.102.

Gaur, A.C. (1985). In: *Proc. of Natioanl Seminar on Development and Use of Biofertilizers,* Ministry of Agriculture, New Delhi

Gaur, A.C. (1986). *Zentralbl. Microbiology* 14:103-05.

Gaur, A.C. (1990). *Phosphate Solubilizing Microorganisms as Biofertilizers.* Omega Scientific Publishers, New Delhi pp. 176.

Gaur, A.C. and Gaind, S. (1983). *Science and Culture* 49:110-12.

Gaur, A.C. and Gaind, S. (1984). *Geobios* 11:227-29.

Gaur, A.C. and Ostwal, K.P. (1972). *Indian J. Exp. Biol.* 10:393–394.

Gaur, A.C. and Sachar, S. (1980). *Current Sci.* 49:553-54.

Gaur, A.C.; Madan, M.; and Ostwal, K.P.(1973). *Indian Journal of Experimental Biology* 11:427-29.

Ghosh, A. (1982a). *FAI.- Training Programme for Fertilizer Promotion Executives,* New Delhi.

Ghosh, A.B. and Hassan, R. (1979). *Bulletin of Indian Society Soil Science* 12:1-8.

Glick, B.R. (1995). *Can. J. Microbiol.* 41:109–117.

Goldstein, A.H. (1986). *Am. J. Alt. Agric.,* 1: 57-65.

Goldstein, A.H. (1994). *In: Torriani-Gorini A, Yagiland E, Silver S (eds) Phosphate in microorganisms: Cellular and molecular biology.* ASM Press, Washington (DC), pp. 197–203.

Goldstein, A.H. and Liu, S.T. (1987). *Biotechnology* 5:72–74.

Goswami, K.P. and Sen, A. (1962). *Indian Joural of Agricultural Science* 32: 96-101.

Grierson, P.F.; Comerford N.B. and Jokela E.J. (1998). *Soil Biol. Biochem.* 30:1323–1331.

Gunes, A.; Ataoglu, N.; Turan, M.; Esitken,A. and Ketterings, Q.M. (2009). *J. Plant Nutr. Soil Sci.* 172:385–392.

Gupta, R.P. and Mishra, B. (1978). *Indian Journal of Agricultural Science* 48:239.

Gupta, R.R.; Singal, R.; Shanker, A.; Kuhad, R.C. and Saxena, R.K. (1994). *J. Gen. Appl. Microbiol.* 40:255–260.

Gyaneshwar, P.; Naresh, K.G.; Parekh, L.J. and Poole, P.S. (2002). *Plant and Soil* 245:83–93.

H, L.Y.; Zhang. Y.F.; Ma, H.Y.; Su, L.N.; Chen, Z.J.; Wang, Q.Y.; Meng, Q. and Fang, S.X. (2010). *Appl. Soil Ecol.* 44:49–55.

Hamdali, H.; Bouizgarne, B.; Hafidi, M.; Lebrihi, A, Virolle, M.J. and Ouhdouch, Y. (2008a). *Appl. Soil Ecol.* 38:12–19.

Hamdali, H.; Hafidi, M.; Lebrihi, A, Virolle, M.J. and Ouhdouch, Y. (2008b). *Appl. Soil Ecol.* 40:510–517.

Havlin, J.; Beaton, J.; Tisdale, S.L. and Nelson,. W (1999). *Soil fertility and fertilizers. An introduction to nutrient management.* Prentice Hall, Upper Saddle River, NJ.

Hayat, R. S.; Ali, U.; Amara, R.; Khalid and Ahmed, I.(2010). *Ann. Microbiol.,* 1-20.

He, Z.L.; Wu J.; O'Donnell, A.G.; Syers, J.K. (1997). *Biol. Fertil. Soils* 24:421–428.

Illmer, P.A. and Schinner, F. (1992). *Soil Biol. Biochem.* 24:389–395.

Illmer, P.A.; Barbato, A. and Schinner, F. (1995). *Soil Biol. Biochem.* 27:260–270.

Illmer, P.A.; Schinner, F. (1995). *Soil Biol. Biochem.* 27:257–263.

Jacobs, H.; Boswell, G.P.; Ritz, K.; Davidson, F.A.; Gadd, G.M. (2002). *FEMS Microbiol. Ecol.* 40:65–71.

Jinhee Park, Nanthi Bolan, Mallavarapu Megharaj and Naidu, R. (2010). *Pedologist.* 67-75.

Jorquera, M.A.; Crowley, D.E.; Marschner, P.; Greiner, R.; Ferna'ndez, M.T.; Romero, D.; Menezes-Blackburn, D. and De La Luz Mora, M. (2011). *FEMS Microbiol. Ecol.* 75:163–172.

Jorquera, M.A.; Hernandez, M.T.; Rengel, Z.; Marschner, P. and Mora, M.D. (2008). *Biol. Fertil. Soils* 44:1025–1034.

Juma, N.G. and Tabatabai, M.A. (1998). *Plant and Soil* 107:31–38.

Kang, S.M.; Khan, A.L.; Hamayun, M.; Shinwari, Z.K.; Kim, Y.H.; Joo, G.J. and Lee. I.J. (2012). *Pak. J. Bot.* 44(1): 365-372.

Karpagam, T. and Nagalakshmi, P.K. (2014). *Int. J. Curr. Microbiol. App. Sci* 3(3): 601-614.

Katznelson, H.; Peterson, E.A. and Rouatt, J.W (1962). *Cannadian J. Bot.* 40:1181-86.

Khan, A.A.; Jilani, G.; Akhtar, M.S.; Naqvi. S.M.S. and Rasheed, M. (2009a). *J. Agric. Biol. Sci.* 1(1):48–58.

Khan, M.r. and Khan, S.M. (2002). *Bioresour. Technol.* 85(2):213–215.

Khan, M.S.; Zaidi, A. and Wani, P.A.(2007). *Agron Sustain. Dev.* 27:29-43.

Khan, M.S.; Zaidi, A. and Wani, P.A. (2009b). *In: Lictfouse et al (Eds) Sustainable Agriculture.* Springer, pp. 552, *DOI: 10.1007/978-90-481-2666-8_34*

Khan, M.S.; Zaidi, A.; Ahemad, M.; Oves, M. and Wani, P.A. (2010). *Arch. Agron. Soil Sci.* 56:73–98.

Khanna, S.S.; Chaudhary, M.L.; Bathla, R.N. (1979). *Bulletin of Indian Society Soil Science* 12:545-549.

Kim, K.Y.; Jordan, D. and McDonald, G.A. (1998). *Soil Biol. Biochem.* 30:995–1003.

Kim, K.Y.; McDonald, G.A. and Jordan, D. (1997). *Biol. Fertil. Soils* 24:347–352.

Koele, N.; Thomas, W.; Kuyper, P. and Bindraban, S. (2014). *Beneficial Organisms for Nutrient Uptake,* Virtual Fertilizer Research Center (VFRC) Report, 2014/1, Washington, D.C, USA.

Krishnaraj, P.U.; Khanuja, S.P.S. and Sadashivam, K.V. (1998). *Mineral phosphate solubilization (MPS) and mps genes -components in eco-friendly P fertilization. Abstracts of Indo US Workshop on Application of Biotechnology for Clean Environment and Energy,* National Institute of Advanced Studies, Bangalore, pp 27.

Kucey, R.M.N. (1983). *Can. J. Soil Sci.* 63:671–678.

Kumar, V.; Behl, R.K. and Narula, N. (2001). *Microbiol. Res.* 156:87–93.

Lindsay, W.L.; Vlek, P.L.G. and Chien, S.H. (1989). *In: Dixon JB, Weed SB (Eds) Minerals in soil environment, 2nd edn.* Soil Science Society of America, Madison, WI, USA. Pp. 1089–1130.

Lipman, JG.; McLean, H.C. and Lint. (1916). *Soil Science* 2:499-538.

Lockhead, A,G. and Thexton, R.H. (1952). *Journal of Bacteriology* 63:219-26.

Lockhead, A.G. (1957). *Soil Science* 84:395-403.

Louw, H.A. and Webley, D.M. (1958). *Nature,* London 182:1317-18.

Maliha, R.; Samina, K.; Najma, A.; Sadia, A. and Farooq, L. (2004). *Pak. J. Biol. Sci.* 7:187–196.

McGill, W.B. and Cole, C.V. (1981). *Geoderma* 26:267–268.

Mishra, B.; Sharma, RD and Mishra, N.P. (1981). *Potatao Research* 24:183.

Mishustin, E.N. and Naumova, A.N. (1962). *Microbiologiya* 31:553-55.

Nair, S.K. and Rao, N.S.S. (1977). *Journal of Plantation Crops* 5:67-70.

Nannipieri, P.; Giagnoni, L.; Landi, L. and Renella, G. (2011). *Soil Biology*, 26. Springer, Heidelberg, pp. 251–244

Naumova, A.N.; Mishustin, E.N. and Marienko, V.M.(1962). *Bulletin of Academy Sciences. USSR* 5:709-17.

Nautiyal, C.S. (1999.) *FEMS Microbiol. Lett.* 170:265–270.

Neller, J.R. (1956). *Soil Science* 82:129.

Norrish, K. and Rosser, H. (1983). *In: Soils: an Australian viewpoint.* Academic Press, Melbourne, CSIRO/London, UK, Australia, pp 335–361.

Oliveira, C.A.; Sa, N.M.H.; Gomes, E.A.; Marriel, I.E.; Scotti, M.R.; Guimaraes, C.T.; Schaffert, R.E. and Alves, V.M.C. (2009). *Appl. Soil Ecol.* 41:249–258.

Omar, S.A. (1998). *World J. Microbiol. Biotechnol.* 14:211–219.

Ortuno, A.; Hernansaez, A.; Nogurea, J.; Morales, V. and Armero, T. (1978). *Microbiologiya Espanola* 31 : 113-20.

Ostwal, K.P. and Bhide, V.P. (1972). *Indian Journal of Experimental Biology* 10:153-154.

Park, K.H.; Lee, C.Y. and Son, H.J. (2009). *Lett. Appl. Microbiol.* 49:222–228.

Parker, D.R.; Reichmann, S.M. and Crowley, D.E. (2005). *In: Zobel RW (Ed) Roots and Soil Management: Interactions Between Roots and the Soil. Agronomy Monograph No. 48.* American Society of Agronomy, Madison, pp 57–93.

Parks, E.J.; Olson, G.J.; Brinckman, F.E. and Baldi, F. (1990). *J. Ind. Microbiol. Biotechnol.* 5:183–189.

Peix, A. Velazquez , E. and Martýnez-Molina, E. (2007). *In: Velazquez E, Rodrguez-Barrueco C (Eds) First International Meeting on Microbial Phosphate Solubilization.* Springer, Berlin, pp. 97–100.

Pikovskaya, R.I. (1948). *Microbiology* 17:362–370.

Ponmurugan and Gopi (2006). *Biol. Fert. Soils* 30:460-468.

Puente, M.E.; Bashan, Y.; Li. C.Y. and Lebsky, V.K. (2004a). *Plant Biol.* 6:629–642.

Puente, M.E.; Li, C.Y. and Bashan, Y. (2004b). *Plant Biol.* 6:643–650.

Puente, M.E.; Li, C.Y. and Bashan, Y. (2009a). *Environ. Exp. Bot.* 66:389–401.

Quiquampoix, H . and, Mousain, D. (2005). *Enzymatic Hydrolysis of Organic Phosphorus. In: Turner BL, Frossardand E, Baldwin DS (Eds) Organic Phosphorus in the Environment.* CAB International, Wallingford UK, pp. 89–112.

Rajan, S.S.S. (1982). *N.Z.J. of Experimental Agriculture* 25:355.

Rajan, S.S.S. (1983). *Fertilizer Research* 4:287.

Rajan, S.S.S. and Edge, E.A. (1980). *N.Z.J. Agriculture Research* 25:355.

Rajan, S.S.S. and Gillingham, A.G. (1986). *N.Z.J. of Experimental Agriculture* 14:313.

Rao, SS.; Saisree, N.S.; Sheth, L.V.; Reddy, G.S.N. and Bhargawa, P.M. (1989). *Applied Environmental Microbiology* 55:767-70.

Rastogi, R.C.; Mishra, B. and Ghildyal, B.P.(1976). *Journal of Indian Society Soil Science* 24:175-81.

Renella, G.; Egamberdiyeva, D.; Landi, L.; Mench, M. and Nannipieri, P. (2006). *Soil Biol. Biochem.* 38:702–708.

Rengel, Z. and Marschner, P. (2005). *New Phytol.* 168:305–312.

Reyes, I.; Bernier, L. and Antoun, H. (2002). *Microb. Ecol.* 44:39–48.

Reyes, I.; Bernier, L.; Simard, R.R. and Antoun, H. (1999). *FEMS Microbiol. Ecol.* 28:281–290.

Richardson, A.E. (1994). *In: Pankhurst CE, Doubeand BM, Gupta VVSR (Eds) Soil biota: management in sustainable farming systems.* CSIRO, Victoria, Australia, pp. 50–62.

Richardson, A.E. (2001). *Aust. J. Plant Physiol.* 28:897–906.

Richardson, A.E. and Simpson, R.J. (2011). *Plant Physiol.* 156:989–996.

Richardson, A.E.; Barea, J.M.; McNeill, A.M. and Prigent-Combaret, C. (2009a). *Plant and Soil* 321:305–339.

Richardson, A.E.; Hadobas, P.A.; Hayes, J.E.; O'Hara, C.P. and Simpson, R.J. (2001). *Plant and* Soil 229:47–56.

Richardson, A.E.; Hocking, P.J.; Simpson, R.J. and George, T.S. (2009b). *Crop Pasture Sci.* 60:124–143.

Rodriguez, H. and Fraga, R. (1999). *Biotechnol. Adv.* 17:319–339.

Rodriguez, H; Fraga, R.; Gonzalez, T. and Bashan, Y. (2006). *Plant Soil* 287:15–21.

Rogers, R.D. and. Wolfram, J. H. (1993). *Phosphorus, Sulphur and Silicon Related Elements.* 77:1-4.

Rose, R.E. (1957). *N.Z.J. of Science and Technology* 38:773-780.

Rudolph, W. (1922).*Soil Sci.* 14:247–263.

Saber, K.; Nahla, L.D. and Chedly, A. (2005). *Agron. Sustain. Dev.* 25:389–393.

Sadia, Alam.; Samina, Khalil.; Najma, Ayub and Maliha, Rashid. (2002). *International Journal of Agriculture and Biology.*4:454-458.

Sagervanshi, A.; Kumari,P.; Nagee, A. and Kumar, A. (2012). *International Journal of Life Sciences and Pharma Research.* 23:256-266.

Sanchez, P. and Logan, T. (1992). *Soil Science Society of America,* Madison, WI, pp 35–46.

Sankram, A. (1960). *Rhizosphere Effect on Soil Management PhD Thesis, PG School,* IARI, New Delhi.

Sattar, M.A. and Gaur, A.C. (1985). *Bangladesh J. Microbiol.*2:22-28.

Sattarand, M.A. and Gaur, A.C. (1987). *Zbl. Microbiology* 142:393-95.

Sharma, S. B.; Trivedi, M. H.; Sayyed R. Z., and Thivakaran, G. A. (2014). *Annual Research & Review in Biology* 4(18): 2901-2909.

Sharma, SN.; Ray, Sb.; Pandey, Sl.; and Prasad, R. (1983). *Journal of Agricultural Sciences* 101:467.

Sims, J.T. and Pierzynski, G.M. (2005). *In: Tabatabai AM, Sparks DL (Eds) Chemical Processes in Soil,* SSSA Book Series 8. SSSA, Madison, pp. 151–192.

Singh, D.; Mannikar, N.D. and Srivas, N.C.(1979). *Journal of Indian Society Soil Science* 37:334.

Singh, H. and Reddy, M.S. (2011). *Eur. J. Soil. Biol.* 47:30-34.

Singh, Y.; Ramteke, P.W. and Shukla, P.K. (2013). *.Adv. Appl. Sci. Res.* 4:269–272.

Son, H.J.; Park, G.T.; Cha, M.S. and Heo, M.S. (2006.). *Bioresour. Technol.* 97:204-210.

Song, O.R.; Lee, S.J.; Lee, Y.S.; Lee, S.C.; Kim, K.K..and Choi, Y.L. (2008). *Braz. J. Microbiol.* 39:151–156.

Sperber, J.I. (1958a). *Aust. J. Agr. Res.* 9:778–781.

Sperber, J.I. (1958b). *Aust. J. Agr. Res.* 9:782–787.

Sridevi, M.; Mallaiah, K.V. and Yadav, N.C.S. (2007). *J. Plant Sci.* 2:635–639.

Srivastava, S.; Yadav, K.S. and Kundu, B.S. (2004). *I. J. Microbiol.* 44:91-94.

SubbaRao, N.S. (1982). *Advances in Agricultural Microbiology.* Oxford and IBH Publications Company, India, pp. 229–305.

Sujatha, E.; Girisham, S. and Reddy, S.M. (2004). *Phosphate Solubilization by Thermophillic Micro-organisms* 44: 101-104.

Sutherland, I.W. (2001). *Microbiology* 147:3–9.

Swaby, R .and Sperber, J.I. (1958). *Soils & Fert.* 22:289–294.

Swaby, R.J. (1975). *Sulphur in Australian Agriculture,* Sydney University Press, Sydney pp.213.

Tarafdar, J.C.; Yadav, R.S. and Meena, S.C, (2001). *J. Plant Nutr. Soil Sci.* 164:279–282.

Tilman, D.; Fargione, J.; Wolff, B.; D'Antonio, C.; Dobson, A.; Howarth, R.; Schindler, D.; Schlesinger, W.H.; Simberloff, D. and Wackhamer, D. (2001). *Science* 292:281–284.

Tinker, P.B. and Nye, P.H. (2000). *Solute movement in the rhizosphere.* Oxford University Press.

Torma, A.E. and Binhegy, I.G. (1984). *Trends of Biotechnolohy* 2:13-15.

Torsvik, V. and Ovreas, L. (2002). *Aust. J. Soil Res.* 41:471–499.

Upadhayay, A. and Srivastava, S. (2012). *Indian J. Exp. Biol.* 48:601–609.

Van Kauwenbergh, S.J.(2010). *World Phosphate Rock Reserves and Resources.* IFDC, Muscle Shoals, AL, USA.

Vassilev, N.; Vassileva, M. and Nikolaeva, I. (2006). *Appl. Microbiol. Biotechnol..* 71:137–144.

Vassilev, N.; Vassileva, M.; Azcon, R. and Medina, A. (2001). *Biotechnol. Lett.* 23:907–909.

Vazquez, P.; Holguin, G.; Puente, M.; Elopez Cortes, A. and Bashan, Y. (2000). *Biology and Fertility of Soils* 30:460-468.

Venkateswarlu, B.; Rao, A.V., Raina, P. and Ahmad, N. (1984). *J. Indian Soc. Soil Sci.* 32:273–277.

Vessey, J.K. and Heisinger, K.G. (2001). *Can. J. Plant Sci. 81:361-366.*

Wakelin, S.; Warren, R.; Harvey, P. and Ryder, M. (2004). *Biol. Fert. Soils* 40:36-43.

Wani , P.V. and Patil L.K. (1979). *National Academy of Science Letters* 2:360-62.

Wani , P.V.; More, B.B. and Patil, P.L. (1979). *I. J. Microbiol.* 19:23-25.

Wani P.A.; Zaidi, A.; Khan, A.A. and Khan, M.S. (2005). *Annals Plant Protection Sci.* 13:139-144.

Watanabe, K. and Hayano, K. (1993). *Canadian Journal of Microbiology* 39:674-80.

Whitelaw, M.A. (2000). *Adv. Agron.* 69:99–151.

Whitelaw, M.A.; Harden, T.J. and Helyar, K.R. (1999). *Soil Biol. Biochem.* 32:655–665.

Widada, J.; Damarjaya, D.I. and Kabirun, S. (2007). *In: Rodriguez-Barrueco C (Ed) Velazquez E. First International Meeting on Microbial Phosphate Solubilization,* Springer, pp 173–177.

Xuan, Yu.; Xu Liu, Tian Hui Zhu, Guang Hai, Liu. and Cui, Mao. (2011). *Biology and Fertility of Soils. doi:10.1007/s00374-011-0548-2.*

Yi, Huang, W. and Ge, Y. (2008). *World J. Microbiol. Biotechnol.* 24:1059–1065.

Young, C.C.; Shen, F.T.; Lai, W.A.; Hung, M.H.; Huang, W.S.; Arun, A.B. and Lu, H.L. (2003). *In: Proceeding of 2nd International Symposium on Phosphorus Dynamics in the Soil-Plant Continuum. Perth, Australia.* pp. 44-45.

Zaidi, A.; Khan, M.S.; Ahemad, M.; Oves, M. and Wani, P.A. (2009). *In: Khan MS et al (Eds) Microbial Strategies for Crop Improvement.* Springer-Verlag, Berlin Heidelberg, pp 23–50.

Zhou, K.; Binkley, D. and Doxtader, K.G. (1992). *Plant and Soil* 147:243–250.

Zhu, F.; Qu, L.; Hong, X. and Sun, X. (2011). *Evid Base Compl. Alternative Med. 615032:6 doi:10.1186/2193-1801-2-*

Chapter 11 : Mycorrhizae

Abbott, L.K. and Robson, A.D. (1978). *New Physiol.* 8: 575-587.

Abbott, L.K. and Robson, A.D.(1984). *The Effect of Mycorrhizas on Plant Growth. [In] VA Mycorrhizae,* ed. C.L. Powell, D.J. Bagyaraj, pp. 113-130. Boca Raton, Fla: CRC Press pp. 234.

Abbott, L.K., and Robson, A.D. (1982). *J. Agric. Res.* 33:389-408.

Abbott, L.K., and Robson, A.D. (1991). *In: The Rhizosphere and Plant Growth, pp.* 355-362, D.L. Keister and P.B. Cregan (eds.), Kluwer Academic Publishers, Dordrecht.

Abbott, LK. and Robson, AD (1985). *New Phytology* 99:255-65.

Abdel-Fattah, G.M. and Shabana, Y.M.(2002). *J. Plant Diseases and Protection* 109 (2): 207-215.

Akhter, A.; Khan, N.A. and Inam, A. (2003). *Indian J.Pulses Res.*16(1): 25-30.

Akiyama, K., Matsuzaki, K., Hayashi, H. (2005). *Nature* 435:824-827.

Alexander, I.L. and Fairley, R.I. (1983). *Plant and Soil* 71:49-53.

Allen, M.F. (1982). *New Phytol.* 91: 191-196.

Allen, M.F. (1988). *In: Proc. of Royal Society of Edinberg* 94:63-71.

Allen, M.F. (1991). *The Ecology of Mycorrhizae.* Cambridge University Press, Cambridge, pp. 184.

Allen, M.F. and Boosalis, M.G. (1983). *New Phytol,* 93: 67-76.

Allen, M.F.; Moore, J.S .Jr. and Christensen, M. (1982). *Canadian J. Bot.* 60:468.

Allen, M.F.; Smith, W.K.; Moore, T.S. and Christensen, M. (1981). *The New Phytologist.* 88: 683-693.

Allison, L.E.(1965). *Organic Carbon. In Methods of Soil-Analysis* (Eds C.A. Black, D.D. Evans, J.L. White, L.E Enminger and F.E. Clark) Madison: Wisconsin: American Society of Agronomy.

Almagrabi, O.A. and Abdelmoneim, T.S. (2012). *Life Science Journal* 9 (4):1648-1654.

Ames, R.N.; Porter, L.; St. John, T.V. and Reid, C.P.P. (1984). *The New Phytologist* 97: 269-276.

Ames, R.N.; Reid, C.P.P.; Porter, L. and Cambardell, C. (1983). *The New Phytologist* 95: 381-396.

Angelard, C.; Colard, H.; Niculita-Hirzel, Croll, D. and Sanders, I.R. (2010). *Curr. Biol.* 20:1216-1221.

Arocena, J.M.; Velde, B. and Robertson, S.J. (2011). *Appl. Clay Sci.* 64:1-94.

Artursson, V.; Finlay R.D. and Jansson, J.K. (2006). *Environ. Microbiol.* 8:1-10.

Asensio, D.F.; Rapparini, J. and Peñuelas, J, (2012). *Phytochemistry* 77:149–161.

Aseri, G.K.; Jain, N.; Panwar. J.; Rao, A.V. and Meghwal, P.R. (2008). *Scientia Horticulturae* 117: 130-135.

Ashton, D.H. (1956). *Autoecology of Eucalyptus Regnans, PhD Thesis,* University Melbourne, Austrailia.

Audet, P., and Charest, C. (2007). *Environ. Pollut.* 147:609-614.

Auge, R.M. (2001). *Mycorrhiza* 11:3-42.

Auge, R.M. and Duan, X. (1991). *Plant Physiol.* 97:821-824.

Auge, R.M.; Stodola, A.J.W.; Brown, M.S.; Bethlenfalvay, G.J. (1992 b). *New Phytol.* 120:117-125.

Ayyappan, S (2011). *Way forward –ICAR News.* 17(3): 20.

Azaizeh, A.; Marschner, H.; Romheld, V. and Wittenmayer, L. (1995). *Mycorrhiza* 5: 321-327.

Azcon, R. and Ocampo, J.A. (1981). *New Phytol.* 87: 677-685.

Azcon, R.; Azcon-Gide Aguilar, C. and Barera, JM. (1978). *New Phytol.* 80:359-64.

Azcon, R.; Barea, J.M. and Hayman, D.S. (1976). *Soil Biology and Biochemistry* 8:135-38.

Azcon, R; Gomez, M. and Tobar, R.M. (1996). *Biol. Fertil. Soils.* 22:156-161.

Azcon-Aguilar, C. and Barea, J.M. (1996). *Mycorrhiza* 6: 457-464.

Azcon-Aguilar, C. and Barea, J.M. (1997). *Scientia Horticulture.*68:1-24.

Bago, B.; Pfeffer, P.E.; Abubaker, J.; Jun, J.; Allen, J.W.; Brovillette, J.; Douds, D.D.; Ammers, P.J. and Shachar-Hill, Y. (2003). *Plant Physiol.* 131: 1496-1507.

Bagyaraj, D.J. (1992). *Methods in Microbiology* 24:359.

Bagyaraj, D.J. and Hedge, S.V. (1978). *Curr. Sci.* 47: 543-549.

Bagyaraj, D.J.; Manjunath, A. and Patil, R.B. (1979). *New Phytol.* 82: 141-145.

Bagyaraj, D.J.; Manunath, A. and Govinda Rao, Y.S. (1988). *Journal of Soil Biology and Ecology* 8:98-103.

Bahr, A.; Ellström, M.; Akselsson, C.; Ekblad, A.; Mikusinska, A. and Wallander, H. (2013). *Soil Biol. Biochem.* 59:38-48.

Ballard, S.S. and Dean, L.A. (1941). *Soil Science.* 52:162-173.

Balyan, S.K.; Chandra, R. and Pareek, R.P. (2002). *Agricultural Research Communication Centre, Karnal, India:* 25(3): 160-164.

Bansal, M.; Chamola, B.P.; Sarwar, N. and Mukerji, K.G. (2000). *In: "Mycorrhizal Biology"* (Eds. Mukerji, K.G, Chamola, B.P. and Jagjit Singh), Kluwer Academic Publishers, NY, pp. 143-152.

Barea, J.M .and Azcon-Aguilar, C. (1983). *Advanced Agronomy* 3:1-54.

Barea, J.M. (1991). *Advanced Soil Science* 15:1-15.

Barea, J.M. and Azcon-Aguilar, C. (1981). *Appl. Environ. Microbiol.* 43:810-813.

Barea, J.M. and Azcon-Aguilar, C. (1987). *New Phytology* 106:717-25.

Barea, J.M., Azcon, R. and Azcon-Aguilar, C (2002). *Antonie van Leeuwenhoek* 81:343-351.

Barea, J.M.; Navare and Montoya, E. (1976). *Journal of Applied Bacteriology* 40:129-34.

Barrow, N.J. (1975). *Aust. J. Agric. Res.* 26:137-143.

Bartels, M. and Caesar, K. (1987). *J. Agron. Crop. Sci.* 158:346-352.

Bayne, H.G. and Bethlenfalvay, G.J. (1987). *Plant Cell Environment* 10:607-612.

Becard, G. and Fortin, J.A. (1988). *New Phytol.* 108:211-218.

Bedini, S.; Pellegrino, E.; Avio, L;, Pellegrini, S.; Bazzoffi, P.; Argese, E. and Giovannetti, M. (2009). *Soil Biol. Biochem.* 41:1491-1496.

Beltrano, J; Ronco, M.; Salerno, M.I. Ruscitti, M.; Peluso, O. (2003). *Revista Cienciay Tecnología* (8):1-7.

Berta, G., Fusconi, A. and Trotta, A. (1993). *Environ. Expt. Bot.* 33: 159-173.

Berta, G.; Fusconi, A.; Trotta, A. and Scannerini, S. (1995). *New Phytol.*114:207.

Bethlenfalvay, G.J. and Yoder, J.F. (1981). *Physiol.Plant.* 52:141-145.

Bethlenfalvay, G.J., Bayne, H.G. and Pacovsky, R.S. (1983). *Physiol. Planta.* 57:543-548,

Bethlenfalvay, G.J., Brown, M.S. and Stafford, A.E.(1985). *Plant Physiol.* 79:1054-1059.

Bethlenfalvay, G.J.; Brown, M.S.; Mishra, M.L. and Stafford, A.E. (1987). *Plant Physiol.* 85:115-119.

Bethlenfalvay, G.J.; Pacovsky, R.S. and Brown, M.S. (1982b). *Phytopathol.* 72:894-897.

Bethlenfalvay, G.J;. Brown, M.S.; Ames, R.N.and Thomas, R.S. (1988). *Physiol. Plant.*72:565-571.

Bhandari , N.N.; and Mukerji , K.G. (1993). *The Haustorium, Research Studies Press Ltd. UK pp.* 235-246.

Bhardwaj, A, and Sharma, S. (2006b). *Journal of Agronomy* 5(3):471-474.

Bhardwaj, A. and Sharma, S. (2006a). *Pathology Journal* 5(2): 166-172.

Bidondo, L.F.; J. Bompadre, M.; Pergola, V. Silvani, R.; Colombo, F.; Bracamnte and Godeas, A. (2012). *Pedobiologia* 55:227-232.

Bieleski, R.L. (1973). *Plant. Physiol.* 24:225-252.

Bierman, N.B. and Linderman, R.G. (1983). *J. Amer. Soc. Hort. Sci.* 108:962-971.

Birhane, E.; Sterck, F.j.; Fetene, M.; Bongers, F. and Kuyper, K.W. (2012). *Oecologia* 169:895-904.

Biro, B.; Pechy, K.K.; Voros, I.; Takacs, T.; Eggenberger, P. and Strasser, R.J. (2000). *Applied Soil Ecology* 15: 159-168.

Blee, K. A., and Anderson, A. J. (1996). *Plant Physiol.* 110:675-688.

Bonfante, P. and Perotto, S. (1995). *New Phytol.* 130:3-21.

Bonfante, P., and Anca, I.A. (2009). *Ann. Rev. Microbiol.* 63:363-383.

Bonfante-Fasolo, P. (1984). *New Phytol.* 99: 5-32.

Boudarga, K. and Dexheimer, J. (1988). *Bulletin de la Societe Botanique de France* 135:11-121.

Brockwell, J. (1977). *In: Hardy, R.W.; Gibon, A.H. (ear), A treatise on dinitrogen fixation sec. IV. Agronomy and Ecology.* Cambridge Univ. Press, Cambridge. U.K.

Brown, M.E. (1974). *Annual Revenue of Phyopathology* 12:181-98.

Brown, R.W.; Brown, C.L. and Kormanil, P.P. (1981). In: *Absts. of 5th North Amer. Conf. on Mycorrhizae, Quebec.* pp. 12.

Buee, M.; Rossingnol, M.; Jauneau, A.; Ranjeva, R. and BeCard, G. (2000). *Mol. Plant. Microbe Interact.* 13: 693-698.

Burges, T. and Malajczuk, N. (1989). *Agricultural Ecosystems Environment* 28:181-98.

Busse, M.D. and Ellis, J.R. (1985). *Can. J. Bot.* 63: 2290-2294.

Buwalda, J.G.; Stribley, D.P. and Tinker, P.B. (1983). *Plant and Soil.* 71:463-467.

Buwalda, J.G.; Stribley, D.P. and Tinker, P.B. (1984). *NewPhytol.* 96:411.

Byra Reddy, M.S. and Nalini, P.A. *et al.* (1989). *Forest Ecology Manage* 27:791-801.

Calvaruso, C.; Turpault, M.P.; Leclerc, E. and Frey-Klett, P. (2007). *Microb. Ecol.* 54:567-577.

Cameron, D.D., Neal, A.L..; Van Wees, S.C.M. and Ton, J. (2013). *Trends Pl. Sci.* 18:539-545.

Campbell, W.J. (1988). *Physiol. Plant* 74:214-219.

Cardoso, I.M. and T.W. Kuyper, (2006). *Agric. Ecosyst. Environ.* 116:72-84.

Cardoso, I.M., Boddington, C.L.; Janssen, B.H.; Oenema, O. and Kuyper, T.W.(2006). *Comm. Soil Sci. Pl. Anal.* 37:1537-1551.

Carvalho, A.M.X. de, R. De C. Tavares, Cardoso, I.M. and Kuyper, T.W. (2010). In: *Soil biology and agriculture in the tropics, pp. Springer, Berlin. 185-208, P. Dion (ed.).*

Cavagnaro, T.R. (2008). *Plant and Soil* 304:315-325.

Cavagnaro, T.R.; Smith, F.A.; Smith, S.E. and Jakobsen, I. (2005). *Plant Cell Environ.* 28:642-650.

Chagnon, P.L., Bradley, R.L.; Maherali, H. and Klironomos, J.N. (2013). *Trends Pl. Sci.*18(9): 484-491.

Chalermpongse, A. (1987). *In: Proccedings of 3rd Round Table Conference on Dipterocarpus* (Ed. AJHG Kostermans, UNESCO).

Cheng, L.; Brooker, F.L.; Cong, T.; Burkey, K.O. Zhou, L.; Shew, H.D.; Rufty T.W. and Hu, S. (2012). *Science* 337:1084-1087.

Chilvers, G.A.; Lapeyrie, F.F. and Horam, D.P. (1987). *New Phytology* 107:441-48.

Clarkson, D.T. (1985). *Ann. Rev. Plant Physiol.* 36: 77-115.

Colla, G; Rouphael, Y.; Cardarelli, M.; Tullio, M.; Rivera, C. and Rea, E. (2007). *Biology and Fertility of Soils*: 44 (3): 501-509.

Cooper, K.M. (1984). *In: VA mycorrhiza. Ed. C.L. Powell, D.J., Bagyaraj;* pp. 113-130. Boca Roton, Fla: CRC Press,London.

Cooper, K.M. and Tinker, P.B. (1978). *New Phytol.* 81:43-52.

Cooper, K.M. and Tinker, P.B. (1981). *New Phytol.* 88: 327-339.

Cordier, C.; Gianinazzi, S. and Gianinazzi-Pearson, V. (1996). *Plant Soil* 185: 223-232.

Cornejo, P.; Meier, S.,; Borie, G; Rilling, M.C.; Borie, F. (2008a). *Sci. Total Environ.* 406:154-160.

Cox, G and Sanders, F.E. (1974). *New Phytol.* 73: 901-912.

Cranen Brouck, S.; Voets, L. and Bivort, C. (2005). *In: Vitro Culture of Mycorrhizas, Declerck S, Strullu DG and Fortin A(Eds.)* Springer Verlag Berlin Heidelberg pp. 342-375.

Crush, J.R. (1974). *New Phytol.* 73: 743-749.

Daft, M.J. and El-Giahmi, A.A. (1974). *New Phytol.* 73.

Daft, M.J. and El-Giahmi, A.A. (1978). *New Phytol.*72:1333-1339.

Daft, M.J. and Nicolson, T.H. (1969). *New Phytol.*68: 953-961.

Daft, M.J. and Nicolson,T.H (1966). *New Phytol.*65:343-350.

Daft, M.J. and Okusanya, B.O. (1973). *New Phytol.* 72: 1333.

Danielson, R.M. (1985). *In: Soil Reclamation Practices* (R.L .Date and D.A .Kelin, Eds) Marcel Dekkar, NY pp.173-201.

Dart, P.J. and Krantz, B.A. (1977). *Agric. Misc. Publ.* 145: 119-154.

Davies, F.T. Potter, J.R. and Linderman, R.G (1993). *Physiol. Plant* 87:45–53.

Day, L.D.; Sylvania, D.M. and Collins, M.E. (1987). *Journal of Soil Science Society,* Amesterdam 51:635-39.

Dehne, H.W. and Schonbeck, F. (1978). *J. Plant Dis. Protec.* 85: 666-678.

Dehne, H.W.(1982). *Phytopath.*72:1115-1119.

Dehne, H.W.; Schonbeck, F. and Baltruschat, H. (1978). *Z.Pflanzenkrankh. pflanzenschutz* 85:666-678.

Deka, A.K. and Azad, P.(2006). *Indian J. Pulses Res.* 19:79-82.

Dickie, I.A.; Martínez-García, l.B.; Koele, N.; Grelet, G.A.; Tylianakis, J.M.; Peltzer, D.A. and Richardson, S.J. (2013). *Plant and Soil* 367:11-39.

Dixon, R.K. and Buchesna , C.A. (1988). *Plant and Soil* 105:265-71.

Douds, D.D.; Pfeffer, P.E. and Shachar-Hill, Y. (2000). *In: Douds DD, Kapulnik Y,eds. Arbuscular mycorrhizas physiology and function.* Dordrecht, the Netherlands: Kluwer Academic Publishers, pp. 107-130.

Douds, J.R.; Nagahashi, D.D.; Pfeffer, G; Reider, P.E. and Kayser, W.M. (2006). *Biores. Technol.* 97:809-818.

Duan, X., Neuman, D.S., Reiber, J.M., Green, C.D., Saxton, A.M. and Auge R.M. (1996). *Journal of Experimental Botany* 47: 1541-1550.

Dubey, S.C. (2006). *Archives. Phytopath. Pl. Protec.* Taylor & Francis, Abingdon,UK : 39(5):341-351.

Duchesne, L.C.; Peterson, R.L. and Ellis, B.E. (1987). *Canadian Journal of Botany* 66:471-76.

Duchesne, L.C.; Peterson, R.L. and Ellis, B.E. (1988). *New Phytol.* 108:471-76.

Dugassa, G.D., Von Alten, H. and Schonbeck, F. (1996). *Plant and Soil* 185: 173-182.

Duke, ER.; Johnson, C.R. and Koch, K.E. (1986). *New Phytol.* 104:583-90.

Dumas-Gaudot, E.; Gollotte, A.; Cordier, C.; Gianinazzi, S. and Gianinazzi-Pearson, V.(2000). *In: Arbuscular Mycorrhizas: Physiology and Function.* Eds Kapulnik Y , DD Douds Jr Y (Kluver Academic Press, Dordrecht, The Netherlands), pp. 173-200.

Dumas-Gaudot, E.; Slezack, S;, Dasi. B.,Pozo, M.J.; Gianinazzi-Pearson, V. and Gianinazzi, S. (1996). *Plant and Soil* 185:211-221.

Dumbrell, A.J.; Nelson, M.; Helgason, T..; Dytham, C. and Fitter, A.H. (2010). *ISME J.* 4:337-345.

Duponnois, R., and J. Garbaye, (1991). *Plant and Soil* 138:169-176.

E Lad, Y.; Chet, I. and Ketan, J. (1980). *Phytopath-* 70:119-21.

Echave, M.; Martin, C.; Ariel, C.; Marcela, R. and José, B. (2005). *Newsletter* 35 (2):182-186.

Ekwebelam, S.A. (1979). *Indian Forester* 105:750-57.

El Ghachtouli, N.; Paynot, M.; Morandi, D.; Martin-Tanguy, J. and Gianinazzi, S. (1995). *Mycorrhiza* 5: 189-192.

Ellis, J.R.; Larsen, H.J. and Boosalis, M.G. (1985). *Plant and Soil* 86:369-378.

Evans, D.G. and Miller, M.H. (1988). *New Phytol.* 110:67-74.

Fernandez, D.S. and Ascencio, J. (1994). *Journal Plant Nutri.* 17(2 &3):229-241.

Fester, T., and Sawers, R. (2011). *Crit. Rev. Pl. Sci.* 30:459-470.

Fischer, R.A. and Turner, N.C. (1978). *Ann. Rev. Plant Physiol.*29: 277-317.

Fitter, A.H. (1977). *New Phytol .* 79:119-125.

Fortin, J.A.; Becard G.; Declerck, S.; Dalpe, Y.; St. Arnaud, M.; Coughlan, A.P. and Piche, Y. (2002). *Can. J. Bot.* 80:1-20.

Frank, A.B. (1985). *Ber. distch Bot. Ges.* 3: 128-45.

Frank, A.B. (1985). *Plize. Ber. Dtsch. Bot. Ges.*3: 128-145.

Franken, P. and Gnadinger, F. (1994). *Mol. Plant. Microbe Interact.*7: 612-620.

Frey-Klett, P.; Burlinson, P.; Deveau, A.; Barret, M..; Tarkka, M. and Sarniguet, A. (2011). *Microbiol. Mol. Biol. Rev.* 75:583-609.

Frey-Klett, P.; J. Garbaye and Tarkka, M. (2007). *New Phytol.* 176:22-36.

Gadkar, V. and Rillig, M.C. (2006). *FEMS Microbiol. Lett.* 263:93-101.

Galván, G.; Kuyper, T.W.; Burger, K.; Keizer, L.C.P. Hoekstra, R.; Kik, C. and Scholten, O. (2011). *Theor. Appl. Genet.* 122:947-960.

Gao, X.P,; Kuyper, T.W.; Zhang, T.; Zou, C. and Hoffland, E. (2008). *In: Development and uses of biofortified agricultural products, pp. 153-170, G.S. Banuelos and Z.-Q. Lin (Eds.),* CRC Press, Boca Raton, FL.

Gao, X.P.; Kuyper, T.W. ; Zou, C.Q.; Zhang, F. and Hoffland, E. (2007). *Plant and Soil* 290:283-291.

Garbaye, J. (1994). *New Phytol.* 128:197-210.

Garcia-Garrido, J.M. and Ocampo, J.A. (2002). *Journal of Experimental Botany*53: 1377-1386.

Gaur, A.C. and Rana, JPS (1990). *In: Role of VAM and their interaction on growth and uptake of nutrients by wheat crops. National Conference on Mycorrhizae,* HAU, Hissar Feb 1990.

Gerdemann, J.W. and Trappe, J. M. (1974). *Mycol. Mem.* 5: 1-76.

Gianianazzi, S.; Gianinazzi – Pearson, V. and Trouvelot A. (1989). *In: Bacteriology of Fungi for Improving Plant Growth,* Cambridge Uni. Press Cambridge pp. 41.

Gianinazzi, S., Golotte, A.; Binet, M.N.; Van Tuinen, D.; Redecker, D. and Wipf, D. (2010). *Mycorrhiza* 20:519-530.

Gianinazzi-Pearson, V. (1984). *In: Proc. 6th North Amer. Conf. on Mycorrhizae Bend. Oregon* pp. 150-154.

Gianinazzi-Pearson, V.; Smith, S.E.; Gianinazzi, S. and Smith, F.A. (1991). *New Phytol.* 117:61-76.

Gildon, A. and Tinker, P.B. (1983). *New Phytol.*95:363-368.

Giltrap, A.E. (1979). *Experimental Studies on the establishment and stability of ectomycorrhizae, PhD Thesis* Sheffield Univ. Sheffield.

Giovannetti, M. and Mosse, B. (1980). *New Phytol.* 84:489-500.

Goicoechea, N.; Dolezal, K.; Antolin, M.C.; Strnad, M. and Sánchez- Díaz, M. (1995). *J. Exp. Bot.* 46:1543–1549.

Goicoechea, N.; Szalai, G; Antol, M.C.; Sachez-Daz, M. and Paldi, E. (1998). *J.Plant Physiol.* 153:706-711.

Gosling, P., Hodge, A.; Goodlass, G and Bending, GD. (2006). *Agric. Ecosyst. Environ.* 113:17-35.

Goss, M.J. and Varennes, de A. (2002). *Soil Biol. Biochem.* 34:1167-1173.

Govinda Rao, Y.S.; Bagyaraj, D.J. and Rai, P.V. (1983). *Zentralbl. Microbiol.* 138:415.

Grace, E.J.; Cotsaftis, O.; Tester, M.; Smith, F.A. and Smith, S.E. (2009). *New Phytol.* 181:938-949.

Graham, J.H.; Leonard, R.T. and Menge, J.A. (1981). *Plant Physiol.* 64:548-52.

Grandmaison, J.; Gyorgy, M.; Olah, R.C. and Valentin, F. (1993). *Mycorrhiza* 3(4):.155-165.

Gray, L.E. and Gerdemann, J.W. (1973). *Plant and Soil* 39:687-689.

Green, C.D.; Stodola, A. and Auge, R.M. (1998). *Mycorrhiza* 8:93-99.

Grierson, P.F. and Comerford, N.B.(2000). *Plants Soil* 218:49-57.

Grimoldi, A.A.; Kavanova, Ma.; Attanzi, F.A.; Schaufele, R. and Schnyder, H. (2006). *New Phytol.* 172:544-553.

Gunasekaran, S.; Balchardar, D. and Sudaram, K.M. (2004). *Studies on Synergism between Rhizobium, plant growth promoting rhizobacteria and Phosphate solubilising bacteria in blackgram . Biofertilizers Technology.* Scientific Publishers (India), Jodhpur, India:269-272.

Gupta, R.P.; Kalra,M.S. and Bhandari,S.C.(1982). *Studies on chickpea-Rhizobium interaction. Biological Nitrogen Fixation. Proc. Nat. Symposium* IARI, New Delhi,pp 544-555.

Guttenberger, M. (2000). *Planta.*211:299-304.

Habte, M. and Manjunath, A. (1991). *Mycorrhiza* 1:3-12.

Hafeel, K.M. and Gunatilake, I.A.V.N. (1988). In: *First Asian Conference on Mycorrhiza,* Madras, India pp. 37-45.

Hall, I.R. (1976). *Trans., Br. Mycol. Soc.* 97:409-411.

Hall, I.R. (1978). *N.Z. J. Agric. Res.*21:509-515.

Hall, I.R. (1987). *New Zealand Journal of Botany* 15:481-84.

Harison, M.J. and Dixon, R.A. (1993). *Mole. Plant Microbe Interac.* 6:643-654.

Harley, J.L. and Smith, S.E. (1983). *Mycorrhizal Symbiosis.* Academic Press, New York pp. 483.

Harris,D.; Pacovsky, R.S. and Paul, E.A. (1985). *New Phytol.*101:427-440.

Harrison, M..J. (2005). *Ann. Rev.Microbiol.*59:19-42.

Harrison, M..J.; Dewbre, G.R. and Liu, J. (2002). *Plant Cell.* 14:2413-2429.

Hart, M.M.; Reader, R.J. and Klironomos, J.N. (2001). *Mycologia* 93:1186-1194.

Hattingh, M.J. and Gerdemann, J.W. (1975). *Phytopathol.* 65:1013-25.

Hayman, D.A. (1974). *New Phytol.*73:71-80.

Hayman, D.S. (1975). *Soil Microbiology.* pp. 67-92.

Hayman, D.S. (1982). *Advances in Agricultural Microbiology* (Ed NS Subba Rao) pp. 328.73.

Hayman, D.S. (1983). *Can. J. Bot.* 61:944.

Hayman, D.S. (1986). *Mircens J.* 2:121-45.

Hayman, D.S. and Mosse, B. (1972). *New Phytol.* 71:41-47.

Hayman, D.S.; Morris, E.J. and Page, R.T. (1981). *Annal of Applied Biology* 99:247-53.

Heinrich, P.A. and Patrick, J.W. (1985). *Austrain Journal of Soil Science* 23:222-25.

Helal, H.M. (1990). *Plant Soil.* 123: 161-163.

Hemanatrajan, A.; Tridevi, A.K.; Kumar, S. and Kumar, A. (2003). *National Seminar on Physiological Interventions for Improved Crop Productivity and Quality.* Dec.12-14, 2003 at Tirupati, pp 154.

Henderson, J.C. and Davies, F.T. (1990). *New Phytol.* 115:503-510.

Hepper, C.M. (1984). *Trans. Br. Mycol. Soc.* 83: 154-156.

Hetrick, B.A.D.(1991). *Mycorrhizas and Root Architecture.Experientia* 47:355-62.

Hetrick, B.A.D.; Hetrick, J.A. and Bloom, J. (1984). *Can. J. Bot.* 62: 2267-2271.

Heys, R.; Reid, C.P.P.; St. Johan, T.V. and Coleman, (1982). *Oecologia* 54:260-265.

Hildebrandt, U., Regvar, M. and Bothe, H. (2007). *Phytochemistry* 68:139-146.

Hildebrandt, U.; Schemelzer, E. and Bothe, H.(2002). *Physiol. Plant.* 115:125-136.

Hirrel, M.C., Meharavaran, H. and Gerdemann, J.W. (1978). *Can. J. Bot.*56: 2813.

Ho, l. and Trappe, J.M. (1973). *Mycology* 67: 886-888.

Hodge, A. (2000). *FEMS Microbiology Ecology* 32: 91-96.

Hodge, A., and Fitter, A.H. (2010). *In: Proc. Nat. Acad. Sci. USA* 107:13754-13759.

Hoeksema, J.D.; Chaudhary, V.B.; Gehring, C.A.; N.C. Johnson, J. Karst, R.T. Koide, A. Pringle, C. Zabinski, J.D. Bever, J.C. Moore, G.W.T. Wilson, J.N. Klironomis and Umbanhowar, J. (2010). *Ecol. Lett.* 13:394-407.

Hoffland, E.; Kuyper, T.W.; Wallander, H.; Plassard, C.; Gorbushina, C.C.; Haselwandter, K.; Holmström, S.; Landeweert, R.; Lundström, U.; Rosling, A.; R. Sen, M.M. Smits, P.A.W. Van Hees and Van Breemen, N. (2004). *Front. Ecol. Environ.* 2:258-264.

Horn, R. and Dexter, A.R. (1989). *Soil and Tillage Research.* 13: 253-266.

Horn. R.; Taubner, H.; Wuttke, M. and Baumgartl, T. (1994). *Soil and Tillage Research.* 30: 187-216.

Huang R.S.; Smith, W.K. and Yost, R.S. (1985). *De wit. New Phytol.* 99:229-243.

Hung, L.L. and Sylvia, D.M. (1998). *Appl. Environ. Microbiol.* 54: 353-357.

Hussey, R.S. and Roncadori, R.W. (1982). *Plant Dis.* 66: 9-14.

Hyppel, A. (1968a). *Studies for Suecica* 64:1-18.

Hyppel, A. (1968b). *Studies for Suecica* 64:3-16.

IJdo, M., Cranenbrouck, S. and Declerck, S. (2011). *Mycorrhiza* 21:1-16.

Ikaram, A., Mahmud, A.W., Chong, K. and Faizah, A.W. (1993). *Field Crops Research.* 31: 131-144.

Jacquelinet-Jeanmougin, J.; Gianinazzi-Pearson, V. and Gianinazzi, S. (1987). *Symbiosis.* 3: 269-286.

Jadhav, S.D.; Kadam, S.B.; Deshmukh, R.K. and Manorama. (2006). *Journal Soils Crops.* 16:1, 88-94.

Jagpal, R. and Mukerji, M.G. (1987). *In: Mycorrhize Roundtable* (Eds. Verma, A.k.; Oka, A.K., Mukjerji, K.G.; Tilak, KVBR and Janakraj) pp. 257.

Jain, V. and Gupta, V.K. (2002). *Ind. Phytopath..* 55(3): 323-324.

Jakobsen, I.; Abbott, L.K. and Robson, A.D. (1992). *New Phytol.* 120:371-380.

Jalali, B.L. (1988). In: *First Asian Conf. Mycorrhizae Centre for Advance Study Bot.* Madras. Abst:27.

Janos, D.P. (2007). *Mycorrhiza* 17:75-91.

Jansa, J., Mozafar, A. and Frossard, E. (2005). *Plant and Soil.* 276:163-176.

Jansa, J.; Mozafar, A. and Frossard, E. (2003). *Agronomie.* 23:481-488.

Janse, J.M. (1986). *An Jard, Bot. Buitenz.* 14:53.

Jasper, D.A.; Robson, A.D. and Abbott, L.K. (1979). *Soil Biol. Biochem.* 11:501-505.

Jeffries, P.; S. Gianinazzi, S.; Peto, K.; Turnau and Barea, J.M. (2003). *Biol. Fertil. Soils* 7:1-16.

John, T.V.; Hay, R.T. and Reid C.P.P. (1981). *New Phytology* 89:81-86.

Johnson D.; Leake J.R. and Read D.J. (2005). *Plant and Soil.* 271:157-164.

Johnson, C.R. and Hummel, R.L. (1985). *Hort. Sci.* 20:754-755.

Johnson, N.C. (1993). *Ecol. Appl.* 3:749-757.

Johnson, N.C. and Graham, J.H. (2013). *Plant and Soil* 363:411-419.

Johnson, N.C.; Graham J.H. and Smith, F.A. (1997). *New Phytol.* 135:575-586.

Johnson, N.C.; Copeland, P.J. ; Crookston, R.K. and Pfleger, F.L. (1992). *Agron. J.* 84:387-390.

Johnson, P.N. (1976). *Nz. J. Bot.* 14:333-340.

Johri, B.N. and Mathew, J. (1989). *Plant Microbe Interaction* (Ed Bilgrami, K.S.) NPH Delhi pp. 269-303.

Jones, D.L.; Hodge, A. and Kuzyakov, Y. (2004). *New Phytol.* 163:459-480.

Jung, K.A. and Claassen, N. (1989). *Zeitschrift fiir Pflanzener Nahrung and Bodenkunde* 152:151-157.

Kabir, Z.; O'Halloran, I.P.; Fyles, J.W. and Hamel, C. (1998). *Agric. Ecosyst. Environ.* 68:151-163.

Kaldorf, M.; Schemelzer, E. and Bothe, H. (1998). *Mol. Plant-Microbe Interact* 11:439-448.

Kaldorf, M.; Zimmer, W. and Bothe, H. (1994). *Mycorrhiza.* 5:23-28.

Kalita, R.; Deka, A.K. and Azad, P. (2006). *Legume Research:*29(3):157-162.

Kapoor, R., Sharma, D. and Bhatnagar, A.K. (2008). *Sci. Horticult.* 116:227-239.

Kapoor, R.; Chaudhary, V. and Bhatnagar, A.K. (2007). *Mycorrhiza* 17:581-587.

Karunaratne, P.S.; Baker, J.H. and Barker, A.V. (1987). *J. Plant. Nutrition.* 10:871-886.

Kaschuk, G.; Kuyper, T.W.; Leffelaar, P.A.; Hungria, M. and Giller, K.E. (2009). *Soil Biol. Biochem.* 41:1233-1244.

Kashyap, S. and Sharma, S. (2006). *Horticulture Science (Prague)* 33:77-86.

Kashyap, S.; Sharma, S. and Vasudevan, P. (2004). *Scientia Horticulturae* 100: 291-307.

Kashyap, S.; Sharma, S. and Vasudevan, P. (2005). *Journal of New Seeds* 7(3): 75-90.

Khalifa, F.M. (1987). *Agric. Sci.* 108:259-265.

Khan, A.G (1972). *New Phytol.* 71: 613-619.

Kiers, E.T.; Duhamel, M.; Beesetty, Y.; Mensah, J.A.; Franken, O.; Verbruggen, E.; Fellbaum, C.R.; Kowalchuk, G.A.; Hart, M.M., Bago, A., Palmer, T.M., West, S.A.; Vandenkoornhuyse, P.; Jansa, J. and Bucking, H. (2011). *Science* 333:880-882.

Kiers, E.T.; West, S.A. and Denison, R.F. (2002). *J. Appl. Ecol.* 39:745-754.

Kinden, D.A. and Brown, M.R. (1975). *Can. J. Microbiol.*21:1768-1780.

Kivlin, S.N.; Emery S.M. and Rudgers, J.A. (2013). *Am. J. Bot.* 100:1445-1457.

Koch, E. and Johnson, C.R. (1984). *Plant Physiol.* 75:26-30.

Koele, N.; Turpault, M..P.; Hildebrand, E.E.; Uroz, S. and Frey-Klett, P. (2009). *Soil Biol. Biochem.* 41:1935-1942.

Koid, R.T. and Mooney, H.A. (1987). *New Phytol.* 107:173-82.

Koide, R. (1985). *New Phytol.* 99:449-462.

Koske, R.E. (1987). *Mycologia* 79: 55-68.

Krajinski, F.; Biela, A.; Schubert, D.; Gianinazzi-Pearson, V.; Kaldenhoff, R. and Franken, P. (2000). *Planta.* 211:85-90.

Krishna, K.R. and Bagyaraj, D.J. (1982). *Legume Res.* 5:18-22.

Krishna, K.R. and Bagyaraj, D.J. (1984). *Experientia.* 40:85-86.

Krishna, K.R. and Bagyaraj, D.J.(1984). *In: Proc. 6th North Amer. Conf. on Micorrhizal. Blend. Oregon.* pp. 372.

Kubikova, E.; Moore, J.L.; Ownlew, B.H.; Mullen, M.D. and Auge, R.M. (2001). *J. Plant Physiol.* 158:1227-1230.

Kucey, R.M.N. and Paul, E.A. (1982). *Soil Biol. Biochem.* 14:407-412.

Kucey, R.M.N.; Janzen, H.H. and Leggett, M.E. (1989). *Adv. Agron.* 199-228.

Kulshreshtha, M. and Khan, M.W. (2002). *Indian Phytopath.* 55 (3): 264-268.

Kumar, A. and Elamathi, S. (2007). *Intern J. Agric. Sci.* 3(1): 179-180.

Kumar, R. and Sharma, C.P. (2006). *Farm Science J.* 15(1): 93-94.

Kumar, R. and Chandra, R. (2008). *World J. Agric. Sciences* 4(3):297-301.

Kuo, C.G and Huang, R.S. (1982). *Plant and Soil* 64:325-30.

Kuo, S,C. and Tang, F. (1981). *Acta Botanica Yunnanica* 3:359-66.

Kuyper, T.W. and Giller, K.E. (2011). *In: Agrobiodiversity Management for Food Security,* pp. 134-149, J.M. Lenné and D. Wood (eds.), CABI, Wallingford.

Kuyper, T.W.. (2013). *Die Auswirkungen von Stickstoffeinträgen auf Artengemeinschaften von Pilzen. Z. Mykol.* 79:565-581 *[in German].*

Laiho, O. (1970). *Acta for Fenn.* 106:1-65.

Lambers, H.; Raven, J.A.; Shaver, G.R. and Smith, S.E. (2008). *Trends Ecol. Evol.* 23:95-103.

Lambert, D.H.; Baker, D.E. and Cole, H. (1979). *Soil Sci. Soc. Am. J.* 43:976-980.

Landis, F.C. and Fraser, L.H. (2008). *New Phytol.* 177:466-479.

Lanowska, J. (1966). *Pamiet Pulawski.* 21:365.

Lee, J.E. and Eom, A.H. (2009). *Mycobiology* 37(4): 272-276.

Lee, K.G (1982). *Korean Journal of Mycology* 10:134-35.

Lee, Y.H. and Yun, H.D. (2011). *J. Korean Soc. Appl. Biol. Chem.* 54(3):434-441.

Lee, Y.J. and George, E. (2005). *Hort. Sci.* 40:378-380.

Lehri, L.K.; Gangwar, B.R. and Mehrotra, C.L. (1974). *J. Indian Soil Sci.*22:66-69.

Lekberg, Y. and Koide, R.T. (2005). *New Phytol.* 168:189-204.

Lendzemo, V.W.; Vast Ast, A. and Kuyper, T.W. (2006). *Outl. Agricult.* 35:307-311.

Levy, I. and Krikun, J. (1980). *New Phytol.* 85:25-32.

Lilleskov, E.A.; Hobbie, E.A. and Horton, T.R. (2011). *Fung. Ecol.* 4:174-183.

Lilly, V.G. (1975). *Cur. Sci.* 44:201-02.

Linderman, R.G. (1994). *In: Pfleger FL, Linderman RG (Ed.) Mycorrhizae and Plant Health.* APS Press, St. Paul, Minn, pp.1–26.

Linderman, R.G (2000). *In: Arbuscular Mycorrhizas: Physiology and Function".* (Eds. Kapulnick, Y. and Douds Jr., D.D.), Kluwer Academic Press pp. 345-366.

Liu, Y.; Mi G.; Chen F.; Zhang, J. and Zhang, F. (2004). *Plant Science.* 167: 217-223.

Lopazaguillon, R. and Garbaye, J. (1989). *Ecosystems Environ.* 29:263-66.

Lovelock, C.E.; Wright, S.F.; Clark, D.A. and Ruess, R.W. (2004). *J. Ecol.* 92:278-287.

Macdonald, R.M. (1981). *New Phytol.* 89:87-93.

Mahadi, A. and Atabani, I.M.A (1992). *Expl. Agric.*28:399-407.

Mahoney, M.J.; Chevone, B.I.; Skelly, J.M. and Moore, L.D. (1985). *Phytopathology* 76:679-82.

Manivannan, V., Thanunathan, K., Imayavaramban,V. and Ramanathan, N. (2003). *Legume Research* 26(4): 296-299.

Manjunath, A. and Bagyaraj, D.J. (1984). *Trop. Agric.* 61(1):48-52.

Marschner, H. and Dell, B. (1994). *Plant and Soil*159:89-102.

Marschner, P.; Solaiman, Z. and Rengel, Z. (2006). *Plant and Soil* 283: 11-24.

Marx, D.H. (1969a). *Phytopathol.* 59:153-63.

Marx, D.H. (1969a). *Phytopathol.* 60:472-473.

Marx, D.H. (1969b). *Phytopathol.* 59:1128-32

Marx, D.H. (1972). *Annual Review Phytopathology* 10:429-54.

Marx, D.H. and Bryan, W.E. (1969). *Phytopathol.* 59:1128-32.

Marx, D.H. (1980). *Tropical Mycorrhiza Research* 13-71.

Mathew, J. and Johari, B.N. (1989). *Mycol. Res.*92(4): 491-493.

Mayerhofer, M.S..; Kernaghan, G. and Harper, K.A. (2013). *Mycorrhiza* 23: 119-128.

McGonigle, T.P., M. Hutton, A. Greenley and Karamanos, R. (2011). *Communications in Soil Science and Plant Analysis* 42: 2134-2142.

McLachlan, K.D. (1980). *Aust. J. Agric. Res.* 31: 441-448.

McLachlan, K.D. and De Marco, D.G. (1982). *Aust. J. Agric. Res.* 33: 1-11.

Menge, J.A.; Lembright, H. and Johnson, E.L.V. (1977). *In: Proc. Int. Soc. Citric.* 1:129-132.

Miller, M., McGonigle, T. and Addy, H. (1994). *Plant and Soil* 159:27-35.

Miller, M.H. (2000). *Can. J. Pl. Sci.* 80:47-52.

Minchin, F.R and Patel, J.S. (1973). *J. Exp. Bot.* 24:259.

Miransari, M., (2010). *Plant Biol.* 12:563-569.

Mishra, S. (2008b). *In:Handbook on Biofertilizers Based Fodder Production Technology.* Sharma, V. (ed.) Pub. by-Director, Science and Society Division, Department of Science and Technology, New Delhi, pp. 91.

Mishra, S.K. and Tiwari, V.N. (2002). *Indian J. Pulses Res.* 15(2): 136-138.

Molina, R. (1979). *Canadian Journal of Botany* 57:1223-25.

Molina, R. and Trappe J.M. (1982). *Forest Science* 28:423-57.

Morales, A.; Castillo, C.G.; Rubio, R.; Godoy, R.; Rouanet, J.L and Borie, F. (2005). *Rev. Cienc. Suelo .Nutr. Veg.* 5(1): 37-45.

Morandi, D. (1996). *Plant and Soil* 185:241-251.

Mosse, B. (1953). *Nature* 171:974.

Mosse, B. (1962). *Journal of General Microbiology* 25:209-50.

Mosse, B. (1972). *Nature* 239:221-223.

Mosse, B. (1973). *Annual Review Phytopathology* 11:171-176.

Mosse, B. (1977). *In: J.M. Vincent, A.S Whiteny and J.Bose (Editors), Exploiting the legume Bradyrhizobium symbiosis in Tropical Agriculture,* University of Hawaii, Kabuli, Maui, pp. 275-292.

Mosse, B. and Thompson, J.P. (1984). *Can. J. Bot.* 62: 1523-1530.

Mosse, B.; Powel, CLi and Hayma, D.S. (1976). *New Phytol.* 76:331-42.

Mosses, B.; Stribley, D.P .and Le Tacon, F. (1981). *Adavnced Microbial Ecology* 5:137-210.

Mukerji, K.G. (1999). *In: "Biotechnological Approaches in Biocontrol of Plant Pathogens".* (Eds. Mukerji, K.G., Chamola, B.P. and Upadhyay, R.K.), Kluwer Academic and Plenum Publishers, New York, pp. 135-155.

Munns, D.N. and Mosse, B. (1980). In: *Advances in Legume science* (Ed. by R.J. Summerfield and A.H. Bunting), pp. 115-125. *Proceedings of the International Legume Conference,* Kew.

Muriithi-Muchane, M.N. (2013). *Influences of Agricultural Management Practices on Arbuscular Mycorrhizal Fungal Symbioses in Kenyan Agroecosystems.* PhD Thesis, Wageningen University.

Murkute, A.A.; Sharma, S., Singh, S.K. (2006). *Horticulture Science.*33(2):22-26.

Murphy, P.J.; Langridge, P. and Smith, S.E. (1997). *New Phytol.* 135:291-301.

Nagrajan, P. and Balchander, D. (2001). *Madras Agric. J.* 88(10/12): 703-705.

Nair, M.G.; Safir, G.R. and Siqueira, J.O. (1991). *Applied Environ. Microbiol.* 57:434-439

Natarajan, K. and Govindasamy, V. (1990). *In: Proceedings of National conference on Mycorrhizae,* HAU, Hissar pp. 98-99.

Nelsen, C.E. (1987). *The Water Relations of Vesicular- Arbuscular Mycorrhizal Systems.* pp.71 – 79. In: *Safir GR (Ed.), Ecophysiology of VA mycorrhizal plants.* CRC Press, Boca Raton, Fla.

Nelson, C.E. and Safir, G.R. (1982a). *J. Amer. Soc. Hortic. Sci.* 107:271-274.

Nelson, C.E. and Safir, G.R. (1982b). *Planta* 154:407-413.

Nemec, S. and Meredith, F.I. (1981). *Aust. Bot.* 47:351-358.

Newman, E.I. (1985). *In: Nuclear Techniques to Study the Role of Mycorrhizae in Increasing Food production* (Ed: danso, SK) Intern. Atomic Energy Agency, Vienna TECDOE-338. pp. 63-68.

Newsham, K.K.; Fitter, A.H. and Watkinson, A.R. (1995). *Trends Ecol. Evolut.* 10:407-411.

Nichols, K.A. and Wright, S.F. (2005). *Soil Sci.* 170:985-987.

Nyabyenda, P. (1977). *Einfluss der Bodentemeratur und organischer staffe auf die wirkung der vesikular arburskularen mykorrhiza.* Ph.D. Dissertation, Univ. Gottingen,Gottingen, (West Germany).

Nye, P.H. and Tinker, P.B. (1977). *Solute Movement in the Soil-Root Systems.* Blackwell Scientific Publications, London.

Oblisami, G.; Balalraman, K. and Natarajan, T. (1976). *Madras Agric. J.*63.587-589.

Ocampo, J. A.; Martin, J. and Hayman, D.S. (1979). *New Phytol.* 85: 27-35.

Oehl, F.; E. Sieverding, P.; Mäder, D.; Dubois, K.; Ineichen, T.; Boller and Wiemken, A. (2004). *Oecologia* 138:574-583.

Oehl, F.; Jansa, J.; Ineichen, K.; Maeder, P. and. Van der Heijden, M.G.A. (2011). *Agrarforschung Schweiz* 2:304-311.

Ohtsuka, T. and Saka H.(1988). *Japan Jour. Crop Sci.* 57 (4):722-727.

Oliver, A.J.; Smith, S.E.; Nicholas D.J.D.; Wallace, W. and Smith F.A. (1983). *New Phytol.* 94:63-79.

Olsson, P.A.; Thingstrup, I.; Jakobsen, I. and Bååth, E. (1999). *Soil Biol. Biochem.* 31:1879-1887.

Öpik, M.; Zobel, M.; Cantero, J.J.; Davison, J.; Facelli, J.M.; Hiiesalu, I.; Jairus, T.; Kalwijs, J.M.; Koorem, K.; Leal, m.E.; J. Liira, M. Metsis, V. Neshataeva, J. Paal, C. Phosri, S. Polme, U. Reier, Saks, H.; Schimann, H.; Thiery, O.; Vasar, M. and Moora, M. (2013). *Mycorrhiza* 23:411-430.

Osonubi, O.; Bakare, O.N. and Mulongoy, K. (1992). *Biol. Fertil. Soils* 14:159-165.

Pacovsky, R.S.; Paul, E.A. and Bethenfalvay, G.J.(1986). *Crop. Sci.* 26:145-150.

Pairunan, A.K.; Rabson, A.D. and Abbott, L.K. (1980). *New Phytol.* 84:327-338.

Pandey, R. (2006). *Journal Nuclear Agric. Biol.* 35 (3-4): 168-179.

Pang, P.C. and Paul, E.A. (1980). *Can. J. Soil Sci.* 60:241-250.

Panozzo, J. and Eagles, H. (1999). *Australian Journal of Agricultural Research* 50:1007-1015.

Panwar, J.D.S. (1991). *Indian J. Plant Physiol.* 34: 357-361.

Panwar, J.D.S. and Thakur, A.K. (1995). *In: Proc. Mycorrhizae Biofertilizers for the Future. Tata Energy Res. Institute.* pp 471-477.

Panwar, J.D.S.; Abbas , S. and Sita Ram, (1990). *Indian J. Plant Physiol.* 33(1):16-20.

Pareek, R.P.; Kumar, N. and Chandra, R. (1990). *Journal of Indian Society of Soil Science* 38:186-88.

Park, J. (1960). *The Ecology of Soil fungi.* Liverpool University Press, Liverpool pp-324.

Park, JY (1970). *Canadian Journal of Microbiology* 16:798-800.

Paszkowski, U.; Kroken, S.; Roux, C. and Briggs, S.P.(2002). *Proc.Natl.Acad.Sci.U.S.A* 99:13324-29

Paszkowski, V. (2006). *New Phytologist* 172:35-46.

Paul, E.A. and Clark, F.E. (1996). *Soil Microbiology and Biochemistry.* San Diego, CA: Academic Press. pp. 340.

Peiffer, C.M. and Bloss, H.E. (1988). *New Phytol.* 108:315-21.

Peretto, R.; Bettini, V.; Favaron, F.; Alghisi, P. and Bonfante, P. (1995). *Mycorrhiza.* 5:157-163.

Perry, D.A., Molina, R. and Amaranthus, M.P. (1987). *Can. J. For. Res.* 17.8:929-940.

Peterson, R.L. and Bonfante, P. (1994). *Plant and* Soil159:79-88.

Plenchette, C. (1982). *Recherches surles endomycorrhizes a vesicules et arbuscules. Influence de la plante-hole du champign on et du phosphore sur 1. expression de la symbiose endomycorrizienne. Ph.D. Thesis,* pp. 170. of Level Canada.

Pond, E.C.; Menge, J.A. and Jarrel, W.M. (1984). *Mycologia* 76:74-84.

Ponder, F. (1983). *Commun. Soil. Sci. Plant Anal.* 14: 507-511.

Powel, C.L. (1979). *New Phytol.* 83:81-85.

Powell, C.L. (1975). *New Phytol.* 75:563-566.

Powell, C.L. (1977). *N.Z.J. Agric.Res.* 20:59-62.

Powell, C.L. (1982). *N.Z.J. Agric. Res.* 25:461-464.

Powell, C.L. and Daniel, J. (1978). *New Phytol.* 80:351-358.

Pozo, M.J. and Azcon-Aguilar, C. (2007). *Curr. Opin. Plant Biol.* 10:393-398.

Prasad, H.; Chandra, R.; Pareek, R.P. and Kumar, N. (2002). *Indian J. Pulses Res.* 15(2):131-135.

Preeti, (2011). *Effect of Rhizobium and VAM on growth and yield of urd bean (Vigna mungo (L.) Hepper) under rain fed conditions.* Ph.D Thesis in Botany, C.C.S.Univ. Meerut.

Price, N.S.; Roncadori, R.W. and Hussey, R.S. (1989). *New Phytol.* 111:61-66..

Pryor, L.D. (1956). *Nature* 177:587-88.

Quintos, M and Valder, M. (1987). *Riv Latinoamericana de Microbiologica* 29:189-92.

Radersma, S. and Grierson, P.F. (2004). *Plant and Soil.* 259: 209-219.

Radin, J.W. (1984). *Plant Physiol.* 76:392-394.

Radin, J.W. and Eidenbach, M.P. (1984). *Plant Physiol.* 75:372-377.

Raghupathy, S. and Mahadevan, A. (1993). *Mycorrhiza* 3:123-26.

Rainbird, R.M.; Atkins, C.A. and Pate, J.S. (1983). *Plant Physiol.* 72:308-312.

Raj, J.; Bagyaraj, D.J. and Manjunath, A. (1981). *Soil Biology Biochemistry* 13:105.

Rajendran, K. and Devraj, P. (2004). *Biomass and Bioenergy* 26:235-249.

Ramchandaran, K.; Menon, M.R. and Aiyer, R.S. (1980). *Indian J. Microbiol.* 20: 220-224.

Rao, C.S., Sharma, G.D. and Shukla, A.K.(1999). *Indian J. Forest* 22(1):7-13.

Rao, Y.S.G.; Suresh, C.K.; Suresh N.S. and Bagyaraj D.J. (1987). *Cur. Sci.* 16:55-59.

Ratnayke, M.; Leonard, R.T. and Monge, J.A. (1978). *New Phytol.* 81:543-52.

Raven, J.A. and Smith, F.A. (1976). *New Phytol.*76:415-431.

Raven, J.A.; Smith, S.E. and Smith, F.A. (1978). *Trans Bot. Soc. (Ediburgh).* 43:27-35.

Rawat, A.K.; Khara, A.K. and Vaishya, U.K. (1991). *Legume Research* 14(3):150-152.

Read, D.J. and Perez-Moreno, J. (2003). *New Phytol.* 157:475-492.

Readhead, J. (1975). *In: Sanders, F.E., Mosse, B. and Tinker, P.B. (Eds.). Endomycorrhizas,* Academic Press, New York and London, pp. 469-484.

Reddy, A.S. and Mallaiah, K.V. (2002). *In:Bionoculants for Sustainable agriculture and forestry: Proceedings of National Symposium* held on February 16-18, 2001. Scientific Publishers (India), Jodhpur, India: 93-98.

Redhead, J.F. (1977). *Translation British Mycological Society* 69: 275-80.

Reena, J. and Bagyaraj, D.J. (1990a). *World J. Microbiology and Biotechnology* 6:59.

Reena, J. and Bagyaraj, D.J. (1990b). *Arid Soil Research and Rehabilitation.* 4:261.

Reid. J.B. and Goss, M.J. (1982). *Journal of Soil Science.* 33:47-53.

Requena, N.; Serrano, E.; Ocon, A. and Brequeninger, M. (2007). *Phytochem.* 68:33-40.

Rhodes, L.H. and Gerdemann, J.W. (1975). *New Phytol.* 75: 555-561.

Rhodes, L.H. and Gerdemann, J.W. (1978a). *Soil Science.* 126: 125-126.

Rhodes, L.H. and Gerdemann, J.W. (1978b). *Soil Biol. Biochem.*10:355-360.

Rhodes, L.H., and Gerdemann, J.W. (1980). *In: Cellular interactions in symbiosis and parasitism, pp. 173-195, C.B. Cook, P.W. Pappas and E.D. Rudolph (Eds.),* Ohio State University Press, Columbus.

Rillig, M.C. (2004). *Can. J. Soil Sci.* 84:355-363.

Rillig, M.C. and Mummey, D.L . (2006). *New Phytol.* 171:41-73.

Rillig, M.C.; Ramsey, P.W.; Morris, S. and Paul, E.A.(2003). *Plant and Soil.* 253:293-299.

Rillig, M.C.; Wright, S.F.; Nichols, K.A.; Schmidt, W.F. and Torn, M,S. (2001). *Plant and Soil*. 233:167-177.

Rilling, M.C.; Lutgen, E.R.; Ramsey, P.W.; Klironomos, J. N. and Gannon, J.E. (2005). *Pedobiologis* 49:251-259.

Rorison, I.H. (1968). *New Phytol.*67:913-923.

Rose, S.L. and Youngberg, C.T. (1981). *Can. J. Bot.* 59:34-39.

Ruiz-Lozano, J.M.,;Azcon, R. and Gomez, M. (1996). *Physiol.Plant.* 98:767-772.

Ruiz-Lozano, J.M.; Azcon, R. and Gomez, M. (1995). *Appl. Environ. Microbiol.* 61:456-460.

Ryan, M.H. and Graham, J.H. (2002). *Plant and Soil.* 244:263-271.

Ryan, M.H.; Tibbett, M.; Edmonds-Tibbett, T.; Suriyagoda, L.D.B.; Lambers, H.; Cawthray, G.R. and Pang, J. (2012). *Plant Cell Environ.* 35:2170-2180.

Ryan, M.H.; McInerney, J.K.; Record, I.R. and Angus, J.F. (2008). *J. Sci. Food. Agric.* 88:1208-1216.

Ryan, M.H.; Van Herwaarden, A.F.; Angus, J.F. and Kirkegaard, J.A. (2005). *Plant and Soil* 270:275-286.

Rygiewicz, P.T., Johnson, M.G., Storm, M.J. and Tingey, D.T. (1994). *Life history and carbon partitioning of ectomycorrhizae on the roots under elevated CO_2 Abst.* First GCTE Science Conference held at Wood Hole, Massachusctts, U.S.A. 23-27, May 1994, pp105.

Safir, G.R. and Nelson, C.E. (1985). *In: Monila, R. (Ed), Proc. of 6th North Amer. Conf. on Mycorrhizae.* Forest Res. Laboratory, Corvallis, Oregon, pp. 161-164.

Safir, G.R.; Bover, J.S. and Gerdemann, J.W. (1971). *Science* 172:581-583.

Safir, G.R.; Bover, J.S. and Gerdemann, J.W. (1972). *Plant Physiol.*49:700-703.

Sahu, S. and Behera, B. (1972). *Indian J. Agron.*17:359-360.

Saker, M. L.; Congdon, R.A. and Mayeok, C. R. (1999). *Tropical Ecology.* 40(2): 261-267.

Same, B.I.; Robson, A.D. and Abbott, L.K. (1983). *Soil Biol. Biochem.* 15:593-597.

Sanchez-Diaz, M. and Aguirreolea, J. (1993). *Phytoma.*51:26-36.

Sanders, F.E. and Sheikh, N.A. (1983). *Plant and Soil.* 71:223-246.

Sanders, F.E. and Tinker, P.B. (1973). *Pest. Sci.* 4:385-395.

Sanders, F.E., Tinker, P.B., Black, R.L.B. and Palmerley, S.M. (1977). *New Phytol.*78:257-268.

Sasek, V. and Musilek (1968). *Folia Microbiology.* 13:43-51.

Sattar, M.A. and Gaur, A.C. (1984). *In: Proc: Inter Symposium on Soil Test Crop Responses Correlation Studies.* Dhaka pp. 237-51.

Sattar, M.A. and Gaur, A.C. (1987). *Zentralbl. Microbial.* 142:393-95.

Sattar, M.A. and Gaur, A.C. 1989a). *Bangladesh Journal of Agricultural Science.* 14:233-39.

Sawers, R.J.H.; Gebreseassie, M.N.; Janos D.P. and Paszkowski, U. (2010). *Theor. Appl. Genet.* 120:1029-1039.

Schenck, N.C. and Hielnson, K. (1973). *Agron. J.* 65:849-850.

Schenck, N.C. and Kinlock, R.A. (1974). *Plant Diseases Reporter.* 58:169-73.

Scholten, O.E. and Kuyper, T.W. (2012). *In: Organic crop breeding,* pp. 263-272, E.T. Lammerts van Bueren and J.R. Myers (Eds.), Wiley-Blackwell, Hoboken, USA.

Schreven, D.A. and Van. (1958). *Nutrition of Legumes* (Ed. Hallsworth, E.G.) Butterworths, London pp. 137-63.

Secilia, J. and Bagyaraj, D.J. (1987). *Can. J. Microbiol.* 33:1069.

Secilia, J. and Bagyaraj, D.J. (1994). *Mycorrhizae.* 4:265.

Seguel, A.; Cumming, J.R.; Klugh-Stewart, K.; Cornejo, P. and Borie, F. (2013). *Mycorrhiza* 23:167-183.

Seth, R.D. (1979). *Studies on the development of rhizobial inoculation for groundnut.* Ph. D Thesis, University of Agric Sci., Bangalore.

Shanmugam, V.; Renukadevi, P.; Senthil, N.; Raguchander, T. and Samiyappan, R.(2001). *Plant Disease Research.* 16(1): 23-29.

Shanta Kumari, P. and Sinha, S.K. (1972). *Phosynthetica.* 6:189-194.

Sharma, M.P.; Gaur, A.; Bhatia, N.P. and Adholya, A. (1996). *Mycorrhizae.* 6:441-446.

Sharma, S.; Kashyap, S. and Vasudevan, P. (2005a). *Indian Journal of Biotechnology* 4: 156-160.

Sharma, S.; Kashyap. S. and Vasudevan, P. (2005b). *European Journal of Horticulture Science* 70(2): 68-74.

Sharma, S.; Madan, M. and Vasudevan, P. (1995). *In: Proceedings of National Academy of Sciences, vol.* LXV:345-356.

Shennan, C. (2008). *Phil. Trans. R. Soc. B* 363:717-739.

Shockley, F.W.; MacGraw, R.L. and Garrett, H.E. (2004). *Agroforestry Systems.* 60: 137-142.

Sieverding, E. (1991). *VAM Mangement in Tropical Agrosystems.* Deutch geselshaft for Rechnische Zusammen arbeit Eschborn.

Singh, A.K. and Shrivastava, S.K. (1985). *Phyton (Austria).* 25(2): 213-217.

Singh, B. and Gupta, B.R. (2006). *Farm Science Journal.* 15 (1): 78-80.

Singh, B.K. and Vyas, D. (2009). *In: Proc. Natl. Acad. Sci. India Section B Biological Sciences* 79:110-128.

Singh, C.S. (1996). *Microbiol. Res.* 157:87-92.

Singh, G. and Tilak, K.B.V.R. (1995). *In: Chandra, S., Khhana, Kheri (ed). Microbes and Man.* pp. 46-46. Dehradun, India.

Singh, P. (1991). *Field crops Res.* 28:1-15.

Singh, R.; Adholeya, A. and Mukerji, K.G. (2000). *In: Mycorrhizal Biology (Eds. Mukerji, K.G., Chamola, B.P. and Jagjit Singh).* Kluwer Academic Publishers, NY, pp. 171-196.

Singh, R.S. and Singh, D. (1995). *Mycorhizal biofertilizers for future* (Eds Adholeya A. and Singh S) TERI, New Delhi pp. 119-23.

Sinha, D.D. and Sharma, A. (2001). *Ann. Agric. Res.*22(1): 151-153.

Sinha, S.K.; Bhargava, S.C. and Baldev, B. (1988). *In: Pulse crops.* Eds. Baldev, B., Ramanujam, S. P. and Jain, H. K. Pub. Oxford and IBH, New Delhi.

Siviero, M.A.; Motta. A.M.; Lima D.S.; Birolli, R.R.; Huh, S.Y.; Santinoni, I.A.; Miyauchi, M.Y.H.; Zangaro, W.; Nogueria, M.A. and Andrade, G. (2008). *Applied Soil Ecology* 39: 144-152.

Skujins, J. and Allen, M.F. (1986). *MIRCENS J.* 2:161-76.

Smiley, R.W. and Cook, R.J. (1973). *Phytopathol.* 63:882-90.

Smith, F.A. and Smith, S.E. (2011). *Plant and Soil.* 348:63-79.

Smith, F.A.; Grace, E.J. and Smith, S.E. (2009). *New Phytol.* 182:347-358.

Smith, J.E. (1994). *Can. J. Microbiol.* 40:736-42.

Smith, S.E. (1980). *Biol. Rev.* 55:475-510.

Smith, S.E. (1982). *New Phytol.* 90:293-303.

Smith, S.E. and Brown, G.D. (1979). *Soil Biol. Biochem.* 11:469-473.

Smith, S.E. and Gianinazzi-Pearson,V. (1988). *Ann. Rev. Plnt Physiol. Plant Mol. Biol.* 39:221-24.

Smith, S.E. and Read, D.J. (1997). *Mycorrhizal symbiosis,*Second ed. Academic Press, London. U.K

Smith, S.E. and Son, C.L. (1987). In: *Proc:. 7th North American Conference on Mycorrhizae* pp. 265.

Smith, S.E., and Read, D.J. (2008). *Plant and Soil* 326:3-20.

Smith, S.E.; Nicholas D.J.D. and Smith, F.A. (1979). *Aust. J. Plant Physiol.* 6:305-316.

Smith, S.E.; Facelli, E.; Pope, S. and Smith, F.A. (2010). *Plant and Soil.* 326:3-20.

Smith, S.E.; St. John, B.J.; Smith F.A. and Bromely, J.L (1986). *New Phytol.* 103: 359-373.

Snellgrove, R.C., Stribley, D.P.; Tinker, P.B. and Lawlor, D.W. (1986). In: *Physiological and genetical aspects of mycorrhizae.* (Eds.) Gianinazzi-Pearson, V. and Gianinazzi, S. pp.421-424.

Song, F.; Song, G.; Dong, A. and Kong, X. (2011). *Acta Ecologica Sinica.* 31:322-327.

Stack, R.W. and Sinclair, W.A. (1975). *Phytopathol.* 65:468-72.

Stahl, P.D.; Williams, S.E. and Christensen, M. (1988). *Mycologia.* 76: 261-67.

Steinkellner, S., Lendzemo, V.; Langer, Schweiger, P.; Khaosaad, T.; Toussaint, J.P. and Vierheilig, H. (2007). *Molecules* 12: 1290-1306.

Subba Rao, N.S.; Tilak, K.V.B.R. and Singh, C.S. (1985). *Plant and Soil.* 84:283.

Subramanian, K.S. and Charest, C. (1998). *Physiol. Plant.* 102: 285-296.

Subramanian, K.S.; Charest, C.; Dwyer, L.M. and Hamilton, R.I. (1995). *New Phytol.* 129:643-650.

Swaminathan, K. and Verma, B.C. (1979). *New Phytol.*84:481-487.

Sylvia, D.M .and Sinclair, W.A. (1983). *Phytopathol.* 73: 384-90.

Sylvia, D.M. (1983). *Plant and Soil.* 71: 299-302.

Tanwar, S.P.S., Sharma, G.L. and Chahar, M.S. (2002). *Annals. Agric Res.* 23(3):491-493

Tanwar, S.P.S.; Sharma, G.L. and Chahar, M.S. (2003). *Legume Research* 26(2):149-150.

Tanwar, S.P.S.; Sharma, G.L. and Chahar, M.S. (2003). *Legume Research* 26(1):39-41.

Tarafdar J.C. and Marschner, H. (1994). *Soil Biol. Biochem.* 26:387-395.

Tawaraya, K. (2003). *Soil Sci. Plant Nutr.* 49:655-668.

Tawaraya, K.; Naito, M. and Wagatsuma, T. (2006). *J. Pl. Nutr.* 29:657-665.

Teste, F.P.; Lieffers, V.J. and Strelkov, S.E. (2012). *Plant and Soil* 360:333-347.

Tester, M. and Smith, S.E. (1985). *Soil. Biol. Biochem.* 17:807-810.

Thakur, A.K. (1994). *Physiological studies on Rhizobium- VAM symbionts in mung bean.* M.Sc. Thesis P.G. School, IARI, New Delhi.

Thakur, A.K. and Panwar, J.D.S. (1995). *Indian J. Plant Physiol.* 38:62-65.

Thapar, H.S. and Khan, S.N. (1985). *In: Proc. Indian National Science Academy.* 39:687-94.

Thiagarajan, T.R.; Ames, R.N. and Ahmad, M.H. (1992). *Can. J. Microbiol.* 38:573-576.

Thompson, J.P. (1987). *Australian Journal of Agricultural Research.* 38:847-67.

Thompson, J.P.; Clewett , T.G. and Fiske, M.L. (2013). *Plant and Soil.* 371:117-137.

Tilak, K.V.B.R. and Dwivedi, (1990). In: *Proc. of National Seminar on Mycorrhizas.* Held at HAU, Hisar, 14-16, Feb. 1990, pp. 59.

Timmer, L.W. and Leyden, R.G. (1980). *New Phytol.*85:15-23.

Tinker, P.B. (1975). *In: Sanders, F.E., Mosse, B., Tinker, P.B. (Eds), Endomycorrhizas,* Academic Press, London and New York, pp. 353-371.

Tinker, P.B. and Gildon, A. (1983). *In: Robb, D.A., Pierpoint, W.S. (eds), Metals and Micronutrients: Uptake and Utilization by Plants.* Academic Press, New York. pp. 21-37.

Tinker, P.B. and Nye, P.H. (2000). *Solute Movement in the Rhizosphere.* Oxford University Press.

Tinker, P.B. and Sanders, F.E. (1975). Soil Science. 119:363-68.

Tiwari, P.; Adholya, A. and Prakash, A. (2003). *In: Fungal Biotechnology in Agricultural, Food and Environmental Application,* D.K. Arora (Ed.,), Marcel Dekker Inc., NY, USA, 195-201.

Tomar, A.; Kumar, N.; Pareek, R.P. and Chandra, R. (2003). *Indian J.Pulses Res.* 16(2):141-143.

Trappe, J.M. (1962). *Bot. Rev.* 28:538-06.

Trappe, J.M. (1977). *Ann. Rev. Phytopathol.* 15:203-222.

Treseder, K.K. (2013). *Plant and Soil* 371:1-13.

Valarini, P.; Curaqueo, G.; Seguel, A.; Manzano, K.; Rubio, R.; Cornejo, P. and Borie, F. (2009). *Chilean J. Agric. Res.* 69(3):416-425.

Valdes, M.; Pina, F. and Grada, R. (1983). *Botetin de la sociedad Mexicana di micologia* 18:65-70.

Varma, A.K. and Schuepp, Hans (1994). *Mycorrhiza* 5:29-37.

Vasanthakrishna, M. and Bagyaraj, D.J. (1993). *Arid Soil Research Rehabilitation* 7:377-80.

Veiga, R.S.L.; Jansa, J.; Frossard, E. and Van der Heijden, M.G.A. (2011). *PLoS One 6:e27825.*

Verbruggen, E. and Kiers, E.T. (2010). *Evol. Appl.* 3:547-560.

Verbruggen, E.; El Mouden, C.; Jansa, J.; Akkermans, G; Bücking, H.; West, S.A. and Kiers, E.T. (2012b). *Am. Nat.* 179:E133-E146.

Verbruggen, E.; Kiers, E.T.; Bakelaar, P.N.C.; Roling, W.F.M. and Van der Heijden, M.G.A. (2012a). *Plant and Soil.* 350:43-55.

Veresoglou, S.D.; Chen, B. and Rillig, M.C. (2012). *Soil Biol. Biochem.* 46:53-62.

Verma, C.; Lallu, B. and Yadav, R. S. (2004). *Indian J. Pulses Res.* 17(2): 149-151.

Verma, N.; Chaturvedi, S. and Sharma, A.K. (2008). *Mycorrhiza News* 20(1):11-15.

Vijaylakshmi. (1988). *Influence of benzyladenine on photosynthesis and assimilate partitioning in mungbean under different source-sink relationship.* M.Sc.Thesis, Div. Plant Physiology P.G. School I.A.R.I.New Delhi.

Vogelzang, B.; Parsons, H. and Smith, S. (1993). *Soil Biol. Biochem.* 25(8):1127-1129.

Wallace, L.L.; Mc Naughton, S.J. and Coughenour, M.B. (1982). *Oecologia.* 54:68-71.

Watanarojanaporn, N.; Bonkerd, N.; Tittabutr, P.; Longtonglang, A.; Young, J.P.W. and Teaumroong, N. (2013). *Environ.* 28:316-324.

Whipps, J.M. (2001). *Journal of Experimental Botany.* 52: 487-511.

Wicklow-Howard, M.(1989). *Mycotaxon.* 34:253-57

Wilson, G.W.T.; Rice, C.W.; Rillig, M.C.; Springer, A. and Hartnett, D.C. (2009). *Ecol. Lett.* 12:452-461.

Wood, T. (1991). *In: Handbook of Applied Mycology Vol. 4* (Eds Arora, D.K, Clander, RP and Mykerjii, KG) Marcel-Dekker Inc. N.Y. pp. 823-71.

Woodhouse, H.W. (1975). *In: Endomycorrhizas* (Eds Sanders, F.E.; Mosse, B. and Tinker, P.B.) Academic Press, London and New York pp. 209-40.

Worchel, E.R.; Giauque, H.E. and Kivlin, S.N. (2013). *Microb. Ecol.* 65:671-678.

Wright, S.F. and Upadhyaya, A. (1998). *Plant and Soil* 198:97-107.

Wright, S.F.; Franke-Snyder, M.; Morton, J.B. and Upadhyaya, A. (1996). *Plant and Soil* 181:193-203.

Wu, S.C.; Cao, Z.H.; Li, Z.G.; Cheung, K.C. and Wong, M.H. (2005). *Geoderma* 125:155-166.

Yocom, D.H.; Boosalis, M.G. and Flessner, T.R.(1987). *In: Sylvia, D. M., Hung,L. L. and Graham, J. H. Mycorrhiza in Next Decade. Practical Applications and Research Priorities.* North Amer. Conf. Mycorrhiza, Gainesville, Florida.

Zaidi, A.; Khan, Sagir, Md. and Amil, Md. (2003). *European Jour. Agron.* 19:15-21.

Zeng, Y.; Guo, L.P.; Chen, B.D.; Hao, Z.P.; Wang, J.Y.; Huang, L.Q.; Yang, G.; Cui, X.M.; Yang, L.; Wu,Z.X.; Chen, M.L. and Zhang, Y. (2013). *Mycorrhizae* 23:253-265.

Zeze, A.; Dulieu, H. and Gianiazzi-Pearson, V. (1994). *Mycorrhizae* 4:2521-24.

Zhu, Y.G. and Miller, R.M. (2003). *Trends Plant Sci.* 8:407-409.

Chapter 12 : *Frankia* (Non-Leguminous Symbiosis)

Alexander, M. (1985). *Introduction to Soil Microbiology,* Wiley Eastern Ltd. New Delhi.

Andeke-Lengui, M.A. and Dommergues, Y.R. (1981). In: *Proc. of an International Workshop,* pp. 158–166, Common Wealth Scientific and Industrial Research Organization (CSIRO), Melbourne, Australia.

Bargali, K. (2011). *African Journal of Plant Science.* 5:401-406.

Becking, J.H. (1970). *Inter. J. System. Bacteriol.* 20:201-220.

Belanger, P.A.; Beaudin, J. and Roy, S. (2011). *Journal of Microbiological Methods.* 85(2):92–97.

Benson D.R. and Silvester, W.B. (1993). *Microbiological Reviews.* 57(2): 293–319.

Benson, D.R., and N. A. Schultz. (1990). *The Biology of Frankia and Actinorhizal Plants.* Academic Press, Inc., New York.

Brunck, F.; Colonna, J.P. and Dommergues, M. (1990). *La maýtrise del'inoculation des arbres avec leurs symbioses racinaires. Synthese d'une selection d'essais au champen zone tropicale," Bois et For^ets des Topiques,* Vol. 223, pp. 24–42, 1990.

Bulloch, T. (1994). *New Zealand Journal of Crop and Horticultural Science.* 22(1):39-44.

Carlson P.T. and Dawson, J. O. (1985). *Turrialba.* 35:141-150.

Chaia, E.; Wall, L.G. and Huss-Danell, K. (2010). *Symbiosis.* 51(3):201-226.

Delver P., and Post, A. (1968). *Plant and Soil.* 28:325-326.

Diagne, N.; Diouf, D. and Svistoonoff, S. (2013). *Journal of Environmental Management.* 128:204-209.

Diem H.G. and Dommergues, Y.R. (1985). *Plant and Soil.* 87(1):17-29.

Dommergues, Y.R. (1997). *Soil Biology and Biochemistry* 29:931–941.

Fernandez, M.P., Meugnier, H.; Grimont, P.A.D. and Bardin. R. (1989). *Int. J. Syst. Bacteriol.* 39:424-429.

Forrester, I.; Bauhus, J.; Cowie, A.L.; and Vanclay, J.K. (2006). *Forest Ecology and Management,* 233:211–230.

Franche, C.; Lindstrom, K. and Elmerich, C. (2009). *Plant and Soil.* 321:35–59.

Gaur, A.C. (2010). *In: Biofertilizers in Sustainable Agriculture,* ICAR, New Delhi, pp. 108-113.

Gopinathan, S. (1995). *Biology and Fertility of Soils.* 20(4):221–225.

Gtari M. and Dawson, J.O. (2011). *Functional Plant Biology* 38(8-9):653–661.

Hafeez, Y.; Sohail Hameed, A. and Malik, K.A. (1999). *Pakistan Journal of Botany.* 31(1):173–182.

Hahn, D., Lechevalier, M.P. ; Fischer, A. and Stackebrandt, E. (1989). *System. Appl. Microbiol.* 11:236-242.

He, X.H. and Critchley, C. (2008). *Frankia nodulation, mycorrhization and interactions between Frankia and mycorrhizal fungi in Casuarina plants in Mycorrhizae,* A. Varma, Ed., Biomedical and life Sciences, chapter 6, pp. 767–781, Springer, Berlin, Germany, 3rd edition.

Huang, J.; Hueyou, H.; Shouping, C.; Yuzhu, K.; Duanqin, C. and Chensheng, T. (2011). *Current research on main Casuarina diseases and insects in China," in Improving Smallholder Livelihoods through Improved Casuarina Productivity, C. Zhong, K. Pinyopusarerk, A. Kalinganire, and C. Franche, Eds., Proceedings of the 4th International Casuarina Workshop,* pp. 232–238.

Izquierdo, F.; Caravaca, M.M.; Alguacil, Hernandez, G. and Roldan A. (2005). *Applied Soil Ecology* 30(1):3-10.

Jain J.K. and Mohan, M. (2011). *Casuarina in farm forestry for sustainable livelihoods: the andhra pradesh paper mills experience," in Improving Small holder Livelihoods through Improved Casuarina Productivity, Proceedings of the 4th International Casuarina Workshop,* pp. 232–238.

Kang, L. (1999). *Forest Research* 12(1):42-46.

Karthikeyan, B.; Deepara and Nepolean, P. (2009). *Forests Trees and Livelihoods* 19(2):153-165.

Lalonde, M. (1979). *Botanical Gazette* 140 (Suppl.):S35-S43.

Landis T.D. and Dumroese, R.K. (2006). *Applying the target plant concept to nursery stock quality," in Proceedings of the COFORD Conference on Plant Quality: A Key to Success in Forest Establishment, L. MacLennan and J. Fennessy, Eds.,* pp. 1–10, National Council for Forest Research and Development, Dublin, Ireland.

Lechevalier, M.P. and Lechevalier. H.A. (1989). *In: S. T. Williams, M. E. Sharpe, and J. G. Holt (Ed.), Bergey's Manual of Systematic Bacteriology,* vol. 4. Williams & Wilkins, Baltimore, pp. 2410-2417.

Lechevalier, M.P.; Labeda, D.P. and Ruan. J.S. (1987). *Physiol. Plant.* 70:249-254.

Lechevalier, M.P. (1994). *International Journal of Systematic Bacteriology* 44(1):1-8.

Lechevalier, M.P. and Lechevalier, H. A. (1990). *In :* C. R. Schwintzer and J. D. Tjepkema (Ed.), *The biology of Frankia and actinorhizal plants.* Academic Press, New York.Systematics Isolation and culture of *Frankia*, pp. 35-60.

Lefrancois, A. Quoreshi, D. Khasa *et al.,* (2007). *Alder-Frankia symbionts enhance the remediation and revegetation of oil sands tailings, http://esaa-events.org/.*

Mahdi, S.S.; Hassan, G.; Samoon, S. A.; Rather, H.A. Dar, S.A. and Zehra, B. (2010). *Journal of Phytology* 2:42–54.

Mailly, P.; Ndiaye, H.; Margolis, A. and Pineau, M. (1994). *Forestry Chronicle* 70(3):282-290.

Mansour, S.R.; Dewedar, A. and Torrey, J.G. (1990). *Botanical Gazette* 151(4):490-496.

Mathish, N.V.; Johnson Prabhu, S. and M. Roopesh *et al.,* (2010). *Development of an In silico gene bank for plant abiotic stresses: towards its utilization for molecular analysis of salt tolerant and susceptible pp.* 144– 151.

Mort, A.; Normand, P. and Lalonde, M. (1983). *Can. J. Microbiol.* 29:993-1002.

Ng, H. (1987). *Plant and Soil* 103(1):123–125.

Normand, P.; Lapierre, P. and Tisa *et al.,* (2007). *Genome Research* 17 (1):7–15.

Normand, P.; Orso, S. and Cournoyer *et al.,* (1996). *International Journal of Systematic Bacteriology* 46(1):1-9.

Obertello, M.; Sy, M.O.; L. Laplaze *et al.,* (2003). *African Journal of Biotechnology* 2(12):528–538.

Oliveira, R.S.; Castro, P.M.L.; Dodd, J.C. and Vosatka, M. (2005). *Chemosphere* 60(10):1462-1470.

Pawlowski, K. and Demchenko, K.N. (2012). *Protoplasma* 249(4):967-979.

Perrine-Walker, F.; Gherbi, H.; Imanishi *et al.*, (2011). *Current Protein and Peptide Science* 12(2):156-164.

Persson, T.; Benson, D.R.; P. Normand *et al.*, (2011). *Journal of Bacteriology* 193(24):7017-7018.

Quispel, A., Burggraaf, A. J. P., Borsje, H. and Tak, T. (1983). *Can. J. Bot.* 61:2801-2806.

Quoreshi, M.; Roy, S.; Greer, C.W.; Beaudin, J.; McCurdy D., and Khasa, D.P. (2007). *Native Plants Journal* 8:271-281.

Roy, S.; Khasa, D. P. and C. W. Greer, C.W. (2007). *Can. J. Bot.* 85(3): 237-251.

Sayed, W. F. (2011). *Folia Microbiologica* 56(1):1–9.

Shah, S. K.; Shah, R.P.; Xu, H.L. and Aryal, U.K. (2007). *Journal of Sustainable Agriculture* 29 (2):85–95.

Soltis E. D. and Soltis, P. S. (2000). *Plant Molecular Biology* 42(1):45–75.

Sougoufara, B.H.; Diem, G. and Dommergues, Y.R. (1989). *Plant and Soil* 118(1-2):133–137.

Srivastava, S. S. Singh, and Mishra, A. K. (2012). *Journal of Basic Microbiology* 52(1):1-2.

St-Laurent, L., Bousquet, J.; Simon, L. and Lalonde, M. (1987). *Can. J. Microbiol.* 33:764-772.

Tani and Sasakawa, H. (2003). *Soil Science and Plant Nutrition,* 49(2):215–222.

Tripathi, S.B.; Mathish, N.V. and Gurumurthi, K. (2001). *In: Proc. of the 5th Annual Workshop on Casuarina,* K. Gurumurthi, Ed., pp. 126–134, Regional Forest Research Center, Rajahmundry, India, October.

Tromas, B. Parizot, N. and Diagne *et al.*, (2012). *Heart of endosymbioses: transcriptome analysis reveals conserved genetic programme between arbuscular mycorrhizal, actinorhizal and legume-rhizobial symbioses," PloS ONE,* vol. 7 no. 9, Article ID 44742.

Voigt, G.K. and Steucek, G.L. (1969). *Soil Science Society American Proceedings 33:946-49.*

Wall, L.G. (2000). *Journal of Plant Growth Regulation* 19(2):167-182.

Zhang X. and Benson, D.R. (1992). *Archives of Microbiology* 158(4):256-261.

Zheng, C. (1996). *Journal of Fujian Forestry Science and Technology* 23:44-47.

Zhong and Zhang, Y. (2003). *Forestry Science and Technology* 17:3-5.

Zhong, C.; Zhang, Y.; Chen, Y. (2010). *Symbiosis* 50(1-2):107-114.

Chapter 13 : Vermiculture and Vermicomposting

Bansal, S.; and Kapoor, K.K. (2000). *Bioresource Technology* 73:95-98.

Baskar, A.; Macgregor, A.N.; and Krikram, J.H. (1993). *Soil Biology and Biochemistry* 25:1673-77.

Bezborodov, G.A.; and khalbayeva, R.A. (1990). *Soviet Soil Science* 22:30-35.

Binet, F.; Fayolle, L. and Pussard, M. (1998). *Biology, Fertilizer and Soil* 27:79-84.

Darwin, C. (1981). *The Formation of Vegetable mould through the action of worms with observations of their habits.* Murray, London pp. 298.

Dorcas Joy, E.K.; Reddy, S.M.; and Reddy, M.V. (1991). *Jornal of Soil Biology and Ecology* 11:57-61.

Duble, B.M.; Stephens, P.M.; Davoren, C.W. and Ryder, M.H. (1994). *Applied Soil Ecology* 1:3-10.

Edwards , C.A. and Lofty, J.R. (1980). *Journal of Applied Ecology* 17:533-43.

Gandhi, M.; Sangwan, V.; Kapoor, K.K.and Dilbaghi, N. (1997). *Environment and Ecology* 15(2):432–434.

Gaur, A. C. (2010). *In: Biofertilizers in Sustainable Agriculture,* ICAR, New Delhi pp. 213-264.

Gaur, A.C and Mathur, R.S. (1990). *Soil Fertility and Fertilizers use, Vol.* 4 (Eds. Kumar *et al.*) IFFCO, New Delhi.

Gosh, M.; Chattopadhyay, G.N. and Baral, K. (1999). *Bioresource Technology* 69:149-54.

Hallat, L., Reineck, A. J. and V iljoen, S. A., 1990, *Life cycle of the oriental compost worm, Perionyx excavatus* (Oligochaeta). *South J. Zool.* 25:41-45.

Heungens, A. (1969). *Rev. Ecological, Biological Society* 2:131-45.

Hooger Kamp, M.; Pogaar, H. and Eijsackers, H.P. (1983). *In: Earthworm Ecology* (Ed J.E. Satchell) Chapman and Hall, London, pp. 85-105.

Hullegalle, N.R. and Ezumah, H.C. (1991). *Agriculture, Ecosystem and Environment* 35:55-63.

Ismail, S.A. (2000). *Tree World* 9(8):4.

Kale, R.D.; Mallesh, B.C.; Bano, K and Bagyaraj, D.J. (1992). *Soil Biology and Biochemistry* 24:1317-20.

Kalpan, D.L.; Hartstein, R.D. and Nauhoser, F.F. (1980). *Soil Biology and Biochemistry* 12:347-52.

Krishnamoorthy, R.V. and Vajranabhaiah, S.N. (1986). *In: Proceedings, Indian Academy of Science (Animal Science)* 95:341-51.

Lee, K.E. (1985). *Eartworms, Their Ecology and relationship with Soil and land use.* Academic Press, New york pp. 441.

Lee, K.E. (1992). *Soil Biology and Biochemistry* 24:1765-71.

Madsen, E.L.; and Alexander, M. (1982). *Soil Science Society of American Journal* 46:537-60.

Ndegwa, P.M. amd Thompson, S.A. (2000). *Bioresource Technol.* 75:7-12.

NIIR Board. (2008). *In: The complete technology book on vermiculture and vermicompost.* National Institute ofIndustrial Research, New Delhi, India. pp. 1.

Nivedita, G.; Susmita, B. and Behera, B.N. (1989). *Journal of Soil Biology and Ecology* 9:46-50.

Philip, G. (1981). *The University Atlas.* 21st Edtn., Philip, London, pp. 150.

Rebatain, S.C.; and Stinner, B.R. (1988). *Agricultural Ecosystems Environment* 24:135-46.

Reddell, F.; and Spain, A.V. (1991*). Soil Biology and Biochemistry* 23:775-78.

Reinecke, A.J. and Venter, J. M. (1987). *Biological Fertility and Soil* 3:135-41.

Reinecke, A.J.; Viljoen, S.A. and Saayman, R.J. (1992). *Soil Biology and Biochemistry* 24:1295-1307.

Shivsubramanian, K. and Ganeshkumar, M. (2004). *Madras Agricultural Journal* 91: 221-225.

Shweta, K. K and Singh, Y. P. (2006). *Asian Journal of Microbiology, Biotechnology and Environmental Science.* 8: 831-834.

Stephens, P.M.; Davoren, C.W.; Double, B.M.and Ryder, M.H. (1994). *Biological Fertilizers and Soils* 18:150-54.

Stewart, V.I. and Scullion, J. (1988). *In: Earthworms in Water and Environment Management* (Eds c.A. Edwards and F. Neuhauser). The hauge, pp. 263-72.

Stockdill, S.M.J and Cossens G.C. (1996). *Proc.: New Zealand Grassland Association,* pp. 168-83.

Subasashri M. (2004). *Vermiwash an effective biopesticide. The Hindu Newspaper,* 30th September, In: Science and Technology Section.

Tara Crescent. (2003). *Vermicomposting. Development Alternatives (DA) Sustainable Livelihoods. (http://www.dainet.org/livelihoods/default.htm)*

Tomati, U.; Grappelli, A. and Galli, E. (1988). *Biological Fertilizers and Soil* 5:288-94.

Trivedi R. and Bhatt S. A. (2006). *Asian Journal of Microbiology, Biotechnology and Environmental Science.* 8: 303-305

Venter, J. M. and Reinecke, A. J., (1988). *South African J. Zool.*,23 : 161-165.

Vermi Co.(2001). *Vermicomposting technology for waste management and agriculture: an executive summary. (http://www.vermico.com/summary.htm) PO Box 2334, Grants Pass, OR 97528,* USA: Vermi Co.

Viljoen, S. A., and Reinecke, A. R. (1989). *S. Afr. J. Zool.*24, 27–32.

Wollters, J.S and Joergensen, R.G. (1992). *Soil, Biology and Biochemistry* 24:171-77.

Zambare V. P., Padul M. V., Yadav A. A. and Shete T. B. (2008). *ARPN Journal of Agricultural and Biological Science* 3(4): 1-5.

Zhang, H. and Schrader, S. (1993). *Biological Fertilizer and Soil* 15:229-34.

Chapter 14 : Liquid Biofertilizers

Ahemad, M. and Khan, M.S. (2011). *Pest Manag. Sci.* 67: 423-429.

Bot, A. and Benites, J. (2005). *The importance of soil organic matter. Natural resources management and environment department, FAO corporation document repository. http://agricoop.nic.in/Annual%20report2011-12/ARE.pdf.*

Chandra, K., Singh., T. and Srivathsa., H. (1999). *In: Biofertilizer ready recokner Pages.* Published by – Regional Biofertilizer Dev. Centre, Bhubaneswar, Orissa.

Chandra, K.; Greep, S. and Srivathsa, R.S.H. (2005). *Promotion of organic farming in Southern States –role of regional centre of organic farming:* In : *proceeding of the Seminar on organic agriculture in peninsular India – promotion, certification and marketing.* pp 25-27.

Chandra, K.; Greep, S. and Srivathsa, R.S.H. (2005). *Spices India* 18(7): 29-33.

Chandra, K.; Greep, S. and Srivathsa, R.S.H. (2006). *Savayava Krishikarige Sahayahasta in Sujatha Sanchike (Reading Kannada Agriculture Magazine)* pp. 3.

Chandra, K.; Greep, S.; Ravindranath, P. and Srivathsa, R.S.H. (2005). *Liquid Biofertilizers,* Ministry of agriculture department of agriculture & co-operation, Government of India.

Chandra, K.; Mukherjee, P.K. and Karmakar, J.B. (1995). *Journal of the North Eastern Council.* 15(1): 13- 15.

Chandra, K.; Mukherjee, P.K. and Karmakar, J.B. (1995). *Journal of Hill Research* 8(2): 199-202.

Chandra, K.; Mukherjee, P.K. and Karmakar, J.B. (1995). *Journal of Hill Research* 8(2): 242-246.

Chandra, K;, Singh. T. and Srivathsa. H. (1999). In: *Biofertilizer ready recokner* pp. 12-14. Published by Regional Biofertilizer Dev. Centre, Bhubaneswar, Orissa.

Chandra,. K.; Karmakar, J.B. and Singh, J. K.H. (1995). *Environment & Ecology* 13(3): 595-597.

Eckert, D. (2012). *Efficient Fertilizer Use- Nitrogen. Retrieved from http://www.rainbowplant food.com/agromomics/efu/nitrogen.pdf.*

Ghosh N;. *Promoting Bio-fertilizers in Indian Agriculture. Institute of Economic Growth University Enclave, India. http://www.ipni.net/ipniweb/portal.nsf/0/94cfd5a0ed0843028525781c 0065437e/ $ FILE/12%20 South %20 Asia. Ghosh. Promoting % 20 Bio-fertilizers%20in% 20India%20Agri.pdf*

Khan, M.S.; Zaidi, A. and Wani, P.A. (2009). *Biomedical and Life Sciences* 551-570.

Mahdi, S.S.; Hassan, G.I.;' Samoon, S.A.; Rather, H.A. and Dar, S.A. (2010). *J. Phytol* 2: 42-54.

Mia, M.A.B., Shamsuddin, Z.H. (2010). *African J. Biotechnology* 9: 6001-6009.

Muthusamy, K,; Gopalakrishnan, S.; Ravi, T.K. and Sivachidambaram, P. (2008). *Current Science,* 94: 736- 774.

Ngampimol, H. and Kunathigan, V. (2008). *Aus. J. T.* 11: 204-208.

Patel, B.C. (2011). *Advance method of preparation of bacterial formulation using potash mobilizing bacteria that mobilize potash and make it available to crop plant. WIPO Patent Application WO/2011/154961.*

Pindi, P.K. (2012). *J. Bioferil. Biopestici.* 3(3) 1000e102: 1-2.

Revellin, C.; Giraud, J.J.; Silva, N.; Wadoux, P. and Catroux, G. (2001). *Pest Manag. Sci.* 57: 1075-1080.

Rodrýìguez, H. and Fraga, R. (1999). *Biotechnol. Adv.* 17: 319-339.

Zaidia, A.; Khan, M.S. and Aamila, M. (2004). *J. Plant Nutr.* 27: 601-612.

Chapter 15 : Production, Quality and Marketing of Biofertilizers

Anonymous (2012). *In: The complete technology Book on Bio-fertilizer and Organic Farming,* NIIR Project Consultancy Services, Kamla Nagar, New Delhi pp.495-523.

Bhattacharyya, P. and Tandon, H.L.S. (2012*).* In: *Biofertilizer Handbook,* Fertilser Development and Consultation Organisation pp. 143-145.

Kabi, M. and Poi, S.C. (2004). *In: Bhattacharya P.; Dwivedi, V. (Eds). Proc. National Conference Quality Control of Biofertilizers.* NBDC, Ghaziabad, pp. 83-87.

Miles, A.A. and Misra, S.S. (1938). *The Journal of Hygiene* 38(6): 732–49.

Specialty Fertilizer Statistics, FAI (2011).MOA and NCOF, Ghaziabad.

Vessey, J.K. (2003). *Plant Soil* 255:571–586.

Web Link- http://www.kribhco.net/images/pdf/PAPER_2013.pdf

Chapter 16 : Some Important Media

Allen, M.B. (1953). *Bacteriological Reviews* 17:125-173.

Ashby, S.F. (1907). *J. Agric. Sci.* 2: 35–51.

Baker, D. and Torrey, J.G. (1979). *The isolation and cultivation of actinomycetous root nodule endophytes, pp.38-56.* In J.C. Gordon, C. T. Wheeler, and D.A. Perry (Ed.), Symbiotic nitrogen fixation in the management of temperate forests. Forest Research Laboratory, Oregon State University, Corvallis.

Becking, J.H. (1959). *Plant and Soil* 11: 193–206.

Bennett, H. S. (1950). *Methods applicable to the study of both I fresh and fixed materials. In McClung's handbook of microscopical technique.* Edited by R. M Jones. Harper & Row, Publishers, Medical Department, New York. pp. 591-677.

Campelo, A.B. and Dobereiner, J. (1970). *Pesq. Agropec. Bras,* 5:327-332.

Chu, S. (1942). *J. Ecol.* 30:284-325.

Dobereiner, J., Marriel, I.E. and Nery, M. (1976). *Can. J. Microbiol.* 22: 1464–1473.

Dubos, R.J. (1928). *J. Bacterial.* 15(4): 223-234.

Hofer, A.W. (1935). *J. Am. Soc. Agron.* 27:228-230.

Jensen, H. L. (1942). In: *Proc. Linn. Soc., N.S.W.S.,* 57: 205–212.

Katznelson, H. and Bose, B. (1959). *Can. J. Microbiol.* 5:79–85.

King, E.O.; Wood, M.K. and Raney, D.E. (1954). *Journal of Laboratory Clinical Medicine.* 44(2): 301-307.

Lalonde, M. and Calvert, H.E. (1979). *In: Proc.: Workshop Symbiotic Nitrogen Fixation in the management of temperate forests* (J.C. Gordon, C.T. Wheeler and D.A. Perry, Eds) pp.95-110, Forest Research Laboratory, Oregon State University.

Lipman, J.G. (1903). *In: Report on the New Jersey Agricultural Experiment Station,* 24: 217-285.

Martin, J. P. (1950). *Soil Sci.* 69:215.

Murry, M.A.; Fontaine, M.S. and Torrey.J.G. (1984). *Plant and Soil* 78:61-78.

Norris, D.O. and Date, R.A. (1976). *Legume bacteriology in tropical pasture research principles and methods.* C.A.B. (Comm. W. Bur. Past. Field Crops Hurl. Berks. Bull.) 51: 134–174.

Okon, Y., Albrecht S.L. and Burris, R.H. (1977). *Appl. Environ. Microbiol.* 33: 85–87.

Pikovskaya, R.I. (1948). *Microbiol* 17: 362–370.

Prigsheim, E. (1951). *The Vitreoscillidae: a family of colourless, gliding filamentous organisms.* J. *Gen. Microbiol.* 5.

Thornton, H.G. (1922). *Ann. Appl. Biol.* 2: 241–274.

Vincent, J.M. (1970). *A manual for the practical study of the root nodule bacteria.* IBP Handbook No. 15. Oxford and Edinburgh, UK, Blackwell Scientific Publications.

Appendix-I & II

Appendix-I

Most Probable Number (MPN) for use with 10 Fold dilution and 5 Tubes per dilution (Cochran, 1950)

P_1	P_2	Most Probable Number for Indicated Value P_3					
		0	**1**	**2**	**3**	**4**	**5**
0	0	-	0.018	0.036	0.054	0.072	0.09
0	1	0.018	0.036	0.055	0.073	0.091	0.11
0	2	0.037	0.055	0.074	0.092	0.11	0.13
0	3	0.056	0.074	0.093	0.11	0.13	0.15
0	4	0.075	0.094	0.11	0.13	0.15	0.17
0	5	0.094	0.11	0.13	0.15	0.17	0.19
1	0	0.02	0.04	0.06	0.08	0.10	0.12
1	1	0.04	0.061	0.081	0.10	0.12	0.14
1	2	0.061	0.082	0.10	0.12	0.16	0.17
1	3	0.089	0.10	0.13	0.16	0.17	0.19
1	4	0.11	0.13	0.15	0.17	0.19	0.22
1	5	0.13	0.15	0.17	0.19	0.22	0.24
2	0	0.046	0.068	0.091	0.12	0.14	0.16
2	1	0.068	0.092	0.12	0.14	0.17	0.19
2	2	0.093	0.12	0.14	0.17	0.19	0.22
2	3	0.12	0.14	0.17	0.20	0.22	0.25
2	4	0.15	0.17	0.20	0.23	0.25	0.28
2	5	0.17	0.20	0.23	0.26	0.29	0.32
3	0	0.078	0.11	0.13	0.16	0.20	0.23
3	1	0.11	0.14	0.17	0.20	0.23	0.27
3	2	0.14	0.17	0.20	0.24	0.27	0.31
3	3	0.17	0.21	0.24	0.28	0.31	0.35
3	4	0.21	0.24	0.28	0.32	0.36	0.40

Contd...

3	5	0.25	0.29	0.32	0.37	0.41	0.45
4	0	0.13	0.17	0.21	0.25	0.30	0.36
4	1	0.17	0.21	0.26	0.31	0.36	0.42
4	2	0.22	0.26	0.32	0.38	0.44	0.50
4	3	0.27	0.33	0.39	0.45	0.52	0.59
4	4	0.34	0.40	0.47	0.54	0.62	0.69
4	5	0.41	0.48	0.56	0.64	0.72	0.81
5	0	0.23	0.31	0.43	0.58	0.76	0.95
5	1	0.33	0.46	0.64	0.84	1.1	1.3
5	2	0.49	0.70	0.95	1.2	1.5	1.8
5	3	0.79	1.1	1.4	1.8	2.1	2.5
5	4	1.3	1.7	2.2	2.8	3.5	4.3
5	5	2.4	3.5	5.4	9.2	16	-

APPENDIX II

Format for reporting the analysis of Biofertilizer samples as per FCO

No.______________________

Government of_______________

____

(Name of the Laboratory)

Date_______________

To_________________

The Fertilizer Inspector

____

The analysis reports of the biofertilizer sample forwarded vide your ref No._____ dated is as per details given below:

1) Name of Biofertilizer_________________
2) Date of sampling____________________
3) Code number of sample as indicated by the inspector_____________
4) Date of receipt of the sample in the laboratory_________________
5) Laboratory sample No._________________
6) Date of analysis of sample_______________
7) Analysis of biofertilizer (On fresh weight basis)

Sr. No.	**Specification as per FCO (Rhizo, Azot, Azsp, PSM)**	**Composition as per analysis (Rhizo, Azot, Azsp, PSM)**	**Variation**	**Permissible tolerance limit**
1	2	3	4	5

A) Physical characteristics
   i) Moisture content
   ii) Particle size

B) Chemical characteristics
   i) pH

C) Mircobial characteristics
   i) Viable cell unit
   ii) Contamination level

D) Efficiency characteristics
   i) *Nodulation test

ii) **Nitrogen fixed (mg)/g of sucrose consumed

iii) ***Formation of white pellicle in semi-solid nitrogen free bromothymol blue media

iv) +a) Solubilization Zone (mm)

+b) P-phosphorus (%)

Spectro-photometer

Remarks: The sample is/is not according to specification and fails in________

Signature of the incharge (Testing Laboratory)

Copy to:-

Director of Agriculture_______________

---

* *Rhizobium* ****Azotobacter** ***Azospirillum***, +PSM

Source: Bhattacharyya and Tandon (2012)

Source: Bhattacharyya, P. and Tandon, H.L.S (2012) *In:* Biofertilizer Handbook, Fertilizer Development and Consultation Organization, 204-204A Bhanot Corner, 1-2 Pamposh Enclave, New Delhi-110048, pp.184.

FSC
www.fsc.org
MIX
Papier | Fördert gute Waldnutzung
FSC® C083411

Kolibri 360 GmbH
Ferdinand-Jühlke-Straße 7
99095 Erfurt, Deutschland
produktsicherheit@kolibri360.de